Genomics of Tree Crops

R.J. Schnell · P.M. Priyadarshan
Editors

Genomics of Tree Crops

Springer

Editors
R.J. Schnell
USDA - ARS
Miami, FL, USA

P.M. Priyadarshan
Rubber Research Institute of India
Central Experiment Station
Ranni, Kerala, India

ISBN 978-1-4899-9384-7 ISBN 978-1-4614-0920-5 (eBook)
DOI 10.1007/978-1-4614-0920-5
Springer New York Heidelberg Dordrecht London

Printed on acid-free paper

Springer is part of Springer Science+Business Media (www.springer.com)

Preface

Editing a book on Genomics is a difficult task since this branch of science is changing so quickly. However, it is necessary to compile the advancements in this area for the use of students and new scientists interested in tree species. Many technological innovations have occurred in a short time span and understanding these techniques and how they are used is essential to move the science of plant genetics forward. Since the rediscovery of Mendelism during the 1900s, the genetics and breeding of plants has experienced many paradigm shifts such as the discovery of DNA and RNA, unraveling of genes and gene expression, the central dogma, reverse genetics, gene sequencing, molecular genetic markers, polymerase chain reaction (PCR), anti-sense RNA, single nucleotide polymorphisms (SNP), RNA interference (RNAi), epigenetics, and finally functional genomics.

In the last few decades, the pace of biological research has accelerated as we have increased our ability to manipulate genes. In agriculture, this has led to waves of controversy, but in medicine the advances are almost universally applauded. Regardless of one's views on genetic engineering, no one questions that it is changing the science of biology in profound ways.

The completion of the Human Genome Project was reported in the first year of the new millennium, with the full sequence becoming available for research and exploration. The Human Genome Project, initiated in the late 1980s, determined the entire sequence of three billion nucleotides. Major goals of this project were to identify and understand a whole repertoire of human genes. During the next few years, the function of all of the estimated 50,000–100,000 human genes will be identified. Embryonic stem cells were cloned for the first time in 2000, and offer the potential for curing a wide range of ills, from spinal cord injuries to diabetes. Golden rice, a genetically modified crop to which a battery of genes that overcome vitamin A and iron deficiencies have been added, was planted for the first time in Asian fields. Even taxonomy seems to be undergoing a sea change, with molecular phylogenies forcing the redrawing of many family trees, from angiosperms to insects and other arthropods.

Molecular Plant Breeding, the new science that emerged from plant genomics, presents many opportunities: shortening the time it takes to domesticate new crops from semi-wild plants, tailoring existing crops to meet new requirements such as nutritional enhancement or resistance to climate change, rapidly incorporating valuable traits from wild relatives into established crops, allowing plant breeders to work with highly complex traits, such as hybrid vigor and flowering, and making it feasible to work on research-neglected "orphan" crops.

Conceptually, whole genome sequencing represents an ultimate form of reductionism in molecular biology. The complex processes of life cannot be totally explained by the linear sequence of DNA. In experimental reality, DNA sequencing requires drastic reductions from higher to lower dimension – to destroy the cell and to extract the DNA molecules. We do not question how much information is lost in these procedures, but simply accept the common wisdom that the genome, or the entire set of DNA molecules, contains all the necessary information to make up the cell.

Genome projects have transformed biology in many ways, but the most immediate outcome is the emergence of computational biology, also known as bioinformatics. It is no longer possible to make advances in biology without integration of informatics technologies and experimental technologies. There is a distinction between genome informatics and post-genome informatics here. Genome informatics was born of necessity to cope with the vast amount of data generated by the genome projects. In contrast, post-genome informatics represents a synthesis of biological knowledge from genomic information toward understanding the basic principles of life. Post-genome informatics has to be coupled with systematic experiments from a large range of scientific disciplines to understand and manipulate plant metabolism for human advantage. This understanding is essential for mitigation of a host of environmental interactions, including physical and biological stress, that are likely to have a greater influence with the accelerating climate change occurring around the world.

The genetic enhancements made in annual crop species stand in sharp contrast to those achieved for tree species. Trees are genetically recalcitrant and have long generation times. They require large amounts of land for phenotypic evaluation and have much less capital resources provided for plant improvement programs. As a result, much less is understood and progress in tree genomics is meager. It is this fact that made us undertake the editing of a book on Genomics of Tree Crops. Our goal was to bring out a compilation of the progress made in tree crops. We have chapters on: the state of the art, bioinformatics, functional genomics of flowering time, gene flow, spatial structure and local adaptation, genetic transformation of fruit trees, genomics of tropical and temperate fruit trees, papaya genomics, genomics of *Hevea* rubber, and genomics of palms. These chapters are contributed by experts in their respective areas of specialization.

We thank Springer for agreeing to publish this book.

Miami, FL, USA R.J. Schnell
Kerala, India P.M. Priyadarshan

Contents

Contributors

María José Aranzana, Ph.D. Department of Plant Genetics, IRTA – Centre for Research in Agricultural Genomics, CSIC-IRTA-UAB-UB, Barcelona, Spain

Renée S. Arias, Ph.D. National Peanut Research Laboratory, USDA-ARS, Dawson, GA, USA

James W. Borrone, MS, Ph.D. Department of Entomology & Plant Pathology, Oklahoma State University, Stillwater, OK, USA

Jaroslaw Burczyk Department of Genetics, Institute of Experimental Biology, Kazimierz Wielki University, Bydgoszcz, Poland

Igor Chybicki, Ph.D. Department of Genetics, Institute of Experimental Biology, Kazimierz Wielki University, Bydgoszcz, Poland

Gregory Downs, B.Sc. Department of Plant Agriculture, University of Guelph, Guelph, ON, Canada

Iban Eduardo, Ph.D. Department of Plant Genetics, IRTA – Centre for Research in Agricultural Genomics, CSIC-IRTA-UAB-UB, Barcelona, Spain

Reiner Finkeldey Büsgen-Institute, Forest Genetics and Forest Tree Breeding, Georg-August University of Göttingen, Göttingen, Lower Saxony, Germany

Henryk Flachowsky Institute for Breeding Research on Horticultural and Fruit Crops, Julius Kuehn-Institute, Dresden, Saxony, Germany

Matthias Fladung, Ph.D. Institute of Forest Genetics, Johann Heinrich von Thünen Institute, Grosshansdorf, Germany

Magda-Viola Hanke Institute for Breeding Research on Horticultural and Fruit Crops, Julius Kuehn-Institute, Dresden, Saxony, Germany

Hans Hoenicka Institute of Forest Genetics, Johann Heinrich von Thünen Institute, Grosshansdorf, Germany

Brian M. Irish, BS, MS, Ph.D. Tropical Agriculture Research Station, USDA-ARS, Mayaguez, PR, USA

Maizura Ithnin, Ph.D. Advanced Biotechnology and Breeding Centre, Malaysian Palm Oil Board, Kajang, Selangor, Malaysia

Robert R. Krueger, Ph.D. USDA-ARS National Clonal Germplasm Repository for Citrus & Dates, Riverside, CA, USA

Konstantin V. Krutovsky, Ph.D. Department of Ecosystem Science and Management, Texas A&M University, College Station, TX, USA

David N. Kuhn, Ph.D. Subtropical Horticulture Research Station, USDA-ARS, Miami, FL, USA

Richard E. Litz, Ph.D. Tropical Research & Education Center, University of Florida, Homestead, FL, USA

Eng-Ti L. Low, Ph.D. Advanced Biotechnology and Breeding Centre, Malaysian Palm Oil Board, Kajang, Selangor, Malaysia

Lewis Lukens, Ph.D. Department of Plant Agriculture, University of Guelph, Guelph, ON, Canada

Ana Montserrat Martín-Hernández Department of Plant Genetics, IRTA – Centre for Research in Agricultural Genomics, CSIC-IRTA-UAB-UB, Barcelona, Spain

Alan W. Meerow, Ph.D. USDA-ARS-SHRS, National Germplasm Repository, Miami, FL, USA

Ray Ming, Ph.D. Department of Plant Biology, University of Illinois at Urbana-Champaign, Urbana, IL, USA

Paul H. Moore, Ph.D. Hawaii Agriculture Research Center, Kunia, HI, USA

Leslie C.-L. Ooi, M.Sc. Advanced Biotechnology and Breeding Centre, Malaysian Palm Oil Board, Kajang, Selangor, Malaysia

Guillermo Padilla, Ph.D. Agrobiología y Medio Ambiente, Instituto de Productos Naturales y Agrobiología (IPNA-CSIC), San Cristóbal de La Laguna, Santa Cruz de Tenerife, Spain

P.M. Priyadarshan, Ph.D. Rubber Research Institute of India, Central Experiment Station, Ranni, Kerala, India

Tanja Pyhäjärvi, Ph.D. Department of Plant Sciences, University of California, Davis, CA, USA

Juan Jose Robledo-Arnuncio, Ph.D. Departamento de Ecología y Genética Forestal, Centro de Investigación Forestal (CIFOR) – INIA, Madrid, Spain

Carlos Romero Citricultura y Producción Vegetal, Instituto Valenciano de Investigaciones Agrarias, Moncada, Valencia, Spain

Thakurdas Saha, Ph.D. Genome Analysis Laboratory, Rubber Research Institute of India, Kottayam, Kerala, India

Raymond J. Schnell, Ph.D. National Germplasm Repository for Tropical/Subtropical Fruit Crops, USDA-ARS, Miami, FL, USA

Rajinder Singh, Ph.D. Advanced Biotechnology and Breeding Centre, Malaysian Palm Oil Board, Kajang, Selangor, Malaysia

Cecile L. Tondo, Ph.D. Subtropical Horticulture Research Station, USDA-ARS, Miami, FL, USA

Santiago Vilanova Instituto Universitario de Conservación y Mejora de la Agrodiversidad Valenciana, Universidad Politécnica de Valencia, Valencia, Spain

Qingyi Yu, Ph.D. Texas Agrilife Research Center, Department of Plant Pathology and Microbiology, Texas A&M University, Weslaco, TX, USA

Chapter 1
The State of the Art: Molecular Genomics and Marker-Assisted Breeding

P.M. Priyadarshan and Raymond J. Schnell

Abstract Focus on tree biotechnology reflects the challenges posed by the genetic attributes of trees. The genetic attributes of trees stand in stark contrast to those of domesticated annual crops. Trees typically have long generation times and are wind pollinated with out-crossing mating systems. Traditional tree breeding is a lengthy process that cannot efficiently capture nonadditive genetic variation, primarily because inbred lines would suffer from inbreeding depression. Clonal propagation of elite genotypes allows for the capture of both additive and nonadditive genetic variation, and the addition of transgenes can confer new or enhanced traits. The molecular analysis of plants often focused on the single gene level. But the recent technological advances have changed this paradigm. The way the genes and genetic information are organized within the genome and the methods of collecting and analyzing this information and the determination of their biological functionality are referred to as genomics. Genomic approaches are permeating every aspect of plant biology, and since they rely on DNA-coded information, they expand molecular analyses from a single to a multispecies level. Plant genomics is reversing the previous paradigm of identifying genes behind biological functions and instead focuses on finding biological functions behind genes. It also reduces the gap between phenotype and genotype. This introductory chapter overviews two main sections: first, the current understanding of genomes, their genetic structure at the inter- and intra-species level, and how whole genomes are sequenced; and second, on finding the biological and functional significance of DNA sequence. It is also

P.M. Priyadarshan, Ph.D. (✉)
Rubber Research Institute of India, Central Experiment Station, Chethackal, Thompikandom, Ranni, Kerala 689676, India
e-mail: rriipriya@gmail.com

R.J. Schnell, Ph.D.
National Germplasm Repository for Tropical/Subtropical Fruit Crops, USDA-ARS, Miami, FL, USA
e-mail: Ray.Schnell@effem.com

R.J. Schnell and P.M. Priyadarshan (eds.), *Genomics of Tree Crops*,
DOI 10.1007/978-1-4614-0920-5_1, © Springer Science+Business Media, LLC 2012

worthwhile to note that these technologies, though extensively used in agricultural species, are only used in forest tree species research. Except for some tropical (avocado, mango, and papaya) and temperate (apple, *Prunus*, and *Pyurus*) fruit species, these techniques are not extensively used in other tree crops.

Keywords Breeding • DNA • Genetic markers • Genomes • Genetic maps • Reverse genetics • Transcriptional profiling

Introduction

Focus on tree biotechnology reflects the challenges posed by the genetic attributes of trees and the limitations of extending available biotechnologies developed for agricultural crop species to trees. The genetic attributes of trees stand in stark contrast to those of domesticated annual crops. Trees typically have long generation times and are wind pollinated with out-crossing mating systems. Individual trees are highly heterozygous and thus carry a high genetic load, such that mating between related individuals results in inbreeding. Furthermore, unlike crop species, trees are expected to have minimal population substructure and low linkage disequilibrium (Gonzalez-Martinez et al. 2006; Ingvarsson 2005). A practical consequence of low linkage disequilibrium is that linkage relationships between markers and alleles of genes controlling phenotypic traits are not consistent among individuals, which limits the application of marker-assisted selection and breeding.

Traditional tree breeding is a lengthy process that cannot efficiently capture non-additive genetic variation, primarily because inbred lines would suffer from inbreeding depression. Clonal propagation of elite genotypes allows for the capture of both additive and nonadditive genetic variation, and the addition of transgenes can confer new or enhanced traits. For example, damage from introduced diseases and insects for which there is no natural genetic basis for resistance could be mitigated through introduction of transgenes conferring resistance (Adams et al. 2000). However, research on the strategies and risks of introducing transgenics into natural populations is still in its infancy (DiFazio et al. 2004; van Frankenhuyzen and Beardmore 2004). Political, societal, and regulatory restrictions make the application of transgenics to trees in the near future uncertain (Herrera 2005). The lengthy traditional tree breeding process typically relies on identifying trees with desirable attributes, followed by indirectly evaluating their breeding potential by measuring phenotypic traits in their progeny. Most traits of interest to forest industry are quantitative in nature, can be costly to measure, and occur later in development (e.g., wood quality). To better understand the genetic regulation of quantitative traits and speed up the progeny testing process, research has focused on the ability to detect chromosomal regions carrying favorable alleles controlling quantitative traits, so called quantitative trait loci (QTL). Studies on tree species have demonstrated the feasibility of this approach within pedigrees and have identified quantitative trait loci influencing traits ranging from wood properties to adaptive traits (Jermstad et al. 2003). Marker-assisted selection is an extension of QTL technology, in which progeny with desired genotypes within a given pedigree are identified using molecular markers linked to favorable QTL alleles. However, QTL and marker-aided selection have limited application outside of pedigreed material. Limitations to QTL and marker-aided selection are exposed when consideration is given to the low linkage disequilibrium and high allelic variation present especially in forest tree populations (Brown et al. 2004; Neale and Savolainen 2004). Although linkage relationships between markers and QTLs can be established within pedigrees resulting from controlled crosses, historical recombination between markers and the QTL within populations means that QTL marker relationships must be reestablished in each new pedigree examined, and are completely uncertain in unrelated individuals taken from natural breeding populations.

The molecular analysis of plants often focused on the single gene level. But the recent technological advances have changed this paradigm. The way the genes and genetic information are organized within the genome, the methods of collecting and analyzing this information, and the determination of their biological functionality is referred to as genomics. Genomic approaches are permeating every aspect of plant biology, and since they rely on DNA-coded information, they expand molecular analyses from a single to a multispecies level. Plant genomics is reversing the previous paradigm of identifying genes behind biological functions and instead focuses on finding biological functions behind genes. It also reduces the gap between phenotype and genotype. This introductory chapter overviews two main sections: first, the current understanding of genomes, their genetic structure at the inter- and intraspecies level, and how whole genomes are sequenced; and second, on finding the biological and functional significance of DNA sequence. It is also worthwhile to note that these technologies, though extensively used in agricultural species, are only used in forest tree species research. Except for some tropical (avocado, mango, and papaya) and temperate (apple, *Prunus*, and *Pyurus*) fruit species, these techniques are not extensively used in other tree crops.

Genetic Markers and Population Genetics

Molecular genetic markers have been extremely useful for tree population genetics, a discipline supporting basic research on the evolution of species and populations, and supporting applications ranging from tree improvement to conservation and restoration. Molecular markers have been used to estimate population parameters including population structure, gene flow, hybridization, migration, mating systems, and inbreeding. Knowledge of these attributes can be used to guide applications for management and conservation. For example, existing marker technologies can be used to determine levels of genetic diversity and inbreeding, two factors indicative of adaptive potential, which can help identify populations at risk. Existing markers can determine taxonomic relationships, a crucial component of establishing the legal basis for protection of endangered plant species. Contamination by nonlocal seed sources can erode the local adaptation of a population, and can potentially be detected using existing marker technology.

A major limitation of currently available markers is that they are neutral, meaning they are not within the actual genes that play a causative role in determining traits of interest. In addition, recombination and low linkage disequilibrium in tree populations means that linkage relationships between markers and alleles of genes controlling phenotypic traits are not consistent among individuals. This is a limiting factor for the application of marker technology to conservation and restoration applications because the markers have little or no predictive value for evaluating adaptive genetic attributes.

Significant sequence resources in the form of expressed sequence tags are available for numerous trees (http://plantta.tigr.org), with the largest conifer resource being >78,000 transcript assemblies for *Pinus taeda* (loblolly pine). Notably, these resources are being expanded through resequencing of alleles in support of association genetic studies (see below). Examples of additional resources for forest genomics include microarray resources (Abbott et al. 2008), proteomics (Lara et al. 2009), gene tagging and mutant collections (Layne and Bassi 2008), and ecotilling (Barkley and Wang 2008). For several angiosperms, transformation systems have been established that enable assessment of gene function using various strategies, including knockdown using RNAi (Enrique et al. 2011) or synthetic miRNAs (Song et al. 2010), or introduction of mutations into the amino acid sequence (Pillitteri et al. 2004).

Currently, association genetic studies in trees require a survey sequencing of alleles of candidate genes within a population to identify single nucleotide polymorphisms (SNPs) that define unique gene alleles. SNP genotypes and phenotypes are then measured for individuals sampled from the population, enabling testing for statistical association between SNP genotypes and phenotypes. Furthermore, linkage disequilibrium decays rapidly within a few hundred base pairs in both pine (Gonzalez-Martinez et al. 2006) and aspen (Ingvarsson 2005). As a result, an SNP with significant association with a phenotypic trait is likely to be close to or in the gene influencing the phenotype. This allows knowledge of gene function to be considered in understanding the genetic mechanisms regulating the trait being evaluated.

Genetic Structure of Plant Genomes

Plant genomes are best described in terms of genome size, gene content, extent of repetitive sequences, and polyploidy/duplication events. Although plants also possess mitochondrial and chloroplast genomes, their nuclear genome is the largest and most complex. There is extensive variation in nuclear genome size (Table 1.1) without obvious functional significance of such variation (Rafalski 2002).

Plant genomes contain various repetitive sequences and retrovirus-like retrotransposons containing long terminal repeats and other retroelements, such as long interspersed nuclear elements and short-interspersed nuclear elements (Kumar and Bennetzen 1999). Retroelement insertions contribute to the large difference in size between collinear genome segments in different plant species and to the 50% or

Table 1.1 Nuclear genome size in plants

Common name	Nuclear genome size[a]
Wheat	15,966
Onion	15,290
Garden pea	3,947
Corn	2,292
Asparagus	1,308
Tomato	907
Sugar beet	758
Apple	743
Common bean	637
Cantaloupe	454
Grape	483
Man	2,910

[a]Expressed in megabases (1 Mb = 1,000,000)

more difference in total genome size among species with relatively large genomes, such as corn. They contribute a smaller percentage of genome size in plants with smaller genomes such as *Arabidopsis* (The *Arabidopsis* Genome Initiative 2000). If other repetitive sequences are accounted for, then corn genome is comprised of over 70% repetitive sequences and 5% protein-encoding regions (Meyers et al. 2001).

It is widely accepted that 70–80% of flowering plants are the product of at least one polyploidization event (Barnes 2002). Many economically important plant species, such as corn, wheat, potato, and oat, are either ancient or more recent polyploids, comprising more than one, and in wheat three different, homologous genomes within a single species. Duplicated segments also account for a significant fraction of the rice genome. About 60% of the *Arabidopsis* genome is present in 24 duplicated segments, each more than 100 kilobases (kb) in size (Bevan et al. 2001). Ancestral polyploidy contributes to create genetic variation through gene duplication and gene silencing. Genome duplication and subsequent divergence is an important generator of protein diversity in plants.

Model Plant Species

Model organisms (*Drosophila melanogaster, Caenorhabditis elegans, Saccharomyces cerevisiae*) provide genetic and molecular insights into the biology of more complex species. Since the genomes of most plant species are either too large or too complex to be fully analyzed, the plant scientific community has adopted model organisms. They share features such as being diploid and appropriate for genetic analysis, being amenable to genetic transformation, having a (relatively) small genome and a short growth cycle, having commonly available tools and resources, and being the focus of research by a large scientific community. Although the advent of tissue culture techniques fostered the use of tobacco and petunia, the species now used as model organisms for mono- and dicotyledonous plants are rice (*Oryza sativa*) and *Arabidopsis* (*Arabidopsis thaliana*), respectively.

Arabidopsis, a small Cruciferae plant without agricultural use, sets seed in only 6 weeks from planting, has a small genome of 120 Megabases (Mb), and only five chromosomes. There are extensive tools available for its genomic analysis, whole genome sequence, Expressed Sequence Tags (ESTs) collections, characterized mutants, and large populations mutagenized with insertion elements (transposons or the T-DNA of *Agrobacterium*). *Arabidopsis* can be genetically transformed on a large scale with *Agrobacterium tumefaciens* and biolistics. Other tools available for this model plant are saturated genetic and physical maps.

Unlike *Arabidopsis*, rice is one of the world's most important cereals. More than 500 million tons of rice is produced each year, and it is the staple food for more than half of the world's population. There are two main rice subspecies. *Japonica* is mostly grown in Japan, while *indica* is grown in China and other Asia-Pacific regions. Rice also has very saturated genetic maps, physical maps, whole genome sequences, as well as EST collections pooled from different tissues and developmental stages. It has 12 chromosomes, a genome size of 420 Mb, and like *Arabidopsis*, it can be transformed through biolistics and *A. tumefaciens*. Efficient transposon-tagging systems for gene knockouts and gene detection have not yet become available for saturation mutagenesis in rice, although some recent successes have been reported.

Maps

Genetic Maps

The development of molecular markers has allowed for constructing complete genetic maps for most economically important plant species. They detect genetic variation directly at the DNA level. A myriad of molecular marker systems are available, yet their description lies beyond the scope of this paper. A genetic map represents the ordering of molecular markers along chromosomes as well as the genetic distances, generally expressed as centiMorgans (cM), existing between adjacent molecular markers. Genetic maps in plants have been created from many experimental populations, but the most frequently used are F_2, backcrosses, and recombinant inbred lines. Although longer to develop, recombinant inbred lines offer a higher genetic resolution and practical advantages. Once a mapping population has been created, it takes only few months to produce a genetic map with a 10-cM resolution. Genetic maps contribute to the understanding of how plant genomes are organized, and once available, they facilitate the development of practical applications in plant breeding, such as the identification of Quantitative Trait Loci and Marker-Assisted Selection. Most economically important plant traits such as yield, plant height, and quality components exhibit a continuous distribution rather than discrete classes and are regarded as quantitative traits. These traits are controlled by several loci each of small effect, and different combinations of alleles at these loci can give different phenotypes.

Quantitative Trait Loci analysis refers to the identification of genomic regions associated with the phenotypic expression of a given trait. Once the location of such genomic regions is known, they can be assembled into designer genotypes, that is, individuals carrying chromosomic fragments associated with the expression of a given phenotype. The most important feature of Marker-Assisted Selection is that once a molecular marker genetically linked to the expression of a phenotypically interesting allele has been detected, an indirect selection for such allele based upon the detection of the molecular marker can be accomplished, since little or any genetic recombination will occur between them. Therefore, the presence of the molecular marker will always be associated with the presence of the allele of interest.

Genetic maps are also an important resource for plant gene isolation, as once the genetic position of any mutation is established, it is possible to attempt its isolation through positional cloning (Campos de Quiroz et al. 2000). Furthermore, genetic maps help establish the extent of genome collinearity and duplication between different species.

Physical Maps

Although genetic maps provide much-needed landmarks along chromosomes, they are still too far apart to provide an entry point into genes, since even in model plants the kilobases per centiMorgan (kb/cM) ratio is large, from 120 to 250 kb/cM in *Arabidopsis* and between 500 and 1.500 kb/cM in corn. Therefore, a 1-cM interval may harbor ~30 to 100 or even more genes. Physical maps bridge such gaps, representing the entire DNA fragment spanning the genetic location of adjacent molecular markers.

Physical maps can be defined as a set of large insert clones with minimum overlap encompassing a given chromosome. First-generation physical maps in plants were based on YACs (Yeast Artificial Chromosomes). Chimerism and stability issues, however, dictated the development of low copy, *E. coli*-maintained vectors such as Bacterial Artificial Chromosomes (BACs) and P1-derived artificial chromosomes. Although BAC vectors are relatively small (molecular weight of BAC vector pBeloBAC11 is 7.4 kb for instance), they carry inserts between 80 and 200 kb on average and possess traditional plasmid selection features such as an antibiotic resistance gene and a polycloning site within a reporter gene allowing insertional inactivation. BAC clones are easier to manipulate than yeast-based clones. Once a BAC library is prepared, clones are assembled into contigs using fluorescent DNA fingerprint technologies and matching probabilities. Physical and genetic maps can be aligned, bringing along continuity from phenotype to genotype. Furthermore, they provide the platform clone-by-clone sequencing approaches rely upon. Physical maps provide the bridge needed between the resolution achieved by genetic maps and that needed to isolate genes through positional cloning.

Genome Collinearity/Genome Evolution

A remarkable feature of plant genomics is its ability to bring together more than one species for analysis. The comparative genome mapping of related plant species has shown that the organization of genes is highly conserved during the evolution of members of taxonomic families. This has led to the identification of genome collinearity between the well-sequenced model crops and their related species (e.g., *Arabidopsis* for dicots and rice for monocots). Collinearity overrides the differences in chromosome number and genome size and can be defined as conservation of gene order within a chromosomal segment between different species. A related concept is synteny, which refers to the presence of two or more loci on the same chromosome, regardless of whether they are genetically linked or not.

Collinear relationships have been observed among cereal species (corn, wheat, rice, barley), legumes (beans, peas, and soybeans), pines, and *Cruciferae* species (canola, broccoli, cabbage, *Arabidopsis thaliana*). Recently, the first studies at the gene level have demonstrated that microcollinearity of genes is less conserved; small-scale rearrangements and deletions complicate microcollinearity between closely related species. For instance, although a 78-kb genomic sequence of sorghum around the locus *adh1* and its homologous genomic fragment from maize showed considerable microcollinearity and the fact that they share nine genes in perfect order and transcriptional direction, five additional, unshared genes reside in this genomic region (Tikhonov et al. 1999).

Comparing sequences of soybean and *Arabidopsis* demonstrated partial homology between two soybean chromosomes and a 25-cM section of chromosome 2 from *Arabidopsis* (Lee et al. 2001). Although such relationships need to be assessed on a case-by-case basis, they reflect the value *Arabidopsis* and other model species offer to economically important species.

Collinearity has also been established between rice and most cereal species, allowing the use of rice for genetic analysis and gene discovery in genetically more complex species, such as wheat and barley (Shimamoto and Kyozuka 2002). A comparison of rice and barley DNA sequences from syntenic regions between barley chromosome 5H and rice chromosome 3 revealed the presence of four conserved regions, containing four predicted genes. General gene structure was largely conserved between rice and barley (Dubcovsky et al. 2001). A similar comparison between corn and rice, based on 340 kb around loci *adh1* and *adh2*, showed five collinear genes between the two species, as well as a possible translocation on *adh1*. Rice genes similar to known disease-resistant genes showed no cross-hybridization with corn genomic DNA, suggesting sequence divergence or their absence in maize (Tarchini et al. 2000). There are even reports of collinearity across the mono-dicotyledoneous division involving *Arabidopsis* and cereals, which diverged as far back as 200 million years ago (Mayer et al. 2001) Exploiting collinearity helps to establish cross-species genetic links and also aids in the extrapolation of information from species with simpler genomes (i.e., rice) to genetically complex species (corn, wheat). Furthermore, it reflects the power of genomics to integrate genetic information across species.

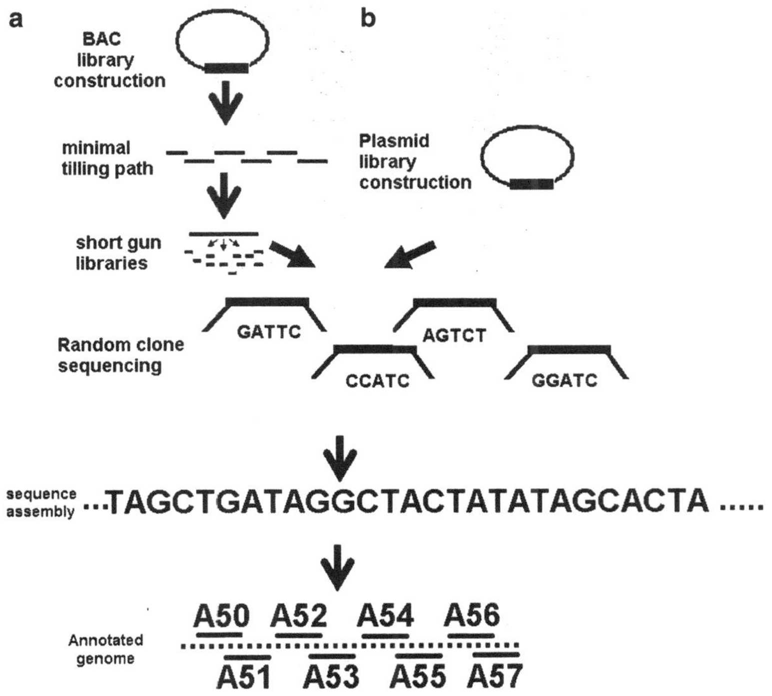

Fig. 1.1 Approaches of large scale sequencing (**a**) clone-by-clone strategy and (**b**) short gun strategy

Whole Genome Sequencing

Genetic and physical maps at the inter- or intraspecies level represent a key layer of genomic information. However, sequence data represents the ultimate level of genetic information. Three major breakthroughs have allowed the sequencing of complete genomes: (1) The development of fluorescence-based DNA sequencing methods that provide at least 500 bases per read; (2) The automation of several processes such as picking and arraying bacterial subclones, purification of DNA from individual subclones, and sample loading among others; and (3) The development of software and hardware able to handle massive amounts (gigabytes) of data points.

There are two main approaches to large-scale sequencing (Fig. 1.1). In clone-by-clone strategies (Fig. 1.1a), large insert libraries, such as those based on BAC clones, are used as sequencing templates, and inserts are arranged into contigs using diverse fingerprinting methods to establish minimal tiling paths. Sequence-tagged connectors extracted from large insert clones as well as FISH (fluorescence in situ hybridization) and optical mapping are used to extend contigs and close gaps (Marra et al. 1997). BAC clones from sequence-ready contigs are then fragmented into plasmid or M13 vector-based shotgun libraries with insert sizes of ~1–3 kb. Using more than one vector system reduces cloning bias issues. Sequencing efforts are tailored to the degree of coverage required. For instance, for a fivefold coverage, and assuming 500 base pairs (bp) per sequencer reading, 800 clones are sequenced to cover an 80-kb BAC clone. Finished sequences are those obtained at a ~8–10-fold coverage and provide >99.99% accuracy, whereas working draft sequences are attained at a ~3–5-fold coverage. It is important to note, however, that even working draft sequences provide an enormous amount of information, and even shotgun approaches rely to some extent on clone-by-clone information.

After sequencing is concluded, the DNA data are used to reassemble BAC clones. Base calling programs assigning quality scores to each read base such as Phred (Ewing et al. 1998), sequence assembly programs such as Phrap (Gordon et al. 1998), and graphical viewing tools are used to achieve such assembly. The finishing of the sequence then ensues, which can be done in part manually or with finishing software, such as Autofinish (Gordon et al. 2001).

Annotation, or the process of identifying start and stop codons and the position of introns that permits the prediction of biological function from DNA sequence, proceeds through three main steps. The first is to use gene finders like Xgrail (Uberbacher and Mural 1991) or others based on generalized hidden Markov models, such as GeneMark.hmm (Lukashin and Borodovsky 1998) and GenScan (Burge and Karlin 1997), specifically developed to recognize *Arabidopsis* genes. In the second step, sequences are aligned to protein and EST databases; and finally, putative functions are assigned to each gene sequence. Successful annotation processes often combine different software and manual inspection.

In shotgun approaches (Fig. 1.1b), which have been successfully used to sequence many microorganisms and *D. melanogaster*, small insert libraries are prepared, and randomly selected inserts are sequenced until a ~5-fold or higher coverage is reached. Sequences are then assembled, gaps are identified and closed, and finally, annotation is conducted. Shotgun sequencing does not rely upon the availability of minimal tiling paths and, therefore, reduces the cost and effort required to obtain whole genome sequences. Nevertheless, they require an enormous amount of computational power to assemble a large number of random sequences into a small number of contigs. Furthermore, the ultimate quality of large genomes that have been shotgun-sequenced may not be as high as that achievable using the clone-by-clone approach. Because of a high content of long and highly conserved repetitive sequences, including retrotransposons, shotgun sequencing of plant genomes may pose special challenges.

Reverse Genetics

Traditional genetic analysis aims to identify the DNA sequences associated with a given phenotype. Reverse genetics determines the function of a gene for which the sequence is known, by generating and analyzing the phenotype of the corresponding knockout mutant (Maes et al. 1999). Unlike yeast, in which gene disruption is available through homologous recombination, transposon and T-DNA tagging are the best methods available for developing mutagenized plant populations suitable for reverse genetics studies (Pereira 2000). There are several mutagenized populations in *Arabidopsis* suited for reverse genetics studies. A European consortium is developing heterologous systems for rice based on the Ac element from corn (Greco et al. 2001). There are also proprietary populations, such as Pioneer Hi-Bred International's Trait Utility System for Corn (TUSC), mutagenized with the high copy Mu element (Multani et al. 1998). Using high copy elements makes it possible to use smaller populations to ensure that tagged mutants will be found for most genes.

There are two main possibilities for identifying tagged genes at insertion sites. For unknown genes, sequences flanking the insertion can be obtained through inverse polymerase chain reaction (PCR) (Ochman et al. 1988) or thermal assimetric interlaced PCR (Liu and Whittier 1995), whereas for insertions in genes of known sequence, it is possible to amplify and clone the sequence of interest through PCR using gene-specific and insertion-specific primers. Since in the latter case it is common to analyze thousands of plants, PCR-based screening is arranged into three-dimensional pools that allow the unequivocal identification of tagged individuals. Large databases of characterized insertion sites are becoming available that will further ease the use of insertion elements to isolate useful genes (Tissier et al. 1999).

Although several genes have been isolated through reverse genetic approaches, two main factors have limited their wider application. First, many genes are functionally redundant, as even species with simple genomes such as *Arabidopsis* carry extensive duplications, and second, mutations in many genes may be highly pleiotropic, which can mask the role of a gene in a specific pathway (Springer 2000). Nevertheless, reverse genetics is considered to be a major component of the functional genomics toolbox, and it plays an important role in assigning biological functions to genes discovered through large-scale sequencing programs. Transposon tagging provides an excellent alternative to isolate tagged genes that exhibit relatively simple inheritance.

Gene traps refer to another application of transposons that responds to regulatory sequences at the site of insertion. Depending on the sequences engineered, they can be classified as reporter traps, enhancer traps, or gene traps. Since they rely on reporter gene expression, mutant phenotypes are not required, and they have been valuable in isolating tissue and cell-specific sequences (Springer 2000).

Transcriptional Profiling

While molecular biology generally analyzes one or a few genes simultaneously, recent developments allow the parallel analysis of thousands of genes. This area of genomics involves the study of gene expression patterns across a wide array of cellular responses, phenotypes, and conditions. The expression profile of a developmental stage or induced condition can identify genes and coordinately regulated pathways and their functions. This produces a more thorough understanding of the underlying biology (Quackenbush 2001).

There are several systems available to analyze the parallel expression of many genes, such as macroarrays (Desprez et al. 1998), microarrays (Schena et al. 1995), and serial analysis of gene expression (SAGE) (Velculescu et al. 1995), which consists of identifying short sequence tags from individual transcripts, their concatenation, sequencing, and subsequent digital quantitation. SAGE provides expression levels for many transcripts across different stages of development.

There are open and closed transcriptional profiling systems. Open technologies survey a large number of transcripts and analyze their levels between different samples, but the identity of the genes involved is not known *a priori*. One example of such a system is the GeneCalling technology (Bruce et al. 2000). Another open system is provided by massively parallel sequence signatures (MPSS), where microbeads are used to construct libraries of DNA templates and create hundreds of thousands of gene signatures (Brenner et al. 2000).

Closed systems, on the other hand, analyze genes that have been previously characterized. They include most of the diverse microarray systems available, and these are based on the specific hybridization of labeled samples to spatially separate immobilized nucleic acids, thus enabling the parallel quantification of many specific mRNAs. It is important to select the system at the onset of any transcriptional profiling study and stay with it.

The focus here is on microarrays. In microarray experiments, DNA samples corresponding to thousands of genes of interest are immobilized on a solid surface such as glass slides in a regular array. The immobilized sequences are usually referred to as probes. RNA samples (or their cDNA derivatives) from biological samples under study are hybridized to the array and are referred to as the target. Labeling with fluorescent dyes with different excitation and emission characteristics allows the simultaneous hybridization of two contrasting targets on a single array (Aharoni and Vorst 2001).

Microarray applications are broadly classified as expression-specific and genome-wide expression studies. In expression-specific studies, they are used as a functional genomics tool to address the biological significance of genes discovered through large-scale sequencing, as well as a means to understanding the genetic networks explaining biological processes or biochemical pathways. The value of using microarrays to identify novel response genes has been demonstrated by studying the gene expression patterns during corn embryo development (Lee et al. 2002), the response to drought and cold stresses (Seki et al. 2001), herbivory (Arimura et al. 2000), and nitrate treatments (Wang et al. 2000).

When addressing a specific pathway or biological process, it is useful to include genes beyond those of apparent interest, since over-specific microarrays would not be able to address genetic interactions with other biological processes. This principle revealed previously unexpected relationships between low soil phosphate levels and cold acclimation in *Arabidopsis* (Hurry et al. 2000). Genes obtained from the transcriptional analysis of plant responses to stress are of particular relevance for transgenic approaches, as thoroughly reviewed by Dunwell et al. (2001).

Genome-wide arrays are mostly designed for model organisms such as *Arabidopsis* or rice, as there are many genes available to select from, either as clones or as annotated genomic sequences for model species. They are also available to species such as corn that have extensive EST collections. This enabling technology is an immediate and direct result of large-scale sequencing projects. It is expected that microarrays covering most of the *Arabidopsis* genes will become available in 2003. Genome-wide expression profiles are the ultimate tool to integrate all genes existing in an organism into a series of experiments. They also help to elucidate the coordinate expression of different genetic networks and document how changes in one would impact others. It is expected that such genome-wide approaches will be particularly useful in identifying new regulatory sequences and master switches that affect distinct but apparently unrelated genetic networks.

Transcriptional profiling technologies play a central role in predicting gene function since sequence comparison alone is insufficient to infer function. They also help to detect phenomena such as gene displacement – nonhomologous genes coding for proteins that serve the same function – and gene recruitment – genes with identical sequences coding for completely different functions (Noordewier and Warren 2001).

Unlike animals, plants cannot move and have developed exquisite mechanisms to cope with changing environmental conditions and biotic challenges, since these directly or indirectly affect most biological processes occurring in plants. Therefore, a significant proportion of the information gathered by specific and genome-wide transcription profiling processes should have practical applications and facilitate the development of plants more resilient to biotic and abiotic stimuli.

Concluding Remarks

The current understanding of plant biology is limited to whole organism biology and its gene functions. Genomics adds another level of understanding to plant biology through the integrated analysis of different species. The large number of genes handled simultaneously by genomics sets a new paradigm in plant biology, since it allows the genetic integration of diverse processes, tissues, and organisms.

Finally, genomics is the ultimate interdisciplinary approach, as it covers the entire spectrum from DNA sequencing to field-based research. The integrated endeavor of genetics, biology, bioinformatics, molecular biology, engineering, microbiology, and related fields will extensively benefit mankind.

Genomics of trees, however, is in its infancy. Concerted efforts by tree biotechnologists will only benefit this paradigm. Except breakthroughs in some temperate and tropical fruit trees, they are to be fully exploited though studies on genomics.

References

Abbott DA, Suir E, van Maris AJ, Pronk JT (2008) Physiological and transcriptional responses to high concentrations of lactic acid in anaerobic chemostat cultures of *Saccharomyces cerevisiae*. Appl Environ Microbiol 18:5759–5768

Adams J et al (2000) The case for genetic engineering of native and landscape trees against introduced pests and diseases. Conserv Biol 16:874–879

Aharoni A, Vorst O (2001) DNA microarrays for functional plant genomics. Plant Mol Biol 48:99–118 [Links]
Arimura G, Tashiro K, Kuhara S, Nishikova T, Ozawa R, Takabayashi J (2000) Gene responses in bean leaves induced by herbivory and by herbivore-induced volatiles. Biochem Biophys Res Commun 277:305–310 [Links]
Barkley NA, Wang ML (2008) Application of TILLING and EcoTILLING as reverse genetic approaches to elucidate the function of genes in plants and animals. Curr Genomics 9(4):212–226
Barnes S (2002) Comparing *Arabidopsis* to other flowering plants. Curr Opin Plant Biol 5:128–133
Bevan M, Mayer M, Mayer K, White O, Eisin JA, Preuss D, Bureau T, Salzberg S, Mewes HW (2001) Sequence and analysis of the *Arabidopsis* genome. Curr Opin Plant Biol 4:105–110
Brenner S, Johnson M, Bridgham J, Golda G, Lloyd DH, Johnson D, Luo SJ, Mccurdy S, Foy M, Ewan M, Roth R, George S, Albrecht G, Vermaas E, Williams SR, Moon K, Burcham T, Pallas M, Dubridge RB, Kirchner J, Fearon K, Mao J, Corcoran NK (2000) Gene expression analysis by massively parallel signature sequencing (MPSS) on microbead arrays. Nat Biotechnol 18:630–634
Brown GR et al (2004) Nucleotide variation and linkage disequilibrium in loblolly pine. Proc Natl Acad Sci USA 101:15255–15260
Bruce W, Folkerts O, Garnaat C, Crasta O, Roth B, Bowen B (2000) Expression profiling of the maize flavonoid pathway genes controlled by estradiol-inducible transcription factors CRC and P. Plant Cell 12:65–79
Burge C, Karlin S (1997) Prediction of complete gene structures in human genomic DNA. J Mol Biol 268:78–94
Campos de Quiros H, Magrath R, McCallum D, Kroymann J, Schnabelrauch D, Mitchell-Olds T, Mithen R (2000) α-Keto acid elongation and glucosinolate biosynthesis in *Arabidopsis thaliana*. Theor Appl Genet 101:429–437
Desprez T, Amselem J, Caboche M, Hofte H (1998) Differential gene expression in *Arabidopsis* monitored using cDNA arrays. Plant J 14:643–652
DiFazio SP, Slavov GT, Burczyk J, Leonardi S, Strauss SH (2004) Gene flow from tree plantations and implications for transgenic risk assessment. In: Forest Biotechnology for the 21st Century (eds. Walter C & Carson M), Research Signpost, Kerala, India. pp. 405–422
Dubcovsky J, Ramakrishna W, Sanmiguel P, Busso C, Yan L, Shiloff B, Bennetzen JL (2001) Comparative sequence analysis of colinear barley and rice Bacterial Artificial Chromosomes. Plant Physiol 125:1342–1353
Dunwell JM, Moya-Leon MA, Herrera R (2001) Transcriptome analysis and crop improvement: a review. Biol Res 34(3–4):153–164
Enrique R, Siciliano F, Favaro MA, Gerhardt N, Roeschlin R, Rigano L, Sendin L, Castagnaro A, Vojnov A, Marano MR (2011) Novel demonstration of RNAi in citrus reveals importance of citrus callose synthase in defence against *Xanthomonas citri* subsp. *Citri*. Plant Biotechnol J 9:394–407
Ewing B, Hillier L, Wendl MC, Green P (1998) Base-calling of automated sequencer traces using phred. I. Accuracy assessment. Genome Res 8:175–185
Gonzalez-Martinez SC et al (2006) DNA sequence variation and selection of tag SNPs at candidate genes for drought-stress response in *Pinus taeda* L. Genetics 172:1915–1926
Gordon D, Abaijan C, Green P (1998) Consed: a graphical tool for sequence finishing. Genome Res 8:195–202
Gordon D, Desmarais C, Green P (2001) Automated finishing with Autofinish. Genome Res 11:614–625
Greco R, Ouwerkerk PBF, Sallaud C, Kohli A, Colombo L, Puigdomenech P, Guiderdoni E, Christou P, Hoge JHC, Pereira A (2001) Transposon insertional mutagenesis in rice. Plant Physiol 125:1175–1177
Herrera S (2005) Struggling to see the forest though the trees. Nat Biotechnol 2:165–167

Hurry V, Strand A, Furbank R, Stitt M (2000) The role of inorganic phosphate in the development of freezing tolerance and the acclimatization of photosynthesis to low temperature is revealed by the pho mutants of *Arabidopsis thaliana*. Plant J 24:383–396

Ingvarsson PK (2005) Nucleotide polymorphism and linkage disequilibrium within and among natural populations of European aspen (*Populus tremula* L., Salicaceae). Genetics 169:945–953

Jermstad KD et al (2003) Mapping of quantitative trait loci controlling adaptive traits in Douglas fir. III. Quantitative trait lociby-environment interactions. Genetics 165:1489–1506

Kumar A, Bennetzen JL (1999) Plant retrotransposons. Ann Rev Genet 33:479–532

Lara MV, Borsani J, Budde CO, Lauxmann MA, Lombardo VA, Murray R, Andreo CS, Drincovich MF (2009) Biochemical and proteomic analysis of 'Dixiland' peach fruit (*Prunus persica*) upon heat treatment. J Exp Bot 60:4315–4333

Layne DR, Bassi D (2008) The peach: botany, production and uses (CABI). CABI 30 Nov 2008, 848 pp. ISBN: 1845933869 PDF 27.1 MB

Lee JM, Grant D, Vallejos CE, Shoemaker RC (2001) Genome organization in dicots II *Arabidopsis* as a ëbridging speciesí to resolve genome evolution events among legumes. Theor Appl Genet 103:765–773

Lee JM, Williams ME, Tingey SV, Rafalski JA (2002) DNA array profiling of gene expression changes during maize embryo development. Funct Integr Genomics 2:13–27

Liu Y, Whittier R (1995) Thermal asymmetric interlaced PCR: automatable amplification and sequencing of insert end fragments from P1 and YAC clones for chromosome walking. Genomics 25:674–681

Lukashin AV, Borodovsky M (1998) GeneMark.hmm: new solutions for gene finding. Nucleic Acids Res 26:1107–1115

Maes T, De Keukeleire P, Gerats T (1999) Plant tagnology. Trends Plant Sci 4:90–96

Marra M, Kucaba T, Dietrich N, Green E, Brownstein WRK, Mcdonald K, Hillier L, Mcpherson J, Waterston RH (1997) High throughput fingerprint analysis of large-insert clones. Genome Res 7:1072–1084

Mayer K, Murphy G, Tarchini R, Wambutt R, Volckaert G, Pohl T, Düsterhöft A, Stiekema W, Entian K-D, Terryn N, Lemcke K, Haase D, Hall C, van Dodeweerd A-M, Tingey S, Mewes H-W, Bevan MW, Bancroft I (2001) Conservation of microstructure between a sequenced region of the genome of rice, multiple segments of the genome of *Arabidopsis thaliana*. Genome Res 11:1167–1174

Meyers B, Tingey SV, Morgante M (2001) Abundance, distribution, and transcriptional activity of repetitive elements in the maize genome. Genome Res 11:1660–1676

Multani D, Meeley RB, Paterson AH, Gray J, Briggs SP, Johal GS (1998) Plant-pathogen microevolution: molecular basis for the origin of a fungal disease in maize. Proc Natl Acad Sci USA 95:1686–1691

Neale DB, Savolainen O (2004) Association genetics of complex traits in conifers. Trends Plant Sci 9:325–330

Noordewier MO, Warren PV (2001) Gene expression microarrays, the integration of biological knowledge. Trends Biotechnol 19:412–415

Ochman H, Gerber A, Hart D (1988) A genetic application of an inverse polymerase chain reaction. Genetics 120:621–623

Pereira A (2000) A transgenic perspective on plant functional genomics. Transgenic Res 9:245–260

Pillitteri LJ, Lovatt CJ, Walling LL (2004) Isolation and characterization of a *TERMINAL FLOWER* homolog and its correlation with juvenility in citrus. Plant Physiol 135:1540–1551

Quackenbush J (2001) Computational analysis of microarray data. Nat Rev Genet 2:418–427

Rafalski AJ (2002) Plant genomics: present state and a perspective on future developments. Brief Fundam Genomics Proteomics 1:1–15

Schena M, Shalon D, Davis RW, Brown P (1995) Quantitative monitoring of gene expression patterns with a complementary DNA microarray. Science 270:467–470

Seki M, Narusaka M, Abe H, Kasuga M, Yamaguchi-Shinozaki K, Carninci P, Hayashizaki Y, Shinozaki K (2001) Monitoring the expression pattern of 1300 *Arabidopsis* genes under drought, cold stresses by using a full-length cDNA microarray. Plant Cell 13:61–72

Shimamoto K, Kyozuka J (2002) Rice as a model for comparative genomics of plants. Annu Rev Plant Biol 53:399–419

Song C, Cao X, Nicholas KK, Wang C, Li X, Wang X, Fang J (2010) Extraction of low molecular weight RNA from *Citrus trifolita* tissues for microRNA Northern blotting and reverse transcriptase polymerase chain reaction (RTPCR). Afr J Biotechnol 9:8726–8730

Springer PS (2000) Gene traps: tools for plant development and genomics. Plant Cell 12:1007–1020

Tarchini R, Biddle P, Winel R, Tingey S, Rafalski A (2000) The complete sequence of 340 kb of DNA around the rice Adh1-Adh2 region reveals interrupted colinearity with maize chromosome 4. Plant Cell 12:381–391

The *Arabidopsis* Genome Initiative (2000) Analysis of the genome sequence of the flowering plant *Arabidopsis* thaliana. Nature 408:796–815

Tikhonov A, Sanmiguel P, Nakajimanina Y, Gorenstein M, Bennetzen JL, Avramova Z (1999) Colinearity and its exceptions in orthologous adh regions of maize and sorghum. Proc Natl Acad Sci USA 96:7409–7414

Tissier AF, Marillonnet S, Klimyuk V, Patel K, Torres MA, Murphy G, Jones JD (1999) Multiple independent defective suppressor-mutator transposon insertions in *Arabidopsis*: a tool for functional genomics. Plant Cell 11:1841–1852

Uberbacher EC, Mural RJ (1991) Locating protein-coding regions in human DNA sequences by a multiple sensor-neural network approach. Proc Natl Acad Sci USA 88:11261–11265

van Frankenhuyzen K, Beardmore T (2004) Current status and environmental impact of transgenic forest trees. Canadian Journal of Forest Res 34(6):1163–1180

Velculescu V, Zhang L, Vogelstein B, Kinzler K (1995) Serial analysis of gene expression. Science 270:484–487

Wang R, Guegler K, Labrie Samuel T, Crawford NM (2000) Genomic analysis of a nutrient response in *Arabidopsis* reveals diverse expression patterns and novel metabolic and potential regulatory genes induced by nitrate. Plant Cell 12:1491–1509

Chapter 2
Bioinformatics Techniques for Understanding and Analyzing Tree Gene Expression Data

Lewis Lukens and Gregory Downs

Abstract There is great interest in enhancing our understanding of the molecular bases of tree biological processes using genomic techniques. One such technique is transcriptional profiling that assays the transcript abundance of thousands of genes. The analyses of these inventories of gene expression help explain the genetic diversity of trees and trees' responses to different developmental stages and environmental conditions. In this chapter, we describe key approaches for collecting transcriptome data and the tree genomic resources available for this data's use and interpretation. We define the factors that cause gene transcript abundances to vary and elucidate how to quantify these factors' effects. We also describe approaches to identify co-regulated genes and to assign functions to genes and groups of genes. Finally, we suggest future directions for tree transcriptome analyses.

Keywords Bioinformatics • Expression analysis • Microarray • High-throughput sequencing • Transcriptome • Gene expression variation • Gene co-regulation • Functional annotation

Introduction

Trees are a critical component of our environment covering a substantial proportion of the earth's land area. They are taxonomically diverse. Although characteristic of the gymnosperms, tree species are found in over 100 plant families, indicating frequent evolution of the tree growth habit. Finally, trees have major economic importance. One key attribute of trees is woodiness. Wood is a major renewable natural resource for the timber, fiber, and bioenergy industries. Tree species are also the

---

L. Lukens, Ph.D. (✉) • G. Downs, B.Sc.
Department of Plant Agriculture, University of Guelph,
50 Stone Road East, Guelph, ON, Canada N1G2W1
e-mail: llukens@uoguelph.ca

R.J. Schnell and P.M. Priyadarshan (eds.), *Genomics of Tree Crops*,
DOI 10.1007/978-1-4614-0920-5_2,

major producers of edible fruits including apples, citrus, peach, pear, plum, olives, and many others.

There is a keen interest in understanding the molecular bases of tree attributes, particularly those of economic and environmental importance, such as drought tolerance, wood development, and dormancy responses. Studies investigating transcriptional changes induced by drought include Street et al. (2006), Bogeat-Triboulot et al. (2007), Caruso et al. (2008), and Wilkins et al. (2009). Because trees and herbaceous plants have relatively few differences in gene content (e.g., Tuskan et al. 2006), gene transcriptional differences rather than novel genes likely cause wood development. A number of analyses have examined the molecular basis of wood formation. Hertzberg et al. (2001) sampled four developmental stages in poplar wood forming tissues and provided molecular signatures for the stages in xylem development. Other studies investigating wood development include Nilsson et al. (2008) and Dharmawardhana et al. (2010). In poplar, short days induce bud formation, dormancy, and acclimation to dehydration and cold. Ruttink et al. (2007) performed a detailed analysis on the response of *Populus tremula* × *Populus alba* to short days using an array with 24,745 cDNA probes corresponding to 16,494 genes. The experiment examined the transcriptional changes of bud development across a 6-week period of exposure to short-day conditions. Other studies examining dormancy include Schrader et al. (2004) who compared the transcriptional profiles of a *P. tremula* cambial meristem during active growth and dormancy, and Hoffman et al. (2010) who examined leaf transcriptional changes in poplar response to short days.

In this chapter, we describe data analysis approaches that have been used and could be used to unlock biological significance from tree gene expression. We first describe how microarrays and high-throughput sequencing technologies assay transcriptomes, and then how existing genomic and expression information facilitate the use and interpretation of transcriptome data. Most large-scale gene expression analyses have been performed in poplar because of its sequenced and annotated genome. Nonetheless, the pace of tree genome and transcriptome discovery is rapid, and such large-scale analyses will advance in other species. Second, we discuss gene expression variation and its causes and detail methods to attribute variation to specific sources. Third, we discuss approaches for functionally characterizing groups of genes. Fourth, we discuss methods to identify novel groups of co-regulated genes, gene modules, and groups of genes that correlate with known traits. Finally, we suggest bioinformatic analytical approaches that will enhance tree expression analyses. We see great potential for advances in tree biology through the use of bioinformatics.

Detecting Transcripts with Microarray and High-Throughput Sequencing Technologies

An assay of gene transcript abundance depends on detecting transcript sequences and mapping the detection information in a quantitative fashion back to individual genes. Microarrays and direct sequencing are the two dominant methods used to perform this task.

Microarrays quantify RNA transcripts by measuring the signal intensity of targets, labeled molecules derived from mRNAs. The targets hybridize to probes on the array. Probes are complementary DNA (cDNA) or oligonucleotides corresponding to different genes. Probes are arranged in a grid on the array substrate, such that tens or hundreds of thousands of probes can fill a centimeter-scale array. Arrays are scanned for target fluorescence because the fluorescent intensity of the target bound to the probe is proportional to the target abundance.

Most arrays are oligonucleotide arrays, although a number of labs have manufactured spotted cDNA arrays (e.g., Brosché et al. 2005; Moreau et al. 2005). Oligonucleotide probes are typically designed from the sequences of unigenes, consensus sets of transcript sequences that appear to come from the same gene, or from genes predicted from genomic DNA sequence. Online databases such as NCBI's UniGene contain these data. The probes are typically composed of sense sequences that bind to labeled antisense target transcripts. Affymetrix, Agilent, and NimbleGen are companies that manufacture oligonucleotide arrays, although oligonucleotides can also be spotted in academic labs. For *Populus*, the NimbleGen array contains three 60-mer probes for each of 48,276 transcripts, including predicted gene models from *P. trichocarpa, P. tremula, P. tremuloides,* and *P. alba* (Drost et al. 2009). The Affymetrix GeneChip Poplar Genome Array contains 25-mer probes in sets of 11 designed from the 3′-end of each of the 61,251 predicted gene models from *P. trichocarpa* (47,835 probe sets) and from EST sequences from 13 other *Populus* species. Researchers can also contract with manufacturers to build custom arrays that have oligonucleotides that query a specific set of genes, specific exons within genes, or single nucleotide polymorphisms (SNPs).

Increasingly, microarrays are being displaced by direct sequencing of cDNA molecules. Sequencing cDNA is not new, and historically utilized Sanger-based sequencing. ESTs, expressed sequence tags, would be generated by single pass sequencing reads from cloned cDNA. The abundance of EST sequences was determined and used as a measure of genes' transcript abundances (e.g., Sterky et al. 2004). A second method for assaying the expression of a large number of samples was the cDNA 3′-fragment sequencing method Serial Analysis of Gene Expression (SAGE; Velculescu et al. 1995). The SAGE method extracted short nucleotide tags (10–14 bp) from the 3′-most position of a restriction enzyme recognition site on a cDNA molecule. These tags were ligated together prior to sequencing, so one sequencing reaction detected a number of gene transcripts. Both EST and SAGE methods were fairly labor-intensive and amplified cDNA through bacterial cloning. A number of sequences failed to replicate in host cells, and thus remained undetected.

High-throughput transcriptome sequencing technologies sequence orders of magnitude more nucleotides and are less labor intensive than were the Sanger EST and SAGE methods. At the time of this writing, the two most widely used approaches are tag-based sequencing approaches and random shotgun RNA sequencing (RNA-Seq). One tag-based method developed by Illumina is the digital gene expression (DGE) profiling system. This procedure is similar to SAGE in a number of ways and generates 20–21-nt sequences from the 3′-end of the transcripts that contain a 4-bp restriction enzyme recognition site. The DGE requires restriction enzyme sites that may not be present on some of the transcripts, meaning that transcripts without sites will not be detected, and the downstream data mining can be problematic. The short sequence tags make the accurate alignment of some reads to genes difficult. RNA-Seq collects information about the entire transcriptome. In one protocol, total mRNA is fragmented. Adapters are added to the RNA fragments. The fragments are converted to cDNA, amplified in a linear fashion, and then sequenced from their ends (Simon et al. 2009). Unlike tag-based methods or microarrays, RNA-Seq can detect expressed polymorphisms, differentially expressed splice variants, fusion genes, different transcriptional start sites, and alternative polyadenylation sites (Wang et al. 2009).

Three technologies currently dominate high-throughput sequencing, and each technology generates different sequence lengths and numbers. The Illumina application of RNA-Seq on a Genome Analyzer II can generate single reads per sample that are 36- or 75-bases long, and paired-end reads of 36 bases each. A paired-end read obtains sequence information from two sides of the same cDNA molecule. The Roche 454 FLX Titanium system can generate 400 base reads. The Applied Biosystems SOLiD system reads up to 50 bases of a fragment. Twenty-five bases at each end of a fragment, e.g., a paired end, can also be sequenced. All three technologies can generate millions of sequences per run, although Illumina and SOLiD have significantly higher throughput than 454. SOLiD is capable of generating 150–200 million reads per single run, and Illumina generates 80–160 million reads per flow cell. In contrast, the current 454 system can generate between 800,000 and 1.2 million 400 nt reads. These technologies are rapidly improving read numbers and read lengths.

To quantify the abundance of each transcript in a sample with confidence using sequencing requires a large number of sequenced cDNA fragments. The number of total sequences necessary to detect differences in cDNA abundance is dependent on the abundance of the transcript. Transcripts that are highly abundant may be confidently quantified with a relatively small number of sequence reads, and transcripts that have low abundance require a large number of sequence reads. A rough estimate is that there are 22,000 different transcripts in tree tissues with 350,000 total mRNA molecules per cell; similar estimates exist for humans. In humans, about two million tags are required to reliably detect low-abundance genes with DGE ('t Hoen et al. 2008), and RNA-Seq requires at least 20 million tags per sample to obtain reasonable coverage of most transcripts (Mortazavi et al. 2008; Wilhelm et al. 2008). Plant genes are about 3 kb and human genes are typically between 10 and 15 kb. Thus, a smaller number of sequence reads should provide good coverage of the tree transcriptome. Roche 454 sequencing experiments typically generate hundreds of thousands of sequences and thus provide exploratory gene expression estimates (Barakat et al. 2009). Nonetheless, even these small-scale experiments can confidently assay the abundance of highly expressed genes (Eveland et al. 2008).

Both tag-based and RNA-Seq-based methods are superior to microarrays for assaying the transcriptome. First, microarrays have low sensitivity for quantifying rare transcripts and detecting differences in the levels of highly expressed transcripts. Low-abundance transcripts are especially important to detect because transcription factors often have low abundance and their levels are important indicators of the biological processes occurring within a tissue. Next-generation sequencing detects rare transcripts, and the dynamic range of detection is very large. For example, digital gene expression detected 10–20% more transcripts than did Affymetrix microarrays in a controlled test (Asmann et al. 2009). Second, the output of sequencing technologies is digital instead of the analog output of microarrays. With sequencing approaches, one estimates transcript abundance by counting the number of sequence reads that map to a transcript. The number of sequence reads that map to a gene is normalized by the length of the gene and the number of total sequence reads in the assay. In contrast, in microarray analyses, one infers transcript changes from changes in image pixel intensities, and one often needs to account for a number of confounding factors in order to compare gene expression values within and between arrays.

Comprehensive unigene sets, sequenced full-length cDNAs, and a complete genome sequence greatly facilitate microarray and sequence-based gene expression analyses. For oligonucleotide microarrays, prior information on gene transcripts is necessary to design probes. Thus, the capacity of the array to assay the transcriptome depends on the breadth of data used for array design. Information about tree gene transcripts is substantial yet also surprisingly sparse, given the importance of these organisms. Most transcript information is from ESTs. A number of tree transcripts are also predicted from genomic DNA sequence. The 485-Mbp *P. trichocarpa* (Black cottonwood) genome has 89,943 ESTs in 14,965 UniGene clusters. Spruce (*Picea glauca* and *Picea sitchensis*), pine (*Pinus taeda*), orange (*Citrus sinensis*), apple (*Malus x domestica*), and cacao (*Theobroma cacao*) all have over 15,000 UniGenes (NCBI in June, 2010). Gene transcripts can also be predicted from genomic sequences that have attributes of a transcribed locus, such as a long open reading frame and conservation with known proteins. The *P. trichocarpa* genome sequence was annotated with over 45,000 gene models, many of which were identified by gene prediction (Tuskan et al. 2006).

Tree transcriptome and genome data are rapidly increasing. For example, a 454 sequencing analysis of American chestnut (*Castanea dentata*) and Chinese chestnut (*C. mollissima*) produced over 40,039 and 28,890 unigenes, respectively (Barakat et al. 2009). Parchman et al. (2010) performed a similar large-scale gene discovery initiative. One difficulty with some tree genomes, especially gymnosperms, is the large genome size. The white spruce (*Picea glauca*) genome is estimated to be 19,796 Mbp, and the loblolly pine (*Pinus taeda*) genome is estimated to be 21,658 Mbp (Sederoff et al. 2009). However, the genomes of angiosperms, such as papaya (*Carica papaya*, 372 Mbp) and peach (*Prunus persica*, 225 Mbp), have sizes that are similar to those of model species such as thale cress (*Arabidopsis thaliana*, 120 Mbp), maize (*Zea mays*, 2,500 Mbp), and rice (*Oryza sativa*, 394 Mbp). There are at least 18 tree genomic DNA sequencing projects underway, most in their early stages. Flooded gum (*E. grandis*) has an 8X draft assembly, River red gum (*E. camaldulensis*) is at the 4–5X draft stage (http://www.eucagen.org), the peach (*Prunus persica*) genome v1.0 includes 7.7X coverage (http://www.rosaceae.org/peach/genome), and the papaya (*Carica papaya*) genome project has a 6X draft assembly (http://asgpb.mhpcc.hawaii.edu/papaya).

Comprehensive atlases of genes are also valuable because it can be difficult to attribute a sequence or probe to a single gene. Transcripts from different genes within the genome may have regions of identical sequences, and a microarray probe may measure more than one mRNA species. Probes on a microarray may be highly similar to a number of genes within a genome, and transcripts from these different genes are difficult to distinguish. A microarray with long oligonucleotides or cDNAs will cross-hybridize to molecules with over 85% identity or with distinct molecules with a long sequence of complementary nucleotides (Tiquia et al. 2004). The problem of confounding gene transcripts is less acute with next-generation sequencing than with microarrays, because no two sequences with nucleotide differences would be counted as coming from the same gene, although SNPs and sequencing errors can confuse the issue. Nonetheless, we have found about 4% of 70-mer oligonucleotides on a spotted maize array have near-perfect identity to more than one annotated gene on the maize genome (Downs and Lukens, unpublished), and in maize, nearly 1% of genes have duplicates that are greater than or equal to 98% identical (Emrich et al. 2007). This large number of highly similar sequences is difficult to distinguish with either an array or sequencing. Comprehensive genomic and EST sequences enable one to identify probe reads that may cross-hybridize and have sequence identity to more than one gene.

A well-annotated genome and transcriptome can also identify probes on a microarray that query the same target transcript. In our analysis of a maize array, we found over 9,000 predicted genes that were represented by multiple probes. If probes are considered individually, the splitting of a gene transcript's signal across more than one probe reduces the estimate of transcript abundance and the power to detect transcript differences between samples. Splitting probe signal can also cause genes that truly have different expression levels to be deemed similar (as measured by Euclidean distance). Grouping genes based on correlation or an expression level normalized relative to a standard solves the issue. For example, Bao et al. (2009) identified groups of genes involved in poplar shoot organogenesis with a 56,000 element Affymetrix array. Groups were identified as those genes with similar relative gene expression changes across samples. For next-generation sequencing approaches, especially RNA-Seq, prior information about a gene transcript is helpful for the identification and quantification of the transcript. The Illumina and SOLiD technologies that generate a large number of sequences also generate relatively short sequences. These short sequences are used to match the sequence with a unique gene in a set of previously characterized transcripts.

Although a complete, correctly annotated genome and a comprehensive set of transcripts is ideal for interpreting transcriptome data, all species' datasets have flaws and gaps. For example, the number of genes estimated by EST sequencing and unigene assembly is sometimes far greater than the number of genes annotated within genomes. Unigenes may not represent an organism's genes because of contaminants in the cDNA libraries or mistakes in cDNA sequence clustering. A genome sequence also does not perfectly represent a genome. Genome sequences typically have large gaps and the annotation of the genome may exclude real genes and include spurious genes. For example, maize (*Zea mays*) has 97,486 UniGenes that appear to be derived from distinct genes; however, the sequencing project reports 39,656 gene models (source: http://maizesequence.org). If genomes are not complete, microarray probes and next-generation sequence reads will not match genes within the genome. In our analysis of a maize microarray, of the probes for which we had expression, close to 50% were not found in a set of predicted transcripts.

The Causes and Measurement of Gene Expression Variation Within Trees

Causes of Gene Expression Variation

Gene transcript abundance in trees, as in other plants, varies for a number of reasons. Transcript abundances differ across tissue types and respond to changes in external stimuli. In addition, transcript levels vary depending on genotype, and genotypes differ in their response to environmental stimulus relative to other genotypes.

A number of studies have used a single genotype to focus on the effect of treatments or developmental stages on transcripts. For example, Bogeat-Triboulot et al. (2007) examined the transcriptional responses of plants derived from vegetatively propagated cuttings to water deficit using a cDNA array with 6,340 distinct ESTs enriched in stress-related genes. The cuttings were taken from a single *P. euphratica* poplar grown in an arid region. Nilsson et al. (2008) examined the transcriptional effects of indole acetic acid (IAA) on wood formation using aspen (*P. tremula* × *P. tremuloides*) clone T89. Quesada et al. (2008) examined gene expression profiles across five clonal *P. trichocarpa* tissues, root, stem node, internode, mature leaf, and young leaf using a NimbleGen array that queried over 40,000 predicted genes, each represented by three 60-nt probes. Bao et al. (2009) studied gene expression across time points over a shoot organogenesis experiment using a clone of *P. tremula* × *P. alba*. To control for genotype, if genetically uniform lines cannot be replicated, one can sample a single individual genotype before and after treatment and replicate across different genotypes. For example, Hoffman et al. (2010) sampled different leaves from the same plants subjected to long day and then short day. A problem with this approach is that transcriptional changes due to the treatment, day length changes, could be confounded by expression changes caused by the earlier sampling such as tissue damage.

Although it is intuitive that external stimuli or developmental stages will influence gene expression, a plant's genotype often has a comparable or greater effect on the transcriptome than does environment or tissue type. For example, Wilkins et al. (2009) sampled different poplar genotypes at different time points and under different water stress conditions. Genotype explained the greatest amount of gene expression variation. The precise genetic basis for transcriptome variation across genotypes is an area of active research. Genes are controlled by both *cis* acting elements such as promoters and *trans* acting factors such as transcription factors. Both *cis* and *trans* factors can vary within a population. Increasingly, studies have shown that both *cis* and *trans* factors explain genetic transcript differences across genotypes. In addition, maternal or paternal effects on the genome seem to be common. If gene expression was *cis*-regulated, an F1 hybrid of two genotypes would be expected to have a gene expression level midpoint between the genotypes. However, Swanson-Wagner et al. (2006), in an analysis of a maize F1 hybrid, found that 22% of 13,999 cDNA array genes exhibited gene expression levels that deviated from the expected mid-parental values. Stupar and Springer (2006), using an Affymetrix array, found similar evidence for nonadditive gene expression. A number of experiments in trees have characterized *cis* and *trans* effects more precisely. An expression quantitative trait locus (eQTL) is a locus segregating within a population that affects a gene's transcript levels. Kirst et al. (2004) mapped eQTL within a population of 91 (*E. grandis* × *E. globulus*) × *E. grandis* pseudo backcross progeny surveyed at 2,608 genes. Kirst et al. (2005) mapped eQTL for all gene transcript levels on two eucalyptus linkage groups. Interestingly, eQTL that mapped to both parental maps often mapped to nonhomologous linkage groups, suggesting that different genotypes have different regulatory loci (Kirst et al. 2005). Recent studies have shown that maternal and paternal effects explain some genetic differences in gene expression. For example, in maize, a number of maternally transmitted genes have a delayed expression within the hybrid endosperm relative to paternally transmitted genes (Guo et al. 2003). Most strikingly, in young plants, regulatory loci inherited from the paternal parent are often dominant to regulatory loci inherited from the maternal parent (Swanson-Wagner et al. 2009).

Finally, gene transcript abundances also vary because of genotypes' specific responses to treatments. Guo et al. (2006) found that alleles within a maize hybrid were differentially expressed depending on the year in which the hybrid was grown. Stress caused a reduction in mid-parent expression; and stress-tolerant genotypes had allele expression values that were more similar to that of their parents than the stress-susceptible genotypes (Guo et al. 2006). Wilkins et al. (2009) showed that water stress affected different gene transcripts in different poplar genotypes. Drost et al. (2010) examined the gene expression abundances across more than 150 individuals from the pseudo backcross progeny of a (*P. trichocarpa* × *P. deltoids)* × *P. deltoides* cross. Gene expression of xylem, leaf, and root samples was assayed on a NimbleGen array. Over 60,000 *cis* and *trans* factors (eQTLs) were detected across the three tissues. Interestingly, *cis* acting factors tended to cause similar differences in gene expression across all tissue types, while *trans* acting factors tended to be specific to a single tissue type. Because of genetic variation and the interactions between genotype and treatments, it would be interesting to quantify the degree to which transcriptional responses to stress observed on a single genotype can be extrapolated to a wide set of genotypes.

Measurement of Gene Expression Variation

The ability to capture and quantify the causes of transcriptional variation depends on experimental design. The first key concept is to ensure that the differences in gene expression that one observes among treatment groups are not due to a confounding factor. An appropriate experimental design randomizes experimental units across known sources of variation, like genotype or growth condition, and estimates the effects of these factors on gene expression. For example, experimental units (e.g., a plant or flats of plants) can be partitioned into spatial or temporal blocks, and the block effect may be removed by the analysis of variance (ANOVA). A second key concept is replication. Replication of experimental units within a treatment group is necessary to quantify the random variation in gene expression. Differences among experimental units within a group are compared with differences between different groups to calculate significance.

Different individuals of the same genotype grown in a controlled environment typically have highly similar transcriptional profiles ('t Hoen et al. 2008). Using technical variation as error variance underestimates the true error variance and may lead to the detection of false positives (e.g., Street et al. 2006). For example, three different genotypes may be subjected to a water-replete and water-deficient growth media in a factorial design with expression measured on three replicates of each treatment-condition combination. One can test for each treatment individually and whether treatments interact with each other by constructing a linear model

$$y_{ijk} = u + \alpha_i + \beta_j + \alpha_i\beta_j + \varepsilon_{ijk}$$

where y_{ijk} is the expression level of a gene in genotype i in growth condition j, where k represents a biological replicate, u is the grand mean, and ε is the experimental error. Different coefficients can be combined to test the comparisons of interest using contrasts. For example, one may determine if two of the genotypes significantly differ from the third, or if two genotypes significantly differ in their response to water stress. As mentioned above, one excellent application of experimental design and ANOVA was recently published in a paper investigating how the transcriptomes of two hybrid poplar clones, *P. deltoides* × *P. nigra* and *P. nigra* × *P. maximowiczii,* responded to drought at four different time points: midnight, predawn, midday, and late day (Wilkins et al. 2009). This experiment was designed as a 4×2×2 completely randomized factorial design, with time of day, genotype, and drought treatment as the main effects. Contrasts were tested to investigate genotype effect, time of day effect, treatment by genotype interactions, and treatment by time of day interactions. Diurnal rhythms are known to be important for transcript abundance changes (Harmer et al. 2000). This study surprisingly discovered that for each genotype, the genes that were differentially expressed at one time point in the well-watered control were often not the same as the genes differentially expressed at the same time point in the drought-stressed plants. Drought stress had altered the timing of transcriptional changes. In well-watered specimens of genotype NM6, transcripts of a large number of genes were downregulated between midnight and predawn; however, the transcripts dropped between predawn and midday in the drought-stressed trees, indicating the timing of expression changes had shifted by approximately one-fourth of the day due to the stress condition. This work is an excellent example of the condition-specific aspects of gene expression.

Transcript profiling experiments will have differences in power, the ability to find significant differences between samples, because of variable sample sizes and experimental heterogeneity. In addition, researchers often apply different criteria to determine if a gene's expression difference between samples is significant. Test statistics may be evaluated by classical and Bayesian approaches. Classical statistics, such as the *t*-test of two treatment groups, calculate the probability of the observed test statistic, given the null hypothesis that the gene is not differentially expressed across treatments. Unlike the classical method, the Bayesian approach calculates a posterior probability for a change in gene expression, given the observed test statistic. The method increases the power of the statistical tests and detects genes with more moderate expression changes than the classical approaches by exploiting the variance information provided by all the genes. Because one tests each gene for differential expression, a number of test statistics from genes that do not differ across treatments would be expected to be high and have low P values by chance. A standard approach is to use the FDR, the false discovery rate. The false discovery rate is the expected proportion of falsely rejected hypotheses, i.e., false positives, out of the total number of rejected hypotheses. The critical value that one uses to declare a test statistic as significant also varies across studies. For example, in Quesada et al. (2008), genes were considered up- or downregulated among organs at an FDR of less than 1% and if they were differentially regulated at least twofold.

Randomization and replication are key concepts for microarray studies. Replication of high-throughput sequencing experiments has often not been possible because of the cost. Most studies sequence a single sample per treatment group. In these cases, when possible, it is important to pool experimental units into a single sample. The transcript abundance in a pooled sample will tend toward the mean. Because there is no replication within a treatment group, we cannot use linear models to identify significantly differentially expressed genes. One approach for evaluating statistical significance is to calculate the probability of observing a certain number of transcript sequences in one sample compared to another, with the null expectation that the numbers are the same between samples. This can be done with the hypergeometric test. However, the test assumes that one sample has the true count of a gene rather than an estimate of that count. Similarly, the probability of the reads differing across samples can be calculated with the chi-square test.

Identifying the Functions of Differentially Expressed Tree Genes

Once a list of genes that differ between treatments is identified, the next question is: What are these genes' functions? The preponderance of gene functional characterization has been done in model species, especially the dicot *Arabidopsis thaliana* (although it is noteworthy that the *A. thaliana* genome still contains a large number of genes with unknown functions). Similar proteins' molecular functions tend to be well conserved (Chervitz et al. 1999). One assumes that a tree gene sequence that is highly conserved with a functionally characterized gene has a similar molecular function.

Homologs to tree genes are typically identified using BLAST (e.g., Kirst et al. 2004) or specialized software for orthologous gene identification such as InParanoid (e.g., Quesada et al. 2008).

Key discoveries in tree bioinformatics have been made by detailed analysis of individual genes that are differentially expressed across treatments. Ruttink et al. (2007) identified genes involved in light signal transduction, ethylene biosynthesis, ABA signal transduction, cell meristem maintenance, and carbohydrate metabolism over a 6-week period of dormancy induction in poplar. Kirst et al. (2004) compared fast-growing plants to slow-growing plants and found the expression of lignin biosynthesis genes and associated methylation pathways to be lower within the former. Hertzberg et al. (2001) found that genes within the lignin biosynthetic pathway tended to be upregulated in late cell maturation during xylem development.

The functions of genes may be difficult to discern on a one-by-one basis when a large number of genes are investigated. Genes may be assigned functions according to Gene Ontology (GO) terms (Ashburner et al. 2000). The GO project assigns gene products to structured terms that define biological processes, cellular components, and molecular functions. For example, proline is instrumental for osmotic adjustment in plants. The Arabidopsis P5C reductase AT5G14800, which degrades proline into 1-pyrroline-5-carboxylate, has the GO biological process term GO:0006561 "proline biosynthetic process," which is a subset of GO:0009084 "glutamine family amino acid biosynthetic process." Approaches can examine the frequency of GO terms to determine if genes with certain molecular functions are over-represented among genes that differ across treatments. One can test the probability that a certain number of genes within a Gene Ontology group would be observed among differentially expressed genes, given an expected frequency using Fisher's Exact test. The expected frequency of a GO category among differentially expressed genes is typically the frequency of the GO term of all the genes within the genome. The over-representation of a GO term means that genes involved in the specific biological process, cellular compartment, or molecular function are differentially expressed more frequently than expected. In the study of transcriptomes from different organ types, Quesada et al. (2008) examined the GO terms associated with organ-specific transcripts. As expected, genes assigned to the chloroplast gene cellular component were high in leaves relative to stem and root. Although a number of sequenced genomes have GO terms assigned to genes, the comprehensiveness of GO terms varies. For example, Knotted1 is a gene discovered in maize, which is important for cell fate commitment and meristem maintenance (Smith et al. 1992). The gene does not have the "cell fate" GO annotation in the recently released maize genome annotation. In contrast, in Arabidopsis, there are seven knotted-like homeobox proteins, and the annotation includes "cell fate commitment," "cell fate specification," "xylem and phloem pattern formation," "nucleus," and "transcription factor activity." Genes can also be assigned to biochemical pathways. The Kyoto Encyclopedia of Genes and Genomes, KEGG (Kanehisa and Goto 2000), has linked many genomic sequences, including poplar genes, to reference biochemical pathway maps. For example, the predicted *P. trichocarpa* gene POPTR_831811 encodes pyrroline-5-carboxylate reductase and is classified into the enzyme category EC 1.5.1.2. This enzyme is assigned to the arginine/proline metabolism pathway 00330.

Increasingly, the functions of genes are inferred from the tissues in which they are preferentially expressed. Ruttink et al. (2007) identified the *Arabidopsis* homologs to poplar genes that were differentially expressed in the apex and cambium and determined the *Arabidopsis* tissues in which these genes were highly expressed. Genes in *Arabidopsis* with more than 10% of the expression signal across all tissues within one tissue were termed as organ-specific. A large number of genes upregulated prior to dormancy were seed-specific in *Arabidopsis*, suggesting parallels in seed and vegetative transitions. Street et al. (2006) compared genes differentially expressed in response to drought with upregulated and downregulated genes found in sequenced poplar EST libraries. Genes that were upregulated in drought stress were far more commonly associated with dormant cambium, dormant buds, and senescing leaves than growing tissues. Zhan and Lukens (2010) described a novel, quantitative method to evaluate genes' tissue specificities in a study of genes mis-expressed in the inflorescences of *A. thaliana* miRNA biogenesis mutants. Microarrays from 23 tissue types were examined. For each tissue type, each gene that was not differentially expressed was ranked in that tissue relative to the 22 other tissues. The counts of gene ranks within each tissue were plotted in a histogram. This histogram represented the expected frequency of genes expressed within each tissue. The genes that were differentially expressed were also assigned ranks across tissues. If mis-expressed genes tended to be expressed at a low level in a particular tissue, they would have low ranks. This analysis showed that inflorescence miRNAs tended to significantly promote reproductive and repress vegetative development (Fig. 2.1).

Identifying Tree Genes Under Shared Regulatory Control

Identifying Co-regulated Genes

Tree transcriptome experiments have measured gene abundances across many distinct conditions, and the number and diversity of experiments are growing. Assaying genes over a number of conditions can find those genes that are under shared regulatory control. Genes that have similar expression patterns likely participate in the same processes. Analyses of sequences upstream of co-regulated genes can identify novel *cis* regulatory elements, and uncharacterized genes that are co-regulated with genes of known function likely also share that function (Ihmels et al. 2002). Finally, clusters of genes whose expression levels vary in parallel with changes in an attribute of interest are good candidates for pathways that respond to that attribute.

A number of tree transcriptome analyses have used clustering methods such as hierarchical methods (Eisen et al. 1998), *k*-means clustering (Tavazoie et al. 1999), and self-organizing maps (SOM) (Tamayo et al. 1999) to identify co-regulated genes. For example, Hertzberg et al. (2001) clustered genes whose abundance changed at least eightfold during the course of wood development using hierarchical clustering. This work found a number of gene clusters, including one cluster that

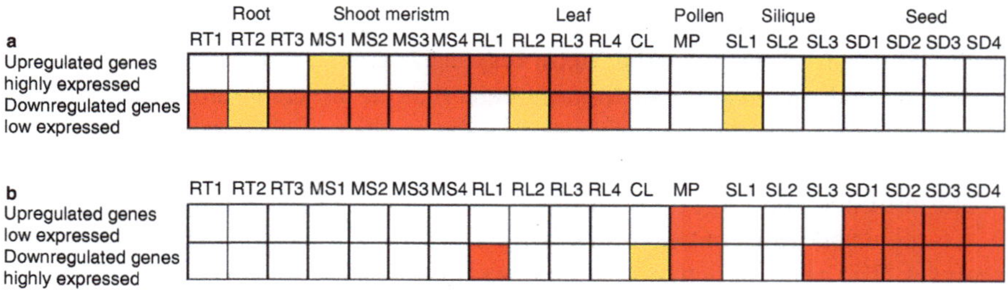

Fig. 2.1 Tissue specificity of genes up- and downregulated in *Arabidopsis* miRNA biogenesis mutant inflorescences. *White*, *yellow*, and *red* indicate no significance, significance at $P<0.05$, and significance at $P<0.01$, respectively. (**a**) The mutants promote vegetative tissue development: tissues in which genes upregulated in the mutants had high abundance in the wild type, and tissues in which genes downregulated in the mutants had low abundance in the wild type. *Colors* in the *top row* indicate if genes upregulated in the mutants were expressed at a significantly high level in the wild-type tissue. *Colors* in the *bottom row* indicate if genes downregulated in the mutants were expressed at a significantly low level in the wild-type tissue. (**b**) The mutants repress reproductive tissue development: tissues in which genes upregulated in the mutants had low abundance in the wild type, and tissues in which genes downregulated in the mutants had high abundance in the wild type. *Colors* in the *top row* indicate if genes upregulated in the mutants were expressed at a significantly low level in the wild-type tissue. *Colors* in the *bottom row* indicate if genes downregulated in the mutants were expressed at a significantly high level in the wild-type tissue. Tissue type abbreviations: *RT1* root, 7 days, *RT2* root, 15 days, *RT3* root, 17 days, *MS1* shoot apex, vegetative plus young leaves, *MS2* shoot apex, vegetative, *MS3* shoot apex, transition, *MS4* shoot apex, inflorescence, *RL1* rosette leaf #4, *RL2* rosette leaf #6, *RL3* rosette leaf #8, *RL4* rosette leaf #10, *CL* cauline leaf, *MP* mature pollen, *SL1* silique, with seeds stage 3, *SL2* silique, with seeds stage 4, *SL3* silique, with seeds stage 5, *SD1* seed, stage 6, *SD2* seed, stage 7, *SD3* seed, stage 8, *SD4* seed, stage 9 (Figure is from Zhan and Lukens 2010)

contained genes with high gene expression in the meristematic cells of the cambium. Hierarchical clustering produces a nested sequence of clusters, represented as a tree. The outermost nodes on the tree, the leaves, are genes. In agglomerative hierarchical clustering, each gene initially represents its own cluster. Larger clusters are iteratively constructed by joining genes to other genes, clusters to genes, or genes to clusters. In each iteration, the two clusters with the highest similarity are grouped together, and this process continues until all clusters and genes have been assembled into a single cluster. Partitioning methods such as *k*-means and SOM used in tree transcriptional profiling differ from hierarchical methods in that these methods subdivide the gene set into a predetermined number of clusters, and there is no implied hierarchical relationship among the clusters. For example, *k*-means subdivides genes into k disjoint clusters with the objective of selecting a partition of clusters and a set of cluster centers that minimizes the sum of squared Euclidean distances of genes to a cluster center. (Genes with small Euclidean distances have similar gene expression levels across treatments.) To group genes into clusters, *k*-means iterates between a cluster-assignment step and a centroid-calculation step. The self-organizing maps (SOM) method is similar to *k*-means but allows clusters to be represented in a grid, and similar clusters are represented in adjacent cells of the grid.

Nilsson used SOM in their analysis of auxin-regulated wood formation (Nilsson et al. 2008). This work found that auxin-responsive gene abundances were higher in expanding and secondary wall-forming cells than in the cambium. Both *k*-means and SOM suffer from the disadvantage that the number of clusters must be chosen before clustering begins, despite the lack of a method for determining the true number of clusters either a priori or post priori.

Co-regulated genes in trees and other plants have also been identified in genetic analyses. The concept is that genes whose transcript levels differ between plants that are genetically distinct at specific loci are under shared regulatory control. As described above, the factors that explain the expression differences between genetically distinct individuals may be mapped to genomic loci. If a locus (e.g., a *trans* eQTL) explains a significant proportion of the expression variation of a large number of genes, it is termed an eQTL hotspot. As mentioned above, Kirst et al. (2004) found that within a population of 91 (*E. grandis* × *E. globulus*) × *E. grandis* pseudo backcross progeny surveyed at 2,608 genes, genes encoding enzymes of the lignin biosynthesis and associated methylation pathways were downregulated in the developing xylem of fast-growing individuals. Interestingly, the RNA abundance of these lignin genes was explained by two *trans* eQTL on different linkage groups. Gene modules can also be identified in a comparison of mutants with wild-type parents. For example, Zhan and Lukens (2010) demonstrated that the expression profiles of plants with mutant alleles at two genes involved in the *A. thaliana* miRNA biogenesis pathway, *hst*15 and *hen*1, influenced a large number of the same genes, an unexpected result given their proteins' predicted roles in miRNA biogenesis.

Within a cluster of genes, gene transcript levels will rise and fall similarly. Groups of genes whose expression patterns are correlated with attributes of the samples from which they are derived can potentially explain those sample attributes. Tree transcriptional analyses have identified groups of genes correlated with sample attributes. Hoffman et al. (2010) investigated diurnal patterns of gene expression in poplar by assaying leaf gene expression four times a day over 2 days in plants grown under long day conditions (16-h light), and again four times a day over 2 days after the plants had been shifted to short day conditions (12-h light). They used a 25,000 spotted cDNA array corresponding to approximately 16,500 genes and identified genes whose transcript abundances followed six diurnal patterns over the 4 days using partial least squares to latent structures regression. Ruttink et al. (2007) clustered genes based on the directionality and the time interval of the maximum log fold change. This approach suggested that light, ethylene, and abscisic acid signal transduction pathways play sequential roles in apical bud development and dormancy.

Another approach for testing if groups of co-expressed genes associate with sample attributes is to examine the absolute value of the correlation between the module's representative gene expression profile, e.g., its eigengene or centroid, and trait values. Relating a group's representative gene to a sample trait instead of relating thousands of genes saves computational time and reduces the loss of power due to multiple testing. For example, in a pooled analysis of transcriptional differences between pairs of six homologous human and chimpanzee brain tissues, Oldham et al. (2006) identified a number of gene expression modules. The value for each module eigengene was plotted for each tissue type. Genes within a number of transcriptional modules had high gene transcript levels in one or a subset of the six different brain regions in both species. A measure of a module's significance for a sample trait can be represented as the average significance of each module gene's correlation with trait values (Langfelder and Horvath 2008). Conceptually, this approach is similar to that used by Nilsson et al. (2008). Nilsson et al. (2008) compared gene expression profiles with an auxin concentration gradient to identify those genes whose expression abundance was highly correlated with auxin abundance within the developing secondary xylem. A gene's module membership can be calculated as the correlation between the gene expression profile and the module eigengene. A gene whose expression pattern is similar to the module's representative expression profile and whose expression is correlated with trait values has a strong likelihood of participation in the specific functions represented by the module.

Future Directions in Tree Transcriptome Analyses

As described above, key objectives in tree transcriptome analyses are to identify and understand those genes and transcription networks that are responsible for traits, particularly traits that are characteristic of trees, such as perennialism and woodiness. A number of novel approaches show promise in the effort to uncover this knowledge.

Based on our understanding of transcriptional control, we expect that genes are co-regulated in a subset of conditions and that they participate in a number of processes. For example, genes may be co-regulated in a subset of samples in which a regulatory factor is active but not in those samples where a regulatory factor is inactive. The involvement of a gene in several processes is reflected in its commonly diverse set of upstream regulatory elements. Hierarchical clustering and partitioning methods have generated a number of important discoveries. However, the methods may cluster each gene into only one group and are not designed to identify genes that are co-expressed in a subset of conditions. Methods can identify subsets of genes that exhibit comparable expression patterns over a subset of treatments and allow genes to be in multiple clusters (Ihmels et al. 2002; Kluger et al. 2003; Prelic et al. 2006; Maere et al. 2008). For example, ISA, the iterative signature algorithm (Ihmels et al. 2002, 2004), performs well in cluster identification using both real and simulated data sets (Prelic et al. 2006). This algorithm is based on the signature algorithm, or SA (Ihmels et al. 2002). In Step 1, the algorithm takes a set of genes as input. In Step 2, it identifies the experimental conditions in which the set of genes are most similarly co-regulated. Threshold values represent the minimal level of co-regulation needed by a gene to be grouped within the set of co-regulated genes. The threshold values can be varied to increase or reduce stringency. When the threshold is reduced, additional genes are included in the set. Genes within the set that are poorly co-regulated are removed. Third, the algorithm examines those genes that were not included in the input set from Step 1 and selects from these new genes those that have a similar pattern of expression changes across the conditions defined

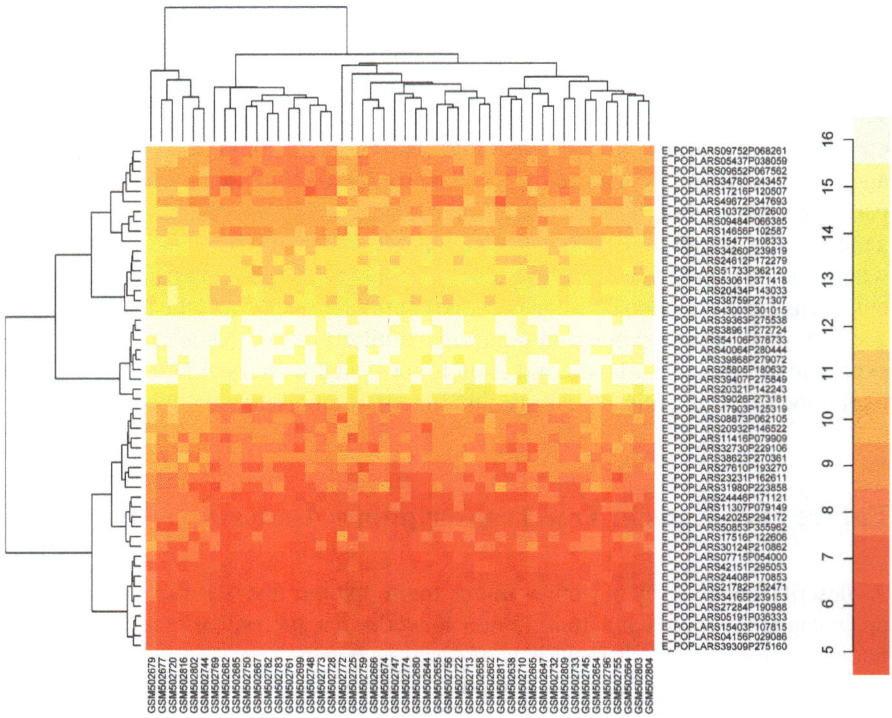

Fig. 2.2 Genes co-regulated in the *Populus* transcriptome. Normalized data from 183 four-plex NimbleGen microarrays (GEO accession number GSE20117) (Drost et al. 2010) was clustered using the Iterative Signature Algorithm (ISA, package eisa in R) (Bergmann et al. 2003). One of five modules produced is shown. The *color scale* represents the $\log_2$ of probe signal intensity values. This bicluster contains 51 genes (*rows*) and 48 tissues (*columns*). Dendrograms at the *top* and *left* indicate the distance between elements of the cluster

in Step 2. In this way, the algorithm identifies a subset of genes that are expressed in a subset of conditions. The ISA generates a large number of random gene sets to provide the SA algorithm for Step 1, and the SA is iteratively applied to each set to obtain two-way clusters. Application of this method to *Saccharomyces cerevisiae* gene expression profiles revealed a number of two-way clusters and a high conservation of putative *cis* regulatory motifs (Ihmels et al. 2004). Analysis of tree transcriptome data with these methods would very likely refine discoveries of tree transcriptional networks (Fig. 2.2).

The weighted gene correlation network analysis (WGCNA) also uses a novel approach to identify groups of co-regulated genes. This method has been powerful in the analysis of mammalian expression data, where WGCNA has identified a number of biologically relevant gene clusters (Oldham et al. 2006; Langfelder and Horvath 2008; Miller et al. 2010; Zhao et al. 2010). The software studies the interconnectedness between all gene pairs and identifies groups of genes that are highly correlated with each other. First, the method calculates the $(n \times n - 1)/2$ correlation coefficients of the *n* genes' expression levels across all samples. Second, the absolute values of the correlation coefficients are weighted by raising them to a power. Third, the topological overlap of each pair of genes is calculated. Two genes have high topological overlap if they both have similar expression patterns with the same group of genes. Two genes that are correlated with each other but are correlated with different sets of other genes will have a low topological overlap score. Scoring gene pairs using topological overlap removes isolated connections between genes. Fourth, genes are clustered according to topological overlap using hierarchical clustering. Individual expression modules are extracted from the hierarchical cluster tree by branch cutting, choosing nodes within the tree that represent true clusters.

The growth in transcriptome data will also enable the analysis of a large number of samples derived from the same tissue type or treatment to identify co-regulated and specific genes that are key to certain processes. The analysis of the inherent variability in gene expression across individuals has proven to be an excellent method to understand transcriptional networks (Zhao et al. 2010). For example, Horvath et al. (2006) examined microarray data extracted from the same cancer type using the WGCNA. One set of co-expressed genes was highly enriched with cell cycle genes. The ASPM gene was highly connected with genes within this module, suggesting the central importance of this gene in cell division. Knockdown of ASPM showed it had specific and dramatic inhibition of cell proliferation. A similar analysis of tree gene expression profiles across a large number of samples undergoing a shared process could identify key modules and genes involved in these processes. For example, key modules and genes involved in wood formation may be identified from a large number of wood development microarray data sets recently published by Drost et al. (2010) (Fig. 2.3).

One interesting aspect of tree molecular research is that traits including perennialism and woodiness evolved multiple times. Thus, the molecular changes that are responsible for these traits are likely due to a limited number of genetic changes. A multi-species comparative transcriptomic approach could determine a shared molecular basis of traits such as dormancy and woodiness. The ISA (Ihmels et al. 2005) may be applied to a data set including transcriptional profiles from a number of species. When the ISA was applied to two fungi, *Candida albicans* and *S. cerevisiae*, some gene expression clusters were shared within both species and some differed. For example, in *C. albicans*, amino acid biosynthesis was associated with protein synthesis genes; while this was not the case for *S. cerevisiae* (Ihmels et al. 2005). WGCNA has also been used for comparing transcriptional networks. As described above, Oldham et al. (2006) examined the similarities and differences in gene expression networks in six matched human and chimp brain regions. Co-regulated genes were identified in each species and compared. A number of modules were highly conserved between the species, which diverged 5–10 Million years ago.

The understanding of the molecular genetic basis for tree development is still in its infancy. Core questions such as "Are transcription programs for wood-formation and dormancy induction largely shared across species?" have not been fully addressed. The answers to this and other questions would represent a

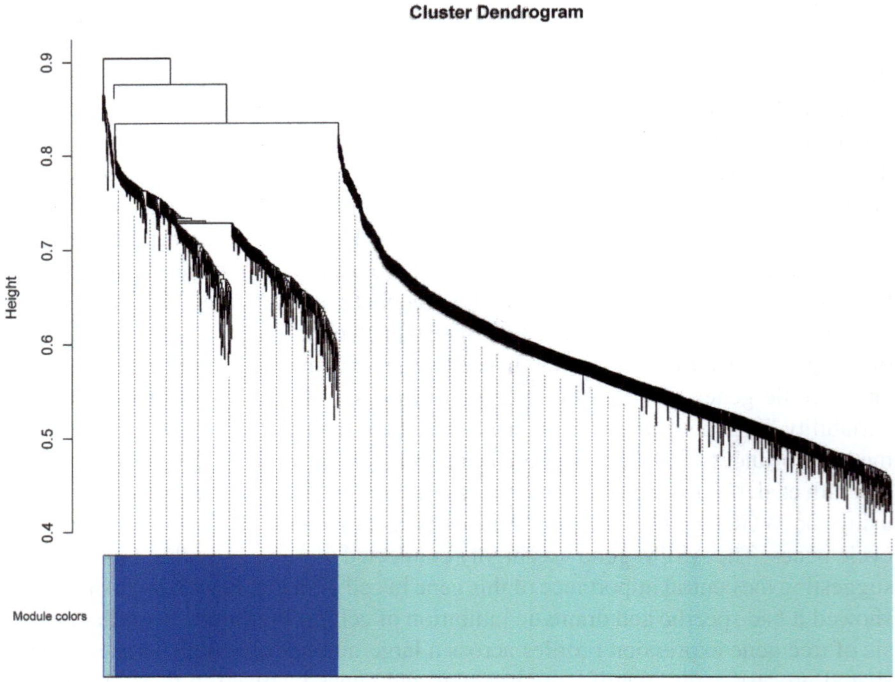

Fig. 2.3 WGCNA analysis of *Populus* transcriptome data. Normalized data from 183 four-plex NimbleGen microarrays (GEO accession number GSE20117) (Drost et al. 2010) was clustered using the Weighted Gene Correlation Network Analysis (WGCNA package in R) (Langfelder and Horvath 2008). This gene dendrogram includes the 1,678 genes classified into co-expression modules by Drost et al. (2010). The *height* in the dendrogram relates the distance between genes in the Topological Overlap Matrix. The *lowest tips* of the dendrogram represent the most highly connected genes in the module. The *colors* along the *bottom* indicate the modules produced by the algorithm. Two large coexpression modules are evident

major advancement in our understanding of tree biology and potentially have great economic impact. For example, genotypes with key alleles of regulatory genes may be selected for cultivation. As sequence data becomes more economical to acquire, and the number of expression microarray and high-throughput RNA sequencing profiles grows, the analysis of tree transcriptome data will continue to yield important discoveries.

Software for Microarrays and for Next-Generation Sequencing

The *R* statistical language (http://www.r-project.org) including the Bioconductor suite is the most common platform for transcriptome statistical analyses and visualizations (Gentleman 2005). There is a strong community of *R* programmers that integrate code and offer support. However, *R* is a command line programming language, and it takes a considerable investment of time to learn. The commercial software SAS is also used for array analysis in addition to the user-friendly software GeneSpring from Agilent technologies. Software applications for high-throughput sequence analysis include ELAND, MAQ, rSeq, SeqMap SSAHA2, TopHat, Cufflinks, and others.

References

Ashburner M, Ball CA, Blake JA, Botstein D, Butler H, Cherry JM, Davis AP, Dolinski K, Dwight SS, Eppig JT, Harris MA, Hill DP, Issel-Tarver L, Kasarskis A, Lewis S, Matese JC, Richardson JE, Ringwald M, Rubin GM, Sherlock G (2000) Gene ontology: tool for the unification of biology. The Gene Ontology Consortium. Nat Genet 25(1):25–29

Asmann YW, Klee EW, Thompson EA, Perez EA, Middha S, Oberg AL, Therneau TM, Smith DI, Poland GA, Wieben ED, Kocher JP (2009) 3′ tag digital gene expression profiling of human brain and universal reference RNA using Illumina Genome Analyzer. BMC Genomics 10:531

Bao Y, Dharmawardhana P, Mockler TC, Strauss SH (2009) Genome scale transcriptome analysis of shoot organogenesis in *Populus*. BMC Plant Biol 9:132

Barakat A, DiLoreto DS, Zhang Y, Smith C, Baier K, Powell WA, Wheeler N, Sederoff R, Carlson JE (2009) Comparison of the transcriptomes of American chestnut (*Castanea dentata*) and Chinese chestnut (*Castanea mollissima*) in response to the chestnut blight infection. BMC Plant Biol 9:51

Bergmann S, Ihmels J, Barkai N (2003) Iterative signature algorithm for the analysis of large-scale gene expression data. Phys Rev E Stat Nonlin Soft Matter Phys 67(3 Pt 1):031902

Bogeat-Triboulot M-B, Brosché M, Renaut J, Jouve L, Le Thiec D, Fayyaz P, Vinocur B, Witters E, Laukens K, Teichmann T, Altman A, Hausman J-F, Polle A, Kangasjärvi J, Dreyer E (2007) Gradual soil water depletion results in reversible changes of gene expression, protein profiles, ecophysiology, and growth performance in *Populus euphratica*, a poplar growing in arid regions. Plant Physiol 143(2):876–892

Brosché M, Vinocur B, Alatalo ER, Lamminmäki A, Teichmann T, Ottow EA, Djilianov D, Afif D, Bogeat-Triboulot M-B, Altman A, Polle A, Dreyer E, Rudd S, Paulin L, Auvinen P, Kangasjärvi J (2005) Gene expression and metabolite profiling of *Populus euphratica* growing in the Negev desert. Genome Biol 6(12):R101

Caruso A, Chefdor F, Carpin S, Depierreux C, Delmotte FM, Kahlem G, Morabito D (2008) Physiological characterization and identification of genes differentially expressed in response to drought induced by PEG 6000 in *Populus canadensis* leaves. J Plant Physiol 165(9):932–941

Chervitz SA, Hester ET, Ball CA, Dolinski K, Dwight SS, Harris MA, Juvik G, Malekian A, Roberts S, Roe T, Scafe C, Schroeder M, Sherlock G, Weng S, Zhu Y, Cherry JM, Botstein D (1999) Using the *Saccharomyces* Genome Database (SGD) for analysis of protein similarities and structure. Nucleic Acids Res 27(1):74–78

Dharmawardhana P, Brunner AM, Strauss SH (2010) Genome-wide transcriptome analysis of the transition from primary to secondary stem development in *Populus trichocarpa*. BMC Genomics 11:150

Drost DR, Novaes E, Boaventura-Novaes C, Benedict CI, Brown RS, Yin T, Tuskan GA, Kirst M (2009) A microarray-based genotyping and genetic mapping approach for highly heterozygous outcrossing species enables localization of a large fraction of the unassembled *Populus trichocarpa* genome sequence. Plant J 58(6):1054–1067

Drost DR, Benedict CI, Berg A, Novaes E, Novaes CR, Yu Q, Dervinis C, Maia JM, Yap J, Miles B, Kirst M (2010) Diversification in the genetic architecture of gene expression and transcriptional networks in organ differentiation of *Populus*. Proc Natl Acad Sci USA 107(18):8492–8497

Eisen MB, Spellman PT, Brown PO, Botstein D (1998) Cluster analysis and display of genome-wide expression patterns. Proc Natl Acad Sci USA 95(25):14863–14868

Emrich SJ, Li L, Wen T-J, Yandeau-Nelson MD, Fu Y, Guo L, Chou H-H, Aluru S, Ashlock DA, Schnable PS (2007) Nearly identical paralogs: implications for maize (*Zea mays L.)* genome evolution. Genetics 175(1):429–439

Eveland AL, McCarty DR, Koch KE (2008) Transcript profiling by 3′-untranslated region sequencing resolves expression of gene families. Plant Physiol 146(1):32–44

Gentleman R (2005) Bioinformatics and computational biology solutions using R and Bioconductor. Springer, New York

Guo M, Rupe MA, Danilevskaya ON, Yang X, Hu Z (2003) Genome-wide mRNA profiling reveals heterochronic allelic variation and a new imprinted gene in hybrid maize endosperm. Plant J 36(1):30–44

Guo M, Rupe MA, Yang X, Crasta O, Zinselmeier C, Smith OS, Bowen B (2006) Genome-wide transcript analysis of maize hybrids: allelic additive gene expression and yield heterosis. Theor Appl Genet 113(5):831–845

Harmer SL, Hogenesch JB, Straume M, Chang HS, Han B, Zhu T, Wang X, Kreps JA, Kay SA (2000) Orchestrated transcription of key pathways in *Arabidopsis* by the circadian clock. Science 290(5499):2110–2113

Hertzberg M, Aspeborg H, Schrader J, Andersson A, Erlandsson R, Blomqvist K, Bhalerao R, Uhlén M, Teeri TT, Lundeberg J, Sundberg B, Nilsson P, Sandberg G (2001) A transcriptional roadmap to wood formation. Proc Natl Acad Sci USA 98(25):14732–14737

Hoffman DE, Jonsson P, Bylesjö M, Trygg J, Antti H, Eriksson ME, Moritz T (2010) Changes in diurnal patterns within the *Populus* transcriptome and metabolome in response to photoperiod variation. Plant Cell Environ 33(8):1298–1313

Horvath S, Zhang B, Carlson M, Lu KV, Zhu S, Felciano RM, Laurance MF, Zhao W, Qi S, Chen Z, Lee Y, Scheck AC, Liau LM, Wu H, Geschwind DH, Febbo PG, Kornblum HI, Cloughesy TF, Nelson SF, Mischel PS (2006) Analysis of oncogenic signaling networks in glioblastoma identifies ASPM as a molecular target. Proc Natl Acad Sci USA 103(46):17402–17407

Ihmels J, Friedlander G, Bergmann S, Sarig O, Ziv Y, Barkai N (2002) Revealing modular organization in the yeast transcriptional network. Nat Genet 31(4):370–377

Ihmels J, Bergmann S, Barkai N (2004) Defining transcription modules using large-scale gene expression data. Bioinformatics 20(13):1993–2003

Ihmels J, Bergmann S, Berman J, Barkai N (2005) Comparative gene expression analysis by differential clustering approach: application to the *Candida albicans* transcription program. PLoS Genet 1(3):e39

Kanehisa M, Goto S (2000) KEGG: Kyoto encyclopedia of genes and genomes. Nucleic Acids Res 28(1):27–30

Kirst M, Myburg AA, De León JPG, Kirst ME, Scott J, Sederoff R (2004) Coordinated genetic regulation of growth and lignin revealed by quantitative trait locus analysis of cDNA microarray data in an interspecific backcross of eucalyptus. Plant Physiol 135(4):2368–2378

Kirst M, Basten CJ, Myburg AA, Zeng Z-B, Sederoff RR (2005) Genetic architecture of transcript-level variation in differentiating xylem of a eucalyptus hybrid. Genetics 169(4):2295–2303

Kluger Y, Basri R, Chang JT, Gerstein M (2003) Spectral biclustering of microarray data: coclustering genes and conditions. Genome Res 13(4):703–716

Langfelder P, Horvath S (2008) WGCNA: an R package for weighted correlation network analysis. BMC Bioinformatics 9:559

Maere S, Van Dijck P, Kuiper M (2008) Extracting expression modules from perturbational gene expression compendia. BMC Syst Biol 2:33

Miller JA, Horvath S, Geschwind DH (2010) Divergence of human and mouse brain transcriptome highlights Alzheimer disease pathways. Proc Natl Acad Sci USA 107(28):12698–12703

Moreau C, Aksenov N, Lorenzo MG, Segerman B, Funk C, Nilsson P, Jansson S, Tuominen H (2005) A genomic approach to investigate developmental cell death in woody tissues of *Populus* trees. Genome Biol 6(4):R34

Mortazavi A, Williams BA, McCue K, Schaeffer L, Wold B (2008) Mapping and quantifying mammalian transcriptomes by RNA-Seq. Nat Methods 5(7):621–628

Nilsson J, Karlberg A, Antti H, Lopez-Vernaza M, Mellerowicz E, Perrot-Rechenmann C, Sandberg G, Bhalerao RP (2008) Dissecting the molecular basis of the regulation of wood formation by auxin in hybrid aspen. Plant Cell 20(4):843–855

Oldham MC, Horvath S, Geschwind DH (2006) Conservation and evolution of gene coexpression networks in human and chimpanzee brains. Proc Natl Acad Sci USA 103(47):17973–17978

Parchman TL, Geist KS, Grahnen JA, Benkman CW, Buerkle CA (2010) Transcriptome sequencing in an ecologically important tree species: assembly, annotation, and marker discovery. BMC Genomics 11:180

Prelic A, Bleuler S, Zimmermann P, Wille A, Buhlmann P, Gruissem W, Hennig L, Thiele L, Zitzler E (2006) A systematic comparison and evaluation of biclustering methods for gene expression data. Bioinformatics 22(9):1122–1129

Quesada T, Li Z, Dervinis C, Li Y, Bocock PN, Tuskan GA, Casella G, Davis JM, Kirst M (2008) Comparative analysis of the transcriptomes of *Populus trichocarpa* and *Arabidopsis thaliana* suggests extensive evolution of gene expression regulation in angiosperms. New Phytol 180(2):408–420

Ruttink T, Arend M, Morreel K, Storme V, Rombauts S, Fromm J, Bhalerao RP, Boerjan W, Rohde A (2007) A molecular timetable for apical bud formation and dormancy induction in poplar. Plant Cell 19(8):2370–2390

Schrader J, Moyle R, Bhalerao R, Hertzberg M, Lundeberg J, Nilsson P, Bhalerao RP (2004) Cambial meristem dormancy in trees involves extensive remodelling of the transcriptome. Plant J 40(2):173–187

Sederoff R, Myburg A, Kirst M (2009) Genomics, domestication, and evolution of forest trees. Cold Spring Harb Symp Quant Biol 74:303–317

Simon SA, Zhai J, Nandety RS, McCormick KP, Zeng J, Mejia D, Meyers BC (2009) Short-read sequencing technologies for transcriptional analyses. Annu Rev Plant Biol 60:305–333

Smith LG, Greene B, Veit B, Hake S (1992) A dominant mutation in the maize homeobox gene, Knotted-1, causes its ectopic expression in leaf cells with altered fates. Development 116(1):21–30

Sterky F, Bhalerao RR, Unneberg P, Segerman B, Nilsson P, Brunner AM, Charbonnel-Campaa L, Lindvall JJ, Tandre K, Strauss SH, Sundberg B, Gustafsson P, Uhlen M, Bhalerao RP, Nilsson O, Sandberg G, Karlsson J, Lundeberg J, Jansson S (2004) A *Populus* EST resource for plant functional genomics. Proc Natl Acad Sci USA 101(38):13951–13956

Street NR, Skogström O, Sjödin A, Tucker J, Rodríguez-Acosta M, Nilsson P, Jansson S, Taylor G (2006) The genetics and genomics of the drought response in *Populus*. Plant J 48(3):321–341

Stupar RM, Springer NM (2006) Cis-transcriptional variation in maize inbred lines B73 and Mo17 leads to additive expression patterns in the F1 hybrid. Genetics 173(4):2199–2210

Swanson-Wagner RA, Jia Y, DeCook R, Borsuk LA, Nettleton D, Schnable PS (2006) All possible modes of gene action are observed in a global comparison of gene expression in a maize F1 hybrid and its inbred parents. Proc Natl Acad Sci USA 103(18):6805–6810

Swanson-Wagner RA, DeCook R, Jia Y, Bancroft T, Ji T, Zhao X, Nettleton D, Schnable PS (2009) Paternal dominance of trans-eQTL influences gene expression patterns in maize hybrids. Science 326(5956):1118–1120

't Hoen PA, Ariyurek Y, Thygesen HH, Vreugdenhil E, Vossen RH, de Menezes RX, Boer JM, van Ommen GJ, den Dunnen JT (2008) Deep sequencing-based expression analysis shows major advances in robustness, resolution and inter-lab portability over five microarray platforms. Nucleic Acids Res 36(21):e141

Tamayo P, Slonim D, Mesirov J, Zhu Q, Kitareewan S, Dmitrovsky E, Lander ES, Golub TR (1999) Interpreting patterns of gene expression with self-organizing maps: methods and application to hematopoietic differentiation. Proc Natl Acad Sci USA 96(6):2907–2912

Tavazoie S, Hughes JD, Campbell MJ, Cho RJ, Church GM (1999) Systematic determination of genetic network architecture. Nat Genet 22(3):281–285

Tiquia SM, Wu L, Chong SC, Passovets S, Xu D, Xu Y, Zhou J (2004) Evaluation of 50-mer oligonucleotide arrays for detecting microbial populations in environmental samples. Biotechniques 36(4):664–670, 672, 674–675

Tuskan GA, DiFazio S, Jansson S, Bohlmann J, Grigoriev I, Hellsten U, Putnam N, Ralph S, Rombauts S, Salamov A, Schein J, Sterck L, Aerts A, Bhalerao RR, Bhalerao RP, Blaudez D, Boerjan W, Brun A, Brunner A, Busov V, Campbell M, Carlson J, Chalot M, Chapman J, Chen GL, Cooper D, Coutinho PM, Couturier J, Covert S, Cronk Q, Cunningham R, Davis J, Degroeve S, Dejardin A, de Pamphilis C, Detter J, Dirks B, Dubchak I, Duplessis S, Ehlting J, Ellis B, Gendler K, Goodstein D, Gribskov M, Grimwood J, Groover A, Gunter L, Hamberger B, Heinze B, Helariutta Y, Henrissat B, Holligan D, Holt R, Huang W, Islam-Faridi N, Jones S, Jones-Rhoades M, Jorgensen R, Joshi C, Kangasjarvi J, Karlsson J, Kelleher C, Kirkpatrick R, Kirst M, Kohler A, Kalluri U, Larimer F, Leebens-Mack J, Leple JC, Locascio P, Lou Y, Lucas S, Martin F, Montanini B, Napoli C, Nelson DR, Nelson C, Nieminen K, Nilsson O, Pereda V, Peter G, Philippe R, Pilate G, Poliakov A, Razumovskaya J, Richardson P, Rinaldi C, Ritland K, Rouze P, Ryaboy D, Schmutz J, Schrader J, Segerman B, Shin H, Siddiqui A, Sterky F, Terry A, Tsai CJ, Uberbacher E, Unneberg P, Vahala J, Wall K, Wessler S, Yang G, Yin T, Douglas C, Marra M, Sandberg G, Van de Peer Y, Rokhsar D (2006) The genome of black cottonwood, *Populus trichocarpa* (Torr. & Gray). Science 31(5793):1596–1604

Velculescu VE, Zhang L, Vogelstein B, Kinzler KW (1995) Serial analysis of gene expression. Science 270(5235):484–487

Wang Z, Gerstein M, Snyder M (2009) RNA-Seq: a revolutionary tool for transcriptomics. Nat Rev Genet 10(1):57–63

Wilhelm BT, Marguerat S, Watt S, Schubert F, Wood V, Goodhead I, Penkett CJ, Rogers J, Bahler J (2008) Dynamic repertoire of a eukaryotic transcriptome surveyed at single-nucleotide resolution. Nature 453(7199):1239–1243

Wilkins O, Waldron L, Nahal H, Provart NJ, Campbell MM (2009) Genotype and time of day shape the *Populus* drought response. Plant J 60(4):703–715

Zhan S, Lukens L (2010) Identification of novel miRNAs and miRNA dependent developmental shifts of gene expression in *Arabidopsis thaliana*. PLoS One 5(4):e10157

Zhao W, Langfelder P, Fuller T, Dong J, Li A, Hovarth S (2010) Weighted gene coexpression network analysis: state of the art. J Biopharm Stat 20(2):281–300

Chapter 3
Functional Genomics of Flowering Time in Trees

Magda-Viola Hanke, Henryk Flachowsky, Hans Hoenicka, and Matthias Fladung

Abstract Genomics of floral induction and flower development in trees is much more complex than in annual/biennial herbaceous plants. Detailed genetic models explaining different steps of these developmental processes already exist in model plants like *Arabidopsis thaliana* and *Antirrhinum majus*. Most of this basic knowledge can be easily applied to many annual/biennial angiosperm plants and, at least in part, to perennial, polycarpic plants such as trees. The transfer of knowledge from the annual to perennial plants and from model plants to economically important crops such as fruit and forest trees is currently underway. A number of major traits have been mapped in different tree species, and individual flowering gene homologues have been isolated and functionally characterized. The establishment of methods for reverse genetics studies and the development of next-generation sequencing technologies were milestones which lead to an acceleration in this field of research. The identification of candidate genes will thereby be dramatically accelerated by the availability of entire genome sequences of different tree species. The functional characterization of such candidate genes will help to complete the picture on genetics of flowering in trees bit by bit.

Keywords *Arabidopsis* • Floral induction • Forest trees • Fruit trees • Climate change

M.-V. Hanke (✉) • H. Flachowsky
Institute for Breeding Research on Horticultural and Fruit Crops,
Julius Kuehn-Institute, Pillnitzer Platz 3a, Dresden, Saxony, Germany
e-mail: viola.hanke@jki.bund.de

H. Hoenicka • M. Fladung, Ph.D.
Institute of Forest Genetics, Johann Heinrich von Thünen Institute,
Grosshansdorf, Germany

R.J. Schnell and P.M. Priyadarshan (eds.), *Genomics of Tree Crops*,
DOI 10.1007/978-1-4614-0920-5_3, © Springer Science+Business Media, LLC 2012

Introduction

Genetic regulation of floral induction in perennial tree species is much more complex than in annual/biennial herbaceous plants. Annual/biennial plants have to pass a transition phase from the vegetative (juvenile) stage to the generative (adult) stage. During the juvenile stage, the plant is unable to produce flowers. After the plant has reached the adult stage, endogenous and environmental signals such as plant age, growth regulators, temperature, and day length decide whether the plant starts to flower or not. Subsequently, the plant stops its vegetative growth step by step and a floral "promoter" induces all above-ground meristems to flower in the same season (Bangerth 2009). Perennial plants have also to pass from the juvenile to the adult stage, but unlike annual herbaceous plants, juvenility in tree species can last a couple of years until the reproductive stage is reached (Table 3.1). During this time, only vegetative growth takes place and the plant is absolutely unable to produce flowers. Favorable environmental conditions such as photoperiod or temperature are ineffective in breaking the juvenility. The end of the juvenile period is indicated by the attainment of the ability to flower, and the actual production of flowers is the first evidence that plant is in the adult phase (Hanke et al. 2007). However, the end of the juvenile period and the first appearance of flowers may not coincide. Seedlings do not flower necessarily after finishing the juvenile stage because of other factors. Flower production takes place not until the seedlings have attained the ability to flower. This period of transition is also defined as the adult vegetative phase

Table 3.1 Duration of the juvenile stage in different tree species

Species	Common name	Juvenility in years	References
Betula sp.	Birch	10–15	Perala and Alm (1990)
Citrus sp.	Citrus	20	Peña et al. (2001)
Cydonia oblonga	Quince	5–6	Kadir (2003)
Malus × domestica	Apple	5–10	Visser (1964)
Mangifera indica	Mango	6–7	Litz et al. (1991)
Phoenix dactylifera	Date palm	4–10	Rajan and Markose (2007)
Populus sp.	Poplar	7–10	Hsu et al. (2006)
Pouteria sapota	Mamey sapote	8–10	Rajan and Markose (2007)
Prunus armeniaca	Apricot	2–5	Kadir (2003)
Prunus avium	Sweet cherry	3–7	Blazek (1985) and Kadir (2003)
Prunus cerasus	Sour cherry	3–5	Kadir (2003)
Prunus domestica	Plum	3–6	Kadir (2003)
Prunus persica	Peach	3–4	Sosinski et al. (2000)
Pyrus sp.	Pear	4–6	Kadir (2003)
Quararibea asterolepis	Molenillo	10–20	De Steven (1994)
Sclerocarya birrea	Marula	8–10	Akinnifesi et al. (2004)
Strychnos cocculoides	Corky-bark monkey-orange	10	Akinnifesi et al. (2004)
Tetragastris panamensis		10–20	De Steven (1994)
Trichilia tuberculata		10–20	De Steven (1994)

(Poethig 1990). During this phase, most floral-inducing techniques are applied successfully. After the plant has reached the adult stage, flowers are produced yearly or in a biennial rhythm. In perennial plants, a sophisticated regulatory system consisting of many different factors fine tunes floral induction so that only a proportion of meristems will be transformed into flowers at any one time.

Floral induction in mature fruit trees constitutes a morphogenetic transition of stem cells in apical and lateral central meristems into differentiated floral cells (Bangerth 2009). This transition implies the activation and transcription of many genes (Boss et al. 2004) in most annual/biennial plants that are controlled by exogenous/endogenous factors. In contrast, in angiosperm trees endogenous factors prevail under temperate conditions, whereas exogenous and endogenous factors are at work in subtropical and tropical species (Bangerth 2009).

In recent years, several genes were identified that are involved in the transition from the vegetative to the reproductive phase and the onset of flowering in *Arabidopsis*, and homologs were isolated and functionally characterized in a number of plant species (for review, see Jung and Müller 2009).

Floral Initiation in Model Plants

Molecular biology and genetics of flowering is already intensively studied in annual/biennial model plants like *Arabidopsis thaliana* and *Antirrhinum majus*. Basic results obtained on these model plants can be applied to many annual/biennial angiosperm plants and, at least in part, to perennial, polycarpic plants such as trees. The genetics of flower induction and floral organ formation seems to be similar among annual/biennial and perennial plants (Tan and Swain 2006). However, whether this emerging molecular-genetic model of floral induction in annual/biennial plants can fully explain all experimental results obtained for floral induction in perennial angiosperm trees has not been fully answered till date (Bangerth 2009). Existing differences between annual/biennial and perennial plants, which cannot be explained with the general models of floral induction in model plants, are discussed below.

In general, the lifecycle of a plant can be divided into two major phases, the vegetative and the generative phase (juvenile and adult phase in perennials). During the vegetative phase, the plant does not obtain the ability to flower. In contrast, the generative phase is determined by the ability of a plant to flower. To switch from the vegetative to the generative phase, plants have to undergo a transition phase.

Flowering in *Arabidopsis* is controlled by four major pathways: photoperiodic, autonomous, gibberellins (GA), and vernalization (Martinez-Zapater et al. 1994). Endogenous as well as exogenous stimulators perceived in leaves (e.g., daylength or temperature) or in the shoot apical meristem itself (e.g., temperature or hormones) can induce promoting pathways, activating the expression of genes that cause floral transition (floral pathway integrators), and enabling pathways (Fig. 3.1). Repressors antagonizing the activation of floral transition are regulated by the enabling pathways (Boss et al. 2004). This complex interaction of multiple pathways ensures the transition of a plant into the generative phase during favorable environmental conditions.

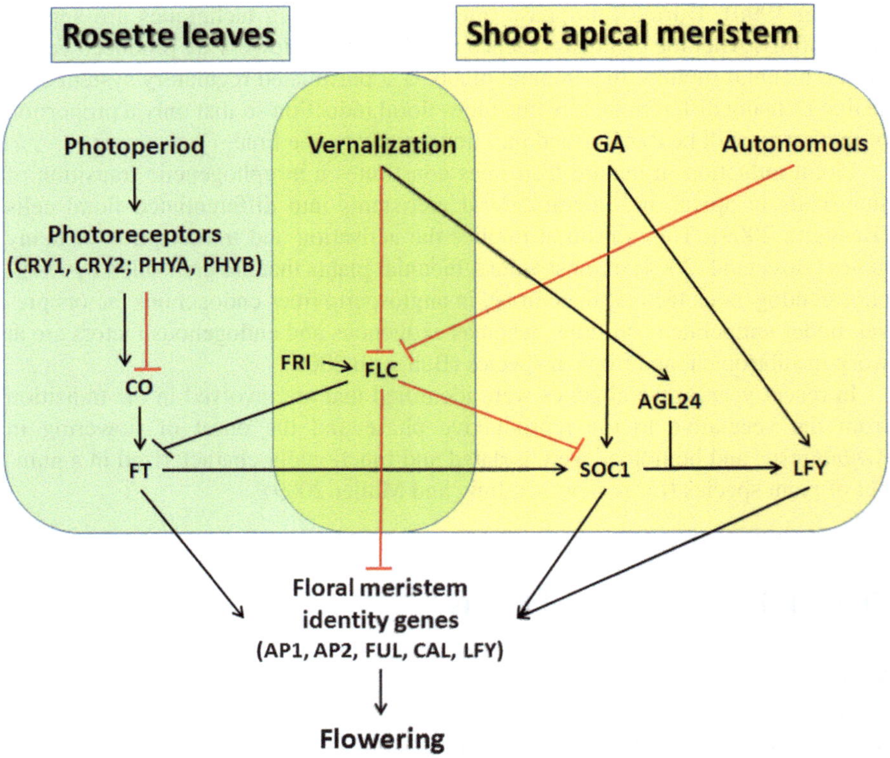

Fig. 3.1 Simplified model of floral induction in the model plant *Arabidopsis thaliana*. Floral induction is controlled by four major pathways (photoperiod, vernalization, gibberellic acid, and autonomous). *Pointed arrows* represent positive regulation (induction), *"T" arrows* represent negative regulation (suppression), *CO* CONSTANS, *FT* FLOWERING LOCUS T, *FRI* FRIGIDA, *FLC* FLOWERING LOCUS C, *GA* gibberellic acid, *SOC1* SUPPRESSOR OF OVEREXPRESSION OF CONSTANS 1 (Modified from Wilkie et al. 2008)

Photoperiodic Induction

Daylength is one of the most important factors in view of flowering in long-day (LD) as well as in short-day (SD) plants. Both, LD and SD plants flower in response to the change in the length of the dark period. LD plants require a short and SD plants a long dark period. The length of dark and light periods is perceived in leaves by photoreceptors such as cryptochrome 1 and 2 and phytochrome A and B. Under LD conditions, CRYPTOCHROMEs (CRY1 and CRY2) and PHYTOCHROME A (PHYA) act antagonistically to PHYB. Valverde et al. (2004) have shown that light stabilizes nuclear CONSTANS (CO) protein in the evening, whereas in the morning or in darkness, the protein is degraded by the proteosome. The *CO* gene is specifically expressed in the vascular tissues of the leaf (Takada and Goto 2003). Its expression is regulated by the circadian clock and the peak expression occurs at the end of the day under LD, but after dark under SD conditions (Suarez-Lopez et al. 2001; Yanovsky and Kay 2002; Valverde et al. 2004). Thus, *CO* is a key factor of floral induction which promotes the expression of the floral pathway integrator genes such as *FLOWERING LOCUS T* (*FT*) and *SUPPRESSOR OF OVEREXPRESSION OF CONSTANS 1* (*SOC1*) (Kardailsky et al. 1999; Kobayashi et al. 1999; Samach et al. 2000; Suarez-Lopez et al. 2001; Hepworth et al. 2002; Valverde et al. 2004). The FT protein is expressed in the vascular tissues of the leaf and then transported via the phloem to the shoot apical meristem (Takada and Goto 2003; Corbesier et al. 2007). In the shoot apical meristem, *FT* forms a protein complex with FLOWERING LOCUS D (FD), a bZIP transcription factor that is preferentially expressed at the shoot apex. The FT/FD protein complex interacts with floral meristem identity genes like *AP1*, which then convert shoot meristems into floral meristems and consequently into flowers (Abe et al. 2003; Wigge et al. 2005).

GA-Dependent Induction

Gibberellic acids (GA) also actively promote flowering in *Arabidopsis*. Their promoting effect to flowering, particularly under short days, has been shown after application of exogenous GA by Chandler and Dean (1994) and Langridge (1957). Grafting experiments showed that endogenous GA also initiates flowering in *Arabidopsis*. Under short day conditions, GA_4, which is most likely produced in the leaves and transported to the shoot apical meristem, upregulates the expression of the floral integrator gene *SOC1* (Fig. 3.1) and/or the floral meristem identity gene *LEAFY* (Blazquez et al. 1997; Blazquez and Weigel 2000; Moon et al. 2003; Bernier and Perilleux 2005; Eriksson et al. 2006). Mutants in GA biosynthesis or signaling fail to flower under short day conditions and show delayed flowering under long day conditions (Wilson et al. 1992; Sun and Kamiya 1994).

Vernalization

Vernalization means the induction of flowering in response to exposure to cold temperatures (−1°C to 10°C) for extended periods (Simpson and Dean 2002). The MADS domain gene *FLOWERING LOCUS C* (*FLC*) is a central repressor of flowering (Michaels and Amasino 1999), which is negatively regulated by the expression of genes induced by the vernalization and the autonomous pathways. *FLC* suppresses flowering by repressing the floral pathway integrators *CO*, *LEAFY*, and *SOC1* (Kobayashi et al. 1999; Blazquez and Weigel 2000; Lee et al. 2000). Cold temperatures are perceived in the shoot apical meristem (SAM). As a consequence of this, the cold-response genes *VERNALISATION 1* (*VRN1*), *VRN2*, and *VRN3* will be activated and changes in DNA methylation take place (Finnegan et al. 1998). These factors suppress *FLC*, which inhibits the floral pathway integrator gene *FT*, and its downstream floral initiation cascade (Michaels and Amasino 2000; Sheldon et al. 2000; Vijayraghavan et al. 2005). Thus, vernalization promotes flowering by suppression of *FLC* and activates the *FT*-controlled transition.

Autonomous Flowering

The autonomous pathway represents a summary of repressor genes that affect flowering under any photoperiod and confer a response to vernalization (Fig. 3.1). This pathway includes the RNA-processing factors *FCA*, *FPA*, *FLOWERING LOCUS K* (*FLK*), and *FY*, the homeodomain protein *LUMINIDEPENDENS* (*LD*), and several histone deacetylation factors like *FVE*, *FLOWERING LOCUS D* (*FLD*), and *RELATION OF EARLY FLOWERING 6* (*REF6*) (Quesada et al. 2005). Mutations in each of these genes lead to an increase in the levels of *FLC* mRNA and FLC protein (Michaels and Amasino 1999; Sheldon et al. 1999; Rouse et al. 2002).

In addition to the four major pathways of flowering time control in *Arabidopsis,* there are other pathways that regulate flowering by more general aspects of plant metabolism (Bernier and Perilleux 2005). Recently, it was found that sucrose is important as a signal and assimilate in the flowering of a starchless mutant (Corbesier et al. 1998). Eriksson et al. (2006) found that sucrose could act synergistically with GA to promote flowering in the absence of long days. In recent time, several micro RNAs (miRNAs) were identified, which are obviously involved in regulation of flowering. miRNAs are small RNA molecules with a length of about 21 nucleotides. In *Arabidopsis,* at least 92 miRNAs have been described, most of them targeting mRNAs coding for transcription factors (Dugas and Bartel 2004). The tree miRNAs, *miR172*, *miR159,* and *miR156* were found to be involved in the regulation of flowering time. They are involved (directly or indirectly) in the regulation of the floral meristem and floral organ identity gene *APETALA2* (*AP2*) and the *AP2*-related genes *TARGET OF EAT1* (*TOE1*), *TOE2*, *SCHLAFMUTZE* (*SMZ*), and *SCHNARCHZAPFEN* (*SNZ*), which are known as repressors of flowering (Aukerman and Sakai 2003; Schmid et al. 2003). They seem furthermore involved in the regulation of *MYB33*, *LEAFY*, and the *SQUAMOSA PROMOTER BINDING PROTEIN LIKE* (*SPL*) genes *SPL3*, *SPL4,* and *SPL5* (Rhoades et al. 2002; Kasschau et al. 2003; Quesada et al. 2005).

Floral Initiation and Flower Induction in Trees

Populus spp.

The genus *Populus* is a genus of 25–35 species of deciduous plants (family Salicaceae) native to the Northern Hemisphere. *Populus* has become a model system for tree and woody plant biology. This has been largely driven by the rapid development of genomic and molecular biology resources for this genus (Jansson and Douglas 2007). Most poplars have a vegetative phase of more than 8 years. However, shorter ones, between 2 and 3 years, have also been reported for *P. tremuloides* Michx. (USDA/NRCS 2010).

In poplars, floral initiation between mid-May and mid-June has been reported (Böhlenius et al. 2006). The role of environmental factors in the timing of tree flowering is not completely clear. However, the central role of photoperiod on flowering time, growth cessation, and bud set in the fall was clearly shown in poplar (Böhlenius et al. 2006). The same authors found that different critical daylengths are necessary for growth cessation in European aspen (*P. tremula*) trees originating from different latitudes (Böhlenius et al. 2006). Variations in temperature and light intensity during the suitable period for flower bud initiation could explain the large year-to-year variations in flowering (i.e., in some years, mature trees do not flower) of aspen and many other tree species (Owens 1995). However, not all poplar species exhibit year-to-year variations in flowering (Brunner and Nilsson 2004). Flowering time seems to be determined by the accumulated amount of heat or heat and light that the plant receives during a period preceding the floral initiation (reviewed in Poethig 2003). A study of adult willow cutting in controlled environments indicated that both photoperiod and temperature affect floral initiation (Junttila 1980). Willow (Salix) and poplars are both members of the family Salicaceae, and similar influence has been proposed for poplars (Brunner and Nilsson 2004).

The association between shoot morphogenesis and flowering was studied in detail in *P. deltoides* (Yuceer et al. 2003). Shoots sequentially initiate early vegetative, floral, and late vegetative buds. Associated with these buds is the formation of three distinct leaf types. In May of the first growing season, the first type begins forming in terminal buds and overwinters as relatively developed foliar structures. These leaves bear early vegetative buds in their axils in the second growing season. The second type forms late in the first growing season in terminal buds. These form floral buds in their axils in the second growing season. The floral bud meristems initiate scale leaves in April and begin forming floral meristems in the axils of the bracts in May. The floral meristems subsequently form floral organs by the end of the second growing season. The floral buds overwinter with floral organs, and anthesis occurs in the third growing season. The third type of leaves forms and develops entirely outside the terminal buds in the second growing season. These leaves bear the late vegetative buds in their axils.

In *Populus alba*, an early flowering genotype that formed female flowers within 1 year after sowing was identified after intensive selection within a breeding program (Meilan et al. 2004). This *P. alba* (clone 6K10), receiving 6 months of root chilling (3°C), 8 weeks of short, cool days (10°C), and 10 weeks of cold temperatures (4°C), formed floral buds 9 months after vegetative propagation. However, the early flowering of this line in the first year was not repeated in a field trial (Strauss, unpublished results) or in greenhouse (Hönicka and Fladung, unpublished results).

Yuceer et al. (2003) conducted experiments with eastern cottonwood (*Populus deltoides* Bartr. Ex Marsch. var. *deltoides*) to evaluate the influence of various treatments on flowering of rooted cuttings from mature and juvenile trees. A combined treatment of water stress, root pruning, and paclobutrazol induced flowering in 3-month-old rooted cuttings from mature trees but no flowering in 1-year-old rooted cuttings from juvenile trees. Development of untreated rooted cuttings from mature trees remained vegetative. Daminozide was informally observed to stimulate flowering in aspen (Meilan 1997). However, treatment with different concentrations of paclobutrazol and daminozide did not induce early flowering in *P. tremula* (clone Brauna 11) (Hoenicka and Fladung, unpublished results).

Eucalyptus spp.

Eucalyptus is a very diverse genus with tropical and temperate species belonging to the family Myrtaceae. There are more than 700 species of *Eucalyptus*, mostly native to Australia. The age at which eucalyptus first produces flowers varies markedly between species and individuals within a species (Eldridge et al. 1993). *Eucalyptus* flowers either biennially or sporadically (Meilan 1997). Most species, including all major commercial species, generally do not flower reliably until 2–6 years of age (Southerton 2007).

The role of environmental factors in the timing of eucalyptus flowering is not completely known. Flower initiation may be probably influenced by the completely different climatic zones inhabited by *Eucalyptus* species ranging from tropical, temperate, dry, and wet regions. The influence of photoperiod on flower initiation has been studied in *Eucalyptus occidentalis* Endl. (Bolotin 1975). In its native habitat, *E. occidentalis* grows at latitudes 31–34°, where the maximum daylengths range from 13.9 to 14.2 h. *E. occidentalis* does not flower until 3–5 years of age (Blakely 1955). However, seedlings of *E. occidentalis* planted in other parts of the world flower precociously at the age of less than 1 year when grown under a long day regime of 16 h or longer (Bolotin 1975). Precocious flowering is more widespread among summer-sown seedlings than among seedlings sown in the winter (Bolotin 1975).

A strong influence of temperature on floral induction was shown in *E. lansdowneana* (Moncur 1992). Transferring seedlings from a heated greenhouse (24°C/19°C) to a cold greenhouse (15°C/10°C) for 5 or 10 weeks and back to the heated greenhouse was sufficient to induce floral buds in *E. lansdowneana*. Bud production was further enhanced when seedlings were transferred to cold conditions during periods of high solar radiation. Under low levels of solar radiation and short duration of cold, 0–5 weeks, the plants reverted to vegetative development, suggesting a low floral induction stimulus (Moncur 1992). Floral response in *E. lansdowneana* seems to be insensitive to daylength as it occurs under both long and short days (Moncur 1992).

Early flowering has been identified as an important trait in several eucalypt breeding programs (Chambers et al. 1997; Wiltshire et al. 1998). Grafting and establishing selections into breeding arboreta or environments more conducive to flowering are common means of stimulating flowering of *Eucalyptus* species (Eldridge et al. 1993). Treatment with the gibberellin biosynthesis-inhibitor paclobutrazol is widely exploited as a means of stimulating flowering in *Eucalyptus* (Moncur 1992; Griffin et al. 1993; Hasan and Reid 1995; Williams et al. 1999). Under specific greenhouse conditions, flowering has been induced after only 19 months in *Eucalyptus globulus* (Hasan and Reid 1995). The application of both nitrogen fertilizer and paclobutrazol substantially increased the occurrence of precocious flowering in reproductive immature *E. globulus* (Williams et al. 2003). Phosphorus shows no influence on flower initiation but on "vegetative phase change" of *E. globulus* (Williams et al. 2004). The physiological and chronological age at which the tree switches from the production of juvenile foliage to adult foliage is called "vegetative phase change" (Williams et al. 2004).

Picea spp.

A spruce is a coniferous evergreen tree of the genus *Picea* consisting of about 35 species (Family Pinaceae), found in the northern temperate and boreal (taiga) regions. In *Picea*, it can take about 20 years before any cones are seen, and heavy seed years may not start until several years later (Longman 1987). Even then, they may only occur at intervals of about 3–5 years (Longman 1987).

Owens and Molder (1976) carried out a very detailed study on cone initiation and induction in *Picea sitchensis* (Bong.). Pollen-cone and seed-cone buds of *P. sitchensis* (Bong.) Carr. are found as either terminal or axillary buds. Pollen cones are most likely to develop from small axillary apices on vigorous distal shoots or small terminal apices on less-vigorous, proximal shoots. Seed cones are most likely to develop from large, distal axillary apices on vigorous shoots or smaller terminal apices on less-vigorous shoots. All apices became mitotically active late in March, passed through a 3.5-month period of bud-scale initiation, and in mid-July became differentiated as vegetative, pollen-cone, or seed-cone apices. Leaf, bract, and microsporophyll initiation began about the end of July. All microsporophylls were initiated by the end of August. Sporogenous cells developed, but meiosis did not occur before the pollen cones became dormant at the end of October. Two thirds of the bracts were initiated by the end of August. The remaining bracts were initiated more slowly until dormancy. Ovuliferous scales were initiated for 3 months beginning in September, and megaspore mother cells appeared, but did not undergo meiosis before seed cones became dormant at the end of November. There was no difference in the time of vegetative, pollen-cone, and seed-cone bud differentiation, which occurred at the end of lateral shoot elongation.

There are several studies on the role of gibberellins on cone induction in *Picea*. Dunberg (1974) first indicated that endogenous gibberellins participate in cone induction in *Picea abies*. There is evidence that endogenous levels of both gibberellins and auxins may change around the time of differentiation of cones (Kulikowska et al. 1978; Ivonis et al. 1982). A positive effect on cone initiation was reported by the application of gibberellin (GA_4, GA_7, GA_9) to *P. abies* (Dunberg 1976, 1980). Treatment with the gibberellin mixture A_4 and A_7 ($GA_{4/7}$) enhanced female cones on

Fig. 3.2 *Picea abies* "*acrocona*" F1-seedling, 6 years old. Flower formation occurred at age of 3–4 years for the first time

mature *Picea glauca* (Marquard and Hanover 1984). Time of treatment, crown position, and $GA_{4/7}$ concentration influenced treatment success. $GA_{4/7}$ treatment initiated before meristematic differentiation began (late June) enhanced female strobilus production, whereas those initiated after late June were ineffective (Marquard and Hanover 1984).

Picea abies "acrocona"

An early flowering variety of *P. abies* named "*acrocona*" is known since 1890, when it was detected in the wild near Uppsala, Sweden (Krüssmann 1983). This variety is characterized by the extremely precocious development of mainly (female) seed cones at terminal positions of shoots (literature in Langner 1954; Fladung et al. 2000). *Acrocona* forms cones abundantly, even on young plants, starting at an age of 3–4 years (while cone formation in wild-type plants usually does not start before a tree age of about 20 years). However, also (male) pollen cones and even hermaphroditic cones with female organs at the top subtended by male organs develop occasionally. Following segregation analyses of the F_1 individuals that originated from *Acrocona,* self-pollinations and crosses with the non-*Acrocona* individual (Fladung, unpublished) confirmed that the *Acrocona* mutation is dominant (Langner 1954). One progeny of about 81 individuals that was used for mapping of the *Acrocona* locus (Acheré et al. 2004) shows a segregation of about 1:1 concerning early cone formation (Fig. 3.2; 42 versus 39 without cones).

So far, the molecular cause for the early flowering in *P. abies* "*acrocona*" is unknown. For conifer species, a number of different putative developmental control genes have been identified (*DAL1*, *DAL2*, *DAL3*, *DAL10*, *DAL11*, *DAL12*, *DAL13*, *DAL14*, *PaLFY*, and *PaNLY*; Tandre et al. 1995; Sundström et al. 1999; Carlsbecker et al. 2003, 2004). All these genes are putative candidates for the "*acrocona*" locus. At present, polymorphisms between the parental trees (TH787F and Sire5 of the above-mentioned cross) are being detected in the *DAL* genes to add these genes to the genetic map established by Acheré et al. (2004).

Fruit Trees

The Effect of Climate Changes on Flowering

The seasonal cycle of growth and dormancy is a distinct feature of perennial plants like fruit trees and represents one of the most basic adaptations of trees to their environment. It requires the timely sensing and processing of a regular and reliable environmental signal. Trees use environmental cues like daylength and temperature to time the dormancy-growth transition properly. Short photoperiods in the autumn are the main dormancy-inducting signal in trees (Garner and Allard 1923). However, an important exception to the short day control of dormancy in tree species are apple and also several other genera in the Rosaceae family (Nitsch 1957). Recently, Heide and Prestrud (2005) demonstrated that low temperature consistently induces growth cessation and dormancy in apple and pear both under short day and long day conditions. The investigation also confirmed that at higher temperatures, short photoperiods have no dormancy-inducing effects in these plants. Different *Prunus* species responded at high temperatures with continuous growth regardless of photoperiod in much the same way as *Malus* and *Pyrus*. However, unlike these genera, most of the *Prunus* species were clearly sensitive to daylength at intermediate temperatures and ceased growing in short day conditions. At low temperatures, *Prunus* species varied in their response to photoperiod (Heide 2008). The process of dormancy in woody perennials is poorly understood at the molecular level (Rohde and Bhalerao 2007). The dormant buds of many plants require a period of cold weather to grow, flower, and develop properly, but requirements vary widely by species. For dormant buds of fruit trees, this is commonly referred to as the chilling requirement. Chilling hours are calculated as a tool for fruit producers to gauge whether their crop has been exposed to cold temperatures for a long enough time period. However, to define chilling requirements for dormancy release in the era of global warming has also gained renewed attention in economical important species, like apple, cherry, and others, when selecting new cultivars. Luedeling et al. (2009a) studied climate change effects on winter chill for tree crops with chilling requirements on the Arabic Peninsula. The studies included species with low chilling requirement, like *Citrus* sp. banana, papaya, guava, and mango, with intermediate chilling requirement, like peach pomegranate, apricot, grape, and fig, and high chilling requirement, like apple, pear, and plum. They stated a decline in the number of chilling hours thus appears to be happening at a rapid pace, with the potential of serious consequences for the traditional cropping pattern. Similar developments are likely to affect other fruit production regions around the world, especially in temperate fruit trees. There are a few studies on overcoming dormancy in flower buds (Labuschagne et al. 2003; Naor et al. 2003; Gariglio et al. 2006; Legave et al. 2008; Mounzer et al. 2008; Tersoglio and Naranjo 2007; Luedeling et al. 2009b; Viti et al. 2010). Fan et al. (2010) recently published the mapping of quantitative trait loci associated with chilling requirement, heat requirement, and bloom date in peach (*Prunus persica*).

In the context of global warming, there is a general trend toward earlier flowering dates of many temperate tree species. Guédon and Legave (2008) analyzed the time-course variation of apple and pear tree dates of flowering stages over the last 40 years in France and Switzerland. The results clearly support the occurrence of a significant abrupt change in the time-course variation of flowering dates at the end of the 1980s toward more frequent early dates. The suddenness in the response to global warming could be explained by changes in rates for completion of chilling and heat requirements, successively essential to the development of floral primordial within buds. Early flowering is likely to result in an increased risk of damage from exposure to frost. Eccel et al. (2009) applied a phenological model of apple flowering to a temperature series from two locations in an important area for apple production in Europe. Flowering dates for the past 40 years and simulated flowering dates for the next 50 years were used in the model. A significant trend toward earlier flowering was clearly detected in the past. The number of frost episodes and flowering dates, on an annual basis, were graphed to assess the risk of spring frost. Risk analysis confirmed a lower risk of exposure to frost at present than in the past, and probably either constant or a slightly lower risk in future, especially given that physiological processes are expected to acclimate to higher temperatures.

As environmental cues, particularly temperature, are subjected to climate changes, all processes of growth cessation, dormancy, dormancy release, and flowering may be affected in temperate fruit trees grown in different geographic regions and environmental conditions. In this respect, it has to be mentioned that General Circulation Models presented by the IPCC (2007) estimate temperature increases of between 1°C and 1.5°C over 1980–1999 and 2020–2029. For 2090–2099, these models predict temperature increases between 2°C and 4°C.

Floral Initiation and Induction

In respect to flowering in temperate trees, some aspects need to be discussed: (1) the juvenile phase and (2) interactions between vegetative growth, flowers, and fruit of the previous year on floral initiation in the current year. The latter may result in biennial bearing.

Hanke et al. (2007) presented a comprehensive review on floral initiation and development in apple. The main phenomenon in perennial trees is the long juvenile phase that lasts for several years, during which time no flowering or fruiting occurs

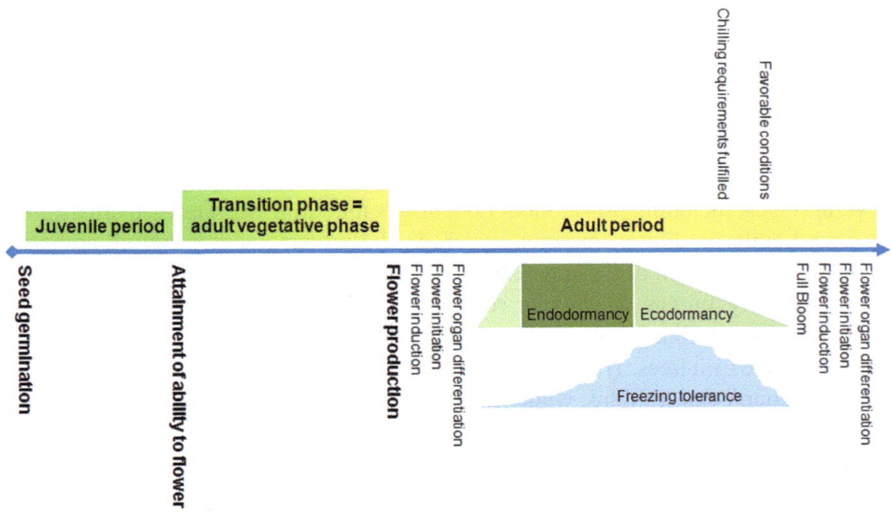

Fig. 3.3 Lifecycle of an apple tree. Apple is a temperate, deciduous tree with a long juvenile phase in seedlings. After several years, the seedling enters the adult phase and obtains the ability to flower depending on exogenous and endogenous factors. Between these two phases, the apple tree passes through a transition phase where floral-inducing techniques are getting increasingly effective. The adult phase of the apple tree is characterized by a sequence of seasonal cycles of growth and dormancy. During endodormancy, the tree will not grow even in growth-conducive conditions, whereas during ecodormancy, the tree will grow if placed in growth-conducive conditions. During endodormancy, the plant accumulates chilling hours. Once the chilling requirement is met, the tree has completed endodormancy and is merely dormant because the weather is too cold for rapid growth. Chilling requirements are usually completed by midwinter. With the return of warm weather, the plant begins to grow. For flower induction, initiation, and flower organ development (see text)

(Table 3.1). As in annual/biennial plants, in trees a phase change is required, which is a transition phase from juvenile to adult. Once fruit trees have passed the juvenile phase and reached an adult phase of reproductive competence, an unknown portion of apical as well as lateral shoot meristems will initiate flowers each year. This seems to be regulated by a sophisticated system consisting of many different factors (Bangerth 2009). As an example for temperate trees, Fig. 3.3 shows the lifecycle of apple. In *Malus* species, the juvenile phase of a seedling can last more than 8 years (Zimmerman 1972), depending on the parent characteristics. Zimmerman (1973) for the first time proposed a transition phase in seedlings between juvenile and adult phases. The end of the juvenile phase is indicated by the attainment of the ability to flower, and the actual production of flowers is the first evidence that the plant is in the adult phase. During the transition, also called adult vegetative phase, most floral-inducing techniques are applied successfully to induce flowering. Due to grafting procedures used to propagate temperate fruit trees, these plants are in an adult state from planting as an adult scion is grafted on a rootstock. Thus, the consequences resulting from a long juvenile phase and a late phase change are important for the breeder dealing with seedlings but less important for fruit production dealing always with mature adult trees.

Once fruit trees, like apple, have reached the adult phase of reproductive competence, part of the meristems in buds will initiate flowers, while part of the buds will initiate leaves giving raise to shoots. The processes of flower bud development and winter dormancy in apple can be followed in Fig. 3.4. Flowering in trees consists of several stages, including flower induction, flower initiation, flower organ differentiation, and anthesis. Floral initiation occurs in the growing season before anthesis. A detailed description of these processes can be found by Hanke et al. (2007). Fruit trees generally initiate flowers in response to either environmental stimulus or autonomously (Wilkie et al. 2008). Two major differences exist between tropical and temperate fruit trees. Tropical species, such as mango, initiate flowers in response to an environmental stimulus, while temperate species, such as apple, initiate flowers autonomously. Temperate trees undergo a period of dormancy and chilling requirement between flower initiation and anthesis, while in tropical species, floral development is continuous from floral induction to anthesis (Wilkie et al. 2008). According to the quantitative "long-distance signal" model involved in floral induction in perennial plants, several exogenous [e.g., temperatures, light (circadian and solar clocks), water shortage, and photosynthesis (carbohydrates)] and endogenous factors [e.g., number of fruits/seeds, vegetative growth (GA concentration), type of rootstock, and localization of buds along the shoot] are involved in the floral induction process (Bangerth 2009; Yeang 2009). Besides, a range of horticultural traits can also promote floral induction, like optimal nutrition (nitrogen); removal of young fruit by thinning, girdling, scoring, shoot bending, shoot pruning in summer; and root pruning. According to Bangerth (2009), there are long-distance hormonal signals (LDS signals) that are likely candidates for the regulation of floral initiation in trees.

Genetic Control of Flowering in Trees

Ectopical Expression of Flowering Genes in Trees

With the isolation of the first flowering genes from model plants in the early 1990s of the last century, it seemed that a wish comes true. Tree breeders were optimistic to find a way for breaking the juvenile stage of trees by ectopic overexpression of such genes in transgenic plants. *LFY* and *AP1*, two flowering genes from *Arabidopsis thaliana*, which were found to be necessary for the determination of the flower meristem identity (Yanofsky 1995), could be used as the key to happiness.

Both genes were transferred into different plant genomes and the initial experiments seemed really to be much promising. Particularly, the *LFY* gene was able to promote flower initiation and flower development in a range of plant species (Fig. 3.5). The overexpression of *LFY* resulted in precocious flowering in *Arabidopsis*, rice, tobacco, and hybrid aspen (Fig. 3.6b; Weigel and Nilsson 1995; Nilsson and

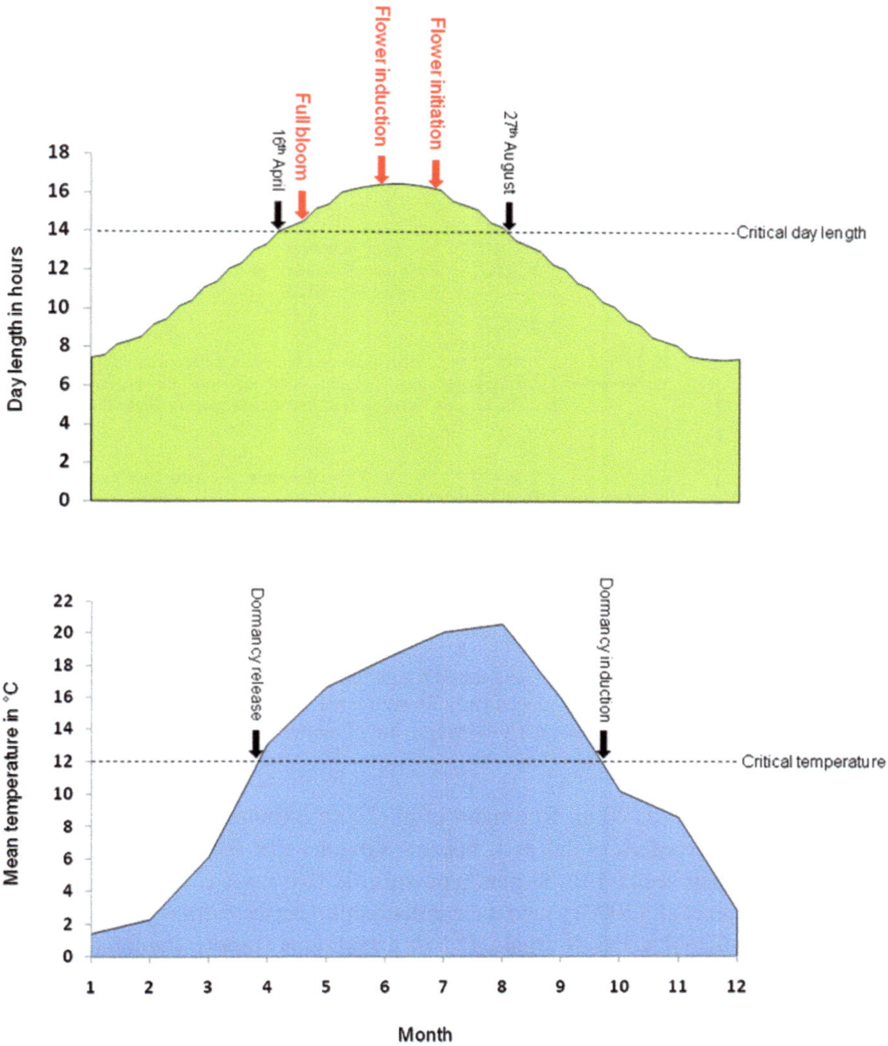

Fig. 3.4 Flower bud development and winter dormancy in apple in dependence of daylength and mean temperature under conditions in Germany. Flower induction and flower initiation in apple are induced in response to a critical daylength (Hanke et al. 2007), whereas growth cessation and dormancy induction are not influenced by photoperiod. Growth cessation and dormancy induction are induced in response to low temperature (<12°C), regardless of photoperiodic conditions. Successive stages of the autumn syndrome (growth cessation, formation of bud scales and winter buds, leaf senescence and abscission, and dormancy induction) occurred in response to low temperature (Heide and Prestrud 2005). Chilling at <7°C for about 700–1,000 h is required for dormancy release and growth resumption

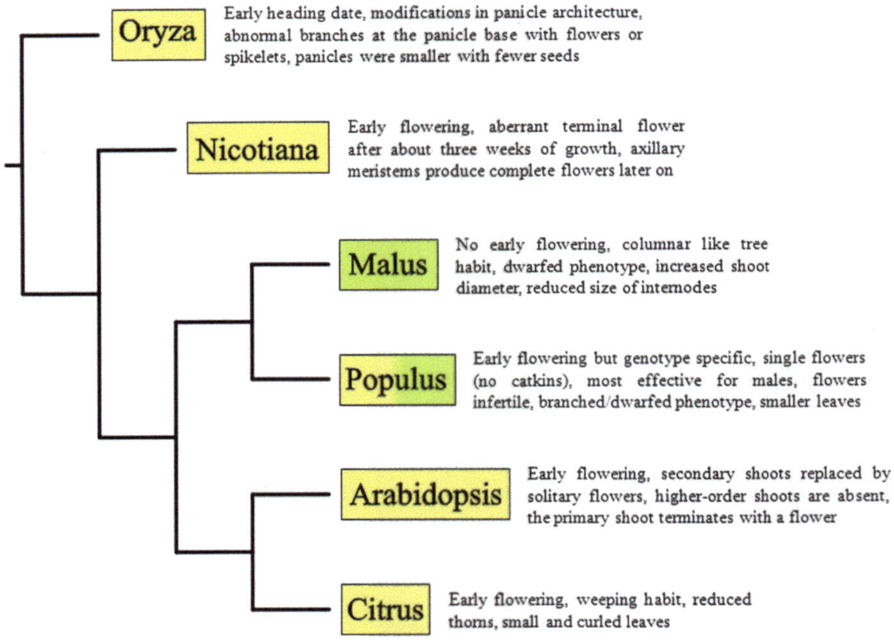

Fig. 3.5 The *LFY* gene of *Arabidopsis* was tested for its function in transgenic plants of different plant genera. Genera in which *LFY* induced early flowering are presented in *yellow boxes*. Genera in which *LFY* was not able to induce early flowering are presented in *green boxes* (for literature, please refer to the text)

Weigel 1997; He et al. 2000; Rottmann et al. 2000; Skinner et al. 2003). A first doubt lurked in breeders' mind as it became obvious that the induction of early flowering in aspen seemed to be genotype-specific (for review, see Strauss et al. 2004). Rottmann et al. (2000) found a clear distinction between transgenic plants of the aspen clones INRA 353-38 (male, *Populus tremula* × *Populus tremuloides*) and INRA 717-1B4 (female, *Populus tremula* × *Populus alba*). Whereas transgenic plants of clone INRA 353-38 showed precocious flowering at very high frequency, plants of clone INRA 717-1B4 flowered rarely. The results obtained were explained with the expression of a poplar gene that probably inhibits the action of *LFY*. Depending on the ability of the inhibitor to interact with *LFY*, the flower development will be more or less repressed. The varying responses of different poplar clones to *LFY* were tempted to explain with a complex set of interactions, including positive and negative regulators (Rottmann et al. 2000).

In the following, the *LFY* gene of *Arabidopsis* was also transferred into the genomes of different fruit tree species. The gene was successfully used to accelerate the juvenile phase in citrus (Peña et al. 2001), but not in apple (Flachowsky et al. 2010). The overexpression of the *LFY* gene resulted in transgenic apple plants having a phenotype that is similar to the columnar phenotype of the "McIntosh Wijcik" mutant (Fig. 3.6a). However, the *LFY* gene was not able to induce the *AP1* homologous genes of apple *MdAP1-1* and *MdAP1-2* (Flachowsky et al. 2010) as it is known

Fig. 3.6 Transgenic apple (**a**) and aspen (**b**) plant constitutively overexpressing the *LFY* gene of *Arabidopsis thaliana*. Transgenic apple plants flowered not earlier than control plants, but their phenotype was changed. The phenotype of these plants is similar to the columnar phenotype of the "McIntosh Wijcik" mutant of apple (Flachowsky et al. 2010). Transgenic aspen plants revealed a dwarf phenotype and single flowers instead of catkins

for *Arabidopsis* (Parcy et al. 1998; Wagner et al. 1999). In vitro and in vivo, it was shown that the LFY protein binds to a proposed CCANTGG consensus binding site present in the regulatory regions of the downstream-acting *Arabidopsis* genes *AP1* (Parcy et al. 1998), *AP3* (Lamb et al. 2002), and *AG* (Busch et al. 1999). The CCANTG motif is several times present in the *AP1* promoter of *Arabidopsis* (Parcy et al. 1998; Wagner et al. 1999; William et al. 2004), but not in the promoter sites of the apple homologous genes *MdAP1-1* and *MdAP1-2*. Whether this is the reason for the ineffectiveness of LFY to induce early flowering in apple or not cannot be stated, because it is not clear if such binding sites are really important for LFY to activate *AP1* genes *in planta* (Parcy 2005). However, the ectopic overexpression of the *Arabidopsis LFY* gene is a good example for the evolution of flower development. It demonstrates well that one and the same gene overexpressed in different plant species leads to different phenotypes. Furthermore, it shows that it is not possible in either case to take any gene and put it into a species of your choice to modify the trait of interest in a predictable direction.

The *AP1* gene of *Arabidopsis* was also transferred to several plant genomes. This gene was effective to induce early flowering in transgenic citrus (Peña et al. 2001) and tomato (Ellul et al. 2004), but not in transgenic hybrid aspen (Nilsson and Weigel 1997) and apple (Zhu et al. 2009). Especially in citrus, early flowering

Fig. 3.7 (**a**) In vitro grown shoot of the transgenic apple clone T1165 overexpressing the *BpMADS4* gene of silver birch (*Betula pendula* Roth.). Several transgenic clones started to flower during in vitro cultivation. Most of the transgenic flowers appeared morphologically normal (Flachowsky et al. 2007). (**b**) *BpMADS4* transgenic aspen did not flower early but induced broad changes in senescence and winter dormancy (Hoenicka et al. 2008) (Photo shows plant with leaves during the winter time)

was obtained using several approaches. *AP1* and *LFY* from *Arabidopsis* were successfully overexpressed in citrange, trifoliate orange, and also sweet orange (Peña et al. 2010). Besides, they overexpressed *FT* and *SOC1* from *Arabidopsis* in citrange and also got early flowering. In all cases phenotypes are different. In addition, *AP1* and *FT* from sweet orange were overexpressed in citrange showing also early flowering.

Inconsistent results were also obtained using *BpMADS4*, the *FUL* homologous gene of silver birch. This gene induced precocious flowering in transgenic birch and apple (Elo et al. 2007; Flachowsky et al. 2007), but not in poplar (Fig. 3.7b; Hoenicka et al. 2008). In apple and birch, the juvenile stage was dramatically reduced. Transgenic shoots overexpressing the *BpMADS4* gene flowered during in vitro culture already (Fig. 3.7a). Other flowering genes such as the *OsMADS1* of rice as well as *CONSTANS* and *AGL20* of *Arabidopsis* were unable to induce precocious flowering in transgenic poplar (Chung et al. 1994; Putterill et al. 1995; Rounsley et al. 1995), but these genes have not been tested in apple so far.

The work which was done with flowering genes ectopically expressed in different tree species is summarized in Table 3.2.

Functional Characterization of Flowering Gene Homologues

In recent years, orthologues of *Arabidopsis* flowering genes have been identified from different economically important tree species (Hanke et al. 2007; Wilkie et al.

Table 3.2 Ectopical expression of flowering genes in trees

Gene construct	Source	Expression in	Early flowering	References
35::*FT*	*Arabidopsis thaliana*	*Populus tremula* × *P. tremuloides* clone T89	Yes	Nilsson (unpublished)
		Populus tremula clone W52	No	Fladung et al. (unpublished)
		Citrus sinensis × *Poncirus trifoliata*	Yes	Peña et al. (2010)
HSP::*FT*	*Arabidopsis thaliana*	*Populus tremula* × *P. tremuloides* clone T89	Yes	Nilsson (unpublished)
		Populus tremula clone W52	Yes	Hoenicka et al. (unpublished)
		Populus tremula × *P. tremuloides* (INRA 353-38)	Yes	Zhang et al. (2010)
		Populus tremula × *P. alba* (INRA 717-1B4)	Yes	Zhang et al. (2010)
35S::*LFY*	*Arabidopsis thaliana*	*Citrus sinensis* × *Poncirus trifoliata*	Yes	Peña et al. (2001)
		Citrus × *sinensis*	Yes	Peña ct al. (2010)
		Malus × *domestica* cv. "Pinova"	No	Flachowsky et al. (2010)
		Malus × *domestica* cv. "Gala"	No	Schaart et al. (unpublished)
		Populus tremula × *P. tremuloides*	Yes	Weigel and Nilsson (1995)
		Populus tremula × *P. tremuloides* (INRA 353-38)	Yes	Rottmann et al. (2000)
		Populus tremula × *P. alba* (INRA 717-1B4)	Yes	Rottmann et al. (2000)
		Populus trichocarpa × *P. deltoides* (184-402)	Yes	Rottmann et al. (2000)
		Populus trichocarpa × *P. deltoides* (189-434)	No	Rottmann et al. (2000)
		Populus trichocarpa × *P. deltoides* (24-305)	No	Rottmann et al. (2000)
		Populus trichocarpa × *P. deltoides* (17-50)	Yes	Rottmann et al. (2000)
		Populus trichocarpa × *P. deltoides* (19-53)	No	Rottmann et al. (2000)
		Populus tremula clone W52	Yes	Hoenicka et al., (2006)
		Populus tremula clone Brauna 11	Yes	Fladung et al. (unpublished)
		Populus tremula × *P. tremuloides* clone Esch 5	Yes	Hoenicka et al. (unpublished)
35S::*AP1*	*Arabidopsis thaliana*	*Citrus sinensis* × *Poncirus trifoliata*	Yes	Peña et al. (2001)
		Citrus × *sinensis*	Yes	Peña et al. (2010)
		Malus × *domestica* cv. "M.26"	No	Zhu et al. (2009)
		Fortunella crassifolia cv. "Meiwa"	Yes	Duan et al. (2010)
		Populus tremula × *P. tremuloides* (INRA 353-53)	No	Mandel and Yanofsky (1995)
		P. tremula × *P. alba* (INRA 717-1B4)	No	Mandel and Yanofsky (1995)
		P. trichocarpa × *P. deltoides* (184-402)	No	Mandel and Yanofsky (1995)

(continued)

Table 3.2 (continued)

Gene construct	Source	Expression in	Early flowering	References
35S::*AGL20*	*Arabidopsis thaliana*	*Populus tremula* × *P. tremuloides* (INRA 353-53)	No	Rounsley et al. (1995)
		P. tremula × *P. alba* (INRA 717-1B4)	No	Rounsley et al. (1995)
		P. trichocarpa × *P. deltoides* (184-402)	No	Rounsley et al. (1995)
		P. trichocarpa × *P. deltoides* (17–50)	No	Rounsley et al. (1995)
		Populus tremula × *P. tremuloides* clone Esch 5	No	Fladung et al. (unpublished)
35S::*CONSTANS*	*Arabidopsis thaliana*	*Populus tremula* × *P. tremuloides* (INRA 353-53)	No	Putterill et al. (1995)
		P. tremula × *P. alba* (INRA 717-1B4)	No	Putterill et al. (1995)
		P. trichocarpa × *P. deltoides* (184-402)	No	Putterill et al. (1995)
		P. trichocarpa × *P. deltoides* (17–50)	No	Putterill et al. (1995)
		P. trichocarpa × *P. deltoides* (19–53)	No	Putterill et al. (1995)
35S::*SOC1*	*Arabidopsis thaliana*	*Citrus sinensis* × *Poncirus trifoliata*	Yes	Peña et al. (2010)
35S::*OsMADS1*	*Oryza sativa*	*Populus tremula* × *P. tremuloides* (INRA 353-53)	No	Chung et al. (1994)
35S::*BpMADS4*	*Betula pendula*	*Malus* × *domestica* cv. "Pinova"	Yes	Flachowsky et al. (2007)
		Populus tremula clone W52	No	Hoenicka et al. (2008)
		Populus tremula clone Brauna 11	No	Hoenicka et al. (2008)
		Populus tremula × *P. tremuloides* clone Esch 5	No	Fladung et al. (unpublished)
35S ::*FUL*	*Arabidopsis thaliana*	*Populus tremula* × *P. tremuloides* clone Esch 5	No	Fladung et al. (unpublished)
35S::*MdFT2*	*Malus* × *domestica*	*Populus tremula* clone W52	Yes	Tränkner et al. (2010)
SUC:*MdFT2*	*Malus* × *domestica*	*Populus tremula* clone W52	No	Tränkner et al. (2010)
35S ::*FPF1*	*Arabidopsis thaliana*	*Populus tremula* × *P. tremuloides* clone Esch 5	No	Fladung et al. (unpublished)
35S ::*MtFT*	*Medicago truncatula*	*Olea europaea* cv. Picual seedlings	Yes	Cerezo et al. (2010)
35S ::*PtFT1*	*Populus trichocarpa*	*Prunus domestica*	Yes	Srinivasan et al. (2010)
35S::*CiFT*	*Citrus unshiu*	*Pyrus communis* cv's. "La France" and "Ballade"	Yes	Matsuda et al. (2009)
		Poncirus trifoliate cv. "Kiyomi"	Yes	Endo et al. (2005)

2008). The function of these genes was mostly studied on the basis of the correlation between gene expression and morphological changes during floral initiation and flower organ development. A number of flowering genes from perennial tree species were ectopically expressed in *Arabidopsis*, but only a few of them have really been characterized by overexpression and/or silencing in their species of origin (Table 3.2). In apple *Malus* × *domestica*, the genes *MdFT1* (homolog to *FT*), *AFL1* and *AFL2* (homolog to *LFY*), *MdAP1* (homolog to *AP1*), and *MdSOC1* (homolog to *SOC1*) were overexpressed (Table 3.3). Only the overexpression of *MdFT1* resulted in early flowering until now. More successful was the downregulation of the *MdTFL1* gene (homolog to *TFL1*) using both the antisense and the RNAi strategy (Kotoda et al. 2006; Szankowski et al. 2009; Flachowsky et al. unpublished).

In poplar, only the *LFY* homologous gene *PTLF* and the *FT*-like genes *PtFT1* and *PtFT2* were tested. No accelerated flowering was obtained using the *PTLF* gene. In contrast, the overexpression of the *FT*-like genes resulted in early flowering in any case (Table 3.3).

The *BpMADS4* gene, which was very effective in breaking the juvenile stage in apple, was also effective in birch itself (Elo et al. 2007).

Future Perspectives

Since more than 140 years are humans in search of the key regulators of flowering in higher plants. In 1865, Julius Sachs was one of the first researchers who concluded, based on the results obtained in his experiments, that leaves in the light produce substances which direct the assimilates to form flowers in darkened shoots (for review, see Zeevaart 2006). Sixty-nine years later, it was found by Knott that in photoperiodically sensitive plants, daylength is perceived by the leaves, but flower development takes place in the shoot apical meristem (Knott 1934). This leads to the conclusion that there is a long-distance transport of a flower-inducing signal from induced leaves to the shoot apex. This graft-transmissible signal was designated by Chailakhyan in 1936 as the so-called "florigen." Since that time, much effort has been made to identify the "florigen" and the gene or genes which are responsible for flower induction, initiation, and flower organ development. Especially, during the last 15 years, a number of flowering genes were identified in model plants like *Arabidopsis thaliana* and *Antirrhinum majus* and the picture of flower development becomes more and more complete in these species. Next step will be the transfer of knowledge from the annual to perennial plants and from model plants to economically important crops such as fruit and forest trees. Several major traits have already been mapped in different tree species using molecular markers. Individual flowering gene homologues have been isolated and functionally characterized. Especially the establishment of methods for reverse genetics studies and the development of the next-generation sequencing technologies were milestones, which lead to an acceleration in this field of research. The identification of

Table 3.3 Functional characterization of native flowering gene homologs and other genes that affect flowering

Plant species/genotype	Gene/construct	Homolog to	Type of expression	Early flowering	References
Pear *Pyrus communis*					
cv. La France	*ACO1*		Antisense	Yes (in vitro)	Gao et al. (2007)
Apple *Malus* × *domestica*					
cv. Orin	*MdTFL1*	*TFL1*	Antisense	Yes	Kotoda et al. (2003, 2006)
cv. Holsteiner Cox	*MdTFL1*	*TFL1*	RNAi	Yes (in vitro)	Szankowski et al. (2009)
cv. Pinova and cv. Galaxy	*MdTFL1*	*TFL1*	RNAi	Yes (in vitro)	Flachowsky et al. (unpublished)
cv. Orin	*MdFT1*	*FT*	Overexpression	Yes (in vitro)	Kotoda et al. (2010)
cv. Pinova	*MdFT2*	*FT*	Overexpression	No	Hättasch et al. (2009)
cv. Pinova	*MdFT2*	*FT*	Overexpression	Yes (in vitro)	Tränkner et al. (2010)
cv. Orin	*AFL1*	*LFY*	Overexpression	No	Kotoda et al. (2003)
cv. Orin	*AFL2*	*LFY*	Overexpression	No	Kotoda et al. (2003)
cv. Rubinola	*MdSOC1*	*SOC1*	Overexpression	Not yet	Zhu et al. (2010)
cv. Rubinola	*MdFT1*	*FT*	Overexpression	Not yet	Zhu et al. (2010)
Poplar *Populus* sp.					
Hybrid aspen clone T89	*PtFT1*	*FT1*	Overexpression	Yes	Böhlenius et al. (2006)
Hybrid aspen 717-1B4	*PtFT2*	*FT2*	Overexpression	Yes	Hsu et al. (2006)
Hybrid aspen 353-38	*PtFT1*	*FT1*	Heat inducible	Yes	Zhang et al. 2010
Hybrid aspen 717-1B4	*PtFT1*	*FT1*	Heat inducible	Yes	Zhang et al. (2010)
Hybrid aspen 353-38	*PTLF*	*LFY*	Overexpression	No	Rottmann et al. (2000)
Hybrid aspen 717-1B4	*PTLF*	*LFY*	Overexpression	No	Rottmann et al. (2000)
Birch *Betula pendula*					
BPM2, JR1/4 and K1898	*BpMADS4*	*FUL*	Overexpression	Yes	Elo et al. (2007)

candidate genes will thereby be dramatically accelerated by the availability of entire genome sequences of different tree species. These genome sequences can be aligned to the genome sequence of model plant species such as *Arabidopsis*. Based on this alignment, it will be much easier to find out sequences (genes) that are homologues or paralogues of known flowering genes. The functional characterization of such candidate genes will help to complete the picture on genetics of flowering in trees bit by bit. Entire genome sequences from apple (Velasco et al. 2010), papaya (Ming et al. 2008), poplar (Tuskan et al. 2006), and grapevine (Jaillon et al. 2007; Velasco et al. 2007; Zharkikh et al. 2008) are available already. For other tree species such as *Citrus sinensis*, *Poncirus trifoliate*, *Eucalyptus grandis*, *Malus* × *domestica*, *Musa accuminata*, *Pinus taeda*, and *Prunus persica*, whole genome sequencing projects are in progress (http://www.ncbi.nlm.nih.gov/genomeprj).

References

Abe M, Kobayashi Y, Yamamoto S, Daimon Y, Yamaguchi A, Ikeda Y, Ichinoki H, Notaguchi M, Goto K, Araki T (2003) FD, a bZIP protein mediating signals from the floral pathway integrator FT at the shoot apex. Science 309(5737):1052–1056

Acheré V, Faivre-Rampant P, Jeandroz S, Besnard G, Markussen T, Aragones A, Fladung M, Ritter E, Favre JM (2004) A full saturated linkage map of *Picea abies* including AFLP, SSR, ESTP, 5S rDNA and morphological markers. Theor Appl Genet 108:1602–1613

Akinnifesi FK, Kwesiga FR, Mhango J, Mkonda A, Chilanga T, Swai R (2004) Domesticating priority for miombo indigenous fruit trees as a promising livelihood option for small-holder farmers in southern Africa. Acta Hortic 632:15–30

Aukerman MJ, Sakai H (2003) Regulation of flowering time and floral organ identity by a microRNA and its *APETALA2*-like target genes. Plant Cell 15:2730–2741

Bangerth KF (2009) Floral induction in mature, perennial angiosperm fruit trees: similarities and discrepancies with annual/biennial plants and the involvement of plant hormones. Sci Hortic 122:153–163

Bernier G, Perilleux C (2005) A physiological overview of the genetics of flowering time control. Plant Biotechnol J 3:3–16

Blakely WF (1955) A key to the eucalypts. Forestry and Timber Bureau, Canberra, 359 pp

Blazek J (1985) Precocity and productivity in some sweet cherry crosses. Acta Hortic 169:105–114

Blazquez M, Weigel D (2000) Integration of floral inductive signals in *Arabidopsis*. Nature 404:889–892

Blazquez M, Soowal L, Lee I, Weigel D (1997) LEAFY expression and flower initiation in *Arabidopsis*. Development 124:3835–3844

Böhlenius H, Huang T, Charbonnel-Campaa L, Brunner AM, Jansson S, Strauss SH, Nilsson O (2006) CO/FT regulatory module controls timing of flowering and seasonal growth cessation in trees. Science 312:1040–1043

Bolotin M (1975) Photoperiodic induction of precocious flowering in a woody species *Eucalyptus occidentalis* Endl. Bot Gaz 136(4):358–365

Boss PK, Bastow RM, Mylne JS, Den C (2004) Multiple pathways in the decision to flower: enabling, promoting, and resetting. Plant Cell 16:S18–S31

Brunner A, Nilsson O (2004) Revisiting tree maturation and floral initiation in the poplar functional genomics era. New Phytol 164:43–51

Busch MA, Bomblies K, Weigel D (1999) Activation of a floral homeotic gene in *Arabidopsis*. Science 285:585–587

Carlsbecker A, Sundström J, Tandre K, Englund M, Kvarnheden A, Johanson U, Engström P (2003) The *DAL10* gene from Norway spruce (*Picea abies*) belongs to a potentially gymnosperm-specific subclass of MADS-box genes and is specifically active in seed cones and pollen cones. Evol Dev 5:551–561

Carlsbecker A, Tandre K, Johanson U, Englund M, Engström P (2004) The MADS-box gene *DAL1* is a potential mediator of the juvenile-to-adult transition in Norway spruce (*Picea abies*). Plant J 40:546–557

Cerezo S, Samach A, Mercado JA, Pliego-Alfaro F (2010) *Agrobacterium*-mediated transformation of olive (*Olea europaea* L.) with a *Medicago truncatula* gene encoding an FT-like protein. In: Abstracts of the 28th international horticultural congress, Lisbon, Portugal, 22–27 Aug 2010, p 278

Chambers PGS, Potts BM, Tilyard PA (1997) The genetic control of flowering precocity in *Eucalyptus globulus* ssp. globulus. Silvae Genet 46:207–214

Chandler J, Dean C (1994) Factors influencing the vernalization response and flowering time of late flowering mutants of *Arabidopsis thaliana*. J Exp Bot 45:1279–1288

Chung YY, Kim SR, Finkel D, Yanofsky MF, An G (1994) Early flowering and reduced apical dominance results from ectopic expression of a rice MADS box gene. Plant Mol Biol 26:657–665

Corbesier L, Lejeune P, Bernier G (1998) The role of carbohydrates in the induction of flowering in *Arabidopsis thaliana*: comparison between the wild type and a starchless mutant. Planta 206:131–137

Corbesier L, Vincent C, Jang S, Fornara F, Fan Q, Searle I, Giakountis A, Farrona S, Gissot L, Turnbull C, Coupland G (2007) FT Protein movement contributes to long-distance signaling in floral induction of *Arabidopsis*. Science 316:1030–1033

Duan Y-X, Fan J, Guo W-W (2010) Regeneration and characterization of transgenic kumquat plants containing the *Arabidopsis APETALA1* gene. Plant Cell Tissue Organ Cult 100:273–281

Dugas DV, Bartel B (2004) MicroRNA regulation of gene expression in plants. Curr Opin Plant Biol 7:512–520

Dunberg A (1974) Occurrence of gibberellin-like substances in Norway spruce (*Picea abies* (L.) Karst.) and their possible relation to growth and flowering. Stud For Suec Nr 111:1–62

Dunberg A (1976) Changes in gibberellin-like substances and indole-3-acetic acid in *Picea abies* during the period of shoot elongation. Physiol Plant 38:186–190

Dunberg A (1980) Stimulation of flowering in *Picea abies* by gibberellins. Silvae Genet 29(2):51–53

Eccel E, Rea R, Caffarra A, Crisci A (2009) Risk of spring frost to apple production under future climate scenarios: the role of phenological acclimation. Int J Biometeorol 53:273–286

Eldridge K, Davidson J, Harwood C, Van Wyk G (1993) Eucalypt domestication and breeding. Oxford University Press Inc., Oxford

Ellul P, Angosto T, Garcia-Sogo B, Garcia-Hurtado N, Martin-Trillo M, Salinas M, Moreno V, Lozano R, Martinez-Zapater M (2004) Expression of *Arabidopsis APETALA1* in tomato reduces its vegetative cycle without affecting plant production. Mol Breed 13:155–163

Elo A, Lemmetyinen J, Novak A, Keinonen K, Porali I, Hassinen M, Sopanen T (2007) *BpMADS4* has a central role in inflorescence initiation in silver birch (*Betula pendula*). Physiol Plant 131:149–158

Endo T, Shimada T, Fujii H, Kobayashi Y, Araki T, Omura M (2005) Ectopic expression of a *FT* homolog from *Citrus* confers an early flowering phenotype on trifoliate orange (*Poncirus trifoliata* L. Raf.). Transgenic Res 14:703–712

Eriksson S, Böhlenius H, Moritz T, Nilsson O (2006) GA_4 is the active gibberellin in the regulation of *LEAFY* transcription and *Arabidopsis* floral initiation. Plant Cell 18:2172–2181

Fan S, Bielenberg DG, Zhebentyayeva TN, Reighard GL, Okie WR, Holland D, Abbott AG (2010) Mapping quantitative trait loci associated with chilling requirement, heat requirement and bloom date in peach (*Prunus persica*). New Phytol 185:917–930

Finnegan EJ, Genger RK, Kovac K, Peacock WJ, Dennis ES (1998) DNA methylation and the promotion of flowering by vernalization. Proc Natl Acad Sci USA 95:5824–5829

Flachowsky H, Peil A, Sopanen T, Elo A, Hanke V (2007) Overexpression of *BpMADS4* from silver birch (*Betula pendula* Roth.) induces early-flowering in apple (*Malus* × *domestica* Borkh.). Plant Breed 126:137–145

Flachowsky H, Hättasch C, Höfer M, Peil A, Hanke MV (2010) Overexpression of *LEAFY* in apple leads to a columnar phenotype with shorter internodes. Planta 231:251–263

Fladung M, Tusch A, Markussen T, Ziegenhagen B (2000) Analysis of morphological mutants in *Picea*. In: Espinel S, Ritter E (eds) Proceedings of the international congress "Applications of biotechnology to forest genetics" (Biofor 99), Vitoria-Gasteiz, Spain, 22–25 Sept 1999, pp 167–170

Gao M, Matsuda N, Murayama H, Toyomasu T, Mitsuhashi W, Dandekar AM, Tao R, Nishimura K (2007) Gene expression and ethylene production in transgenic pear (*Pyrus communis* cv. 'La France') with sense and antisense cDNA encoding ACC oxidase. Plant Sci 173:32–42

Gariglio N, Rossia DEG, Mendow M, Reig C, Agusti M (2006) Effect of artificial chilling on the depth of endodormancy and vegetative and flower budbreak of peach and nectarine cultivars using excised shoots. Sci Hortic 108:371–377

Garner WW, Allard HA (1923) Further studies in photoperiodism, the response of the plant to relative length of day and night. J Agric Res 23:871–920

Griffin AR, Whiteman P, Rudge T, Burgess IP, Moncur M (1993) Effect of paclobutrazol on flower bud production and vegetative growth in two species of *Eucalyptus*. Can J For Res-Revue Canadienne De Recherche Forestiere 23:640–647

Guedon Y, Legave JM (2008) Analyzing the time-course variation of apple and pear tree dates of flowering stages in the global warming context. Ecol Model 219:189–199

Hanke MV, Flachowsky H, Peil A, Hättasch C (2007) No flower no fruit – genetic potential to trigger flowering in fruit trees. Genes Genomes Genomics 1:1–20

Hasan O, Reid JB (1995) Reduction of generation time in *Eucalyptus globulus*. Plant Growth Regul 17:53–60

Hättasch C, Flachowsky H, Hanke M-V, Lehmann S, Gau A, Kapturska D (2009) The switch to flowering: genes involved in floral induction of the apple cultivar 'Pinova' and the role of the flowering gene *MdFT*. Acta Hortic 839:701–705

He ZH, Zhu Q, Dabi T, Li DB, Weigel D, Lamb C (2000) Transformation of rice with the *Arabidopsis* floral regulator *LEAFY* causes early heading. Transgenic Res 9:223–227

Heide OM (2008) Interaction of photoperiod and temperature in the control of growth and dormancy of *Prunus* species. Sci Hortic 115:309–314

Heide OM, Prestrud AK (2005) Low temperature, but not photoperiod, controls growth cessation and dormancy induction and release in apple and pear. Tree Physiol 25:109–114

Hepworth SR, Valverde F, Ravenscroft D, Mouradov A, Coupland G (2002) Antagonistic regulation of flowering-time gene *SOC1* by CONSTANS and FLC via separate promoter motifs. EMBO J 21:4327–4337

Hoenicka H, Nowitzki O, Debener T, Fladung M (2006) Faster evaluation of floral sterility in transgenic early flowering poplar. Silvae Genet 55(6):285–291

Hoenicka H, Nowitzki O, Hanelt D, Fladung M (2008) Heterologous overexpression of the birch *FRUITFULL*-like MADS-box gene *BpMADS4* prevents normal senescence and winter dormancy in *Populus tremula* L. Planta 227:1001–1011

Hsu CY, Liu Y, Luthe DS, Yuceer C (2006) Poplar FT2 shortens the juvenile phase and promotes seasonal flowering. Plant Cell 18:1846–1861

IPCC (2007) Climate change 2007: synthesis report. In: Contribution of working groups I, II and III to the fourth assessment report of the intergovernmental panel on climate change. IPCC, Geneva, p 104

Ivonis IYu, Kyalina LV, Litvinova VB, Khokhlina EV (1982) Annual dynamics of gibberellin-like substances in needles of Norway spruce clones of different sexes. Sov Plant Physiol 28:915–920

Jaillon O, Aury J-M, Noel B, Policriti A, Clepet C, Casagrande A, Choisne N, Aubourg S, Vitulo N, Jubin C, Vezzi A, Legeai F, Hugueney P, Dasilva C, Horner D, Mica E, Jublot D, Poulain J, Bruyère C, Billault A, Segurens B, Gouyvenoux M, Ugarte E, Cattonaro F, Anthouard V, Vico V, Del Fabbro C, Alaux M, Di Gaspero G, Dumas V, Felice N, Paillard S, Juman I, Moroldo M, Scalabrin S, Canaguier A, Le Clainche I, Malacrida G, Durand E, Pesole G, Laucou V, Chatelet P, Merdinoglu D, Delledonne M, Pezzotti M, Lecharny A, Scarpelli C, Artiguenave F, Pè ME, Valle G, Morgante M, Caboche M, Adam-Blondon A-F, Weissenbach J, Quétier F, Wincker P (2007) The grapevine genome sequence suggests ancestral hexaploidization in major angiosperm phyla. Nature 449:463–467

Jansson S, Douglas CJ (2007) *Populus*: a model system for plant biology. Annu Rev Plant Biol 58:435–458

Jung C, Müller AE (2009) Flowering time control and applications in plant breeding. Trends Plant Sci 14:563–573

Junttila O (1980) Flower bud differentiation in *Salix pentandra* as affected by photoperiod, temperature and growth regulators. Physiol Plant 49:127–134

Kadir S (2003) Why fruit trees fail to beer. Hortic Rep MF-2166:1–3

Kardailsky I, Shukla VK, Ahn JH, Dagenais N, Christensen SK, Ngyen JT, Chory J, Harrison MJ, Weigel D (1999) Activation tagging of the floral inducer FT. Science 286:1962–1965

Kasschau KD, Xie Z, Allen E, Llave C, Chapman EJ, Krizan KA, Carrington JC (2003) P1/HC-Pro, a viral suppressor of RNA silencing, interferes with *Arabidopsis* development and miRNA function. Dev Cell 4:205–217

Knott JE (1934) Effect of a localized photoperiod on spinach. Proc Am Soc Hortic Sci 31:152–154

Kobayashi Y, Kaya H, Goto K, Iwabuchi M, Araki T (1999) A pair of related genes with antagonistic roles in mediating flowering signals. Science 286:1960–1962

Kotoda N, Wada M, Masuda T, Soejima J (2003) The break-through in the reduction of juvenile phase in apple using transgenic approaches. Acta Hortic 625:337–343

Kotoda N, Iwanami H, Takahashi S, Abe K (2006) Antisense expression of *MdTFL1*, a *TFL1*-like gene, reduces the juvenile phase in apple. J Am Soc Hortic Sci 131:74–81

Kotoda N, Hayashi H, Suzuki M, Igarashi M, Hatsuyama Y, Kidou S-I, Igasaki T, Nishiguchi M, Yano K, Shimizu T, Takahashi S, Iwanami H, Moriya S, Abe K (2010) Molecular characterization of *FLOWERING LOCUS T*-like genes of apple (*Malus* × *domestica* Borkh.). Plant Cell Physiol 51(4):561–575

Krüssmann G (1983) Handbuch der Nadelgehölze. Paul Parey Verlag, Berlin/Hamburg

Kulikowska HJ, Kopcewicz J, Zatorska Z, Szczesniak T (1978) Auxins and gibberellins in embryonic shoots of Scats pine in relation to flower sex differentiation. Acta Soc Bot Pol 47:403–409

Labuschagne IF, Louw JH, Schmidt K, Sadie A (2003) Selection for increased budbreak in apple. J Am Soc Hortic Sci 128:363–373

Lamb RS, Hill TA, Tan QKG, Irish VF (2002) Regulation of *APETALA3* floral homeotic gene expression by meristem identity genes. Development 129:2079–2086

Langner W (1954) Über die Ursachen einiger phänotypischer Besonderheiten bei Fichtenpfropflingen. Zeitschrift Forstgenetik Forstpflanzenzüchtung 3:83–86

Langridge J (1957) Effect of day-length and gibberellic acid on the flowering of *Arabidopsis*. Nature 180:36–37

Lee H, Suh SS, Park E, Cho E, Ahn JH, Kim SG, Lee JS, Kwon YM, Lee I (2000) The *AGAMOUS*-LIKE 20 MADS domain protein integrates floral inductive pathways in *Arabidopsis*. Genes Dev 14:2366–2367

Legave JM, Farrera I, Almeras T, Calleja M (2008) Selecting models of apple flowering time and understanding how global warming has had an impact on this trait. J Hortic Sci Biotechnol 83:76–84

Litz RE, Mathews VH, Hendrix RC, Yurgalevitch C (1991) Mango somatic cell genetics. Acta Hortic 291:133–140

Longman KA (1987) The significance of juvenility for seed orchard. For Ecol Manage 19:9–16
Luedeling E, Gebauer J, Buerkert A (2009a) Climate change effects on winter chill for tree crops with chilling requirements on the Arabic Peninsula. Clim Chang 96:219–237
Luedeling E, Zhang MH, Luedeling V, Girvetz EH (2009b) Sensitivity of winter chill models for fruit and nut trees to climatic changes expected in California's Central Valley. Agric Ecosyst Environ 133:23–31
Mandel MA, Yanofsky MF (1995) The *Arabidopsis Agl8* MADS box gene is expressed in inflorescence meristems and is negatively regulated by *Apetala* 1. Plant Cell 7:1763–1771
Marquard R, Hanover J (1984) Relationship between gibberellin A4/7 concentration, time of treatment and crown position on flowering of *Picea glauca*. Can J For Res 14:547–553
Martinez-Zapater JM, Coupland G, Dean C, Koornneef M (1994) The transition to flowering in *Arabidopsis*. In: Meyerowitz EM, Somerville CR (eds) *Arabidopsis*. Cold Spring Harbor Laboratory Press, New York, pp 403–434
Matsuda N, Ikeda K, Kurosaka M, Takashina T, Isuzugawa K, Endo T, Omura M (2009) Early flowering phenotype in transgenic pears (*Pyrus communis* L.) expressing the *CiFT* gene. J Jpn Soc Hortic Sci 78:410–416
Meilan R (1997) Floral induction in woody angiosperms. New Forests 14:179–202
Meilan R, Sabatti M, Caiping M, Kuzminsky E (2004) An early-flowering genotype of *Populus*. J Plant Biol 47(1):52–56
Michaels SD, Amasino RM (1999) *FLOWERING LOCUS C* encodes a novel MADS domain protein that acts as a repressor of flowering. Plant Cell 11:949–956
Michaels SD, Amasino RM (2000) Memories of winter: vernalization and the competence to flower. Plant Cell Environ 23:1145–1153
Ming R, Hou S, Feng Y, Yu Q, Dionne-Laporte A, Saw JH, Senin P, Wang W, Ly BV, Lewis KLT, Salzberg SL, Feng L, Jones MR, Skelton RL, Murray JE, Chen C, Qian W, Shen J, Du P, Eustice M, Tong E, Tang H, Lyons E, Paull RE, Michael TP, Wall K, Rice DW, Albert H, Wang M-L, Zhu YJ, Schatz M, Nagarajan N, Acob RA, Guan P, Blas A, Man Wai C, Ackerman CM, Ren Y, Liu C, Wang J, Wang J, Na J-K, Shakirov EV, Haas B, Thimmapuram J, Nelson D, Wang X, Bowers JE, Gschwend AR, Delcher AL, Singh R, Suzuki JY, Tripathi S, Neupane K, Wei H, Irikura B, Paidil M, Jiang N, Zhang W, Presting G, Windsor A, Navajas-Pérez R, Torres MJ, Feltus FA, Porter B, Li Y, Burroughs AM, Luo M-C, Liu L, Christopher DA, Mount SM, Moore PH, Sugimura T, Jiang J, Schuler MA, Friedman V, Mitchell-Olds T, Shippen DE, de Pamphilis CW, Palmer JD, Freeling M, Paterson AH, Gonsalves D, Wang L, Alam M (2008) The draft genome of the transgenic tropical fruit tree papaya (*Carica papaya* Linnaeus). Nature 452:991–996
Moncur MW (1992) Effect of low temperature on floral induction of *Eucalyptus lansdowneana* Muell, F. and Brown, J. Subsp *lansdowneana*. Aust J Bot 40:157–167
Moon J, Suh SS, Lee H, Choi KR, Hong CB, Paek NC, Kim SG, Lee I (2003) The *SOC1* MADS-box gene integrates vernalisation and gibberellins signals for flowering in *Arabidopsis*. Plant J 35:613–623
Mounzer OR, Conejero W, Nicolas E, Abrisqueta I, Garcia-Orellana YV, Tapia LM, Vera J, Abrisqueta JM, Ruiz-Sanchez MD (2008) Growth pattern and phenological stages of early-maturing peach trees under a Mediterranean climate. HortScience 6:1813–1818
Naor A, Flaishman M, Stern R, Moshe A, Erez A (2003) Temperature effects on dormancy completion of vegetative buds in apple. J Am Soc Hortic Sci 128:636–641
Nilsson O, Weigel D (1997) Modulating the timing of flowering. Curr Opin Biotechnol 8:195–199
Nitsch JP (1957) Photoperiodism in woody plants. Proc Am Soc Hortic Sci 70:526–544
Owens JN (1995) Constraints to seed production: temperate and tropical forest trees. Tree Physiol 15:477–484
Owens JN, Molder M (1976) Bud development in Sitka spruce. II. Cone differentiation and early development. Can J Bot 54(8):766–779
Parcy F (2005) Flowering: a time for integration. Int J Dev Biol 49:585–593

Parcy F, Nilsson O, Busch MA, Lee I, Weigel D (1998) A genetic framework for floral patterning. Nature 395:561–566

Peña L, Martin-Trillo M, Juarez J, Pina JA, Navarro L, Martinez-Zapater JM (2001) Constitutive expression of *Arabidopsis LEAFY* or *APETALA1* genes in citrus reduces their generation time. Nat Biotechnol 19:263–267

Peña et al. (2010) Personal communication

Perala DA, Alm AA (1990) Reproductive ecology of birch: a review. For Ecol Manage 32:1–38

Poethig RS (1990) Phase change and the regulation of shoot morphogenesis in plants. Science 259:923–930

Poethig RS (2003) Phase change and the regulation of developmental timing in plants. Science 301:334–336

Putterill J, Robson F, Lee K, Simon R, Coupland G (1995) The *CONSTANS* gene of *Arabidopsis* promotes flowering and encodes a protein showing similarities to zinc finger transcription factors. Cell 80:847–857

Quesada V, Dean C, Simpson GG (2005) Regulated RNA processing in the control of *Arabidopsis* flowering. Int J Dev Biol 49:773–780

Rajan S, Markose BL (2007) Propagation of horticultural crops. In: Horticulture Science Series, ed. Peter KV. New India publishing Agency, pp. 1–275

Rhoades MW, Reinhart BJ, Lim LP, Burge CB, Bartel B, Bartel DP (2002) Prediction of plant microRNA targets. Cell 110:513–520

Rohde A, Bhalerao RP (2007) Plant dormancy in the perennial context. Trends Plant Sci 12:217–223

Rottmann WH, Meilan R, Sheppard LA, Brunner AM, Skinner JS, Ma C, Cheng S, Jouanin L, Pilate G, Strauss SH (2000) Diverse effects of overexpression of *LEAFY* and PTLF, a poplar (*Populus*) homolog of *LEAFY/FLORICAULA*, in transgenic poplar and *Arabidopsis*. Plant J 22:235–245

Rounsley SD, Ditta GS, Yanofsky MF (1995) Diverse roles for MADS box genes in *Arabidopsis* development. Plant Cell 7:1259–1269

Rouse DT, Sheldon CC, Bagnall DJ, Peacock WJ, Dennis ES (2002) FLC, a repressor of flowering, is regulated by genes in different inductive pathways. Plant J 29:183–191

Samach A, Onouchi H, Gold SE, Ditta GS, Schwarz-Sommer Z, Yanofsky MF, Coupland G (2000) Distinct roles of *CONSTANT* target genes in reproductive development of *Arabidopsis*. Science 288:1613–1616

Schmid M, Uhlenhaut NH, Godard F, Demar M, Bressan R, Weigel D, Lohmann JU (2003) Dissection of floral induction pathways using global expression analysis. Development 130:6001–6012

Sheldon CC, Burn JE, Perez PP, Metzger J, Edwards JA, Peacock WJ, Dennis ES (1999) The FLF MADS box gene: a repressor of flowering in *Arabidopsis* regulated by vernalization and methylation. Plant Cell 1:445–458

Sheldon CC, Rouse DT, Finnegan EJ, Pacock WJ, Dennis ES (2000) The molecular basis of vernalization: the central role of *FLOWERING LOCUS C* (*FLC*). Proc Natl Acad Sci USA 97:3753–3758

Simpson GG, Dean C (2002) *Arabidopsis*, the rosetta stone of flowering time? Science 296:285–289

Skinner JS, Meilan R, Ma C, Strauss SH (2003) The *Populus* PTD promoter imparts floral-predominant expression and enables high levels of floral-organ ablation in *Populus*, *Nicotiana* and *Arabidopsis*. Mol Breed 12:119–132

Sosinski B, Gannavarapu M, Beck LE, Rajapakse S, Ballard RE, Abbott AG (2000) Characterization of microsatellite markers in peach [*Prunus persica* (L.) Batsch]. Theor Appl Genet 101:421–428

Southerton S (2007) Early flowering induction and *Agrobacterium* transformation of the hardwood tree species *Eucalyptus occidentalis*. Funct Plant Biol 34:707–713

Srinivasan C, Callahan A, Dardick C, Scorza R (2010) Expression of the poplar FLOWERING LOCUS T1 (FT1) gene reduces generation time in plum (*Prunus domestica* L.). Abstracts of the 28th international horticultural congress, Lisbon, Portugal, 22–27 Aug 2010, p 277

Steven D (1994) Tropical tree seedling dynamics: recruitment patterns and their population consequences for three canopy species in Panama. J Trop Ecol 10:369–383

Strauss SH, Brunner AM, Busov VB, Ma C, Meilan R (2004) Ten lessons from 15 years of transgenic *Populus* research. Forestry 77:455–465

Suarez-Lopez P, Wheatley K, Robson F, Onouchi H, Valverde F, Coupland G (2001) *CONSTANS* mediates between the circadian clock and the control of flowering in *Arabidopsis*. Nature 410:1116–1120

Sun TP, Kamiya Y (1994) The *Arabidopsis Ga1* locus encodes the cyclase ent-kaurene synthetase A of gibberellins biosynthesis. Plant Cell 6:1509–1518

Sundström J, Carlsbecker A, Svensson ME, Svenson M, Johanson U, Theissen G, Engström P (1999) MADS-box genes active in developing pollen cones of Norway spruce (*Picea abies*) are homologous to the B-class floral homeotic genes in angiosperms. Dev Genet 25:253–266

Szankowski I, Waidmann S, El-Din Saad Omar A, Flachowsky H, Hättasch C, Hanke M-V (2009) RNAi-silencing of *MdTFL1* induces early flowering in apple. Acta Hortic 839:633–636

Takada S, Goto K (2003) TERMINAL FLOWER2, a HETEROCHROMATIN PROTEIN 1-LIKE protein of *Arabidopsis*, counteracts the activation of FLOWERING LOCUS T by CONSTANS in the vascular tissues of leaves to regulate flowering time. Plant Cell 15:2856–2865

Tan FC, Swain SM (2006) Genetics of flower initiation and development in annual and perennial plants. Physiol Plant 128:8–17

Tandre K, Albert VA, Sundas A, Engström P (1995) Conifer homologues to genes that control floral development in angiosperms. Plant Mol Biol 27:69–78

Tersoglio E, Naranjo G (2007) Estimation model of winter chilling availability in sweet cherry areas of Mendoza, Argentina. Part II. Itea-Inf Tec Econ Agraria 103:198–211

Tränkner C, Lehmann S, Hoenicka H, Hanke M-V, Fladung M, Lenhardt D, Dunemann F, Gau A, Schlangen K, Malnoy M, Flachowsky H (2010) Over-expression of an *FT*-homologous gene of apple induces early flowering in annual and perennial plants. Planta 232:1309–1324

Tuskan GA, DiFazio S, Jansson S, Bohlmann J, Grigoriev I, Hellsten U, Putnam N, Ralph S, Rombauts S, Salamov A, Schein J, Sterck L, Aerts A, Bhalerao RR, Bhalerao RP, Blaudez D, Boerjan W, Brun A, Brunner A, Busov V, Campbell M, Carlson J, Chalot M, Chapman J, Chen G-L, Cooper D, Coutinho PM, Couturier J, Covert S, Cronk Q, Cunningham R, Davis J, Degroeve S, Déjardin A, dePamphilis C, Detter J, Dirks B, Dubchak I, Duplessis S, Ehlting J, Ellis B, Gendler K, Goodstein D, Gribskov M, Grimwood J, Groover A, Gunter L, Hamberger B, Heinze B, Helariutta Y, Henrissat B, Holligan D, Holt R, Huang W, Islam-Faridi N, Jones S, Jones-Rhoades M, Jorgensen R, Joshi C, Kangasjärvi J, Karlsson J, Kelleher C, Kirkpatrick R, Kirst M, Kohler A, Kalluri U, Larimer F, Leebens-Mack J, Leplé C-J, Locascio P, Lou Y, Lucas S, Martin F, Montanini B, Napoli C, Nelson DR, Nelson C, Nieminen K, Nilsson O, Pereda V, Peter G, Philippe R, Pilate G, Poliakov A, Razumovskaya J, Richardson P, Rinaldi C, Ritland K, Rouzé P, Ryaboy D, Schmutz J, Schrader J, Segerman B, Shin H, Siddiqui A, Sterky F, Terry A, Tsai C-J, Uberbacher E, Unneberg P, Vahala J, Wall K, Wessler S, Yang G, Yin T, Douglas C, Marra M, Sandberg G, Van de Peer Y, Rokhsar D (2006) The genome of black cottonwood, *Populus trichocarpa* (Torr. & Gray). Science 15:1596–1604

USDA/NRCS National Plant Data Center & the Biota of North America Program (2010) http://plants.usda.gov/plantguide/pdf/cs_potr5.pdf

Valverde F, Mouradov A, Soppe W, Ravenscroft D, Samach A, Coupland G (2004) Photoreceptor regulation of CONSTANS protein in photoperiodic flowering. Science 13:1003–1006

Velasco R, Zharkikh A, Troggio M, Cartwright DA, Cestaro A, Pruss D, Pindo M, Fitzgerald LM, Vezzulli S, Reid J, Malacarne G, Iliev D, Coppola G, Wardell B, Micheletti D, Macalma T, Facci M, Mitchell JT, Perazzolli M, Eldredge G, Gatto P, Oyzerski R, Moretto M, Gutin N, Stefanini M, Chen Y, Segala C, Davenport C, Demattè L, Mraz A, Battilana J, Stormo K, Costa F, Tao Q, Si-Ammour A, Harkins T, Lackey A, Perbost C, Taillon B, Stella A, Solovyev V, Fawcett JA, Sterck L, Vandepoele K, Grando SM, Toppo S, Moser C, Lanchbury J, Bogden R,

Skolnick M, Sgaramella V, Bhatnagar SK, Fontana P, Gutin A, Van de Peer Y, Salamini F, Viola R (2007) A high quality draft consensus sequence of the genome of a heterozygous grapevine variety. PLoS One 2(12):e1326. doi:10.1371/journal.pone.0001326

Velasco R, Zharkikh A, Affourtit J, Dhingra A, Cestaro A, Kalyanaraman A, Fontana P, Bhatnagar SK, Troggio M, Pruss D, Salvi S, Pindo M, Baldi P, Castelletti S, Cavaiuolo M, Coppola G, Costa F, Cova V, Dal Ri A, Goremykin V, Komjanc M, Longhi S, Magnago P, Malacarne G, Malnoy M, Micheletti D, Moretto M, Perazzolli M, Si-Ammour A, Vezzulli S, Zini E, Eldredge G, Fitzgerald LM, Gutin N, Lanchbury J, Macalma T, Mitchell JT, Reid J, Wardell B, Kodira C, Chen Z, Desany B, Niazi F, Palmer M, Koepke T, Jiwan D, Schaeffer S, Krishnan V, Wu C, Chu VT, King ST, Vick J, Tao Q, Mraz A, Stormo A, Stormo K, Bogden R, Ederle D, Stella A, Vecchietti A, Kater MM, Masiero S, Lasserre P, Lespinasse Y, Allan AC, Bus V, Chagné D, Crowhurst RN, Gleave AP, Lavezzo E, Fawcett JA, Proost S, Rouzé P, Sterck L, Toppo S, Lazzari B, Hellens RP, Durel C-E, Gutin A, Bumgarner RE, Gardiner SE, Skolnick M, Egholm M, Van de Peer Y, Salamini F, Viola R (2010) The genome of the domesticated apple (*Malus × domestica* Borkh.). Nat Genet 42:833–839. doi:10.1038/ng.654

Vijayraghavan U, Prasad K, Meyerowitz EM (2005) Specification and maintenance of the floral meristem: interaction between positively-acting promoters of flowering and negative regulators. Curr Sci 89:1835–1843

Visser T (1964) Juvenile phase and growth of apple and pear seedlings. Euphytica 13:119–129

Viti R, Andreini L, Ruiz D, Egea J, Bartolini S, Iacona C, Campoy JA (2010) Effect of climatic conditions on the overcoming of dormancy in apricot flower buds in two Mediterranean areas: Murcia (Spain) and Tuscany (Italy). Sci Hortic 124:217–224

Wagner D, Sablowski RWM, Meyerowitz EM (1999) Transcriptional activation of *APETALA1* by *LEAFY*. Science 285:582–584

Weigel D, Nilsson O (1995) A developmental switch sufficient for flower initiation in diverse plants. Nature 377:495–500

Wigge PA, Kim MC, Jaeger KE, Busch W, Schmid M, Lohmann JU, Weigel D (2005) Integration of spatial and temporal information during floral induction in *Arabidopsis*. Science 309:1056–1059

Wilkie JD, Sedley M, Trevor O (2008) Regulation of floral initiation in horticultural tress. J Exp Bot 59:3215–3228

William DA, Su Y, Smith MR, Lu M, Baldwin DA, Wagner D (2004) Genomic identification of direct target genes of *LEAFY*. Proc Natl Acad Sci USA 101:1775–1780

Williams DR, Ross JJ, Reid JB, Potts BM (1999) Response of *Eucalyptus nitens* seedlings to gibberellin biosynthesis inhibitors. Plant Growth Regul 27:125–129

Williams DR, Potts BM, Smethurst PJ (2003) Promotion of flowering in *Eucalyptus nitens* by paclobutrazol was enhanced by nitrogen fertilizer. Can J For Res 33:74–78

Williams DR, Potts BM, Smethurst PJ (2004) Phosphorus fertilizer can induce earlier vegetative phase change in *Eucalyptus nitens*. Aust J Bot 52:281–284

Wilson RN, Heckman JW, Sommerville CR (1992) Gibberellin is required for flowering in *Arabidopsis* under short days. Plant Physiol 100:403

Wiltshire RJE, Reid JB, Potts BM (1998) Genetic control of reproductive and vegetative phase change in the *Eucalyptus risdonii–E. tenuiramis* complex. Aust J Bot 46:45–63

Yanofsky MF (1995) Floral meristems to floral organs – genes controlling early events in *Arabidopsis* flower development. Annu Rev Plant Physiol 46:167–188

Yanovsky MJ, Kay SA (2002) Molecular basis of seasonal time measurement in *Arabidopsis*. Nature 419:308–312

Yeang H-Y (2009) Circadian and solar clocks interact in seasonal flowering. Bioessays 31:1211–1218

Yuceer C, Land SB, Kubiske ME, Harkess RL (2003) Shoot morphogenesis associated with flowering in *Populus deltoids* (Salicaceae). Am J Bot 90(2):196–206

Zeevaart JAD (2006) Florigen coming of age after 70 years. Plant Cell 18:1783–1789

Zhang H, Harry DE, Ma C, Yuceer C, Hsu CY, Vikram V, Shevchenko O, Etherington E, Strauss SH (2010) Precocious flowering in trees: the *FLOWERING LOCUS T* gene as a research and breeding tool in *Populus*. J Exp Bot 61(10):2549–2560

Zharkikh A, Troggio M, Pruss D, Cestaro A, Eldrdge G, Pindo M, Mitchell JT, Vezzulli S, Bhatnagar S, Fontana P, Viola R, Gutin A, Salamini F, Skolnick M, Velasco R (2008) Sequencing and assembly of highly heterozygous genome of *Vitis vinifera* L. cv Pinot noir: problems and solutions. J Biotechnol 136:38–43

Zhu LH, Li XY, Welander M (2009) Can *Arabidopsis AP1* gene shorten the juvenility of apple? Acta Hortic 829:259–264

Zhu et al. (2010) Personal communication

Zimmerman RH (1972) Juvenility and flowering in woody plants: a review. HortScience 10:1–8

Zimmerman RH (1973) Juvenility and flowering of fruit trees. Acta Hortic 34:139–142

Chapter 4
Gene Flow, Spatial Structure, Local Adaptation, and Assisted Migration in Trees

Konstantin V. Krutovsky, Jaroslaw Burczyk, Igor Chybicki, Reiner Finkeldey, Tanja Pyhäjärvi, and Juan Jose Robledo-Arnuncio

Abstract Gene flow is a process of transferring and exchanging genetic material among groups of organisms and is an important evolutionary factor that greatly affects genetic variation and differentiation in trees. Gene flow within and among populations of forest trees plays an important role in forest tree improvement, conservation genetics, and containment of genetically modified trees (GM trees). Although trees are a very diverse group of woody plants, they share many common life traits that affect gene flow, which we discuss in this review. Some unfavorable processes in current forest tree ecosystems, such as habitat loss and fragmentation, increased environmental stress due to global climate change, introgression from domesticated trees into their wild relatives, introduction of maladapted germplasm during reforestation, etc., may badly affect naturally established balance between gene flow, isolation, and local adaptation. Therefore, estimation of gene flow becomes increasingly important for monitoring these processes and developing the

K.V. Krutovsky, Ph.D. (✉)
Department of Ecosystem Science and Management, Texas A&M University, #319 Horticulture and Forest Science Building, 2138 TAMU, College Station, TX 77843-2138, USA
e-mail: k-krutovsky@tamu.edu

J. Burczyk • I. Chybicki, Ph.D.
Department of Genetics, Institute of Experimental Biology, Kazimierz Wielki University, Bydgoszcz 85-064, Poland

R. Finkeldey
Büsgen-Institute, Forest Genetics and Forest Tree Breeding, Georg-August University of Göttingen, 37077 Göttingen, Lower Saxony, Germany

T. Pyhäjärvi, Ph.D.
Department of Plant Sciences, University of California, Davis, CA 95616, USA

J.J. Robledo-Arnuncio, Ph.D.
Departamento de Ecología y Genética Forestal, Centro de Investigación Forestal (CIFOR) – INIA, 28040 Madrid, Spain

R.J. Schnell and P.M. Priyadarshan (eds.), *Genomics of Tree Crops*,
DOI 10.1007/978-1-4614-0920-5_4, © Springer Science+Business Media, LLC 2012

best possible strategy to manage and protect forest ecosystems. Here, we review what has been done in this area recently and the methods and approaches currently used to measure gene flow in forest tree populations.

Keywords Assisted migration • Gene flow • Dispersal • Introgression • Local adaptation • Molecular genetic markers • Spatial structure

Introduction

Gene flow is a process of transferring and exchanging genetic material among groups of organisms, such as subpopulations, populations, races, subspecies, or even species. In trees, gene flow occurs mostly via pollen and seed dispersal, although not all dispersed propagules contribute ultimately to gene exchange. Trees as a group of woody plants are extremely diverse. All major eudicot orders except Gunnerales and Geraniales contain woody or tree-like species (Groover 2005). Their mating and reproduction systems are very different. Even among forest trees, there are great differences between angiosperms and gymnosperms (conifers), boreal and tropical trees, etc. In our review, we will focus mainly on the most common forest tree species constituting major tree crops, trying to take biological differences among them into consideration (see below, for instance, a separate section on tropical trees). At the same time, there are several life history traits that are common for most forest trees and that affect gene flow, such as sedentary life, sexual reproduction with prevailing outcrossing, potentially long-distance migration of pollen (and often seeds), longevity, and mixed age populations. There are tree species with unique features and exceptions that are, for instance, dispersed mainly by clonal propagation (e.g., some aspens) or via young, germinated seedlings rather than seeds, such as viviparous mangrove species (Geng et al. 2008), but we will focus mostly on forest trees with the "typical" life history traits mentioned above.

Gene flow is common in forest trees (see Sork et al. 1998, and more recent reviews by Savolainen et al. 2007 for trees, and by Mitton and Williams 2006 and Bagnoli et al. 2011 specifically for conifers). It affects the distribution of genotypes within and between populations, and, therefore, has a great impact on tree population genetic structure, adaptation, and evolution (e.g., Ellstrand 1992a). In most tree species, populations are subdivided into smaller groups, families, or subpopulations because of geographical (spatial), ecological, and/or phenological isolation. Gene flow depends mainly on mating system, fecundity, and dispersal distances of pollen and seeds, which can vary widely due to geographical, ecological, demographic, and environmental heterogeneity. Tree populations diverge genetically over time and space due to genetic drift that is a random allelic frequency change resulting from stochastic variation in survival and reproduction in populations of finite size and leads finally to incidental loss or fixation of alleles, the probability and the rate of which are functions of allele frequency and effective population size, respectively. Isolated populations with small effective population size diverge faster (e.g., Ellstrand and Elam 1993).

Different types of selection acting on the same alleles and genotypes or the same type of selection acting on different alleles and genotypes in different populations can also promote genetic differentiation between completely or partly isolated populations. In addition, mutations constantly introduce new alleles into populations, also promoting their divergence. Gene flow works effectively against genetic differentiation between populations toward genetic unification across a species' range, and, therefore, counteracts genetic drift, mutation, and selection, depending much on the level of physical and spatial isolation between different populations and parts of a species' range (described in detail in two separate sections below). Low levels of gene flow between populations may not prevent their divergence, while high levels can significantly delay or halt it, and practically homogenize them and maintain their similarity.

Gene flow may either decrease or increase local adaptation in tree populations. If populations are strictly adapted to their locations and have the highest fitness at their home sites and lower fitness outside their range, then gene flow of maladaptive alleles from other populations would decrease relative fitness of local populations. Alternatively, gene flow can also provide local populations with new favorable alleles, increasing adaptive potential and fitness of local populations (e.g., De Carvalho et al. 2010). Therefore, the extent of local adaptation is determined by the balance between gene flow and selection. Adaptation of forest trees to climate change is a very important issue now. The potential for such adaptation depends on genetic variation and dispersal and establishment rates. Assisted migration and dispersal may help to mitigate negative effects of climate change and are described in detail in one of the following sections.

Not every migration or dispersal event leads to gene flow, but only those that result in genetic exchange. It is experimentally challenging to directly estimate levels of gene flow in natural populations. Indirect estimates of gene flow need advanced statistical models and methods and may rely on strong assumptions (described in detail in the following section) and highly polymorphic, selectively neutral molecular genetic markers, such as microsatellites (or simple sequence repeats – SSRs) and single nucleotide polymorphisms (SNPs) in noncoding regions. Recent progress in genomics and high-throughput genome sequencing of multiple individuals greatly facilitates this task and provides researchers with a practically unlimited number of supposedly neutral markers (described briefly below in a separate section).

Gene flow occurs on different spatial and geographic scales (between subpopulations, regions, etc.), and biotic and abiotic factors may greatly affect contemporary and historical gene flow. For example, there may be gene flow between adjacent watersheds, mountain ranges, or valleys, but less than between populations within a watershed, a valley, or a mountain range. Separated populations may have little direct exchange currently (e.g., through occasional long-distance gene flow) but may share a common genetic history. There has been increasing interest in landscape genetics and geographic approaches to estimating historical and contemporary gene flow and local adaptation (Sork et al. 1999; Epperson 2003; Manel et al. 2003; Holderegger and Wagner 2008; Holderegger et al. 2010). Therefore, gene flow is also reviewed in ecological and biogeographic context in one of the separate sections below. The hierarchical or spatial genetic structure (SGS) analysis, gene coalescence, and phylogenetic methods are also often used to infer historical gene flow and understand genetic relationships and differentiation of populations.

It is often useful to study gene flow from different perspectives using genetic data based on different modes of inheritance – maternal, paternal, and biparental (see Jordano 2010 for historic perspective and additional references). Conifers are especially informative in this regard, providing genetic data with strictly paternal (chloroplast genome), maternal (mitochondrial genome), and biparental (nuclear genome) inheritance. Therefore, gene flow estimates based on mtDNA address maternal (seed) gene flow, based on cpDNA address paternal (pollen) gene flow in conifers, and based on nuclear DNA address both (e.g., Mitton and Williams 2006). In most angiosperms, by contrast, both chloroplast and mitochondrial genomes are maternally inherited (McCauley 1995), reducing inferential opportunities to dissect the paternal and maternal contributions to total gene flow.

Gene flow estimates have multiple important practical applications in trees, such as estimation and modeling of transgene flow from genetically modified trees (GM trees) into related wild populations (Katul et al. 2006; Williams et al. 2006; Schoen et al. 2008; Ahuja 2009, 2011; see also a separate section below), seed orchard pollen contamination (e.g., Adams et al. 1997; Kaya et al. 2006; Slavov et al. 2005a, b), gene flow from domesticated populations (e.g., Adams and Burczyk 2000), invasion of nonnative trees into new areas and their hybridization (gene flow) with native taxa (Sax et al. 2007), and genetic rescue or restoration of declined tree populations of endangered species through introducing new variation or promoting gene flow from other populations of the same or closely related species (Tallmon et al. 2005; Edmands 2007).

Statistical Models and Methods for Estimation of Gene Flow in Trees

The estimation of gene flow is not trivial, and nearly 30 years ago Slatkin (1981) skeptically concluded that "there is no way to determine the importance of gene flow in natural populations because there is no direct way to estimate levels of gene flow." Although estimation of gene flow is still a challenging task, it has become more accurate nowadays due to two major achievements in plant population genetics: development of powerful genetic markers available for various kinds of estimation methods (especially for parentage analyses) and significant conceptual advancement in statistical models and methods.

Gene flow is often described in terms of migration rates and distances at which genes are dispersed. In population genetics theory, there are several key parameters that are used to describe the intensity and scale of gene flow. Some of them are introduced here for reference, such as $N_e m$ that defines the number of migrants exchanged between discrete populations per generation, where m is the migration rate (proportion of individuals or propagules that migrated between populations), and N_e is the effective population size. $N_e m$ has a simple relationship with among-population genetic differentiation under Wright's classical island model of population substructure (Wright 1931), although many tree populations are structured following isolation by distance model rather than island model (Wright 1943, 1946). Quite often, m is considered as the proportion of immigrants from outside of the target population. Mean dispersal distance (δ) is the average distance of dispersal of seed or pollen from the point of release to the point of settlement. Variance of the dispersal distance (σ^2) (or average squared axial parent-to-offspring or birth-to-breeding dispersal distance) is used to infer an effective population size for species with a continuous distribution over space using the concept of "neighborhood size" (Wright 1946; Crawford 1984; Rousset 2000): $N_e = 4\pi\sigma^2 D$, where σ is defined as above, and D is effective population density, determining the rate of the decrease of kinship between individuals with increase of distance separating them. While different modes of dispersal may have the same mean dispersal distances, they may follow dispersal kernels with different shapes. Dispersal kernel refers to the probability distribution (probability density function) of dispersal distances from an individual plant. A number of dispersal kernels of varying complexity are commonly used to fit observed dispersal patterns, ranging from thin to fat tailed distributions (Austerlitz et al. 2004), and among those most often used are the normal, exponential, exponential-power, Weibull, two-dimensional Student's *t* (2*Dt*), and others including mixture kernels (e.g., Goto et al. 2006; Slavov et al. 2009).

Pollen and seed dispersal can be measured by means of ecological methods such as tagged seed tracking or frugivore observations: such approaches have been reviewed elsewhere (DiFazio et al. 2004; Ashley 2010). However, tracking dispersal of pollen and seed over broad scales is very difficult, and the evolutionary relevance of these potential dispersal estimates is questionable unless dispersed propagules get effectively established, i.e., unless genes are finally exchanged via mating (pollination) or seedling establishment and future reproduction. Therefore, it is generally acknowledged that the most reliable and useful estimates of gene flow are obtained using genetic markers (Sork et al. 1999).

Indirect Methods

Gene flow can be estimated indirectly from the analysis of genetic structure among subpopulations, assuming that they are in migration–drift equilibrium, and that their differentiation is mainly due to the balance between genetic drift and limited gene flow. Under these assumptions $N_e m \approx (1 - F_{\mathrm{ST}})/(4F_{\mathrm{ST}})$, where F_{ST} is the fixation index of population subdivision relative to the total population, as formulated in a classical Wright's island model of population structure (Wright 1931; Slatkin 1985; Neigel 1997). However, indirect methods have several assumptions that are rarely met in real situations (Whitlock and McCauley 1999; Broquet and Petit 2009). The assumption about the equilibrium between genetic drift and migration is unlikely to be fulfilled in forest tree species because of historical demographic fluctuations, large census numbers of most tree populations, and long generation time (Petit and

Hampe 2006). For example, neighboring populations could be completely isolated but still very similar to each other due to recent fragmentation and sharing ancestral polymorphism, but not due to high gene flow. Therefore, even in the case of the absence of actual gene flow between populations, indirect methods would suggest extensive gene exchange between populations, but this gene exchange should be considered as historic rather than current. Eventually, populations of forest tree species are often structured continuously in space following isolation-by-distance rather than island models.

Trying to overcome the limitations of the classical island model of population structure, new methods for inferring historical migration among populations have been developed, based on coalescent theory (Beerli and Felsenstein 2001). The method implemented in MIGRATE software (Box 4.1) allows one to estimate effective population sizes and historical migration rates between populations assuming a migration matrix model with asymmetric migration rates and different subpopulation sizes and using maximum likelihood or Bayesian inference to jointly estimate all parameters. Indirect gene flow estimates, if they comply with underlying assumptions, can estimate long-term historic averages and contribute to the general knowledge about the biology of the species at evolutionary time scales, but they are not appropriate for inferring contemporary gene flow over a time scale of one or a few generations. If recent migration patterns are of interest, they can be rather inferred with methods known generally as assignment methods (Berry et al. 2004; Manel et al. 2005). Assignment methods (e.g., Pritchard et al. 2000; Wilson and Rannala 2003; Guillot et al. 2005) embrace several statistical approaches (clustering or classification methods) that use multilocus genetic information to ascertain population membership of individuals or groups of individuals (Manel et al. 2005). The real advantage of some of these implementations is that they do not require prior knowledge about population subdivision, and, can, therefore, be efficiently used even in a case of cryptic genetic structure. Provided their underlying assumptions are met, these methods appear to be most effective when differentiation among populations is relatively large ($F_{ST} \geq 0.05$; Faubet et al. 2007), which is often not the case in forest trees (Hamrick et al. 1992).

On a relatively small spatial scale within populations, the extent of gene flow can be inferred from the analysis of SGS (Rousset 2000; Vekemans and Hardy 2004). In general, the method approximates the neighborhood effective population size N_e (*sensu* Wright 1946) based on the relationship between pairwise kinship coefficients or genetic distances and spatial distances between individuals, i.e. $N_b = 1/b$, where b is the slope of regression of pairwise kinship coefficient or genetic distance on the log-transformed pairwise spatial distance in two-dimensional space (Rousset 2000). Then, dispersal parameters (namely, σ^2) might be obtained from $N_b = 4\pi\sigma^2 D$, assuming that effective population density (D) is known (i.e., density measured based on the effective number of individuals, Hardy et al. 2006). Since the actual effective density of a population is difficult to estimate, results are often presented as a list of estimates, given some range of possible D priors (Hardy et al. 2006; Oddou-Muratorio and Klein 2008). Although SGS analysis reflects the joint effect of dispersal of both pollen and seed, careful assessment of the shape of the relationship between kinship and distance might provide some insights into the ratio of seed to pollen dispersal (Heuertz et al. 2003), especially for tree species with contrasting modes of seed vs. pollen dispersal (Chybicki et al. 2009; Craft and Ashley 2010).

Finally, an additional indirect procedure, designed for estimating contemporary pollen flow within a continuous population, has been proposed by Smouse et al. (2001). This method, dubbed TwoGener, is based on the genetic structure of pollen pools "sampled" by a number of mother plants distributed across the landscape. This structure is measured by an intraclass correlation coefficient (Φ_{FT}), which is the correlation of male gametes drawn at random from the same mother relative to those drawn at random from the population as a whole, calculated using analysis of molecular variance (AMOVA, Excoffier et al. 1992). This parameter is related to the probability of paternal identity and thus inversely related to effective number of pollen parents (Austerlitz and Smouse 2001), which can be used to infer jointly dispersal distances and effective density, assuming a dispersal kernel function (Austerlitz and Smouse 2001; Austerlitz et al. 2004). The assessment of the pollen pool genetic structure has been further developed as a normalized measure of correlated paternity between female pairs, which again can be used for the inference of dispersal distance, but in this case independently of effective density (Robledo-Arnuncio et al. 2006). Recently, a similar approach has been developed to dissect the respective contributions of pollen and seed gene flow to naturally established seedlings, which is possible if the maternal and paternal gametic contributions can be unambiguously determined, as for some tree species in which maternal-origin seed tissues remain attached to the seedling after germination (Grivet et al. 2009).

Direct Methods

Direct methods attempt to trace the migration of pollen and/or seeds using parentage-offspring analyses (Smouse and Sork 2004; Ashley 2010). In its simplest form, usually called parental or genotypic exclusion (DiFazio et al. 2004), if for any particular offspring genotype or pollen gamete haplotype (usually inferred via comparison of an offspring and its mother genotypes), there is no genetically compatible parent in a local population, such an offspring or pollen gamete is considered as a putative migrant. Some actual migrants may remain undetected if their genotypes could also be generated by any of the local adults, thus contributing to cryptic gene flow. The estimates of apparent migration rates can be easily adjusted for undetected (cryptic) gene flow in both conifers (Smith and Adams 1983; Adams et al. 1997) and angiosperms (Devlin and Ellstrand 1990), although these methods are sensitive to the assumptions about the allele frequencies in surrounding populations, which are used for inference (Burczyk and Chybicki 2004). Estimates of pollen migration have become important measures for evaluating reproductive isolation of forest seed orchards from surrounding forest stands (Hamrick and Nason 2000).

A more detailed pattern of gene dispersal can be obtained using direct parental (paternal and/or maternal) assignment. In an ideal situation, if for any given offspring it is possible to find one and only one paternal and/or maternal parent, the dispersal distance can be measured, and these data collected over a large number of offspring can be used to calculate mean dispersal distances and to infer dispersal kernels. However, even using a significant number of microsatellite markers, it is difficult, if not impossible, to unambiguously assign all offspring to their parents in forest tree populations, and there are only a few studies where such parentage analyses were relatively successful (Dow and Ashley 1996, 1998a, b; Robledo-Arnuncio and Gil 2005; Hardesty et al. 2006; see also review by Ashley 2010). Among the most common problems are the presence of more than one compatible parent for each offspring (ambiguous parentage), pollen and/or seed migration from unidentified sources, genotyping errors, and mutations (Ashley 2010). Direct parentage methods are susceptible to Type I and Type II inference errors. Type I error occurs when for a given offspring an adult tree, which is not the actual parent, is assigned as a parent (false positive case). This may happen when the actual parent of an offspring is located outside the study population (and is thus unknown), but one of the local adult trees has a genotype compatible with the focal offspring, or when, due to genotyping error or mutation, a wrong parent is assigned. Type II error occurs when the actual parent is excluded (false negative), which happens most often as a result of misgenotyping, mutation, or null alleles (Slavov et al. 2005b).

Some of these problems may be tackled using likelihood-based parentage methods (e.g., Marshall et al. 1998) that assign a single parent to an individual with the highest likelihood, based on population allele frequencies. This kind of method is implemented in the software CERVUS (Marshall et al. 1998; Kalinowski et al. 2007), in which likelihood ratios are calculated allowing for the possibility that the genotypes of parents and offspring may be mistyped, and additionally the level of confidence in the parentages it assigns may be determined via simulation. Although the feature is not available in CERVUS, the frequency distribution of dispersal

Box 4.1 Gene Flow-Related Software

(Mostly updated from the list compiled by Rodney Dyer, Juan Fernandez, and Victoria Sork, which is available at http://www.nceas.ucsb.edu/nceas-web/projects/2057/nceas-paper3/software/index.html; see also Jones et al. 2010 for a review of parentage software that can be also used for assessing gene flow.)

Migrate (http://popgen.sc.fsu.edu/Migrate/Migrate-n.html) by Peter Beerli, Florida State University, estimates effective population sizes and past migration rates between *n* populations assuming a migration matrix model with asymmetric migration rates and different subpopulation sizes. Migrate uses maximum likelihood or Bayesian inference to jointly estimate all parameters. It can use the following data types: (1) nucleotide sequence data using Felsenstein's 84 model with or without site rate variation, (2) single nucleotide polymorphism (SNP) data (sequence-like data input, HAPMAP-like data input), (3) microsatellite data using a stepwise mutation model or a Brownian motion mutation model (using the repeat length input format or the

(continued)

Box 4.1 (continued)

fragment-length input format), and (4) allozyme data using an "infinite" allele model. The output can contain estimates of all migration rates and all population sizes, assuming constant mutation rates among loci or a gamma distributed mutation rate among loci, profile likelihood tables, percentiles, likelihood-ratio tests, and simple plots of the log-likelihood surfaces for all populations and all loci.

Platform: Linux, Mac OS X, Sun, Ubuntu, Windows, source code. It runs on desktop and parallel computers (MPI-parallel, multiple heated chains: multithreading, etc.).

PollenGF (http://www.nceas.ucsb.edu/nceas-web/projects/2057/nceas-paper3/software/pollen.gene.flow.sea.hqx) by John Nason, University of Iowa, uses paternity exclusion techniques to estimate apparent gene flow using progeny arrays from maternal plants. Gene flow can be calculated for individual mothers or local population.

Platform: Mac OS X.

SPAGeDi (Spatial Pattern Analysis of Genetic Diversity; http://ebe.ulb.ac.be/ebe/Software.html) by Olivier Hardy and Xavier Vekemans, Université Libre de Bruxelles, is a computer package primarily designed to characterize the spatial genetic structure (SGS) of mapped individuals and/or mapped populations using genotype data of any ploidy level. It computes various statistics describing relatedness or differentiation between individuals or populations by pairwise comparisons, and analyzes how these values are related to spatial distances. The statistics computed include F_{ST}, R_{ST}, N_{ST}, D_S (Nei's standard genetic distance), and $(\delta\mu)^2$ (Goldstein and Pollok 1997) for analyses at the population level, and, for analyses at the individual level, pairwise kinship, relatedness, and fraternity coefficients (with many different estimators for each). Jackknife over loci gives approximate standard errors. Permutations of locations, individuals or genes provide *ad hoc* tests of spatial structure, population differentiation or inbreeding, respectively. For analyses at the individual level, assuming a two-dimensional population at drift-dispersal equilibrium, SPAGeDi can estimate σ, the historical dispersal rate, by regressing pairwise kinship coefficients on natural logarithm of distance. The procedure requires an estimate of the effective population density.

Platform: Linux, Mac OS X, Windows, source code.

Analyse (http://www.biology.ed.ac.uk/archive/software/Mac/Analyse/index.html) by Nick Barton and Stuart Baird, University of Edinburgh, is a MacOS/PowerPC application for the analysis of hybrid zone data.

(continued)

Box 4.1 (continued)

Platform: MacOSX, PC, Unix.

Cervus (http://www.fieldgenetics.com/pages/home.jsp) by Tristan Marshall, University of Edinburgh, is a computer program for assignment of parents to their offspring using codominant genetic markers such as microsatellites (short tandem repeats [STRs]), SNPs, and allozymes. It uses likelihood to assign parentage, and likelihood ratios are calculated allowing for the possibility that the genotypes of parents and offspring may be mistyped. Cervus determines via simulation the level of confidence in the parentages it assigns.

Platform: Windows.

GENFLOW (http://www.nceas.ucsb.edu/nceas-web/projects/2057/nceas-paper3/software/Genflow.zip) by W. Tom Adams, Oregon State University, and Jaroslaw Burczyk, Bydgoszcz University, is a computer program for estimating levels of pollen contamination in clonal seed orchards using codominant genetic markers.

Platform: PC.

NM+ (http://www.genetyka.ukw.edu.pl/index_pliki/software.htm) by Igor Chybicki and Jaroslaw Burczyk, Bydgoszcz University, is a more advanced and flexible version of *NEIGHBOR* that makes inferences on plant gene dispersal and mating patterns via modeling parentage probabilities based on a spatially explicit parentage model called neighborhood model (Chybicki and Burczyk 2010a). Therefore, NM+ requires a sample of mapped and genotyped candidate parents and offspring; however, offspring may optionally be assigned to single maternal parents (forming so-called half-sib progeny arrays). Using the maximum likelihood approach, NM+ estimates a number of parameters, including self-fertilization rate, immigration rates from outside of a defined study site, parameters of pollen (and/or seed) dispersal kernels (exponential power, Weibull, geometric or 2*Dt*), and selection gradients relating covariates (phenotypic traits) with male (and/or female) reproductive success.

Platform: Linux using WINE emulator, Windows.

POLDISP (http://sites.google.com/site/poldisp2) by Juan José Robledo-Arnuncio and Frédéric Austerlitz, Centro de Investigación Forestal (CIFOR) – INIA and Paris-Sud University, is a free software package to estimate contemporary pollen dispersal using mother-offspring diploid codominant genotypic data. It contains the necessary programs to apply the two available indirect (non-paternity-based) methods of contemporary pollen dispersal estimation: *KINDIST* and *TWOGENER*. POLDISP estimates the following

(continued)

Box 4.1 (continued)

parameters: the average pollen dispersal distance, the variance of the pollen dispersal distribution, the kurtosis of the pollen dispersal distribution, the correlated paternity rate within and among maternal sibships, and the effective male population density. POLDISP requires the spatial coordinates and genotypes of a sample of seed-plants and their respective maternal progenies.

Platform: Mac OS, Windows.

Immanc (http://www.nceas.ucsb.edu/nceas-web/projects/2057/nceas-paper3/software/Immigration-_Slatkin.sea.hqx or http://www.rannala.org/?page_id=13) by Bruce Rannala, UC Davis, and Joanna Mountain, Stanford University, is detecting immigration by using multilocus genotypes. It tests whether or not an individual is an immigrant or is of recent immigrant ancestry. In order to apply the test, one must have genotype data for multiple loci for multiple individuals of two or more populations. The method is appropriate for use with allozyme, microsatellite, or RFLP data.

Platform: Linux, Mac OS X, Windows.

Isolation by Distance Program (http://www.nceas.ucsb.edu/nceas-web/projects/2057/nceas-paper3/software/Slatkins_IBD.sea.hqx or http://ib.berkeley.edu/labs/slatkin/monty/Isolation_by_distance.program) by Montgomery Slatkin, University of California, Berkeley, estimates levels of gene flow between pairs of sampling locations and regresses the estimated level of gene flow on geographic distance.

Platforms: PC, Unix

PMLE (Pseudo Maximum Likelihood Estimator of Gene Flow; http://www.rannala.org/?page_id=258 or http://www.rannala.org/docs/pmle20.html) by Bruce Rannala, University of California, Davis, and J. A. Hartigan. This program estimates the gene flow parameter q for a collection of two or more semi-isolated populations by (pseudo) maximum likelihood using either single or multilocus, haploid or diploid, genotype data. Suitable molecular data types minimally include allozymes, microsatellites, and RFLPs. The method is explicitly based on Wright's (1931) island model of population demographic structure. An exact solution exists for the likelihood function of this model under Wright's diffusion approximation. However, the exact form of the likelihood function appears quite robust to the specific assumptions of the model and other, more general, population demographic models may have an identical likelihood function. For a (haploid) island model with a constant population size, the parameter q can be interpreted as Nm, the

(continued)

Box 4.1 (continued)

expected number of immigrants per island, per generation. This quantity is most often estimated by the widely used formula $Nm = (1 - F_{ST})/(4F_{ST})$. However, the maximum likelihood estimator has less bias and lower variance than this classical estimator over widely varying levels of gene flow and is therefore preferable.

Platform: Linux, Windows.

BayesAss+: Bayesian Inference of Recent Migration Using Multilocus Genotypes (http://www.rannala.org/?page_id=245) by G. A. Wilson, University of California, Berkeley, and Bruce Rannala, University of California, Davis, is a program that estimates recent migration rates between populations using Markov Chain Monte Carlo (MCMC) simulation. It also estimates each individual's immigrant ancestry, the generation in which immigration occurred (i.e., the individual can be assigned as an immigrant from a specific population, a nonimmigrant, or the offspring of an immigrant and a nonimmigrant), the population (as opposed to the sample) allele frequencies, inbreeding (*F*) values within each population, and genotypes at any locus with missing data. Details of the method can be found in Wilson and Rannala (2003). BayesAss requires diploid data, and can be used for genetic markers such as microsatellites, RFLPs, SNPs, and allozymes. The method assumes relatively low levels of migration, and the proportion of migrant individuals into a population cannot exceed one-third of the population total per generation. Loci are assumed to be in linkage equilibrium, but deviations from Hardy-Weinberg equilibrium are allowable. The method does not check for the presence of null alleles, and their occurrence may negatively affect the results. Estimates are most accurate when a large number of variable loci are used, sample size is large, and population differentiation is high.

Platform: Linux, Mac OS X, Windows, source code.

BayesAssNM: Bayesian Inference of Recent Migration Using Multilocus Genotypes (Migrants Not Included In the Sample) (http://www.rannala.org/?page_id=245) by R. Jehle, G. A. Wilson, J. W. Arntzen, and T. Burke is only to be used when genotypes are obtained from samples that are known not to be migrants, such as embryos. It estimates recent migration rates between populations using MCMC. It also estimates each individual's immigrant ancestry, the generation in which immigration occurred (i.e., the individual can be assigned as a nonimmigrant or the offspring of an immigrant and a nonimmigrant), the population (as opposed to the sample) allele frequencies, inbreeding (*F*) values within each population, and genotypes at any locus with missing data. In most respects, BayesAssNM is identical to BayesAss, originally described in Wilson and Rannala (2003). However, BayesAssNM was

(continued)

Box 4.1 (continued)

designed for the situation where first generation migrants are not included in the sample, due to the sampling scheme. BayesAssNM was originally used to estimate gene flow between a number of new populations. As samples were collected as fertilized eggs, no first generation migrants were obtained. BayesAssNM was then designed to relax the assumption that migrants occur in the genotyped sample of individuals (Jehle et al. 2005). It should be stressed that BayesAssNM is only for situations where there are known to be no migrants among the sampled individuals.

Platform: Linux, source code.

distances based on parentage analyses might then be translated into dispersal kernels (Austerlitz et al. 2004; Robledo-Arnuncio and Gil 2005). See also Box 4.1 for software available for gene flow analysis.

If dispersal parameters, rather than parental identities, are of primary interest, it is more adequate to use all available genetic information on genotypes and locations of adults and offspring for designing probability models to estimate dispersal parameters directly, without the need for separate categorical parentage inference for each individual offspring. Such models are often referred to as *full probability models* (Hadfield et al. 2006; Jones et al. 2010), since they combine genetic data defining parentage relationships together with nongenetic data, such as spatial distribution of individuals, which themselves should determine parentage probabilities. For example, under the reasonable assumption that the probability of parentage is inversely related to the distance between parent and offspring, this spatial information should be incorporated when computing parentage likelihoods, but, more interestingly, it allows incorporation of explicit dispersal models (dispersal kernels), the parameters of which can be estimated directly. The simplest model of this kind, considering only pollen dispersal, is presented in Box 4.2.

More advanced models have been designed for joint estimation of seed and pollen dispersal parameters based on established seedlings (Burczyk et al. 2006; Goto et al. 2006; Oddou-Muratorio and Klein 2008; Chybicki and Burczyk 2010b; Moran and Clark 2011), and the same probabilistic scheme can be applied to the specific case of categorical maternal analysis based on maternal-origin seed tissues (Robledo-Arnuncio and García 2007). The estimates of gene dispersal based on established seedlings seem to be of particular interest, because patterns of such "realized" gene dispersal reflect not only dispersal processes but also the impact of factors affecting seedling establishment, either of demographical or adaptive nature (González-Martínez et al. 2006a; Grivet et al. 2009).

Modeling propagule dispersal based on parentage analyses has several advantages. For example, the proportion of migration can be modeled as a function of the assumed dispersal kernel in order to integrate long-distance dispersal (LDD) events

Box 4.2 Simple Model of Pollen Dispersal

In the model of pollen dispersal, the probability of observing a particular multilocus offspring genotype in a progeny array is defined as:

$$P(g_i \mid M) = sP(g_i \mid M, M) + mP(g_i \mid B, M) + (1 - s - m)\Sigma(\varphi_j P(g_i \mid F_j, M))$$

Where $P(g_i \mid M)$ is the probability of observing a particular diploid genotype g_i in the progeny array of the maternal tree of genotype M; s is the self-fertilization rate (in monoecious species), m is the migration rate of a pollen gamete from outside of an area where all potential paternal trees have been genotyped (often referred to as neighborhood of a maternal tree; Adams and Birkes 1991). $P(g_i \mid M, M)$, $P(g_i \mid F_j, M)$, and $P(g_i \mid B, M)$ are genetic Mendelian transition probabilities that a particular genotype g_i from the progeny array of maternal tree M is the result of selfing, outcrossing with the j_{th} father of genotype F_j residing in a local population, or outcrossing with distant, not genotyped fathers located outside a local population, respectively. B is an array of allelic frequencies in background (surrounding) populations (Burczyk and Chybicki 2004); φ_j is a function of factors affecting mating success of *j*th paternal tree and could be related exclusively to distance of pollen dispersal (Chybicki and Burczyk 2010b), or could be extended to include a number of individual paternal tree phenotypic measures as correlates of male mating success (Burczyk et al. 1996; Smouse et al. 1999; Oddou-Muratorio et al. 2005; van Kleunen and Burczyk 2008). The function φ_j could represent any desired dispersal kernel (Austerlitz et al. 2004; Pluess et al. 2009), and its parameters can be estimated using maximum likelihood or Bayesian methods.

in the kernel parameter estimates (Goto et al. 2006; Chybicki and Burczyk 2010b). The problem of misgenotyping can also be accounted for by incorporating genotyping error/mutation rates into the model and even estimating them (Gerard et al. 2006). Finally, the models are capable of accounting for fertility variation directly (Klein et al. 2008) or indirectly through selection gradients of phenotypic traits as covariates of reproductive success (Smouse et al. 1999) in order to provide a more comprehensive description of reproductive patterns. However, parentage-based methods are very laborious and have inherent logistic restrictions resulting from the required exhaustive sampling and genotyping of candidate parents within the study area. The larger the parental population, the higher parentage assignment uncertainty becomes and the larger the offspring sample should be to achieve acceptable precision of dispersal estimates. The need of an exhaustive sampling of parents does indeed limit the spatial scale of analysis. Hence, parentage-based methods poorly handle long-distance dispersal events that are treated usually as migrants from unknown sources.

Molecular Genetic Markers

Gene flow estimation requires appropriate polymorphic genetic markers. Here, we briefly review the most common markers used in gene flow studies (see also Gillet 1999; Namkoong and Koshy 2001; Wang and Szmidt 2001; Lörz and Wenzel 2004; Ziegenhagen and Fladung 2004; Semagn et al. 2006; Dean 2006; Pijut et al. 2007; Burdon and Wilcox 2011 for review of DNA markers in forest trees, and Bagnoli et al. 2011 specifically for selectively neutral markers in conifers).

Allozyme loci were the most popular genetic markers a couple of decades ago, and are still occasionally used to study gene flow (see El-Kassaby and White 1985 for earlier references). Allozymes are allelic forms of enzymes that have a codominant mode of inheritance and can be relatively easily genotyped. They usually have moderate levels of polymorphism; therefore, a relatively large number of allozyme loci are required for accurate gene flow estimation. The great advantage of allozyme loci is that their analysis is relatively inexpensive, reproducible, and transferable, and can be done with minor modifications for the same loci across different tree species (Conkle et al. 1982; Cheliak and Pitel 1984; Jech and Wheeler 1984; Konnert and Maurer 1995; Manchenko 2003). Allozyme loci have been used effectively in gene flow studies for more than three decades, and a large body of data was obtained from indirect and direct estimates. In particular, early studies of pollen migration into artificial populations such as seed orchards revealed a great potential for pollen migration from neighboring stands, accounting for up to 50% of effective pollen in seed orchards isolated by several hundreds of meters from nearest stands (Hamrick and Nason 2000; Adams and Burczyk 2000). However, the relatively low exclusion power of allozyme markers made them generally ineffective in parentage studies. Their low exclusion power promoted though the development of several advanced models of parentage analyses intended to fractionally partition probability of paternity based on Mendelian transition probabilities and/or to estimate dispersal parameters directly in the model (e.g., Adams and Birkes 1991).

Microsatellite loci are now the most common type of markers used in gene flow studies (see Ashley 2010 for review). Microsatellites, *aka* simple sequence repeats (SSRs) or short tandem repeats (STRs) are relatively short stretches of DNA that consists of short sequence motifs (1–6 nucleotides) repeated several times, and the difference in number of repeats between different alleles constitutes the microsatellite polymorphisms. Microsatellites are ubiquitous and widely dispersed in plant genomes (Morgante and Olivieri 1993). They are genotyped following PCR amplification, most often using species-specific primers, but some microsatellites have been successfully amplified across closely related species using the same PCR primers (e.g., Curtu et al. 2009). Their features, such as codominance, high polymorphism, selective neutrality, considerable number of available unlinked loci, and feasibility for high-throughput analyses, make them one of the best, if not the best, tool for gene flow studies.

Amplified fragment length polymorphism (AFLP) loci (Vos et al. 1995) are also useful molecular markers. They can be relatively easily generated in practically any species without *a priori* knowledge of genomic sequence or genetic data and are especially suitable for species that have not been well studied, such as tropical trees (e.g., Stefenon et al. 2008; Cao et al. 2009; Derero et al. 2011), when other genetic markers are not available. AFLPs usually represent restriction site polymorphisms and are typically genotyped as binary dominant-recessive bi-allelic markers (present-absent fragments). Therefore, a single locus provides incomplete information about an individual genotype. However, the number of loci available in a typical assay is very large (up to several hundred), enabling efficient multilocus parentage inference. Although AFLPs usually provide lower exclusion power when compared with SSRs (Gerber et al. 2000), they can still be markers of choice for parentage analysis in specific situations, because they do not require species-specific primers to be developed first.

Single nucleotide polymorphism (SNP; see Chagné et al. 2007; Henry 2008; Pavy et al. 2008; Eckert et al. 2009; Lepoittevin et al. 2010 for review) markers have been recently proposed for parentage analyses (Jones et al. 2009). SNPs are usually bi-allelic polymorphisms and are not as informative as multi-allelic markers, such as microsatellites, but they are very common and often highly polymorphic, especially in noncoding regions (see González-Martínez et al. 2011 for review). Using blocks of linked SNPs (haploblocks) can increase their efficiency in parentage analysis (Jones et al. 2009), but such genotyping has to consider decay of linkage disequilibrium (LD) and take into account the possibility of sequencing and genotyping errors. On the other hand, SNP markers are relatively expensive and not easily available for nonmodel species with limited genome sequence resources. Therefore, SNP markers are not widely used for parentage methods and gene flow estimations yet, but, considering a huge progress in next generation high-throughput massively parallel sequencing and high-throughput SNP genotyping (see respective section on tree genomics below), these markers will become increasingly available and affordable.

Limitation and Challenges to Estimating Gene Flow

Small progeny sample size is a common problem in many parentage studies. Genotyping of large numbers of parents and offspring is laborious and costly, and researchers tend to keep sample sizes at affordable levels, sometimes at the cost of estimation precision. The problem becomes worse when gene flow is extensive, and a substantial proportion of gametes migrate into the study population from external sources. In this case, estimating dispersal kernels using local parentage assignments based on the small fraction of locally sired offspring usually results in very imprecise estimates.

Long-distance dispersal (LDD) of seed and especially pollen seems common in forest trees, as a consequence of their efficient dispersal and high fecundity (Williams et al. 2006, 2010; Robledo-Arnuncio 2011). However, contemporary LDD is particularly difficult to trace precisely with available methods. Most dispersal kernels in use (such as the exponential, normal, power exponential, Weibull, 2D – Student *t*) might oversimplify the mode of propagule dispersal. If short- and long-distance dispersal events are governed by different processes (e.g., gravity and animal dispersal of oak acorns, respectively), it might be difficult to fit actual dispersal patterns to any single simple dispersal kernel. A combination of different kernels at near and far distances can help (Goto et al. 2006; Chybicki and Burczyk 2010b), although it will also increase parameter dimensionality.

Increasing the number of loci (usually microsatellites) increases the exclusion power of the marker set, but, on the other hand, it raises the problems of genotype mismatching at multilocus levels due to genotyping errors, *null* alleles and/or mutations. Handling of these problems becomes an important methodological aspect in parentage studies, but there are several efforts as to how to address them (Marshall et al. 1998; Wang 2004; Hoffman and Amos 2005; Slavov et al. 2005b; DeWoody et al. 2006).

Interspecific Gene Flow, Introgression, and Introgressive Hybridization

A considerable proportion of gene flow in plants (in terms of distance and rate) occurs through pollen dispersal, enabling the process of mating. Most mating occurs within species, but mating between closely related species is frequent in many taxa. The process is known as hybridization, and, when repeated backcrossing of hybrids with parental species cause transfer of genes from one species to another, this process is called introgression, introgressive hybridization, or sometimes "interspecific gene flow" (e.g., Whittemore and Schaal 1991). While the problem of hybridization in forest trees is important for tree conservation management and improvement (White et al. 2007), here we only present some examples on hybridization in natural populations, and how it might facilitate pollen-mediated gene flow and adaptation.

An interesting example of hybridization comes from oak species (Rushton 1993). Oaks are widely distributed over the northern hemisphere, and they often form species complexes, where two (Whittemore and Schaal 1991; Bacilieri et al. 1996; Craft et al. 2002) or more (Curtu et al. 2009; Penaloza-Ramirez et al. 2010) sympatric oak species can hybridize. Petit et al. (2004) formulated the hypothesis that in case of the *Quercus petraea*/*robur* complex, dispersal and adaptation could benefit from asymmetric hybridization (introgression). The dispersal of seeds of *Q. robur* is more efficient in terms of distance and abundance. The other species (*Q. petraea*) might be dispersed into the areas already occupied by *Q. robur* by means of pollen dispersal and hybridization followed by subsequent pollinations in consecutive generations from *Q. petraea*. Backcross of F_1 hybrids with *Q. petraea* as a pollen donor seems quite efficient, so *Q. petraea* "swamps" *Q. robur* population relatively quickly, in a few generations (Petit et al. 2004). Similar situation might exist in other oak complexes (Whittemore and Schaal 1991). Proposed gene flow patterns could partly explain the lack of chloroplast genome differentiation between species pairs occupying the same area, while exhibiting pronounced differentiation between populations of distant regions (Whittemore and Schaal 1991; Petit et al. 2004). While this particular concept generated some controversy (Muir and Schlötterer 2005, 2006; Lexer et al. 2006), it is generally agreed that during the process of hybridization and subsequent backcrossing with parental species, the segregation of QTLs and selection of traits responsible for adaptation might be quite efficient in providing the basis for adaptation and invasion of species (Rieseberg et al. 2003; Arnold 2006), given relatively low LD in forest trees (Krutovsky and Neale 2005a; Kole 2007; Ingvarsson 2010; González-Martínez et al. 2011; but see Pyhäjärvi et al. 2011 for cases of extended LD).

Natural complexes of hybridizing species are found among various forest tree genera, including *Pinus* (Wheeler and Guries 1987; Heuertz et al. 2010), *Picea* (Krutovskii and Bergmann 1995), *Populus* (Eckenwalder 1984; Lexer et al. 2005), *Eucalyptus* (Griffin et al. 1988), *Betula* (Johnsson 1945), and others. Interspecific gene flow as a source of genetic variation fostering adaptation of populations cannot be ignored, and may be more frequent than previously thought. While understanding the mechanisms underlying the persistence and coexistence of separate parental species along with its hybrids, often in a form of hybrid swarm, is a timely debate, it extends well beyond the scope of this chapter.

Gene Flow in Ecological and Biogeographic Context

In this section, we focus on ecological factors that affect contemporary gene flow at the landscape level. Plants in general are sedentary organisms whose dispersal depends greatly on extrinsic factors. Relatedness among individuals is expected to decline along geographical distance, but the physical distance is not the only factor affecting gene flow among trees. Air or pollinator movement, fragmentation, flowering phenology, and population density are examples of other factors affecting gene flow in trees.

Wind-pollinated trees typically exhibit extensive gene flow within and among stands, resulting in broad mating neighborhoods (Slavov et al. 2009; Ashley 2010). For instance, Slavov et al. (2009) investigated pollen-mediated gene flow of wind-pollinated *Populus trichocarpa* in an exceptionally vast geographic area (radius 10 km). In their Vinson site in eastern Oregon, a two-component model of pollen flow explained the observed pattern of pollen flow better than commonly used exponential power dispersal curves. It has been suggested that short-distance portion of pollen is distributed locally following exponentially bounded distributions, while part of the pollen gets updrafted to higher levels of the atmosphere and results in substantially longer dispersal distances (e.g., Lanner 1965; Di-Giovanni et al. 1996; Slavov et al. 2009). Somewhat surprisingly, wind direction has not been found to have a great impact on the direction of gene flow in wind-pollinated trees in the few studies where wind direction has been considered (Dow and Ashley 1998a, b; Dutech et al. 2005; Pluess et al. 2009). Although in these studies it would be necessary and could be helpful to investigate directionality patterns over broader spatial scales, it is practically unfeasible using direct approaches. Correlation still can be poor solely due to great variation in wind direction (daily, weekly, etc.).

Average pollen dispersal distances in animal-pollinated trees vary depending on pollinator size, abundance, and behavior (Dick et al. 2008a, b). As expected, pollination distances for small insects are shorter than for larger insects or mammals. The timing of flowering and the density of flowers in the whole plant community also affect the movement of pollinators. When flowers are abundant, pollinators are not required to visit plants separated by long distances (Dick et al. 2008a, b). Genetic differentiation in tropical trees is higher compared to temperate and boreal trees, and the difference is at least partially caused by different pollination mechanisms (Hamrick et al. 1992; Dick et al. 2008a, b). However, in a recent review on pollination and seed dispersal by Ashley (2010), the mean dispersal distances of 15 insect-pollinated and 17 wind-pollinated trees were not very different, both spanning from tens to thousands of meters. Thus, animal-pollinated trees can still have an extensive amount of contemporary gene flow (Dick et al. 2008a, b). The apparent discrepancy of observing genetic differentiation despite estimates of high contemporary pollen dispersal could result from the fact that genetic structure reflects historical gene flow as well as the amount of drift within population promoted by selfing and/or low density (Dick et al. 2008a, b).

Seed-mediated gene flow in trees is not studied as extensively as pollen-mediated gene flow, despite its importance for effective gene flow and small spatial scale genetic structure (Hardy et al. 2006; Dick et al. 2008a, b). Pollen flow is not accomplished until the pollinated seeds have dispersed, germinated, and matured. On average, seed flow reaches shorter distances compared to pollen flow (Dick et al. 2008a, b; Ashley 2010), but there are some cases where seed flow reaches extensive distances and even exceeds pollen flow, as in *Sorbus domestica* (Kamm et al. 2009) and *Fraxinus excelsior* (Bacles et al. 2006). Seed flow in trees is mediated by gravity, wind, water, autochory, and frugivores. The variety of animals (birds, rodents, bats, monkeys, ants, etc.) that disperse seeds is wider than that of pollinators, which are insects for most plant species. Trees whose seeds are dispersed by gravity have higher genetic differentiation than trees with animal or wind seed dispersal, suggesting restricted seed-mediated gene flow in the former (Hamrick et al. 1992). Survival near the mother tree may change during establishment (Nathan and Casagrandi 2004), which may affect the estimates of seed dispersal distances depending on the life stage of studied offspring, such as in case of *Pinus pinaster*, where saplings had higher estimated seed dispersal distances compared to seeds (González-Martínez et al. 2006a). Abiotic factors such as water flow or wind may also affect seed dispersal in trees. For example, spatial distribution of mature and juvenile seedlings of *Eucalyptus globulus* indicates the effect of western winds on the direction of gene flow (Jones et al. 2007).

An important conclusion derived from estimates of gene flow in trees is its effect on genetic diversity of fragmented populations. It has been suggested that small isolated populations or solitary trees would have low reproductive potential. If they are isolated from other populations, population genetic theory predicts that they would become genetically depauperate due to increased selfing (if possible), inbreeding, genetic drift, and reduced migration (e.g., Franklin 1980; Janzen 1986). A growing number of studies that used parentage assignment to estimate gene flow in trees have shown that this is not always the case (Kramer et al. 2008; Ashley 2010; Robledo-Arnuncio 2011). Populations that are physically isolated from the main part of the species distribution may still be genetically connected to other populations. In fact, trees in fragmented populations can have more long-distance pollen flow than trees in large populations. The estimated mean pollen dispersal distance was 212 m in an undisturbed population of *Dinizia excelsa* – in contrast to 1,509 m in a fragmented ranch population (Dick et al. 2003). In a study that compared pairs of fragmented and continuous populations of *Eucalyptus globulus*, the fragmented populations had more individuals contributing to reproduction, although they had lower outcrossing rates (Mimura et al. 2009). The difference in dispersal distance between fragmented and continuous populations can partly explain these observations: pollen flow is dominated by pollen migration from larger source populations to smaller sink populations (e.g., Ellstrand 1992b). As a consequence of extensive gene flow to fragmented populations, the genetic drift in isolated populations may not be as strong as previously postulated (Kramer et al. 2008; Ashley 2010). Furthermore, the effects of fragmentation on genetic diversity of tree populations take a long time due to the long life span of trees (Hamrick 2004). However, many authors have pinpointed that the effects of fragmentation on gene flow may have strong variation among species (e.g., Ellstrand 1992b; Mimura et al. 2009). For example, self-incompatible species may suffer more from increased inbreeding compared to self-compatible species that have already purged recessive deleterious alleles (Charlesworth and Charlesworth 1987; Kramer et al. 2008).

Flowering tree density and synchrony affect the outcrossing rates and the distance of gene flow in trees. In animal-pollinated species, flowering tree density changes outcrossing rates via its effects on pollinator behavior. In low density, pollinators move within a tree, but do not necessarily go to the nearest neighbor, when they move from one tree to another (Dick et al. 2008a, b). In tropical trees, the outcrossing rate was positively correlated with flowering tree density (Murawski and Hamrick 1991; Dick et al. 2008a, b). However, the effect of density on outcrossing rates depends on species self-incompatibility systems. Self-compatible mating systems may have evolved to ensure reproduction in species that naturally grow in extremely low densities (Murawski and Hamrick 1991). In species with strong self-incompatibility systems, shifting to lower outbreeding rates is difficult. For example in highly outcrossing African mahogany, *Entandrophragma cylindricum*, logged, low-density sites had the same level of outcrossing as an unlogged site (Lourmas et al. 2007). An animal-pollinated *Sorbus domestica* growing in low densities has a high selfing rate (33.5%), and its average pollen flow distance was estimated to be 1.2 km, which is high compared to most previous estimates in trees (Kamm et al. 2009; Ashley 2010).

Overall, the mechanism of pollen flow both in animal- and wind-pollinated trees is a complex stochastic phenomenon that is hard to model: in wind-pollinated species due to complexities in air movement (Kuparinen 2006) and in animal-pollinated species due to intra- and interspecific ecological factors affecting the behavior of pollinators (Ashley 2010). Effects of fragmentation, density, pollination system, and the degree of self-incompatibility are difficult to partition because many of them are correlated. Most tropical tree species are animal-pollinated and grow in low densities. Effects of fragmentation and flowering tree density are also intermingled and hard to distinguish from each other. In the following section, we especially discuss the differences among temperate and tropical tree species.

Temperate Versus Tropical Forest Trees

Forest tree species differ with regard to their life history traits, population biology and history, and spatial distribution patterns, especially when comparing temperate and tropical tree species. Trees, mostly conifers, of the boreal zone typically occur in high density and often in pure stands. Species diversity increases and, accordingly, population density of a given species usually decreases toward the equator. This has far-reaching consequences concerning the dispersal agents of diaspores containing pollen or seed. Most trees of the boreal and temperate zones, in particular those occurring in high density, are wind-pollinated, and their seeds are also dispersed mainly by wind. This applies, for example, to Norway spruce (*Picea abies*), Scots pine (*Pinus sylvestris*), beech (*Fagus sylvatica*), and two oak species (*Quercus robur* and *Q. petraea*), which together comprise more than 75% of the total forest area in Europe. However, there are wind-pollinated species, whose seeds are dispersed mainly by animals. These include some white pines (subgenus *Strobus*), such as Siberian stone pine (*Pinus sibirica*), one of the major forest tree species in the Siberian boreal forest. On the other hand, tropical trees are rarely wind-pollinated (Finkeldey and Hattemer 2007). The pollination of tropical trees by animals is often accompanied by the production of fleshy fruits containing animal-dispersed seeds (Howe and Smallwood 1982). The evolution of different pollen and seed dispersal mechanisms was not a random process, but reflects adaptation of trees to different environments in forests with different structure and species diversity.

Historical Gene Flow in Different Climate Zones

Different dispersal strategies and efficiencies of biparentally and uniparentally inherited genetic information are reflected in the spatial distribution of genetic information with different modes of inheritance. The wide distribution of pollen in wind-pollinated trees of the temperate and boreal zone is regarded as a main factor contributing to low genetic differentiation among even widely separated populations in most forest tree species (Austerlitz et al. 2000).

A well-studied example reflecting the importance of historical gene flow in the temperate zone is the study of the distribution of chloroplast DNA variation in European species of white oaks (*Quercus* spp.), which revealed strong differentiation among populations across the continent (Petit et al. 2002b). The differentiation is obviously related to the postglacial history of oak populations in Europe (Petit et al. 2002a). Different refugial regions and recolonization routes can be identified, and human impact on the gene flow can be detected in oaks (Gailing et al. 2007). The crucial impact of the ice ages on "historic" gene flow and the genetic structure is obvious in boreal and temperate regions that were denuded from forest vegetation (Hewitt 2000). However, founder effects rarely diminished the genetic variation of populations during postglacial remigration, and glacial refugia with a long, continuous evolutionary history in a single region are not always characterized by high levels of variation (Petit et al. 2003). Intermixing of populations from several refugia and migration routes contributed to the high diversity of tree populations after recolonization.

The impact of evolutionary history and in particular "historical" gene flow on populations is much less obvious in the tropics. However, the climatic oscillations during the ice ages had considerable impact on ecosystems throughout the globe including the tropics. For example, low sea levels in Southeast Asia created large suitable habitats for tropical rain forests in the gulf of Siam (Cannon et al. 2009), but low precipitation presumably restricted the extent of tropical rainforests in the Neotropics (Colinvaux and De Oliveira 2001). There is growing evidence that population migration and "historical" gene flow shaped genetic structures of tree species in the tropics as it did in temperate and boreal regions (Ayele et al. 2009; Cavers et al. 2003).

Thus, past climatic events had strong effects on gene flow of forest trees throughout the globe. Human activities, including gene flow facilitated by humans, who transferred forest reproductive material on a large scale, are strongly confounding factors for almost all forest tree species in Europe (Finkeldey et al. 2010) and other densely populated regions of the temperate zone. This is also true for some tropical tree species, which have been used extensively in plantations. Unfortunately, seed movement of tropical trees grown in plantations is usually poorly documented, and the consequences of human-induced gene flow are presumably severe, but difficult to assess (Finkeldey and Hattemer 2007). Humans have never moved reproductive material of most noncommercial tropical trees. Other activities, in particular forest destruction resulting in fragmentation of isolated patches of different size, influence gene flow patterns of most tropical trees much stronger than voluntary seed transfer.

Current Gene Flow in Main Climatic Regions

Current gene flow has been studied in numerous temperate and boreal tree species, both conifers and broadleaved trees. Most studies found low levels of selfing and biparental inbreeding (Finkeldey 1995), preferential mating within neighborhoods

(Bai et al. 2007; Müller-Starck 1977), but also high levels of long-distance gene flow, mainly through pollen. The origin of reproductively effective pollen is difficult to monitor after long-distance travel. However, the observation of high contamination rates in clonal seed orchards of conifers suggests that a large proportion of genetically effective pollen is of distant origin of at least several kilometers (Lindgren et al. 1995). The observation of efficient long-distance gene flow, especially through pollen, matches the result of numerous genetic inventories, proving high diversity within but low differentiation among subpopulations of forest trees (Hamrick et al. 1992).

This view of trees as species with huge population sizes and highly efficient means of gene dispersal is challenged by the fact that most tree species occur in low density in tropical forests. Early studies on gene flow in tropical trees suggested that self-pollination would be more common in tropical trees that grow in low density and depend on animal pollination, which would consequently reduce the overall gene flow (Corner 1954; Fedorov 1966). Indeed, genetic differentiation in tropical trees is higher compared to temperate and boreal trees, and in addition to population history, the difference is at least partially caused by different pollination mechanisms (Hamrick et al. 1992; Dick et al. 2008a, b). Studies on species occurring on Barro Colorado Island, an artificial island in the Panama channel, depicts an impact of population structure on gene flow via pollen and the mating system for predominantly outcrossing tree species (Stacy et al. 1996). The rare tree species *Turpinia occidentalis* occurred in two small clusters of only two and three trees, respectively, and mating was almost exclusively within those clusters.

On the other hand, Murawski and Hamrick (1991) suggested that tropical trees that live in low densities and have a high selfing rate can have a considerable amount of long-distance pollen flow that compensates for the high amount of consanguineous matings. Far-reaching pollen transport was observed for the slightly more frequent and more homogeneously distributed *Calophyllum longifolium* (Stacy et al. 1996). The efficiency of insect pollinators for pollen transport over distances of hundreds of meters has been demonstrated in numerous other studies. For example, average pollination distances of at least 525 m were observed for the bee pollinated dipterocarp *Neobalanocarpus heimii* in the Pasoh Forest Reserve, Malaysia (Konuma et al. 2000). Transport of pollen over distances of many kilometers has been reported for figs (*Ficus* spp., Nason et al. 1998). The African *Ficus sycomorus* represents an extreme case of successful pollen movement, where paternity analyses revealed mean distances of more than 88 km between pollen and seed parent (Ahmed et al. 2009). Figs are pollinated by tiny wasps; passive movement of these insects by the wind plays an important role to ensure successful pollination over large distances (Compton et al. 2000).

Tropical figs are a well-studied example of the importance of plant-pollinator interactions to ensure gene flow via pollen. Most of the more than 800 *Ficus* species live in obligatory symbiosis with a specific fig wasp species (Agaonidae), which are the only pollinators for these species. The wasps completely rely on the inflorescences of figs as breeding habitat of their larvae (Janzen 1979). Disturbance of this symbiosis may result in the loss of both species. However, there are also examples of tropical forest trees with a wide array of potential pollinators and no specific symbioses. For example, pollination by different birds, bats, opossums, and large bees has been reported for the neotropical tree *Mabea fistulifera* (Viera and de Carvalho-Okano 1996).

Endozoochorous dispersal of seeds is the most common in tropical trees producing fleshy fruits. Dispersal distances depend on the migration behavior of dispersing animals, mostly vertebrates. Long-distance seed dispersal by animals promotes low differentiation among populations not only at biparentally inherited genes, but also at maternally inherited markers. On the other hand, the absence of seed dispersing agents for barochorous species (trees with large seeds dispersed by gravity) with heavy diaspores results in family structures within stands, such as in two *Shorea* species in Malaysia (Ng et al. 2004), and in strong differentiation of chloroplast haplotypes even among closely neighboring populations, such as in the neotropical *Corythophora alta* (Hamilton 1999).

The enormous diversity of tree species in the tropics is mirrored in the high diversity of their reproductive biology and gene flow mechanisms. However, most tropical tree species developed specific adaptation mechanisms often associated with plant-pollinator symbioses, ensuring mating contact even among spatially isolated individuals. Thus, long-distance dispersal of pollen and, to a lesser extent, of seed contributes to the maintenance of large effective population sizes in all climatic regions and in forests of very different species diversity. This promotes high levels of genetic diversity also within populations of tropical trees and results in low to moderate differentiation among populations (Hamrick et al. 1992). Thus, contrasting patterns of population densities and spatial distribution patterns between trees of the tropical and other climatic zones with lower species diversity result in neither severely restricted gene flow in tropical forests nor in consistent differences among forest species from different climatic zones with regard to the distribution of genetic diversity within and among populations. However, the significance of plant-pollinator interactions in the tropics and low population densities of most tropical trees suggest that the disturbance and fragmentation of these ecosystems is likely to affect the gene flow mechanisms in many tropical tree species, eventually decreasing effective population sizes and genetic diversity within populations, and, finally, compromising their adaptive potential (Finkeldey and Ziehe 2004).

Effects of Gene Flow on Neutral Genetic Differentiation and Spatial Structure

Given the high fecundities and efficient dispersal of trees, substantial gene exchange can be expected over broad spatial scales. How do these levels of gene flow impact the spatial distribution of genetic diversity? In the case of selectively neutral DNA regions (see next sections for adaptive variation), there is a balance between gene flow, mutation, and genetic drift that all together determine SGS, both within and among populations. For low mutation rates, the expected SGS at equilibrium can be approximated considering solely the interaction between gene flow and drift (Ewens 2004).

Within large continuous populations, the traditional theoretical isolation by distance lattice model predicts that under equilibrium the genetic structure statistic a_r [that contains the same information about genetic differentiation as the $F_{ST}/(1-F_{ST})$] for selectively neutral genes is inversely related to the product $\sigma^2 D$,

$$a_r = \frac{\ln(r)}{4D\pi\sigma^2} + \text{constant}$$

where r is the spatial (geographical) distance, σ^2 is the effective average squared axial parent-offspring dispersal distance ($\sigma^2 = \sigma_s^2 + \sigma_p^2/2$, considering seed and pollen dispersal components; Crawford 1984), and D is the effective population density (Wright 1946; Malécot 1948; Rousset 2000). Then, $4D\pi\sigma^2$ can be estimated from the inverse of the slope of the regression line of a_r vs. logarithm of distance. For tree species, while D ranges from about 10^{-5} trees/m^2 in low-density species to 10^{-1} trees/m^2 in dense monospecific stands (Hardy et al. 2006), σ^2 is usually very large, on the order of 10^4–10^6 m^2, and, consequently, $\sigma^2 D$ is also typically too large to promote significant differentiation and results in weak SGS within populations. That is, gene flow within tree populations is expected to largely counterbalance drift-induced local fluctuations of allelic frequencies in most populations, although the strength of SGS may vary, depending on, among other factors, the mating system, the species dispersal ability, and the spatiotemporal distribution of individuals. Increase in selfing should decrease both σ^2 and D, promoting SGS, and this has been confirmed in an analysis of 47 plant species, which also revealed significantly weaker SGS for trees than for herbaceous species (Vekemans and Hardy 2004). Increase in dispersal ability will enlarge σ^2, decreasing SGS, and this trend has also been confirmed in an analysis of 39 tree species where animal-pollinated temperate trees demonstrated stronger SGS than wind-pollinated temperate trees, and zoochorous tropical trees demonstrated weaker SGS than barochorous/anemochorous tropical trees (Dick et al. 2008a, b). As for demographic dynamics, stronger SGS is predicted in nonuniform populations with spatiotemporal fluctuations in plant local density, which are expected to reduce both σ^2 and D (Robledo-Arnuncio and Rousset 2010).

Measuring gene flow among discrete tree populations is much more difficult than assessing within-population gene movement patterns, especially over long distances. As we have seen before, theoretical predictions based on simple island migration models have found that the balance between gene flow and genetic drift in subdivided populations is determined by the product Nm, where N is the effective population size of each subpopulation or deme and m the effective migration rate between demes per generation (Wright 1931). For values of $Nm > 1$, differentiation among population is predicted to be relatively weak ($F_{ST} < 0.2$ under the island migration model). Both N and m are usually very large in trees (Petit and Hampe 2006) and generate sufficiently large Nm values (although it is very difficult to measure) that should result in weak neutral genetic divergence among population (low F_{ST}) in most tree species. It is confirmed by empirical data that $F_{ST} < 0.05$ for many forest tree populations at neutral markers (McKay and Latta 2002), although substantially higher values have been observed for very small subpopulations (e.g., Ledig et al. 1997; Oline et al. 2000) and are predicted under extinction-colonization dynamics (Barton 1993; Whitlock 2003). It is worth noticing that m is migration rate per generation, and that trees typically reproduce each year during long maturity periods, thus increasing chances of long-distance dispersal among populations. Moreover, estimated tree pollen and seed dispersal curves are very leptokurtic (e.g., Austerlitz et al. 2004), which should further increase the probability of gene flow between distant populations (Klein et al. 2006).

Effects of Gene Flow on Local Adaptation

Does gene flow favor or hamper local adaptation in tree populations inhabiting different environments? Should we consider allochthonous (long-distance) gene migration into native stands as a positive or a negative process? Can gene flow affect tree species' response to climate change? We still lack precise answers for many of these questions, and there may be no simple answer since they depend on multiple factors. Evaluating the role of among-population gene flow is certainly very difficult. On the one hand, estimating long-distance gene migration is a problem of intrinsic experimental and statistical complexity, especially in real time, since it takes place on large spatial scales at low frequency. On the other hand, it is not easy to determine the long-term adaptive consequences of a given gene flow rate, due to multiple interactions between gene flow and other evolutionary forces, such as selection, drift, phenotypic plasticity, and inbreeding depression (Templeton 2006). For this reason, it is considered that gene flow plays an important but ambiguous role in evolutionary biology.

Gene flow can hamper local adaptation in heterogeneous environments by disrupting coadapted gene complexes and modifying allelic frequencies in a direction opposite to natural selection ("migration load" effect; Lenormand 2002). At the same time, gene flow can reduce the deleterious effects of inbreeding and increase the genotypic diversity available for selection, favoring local adaptation ("genetic rescue" effect) this way. The relative importance of these effects depends on gene flow intensity, type of selection, demographic structure, and genetic architecture of the phenotypic traits under selection (Lenormand 2002). For instance, migration load is expected to prevail if gene flow is extensive, resulting in a genetic homogenization and subsequent swamping of locally adapted genotypes. By contrast, a relatively small migration rate may favor local adaptation, providing the necessary genetic variation for natural selection and counteracting genetic drift and local inbreeding without preventing local population adaptive divergence.

Due to the enormous difficulties in conducting evolutionary experiments in long-living species, our knowledge on the interaction between gene flow and local adaptation in forest trees is mainly based on theoretical models and indirect inferences from the observed difference in population differentiation at neutral vs. adaptive traits (Le Corre and Kremer 2003; Savolainen et al. 2007; Lopez et al. 2008). Empirical evidence indicates that forest tree species generally demonstrate (1) high neutral and adaptive diversity within populations, (2) weak neutral genetic differentiation among populations, and (3) strong among-population genetic differentiation in quantitative traits associated with local adaptation (Yeaman and Jarvis 2006; Petit and Hampe 2006; Savolainen et al. 2007). Based on theoretical model predictions, these empirical observations can be interpreted as the presence of (1) a diversifying selection among populations and (2) a significant long-distance gene flow (as suggested by the low neutral genetic divergence among populations) that tends to increase neutral genetic diversity and additive genetic variance within populations, counteracting a potential within-population stabilizing selection. Thus, it seems that in forest tree species gene flow is extensive enough to homogenize neutral variation across populations and increase the additive variance within populations, but not always to the extent of preventing adaptive divergence.

This interpretation is widely accepted, but the empirical data on which it is based are rather limited. Among the strengths of experimental data are highly reliable measures of adaptive genetic divergence among tree populations that have been obtained in multiple available provenance trials, where the contribution of additive genetic variance to total phenotypic variance can be readily determined, avoiding the potential interference that phenotypic plasticity of the target traits may introduce in the evaluation of gene flow-local adaptation balance (Crispo 2008). Moreover, there is abundant evidence that long-distance gene flow is substantial in trees (see previous sections). Now, the above-mentioned hypothesis relies on (1) gene flow being intense enough, relative to stabilizing selection, to maintain ample additive genetic variation within populations and (2) local environmental adaptation occurring on a smaller scale than gene flow, which might not always be the case. Indeed, the indirect methods used for estimating gene flow from neutral spatial genetic differentiation are very sensitive to the underlying demographic model (island model), the simplicity of which is expected to be far from reality in many cases (Whitlock and McCauley 1999). For example, a low genetic differentiation at neutral markers might also be observed in recently fragmented populations not connected by contemporary gene flow (but it would be invariably explained as the result of extensive gene flow when using the island model), while adaptive divergence might have rapidly emerged *ex novo* following fragmentation under strong enough environmental heterogeneity. On the other hand, the high observed levels of intrapopulation additive variation could be due to temporal (Burger and Gimelfarb 2002) or even microspatial environmental (Campbell 1979) heterogeneities within populations, and not only due to gene exchange among populations with different local optima.

The methodological drawbacks of indirect approaches become especially relevant if we are to explore the way in which gene flow interplays with local adaptation in present (not historical) demographic conditions, and on a well-defined geographical framework. This would be the case if we wanted to assess either the effect of potential gene flow reductions subsequent to anthropogenic fragmentation or the risks of gene flow from allochthonous plantations into protected natural populations. In these cases, it would be more adequate to use a more direct approach to the problem, allowing real-time field assessment of gene flow and local selection processes. This kind of approximation is common in herbaceous species (e.g., Willi and Van Buskirk 2005; Sambatti and Rice 2006; Bossuyt 2007), but it is very difficult for forest trees, where the spatial range of seed and pollen dispersal and the long generation time create great problems for estimating gene flow and measuring selection empirically in real time. Future development of more accurate statistical models and advanced molecular tools might bring better chances to study interactions between contemporary gene flow and local adaptation in forest trees.

Clinal Variation: Gene Flow Versus Selection

The interaction between gene flow and selection is frequently observed in species gradually distributed along ecological gradients and latitudinal, longitudinal, and elevational clines. Latitude is associated with steep gradients in temperature and photoperiod, which have imposed strong selective pressure on populations of species such as *Populus tremula* (Hall et al. 2007), *Pinus sylvestris* (Savolainen et al. 2007), *Picea mariana* (Beaulieu et al. 2004), and *Picea sitchensis* (Mimura and Aitken 2010) that expanded northwardly after the last glacial maximum, resulting in clinal adaptive divergence patterns in survival, growth, and phenological traits, even in northern marginal areas colonized only a few generations ago (Mimura and Aitken 2010). Although epigenetics may also play very important role in this pattern (Johnsen et al. 2005, 2009; Kvaalen and Johnson 2008; Skrøppa et al. 2009; Gömöry et al. 2010; Mutke et al. 2010; Yakovlev et al. 2010), it confirms that divergent selection can counterbalance the high levels of gene flow in tree species, at least in strongly heterogeneous environments. However, a careful analysis of the relative fitness of local vs. nonlocal populations in *Pinus sylvestris* revealed that populations in central parts of the range are locally adapted, while those in northern marginal areas appear to be adapted to somewhat more southern conditions (Savolainen et al. 2007). This suboptimal phenotypic value for marginal northern populations is consistent with analytical predictions of the balance between gene flow and local adaptation along clines with decreasing population density toward range margins, inducing a net flow of (maladaptive) genes from central to peripheral populations (Garcia-Ramos and Kirkpatrick 1997). In other species, however, such as *Picea sitchensis*, even the northernmost populations seem to have approached their local optima (Mimura and Aitken 2010). This might be the result of a relatively lower level of gene flow from central populations, since low-to-intermediate migration rates may improve fitness in marginal populations by replenishing the genetic variation and alleviating inbreeding in small populations, without compromising adaptive divergence (Alleaume-Benharira et al. 2006). In the face of climate warming, an interesting question for future research is whether south-north gene flow would facilitate the adaptation of northern populations to their increasingly warm habitats (Aitken et al. 2008).

Gene Flow from Tree Plantations and Containment of Genetically Modified Forest Trees

Assessing gene flow and its potential consequences on local adaptation is especially relevant for native forests exposed to seed or pollen dispersal from allochthonous populations. Anthropogenic landscape alteration has resulted in substantial regional and continental translocation of forest reproductive materials. Nowadays, it is frequent to find mosaic landscape structures, where natural populations coexist in proximity to allochthonous plantations of conspecific or closely related interfertile species or varieties. Allochthonous populations can represent genetically selected, clonal, or even genetically modified (GM) trees, posing additional concerns on the preservation of native forest genetic resources. GM poplar and papaya are already being commercially planted in China and Hawaii, respectively, while field trials of several other GM tree species are underway in the US, Japan, and other countries (Van Frankenhuyzen and Beardmore 2004; Kikuchi et al. 2008). Since trees represent the dominant life form in many ecosystems, with profound implications on global biodiversity, it is especially important to improve our deep understanding of the interplay between allochthonous gene flow and local adaptation, a task that the long generation time of tree species makes experimentally very challenging.

Risk assessment of gene flow from allochthonous or GM tree plantations has two main components: hazard and exposure assessments (Johnson et al. 2006). Hazard assessment deals with the evaluation of ultimate ecological consequences of allochthonous genes for the environment. Exposure assessment evaluates the probability of the environment being exposed to the hazards, that is, the probability that allochthonous genes do actually get incorporated into the gene pool of native populations. The latter will occur if (1) pollen or seed gene flow brings allochthonous genes into native populations and (2) allochthonous genes are not rapidly eliminated by natural selection. The first condition is mostly determined by tree dispersal ability, the second by the relative selective advantage of allochthonous vs. local genes.

Considering the evidence on propagule dispersal patterns in natural tree populations, it is reasonable to assume that the efficient dispersal systems of trees make the movement of transgenes from exotic or GM plantations into natural forests highly probable, although more research is needed to predict at what precise rate it would happen (Robledo-Arnuncio et al. 2010). As a matter of fact, containment strategies based simply on spatial isolation distances of a few kilometers between plantations and native stands seem clearly insufficient to eliminate the risk of gene escape. For instance, pine and spruce pollen is able to move for long distances of up to several thousand kilometers while remaining viable (Lindgren et al. 1995; Campbell et al. 1999; Williams et al. 2006, 2010; Varis et al. 2009; Bohrerova et al. 2009), and effective pollination distances of a few tens of kilometers have been reported for several tree species (see Table 2 in Petit and Hampe 2006; Williams et al. 2006, 2010; Robledo-Arnuncio 2011). Induced sterility and other forms of genetically engineered containment methods are promising, since they could potentially interrupt transgene flow into native stands (Lee and Natesan 2006; Brunner et al. 2007), but fully safe and reliable methods are yet to be developed, and further research is needed to test their stability, considering the long life cycle of trees and potentially changing environmental conditions.

Although maladaptive genes are expected to be rapidly eliminated by selection, they can become fixed in recipient populations with very small effective sizes, in which selection efficiency can be overtaken by genetic drift (e.g., Ellstrand and Elam 1993). High gene flow of maladaptive genes or slightly deleterious transgenes will increase their chances to become established in a wild population (Ellstrand 1992b; Potts et al. 2003). Moreover, if the transgenes are advantageous, they can become incorporated into wild populations even if gene flow is extremely limited

(Slatkin 1976; Takahata 1991). That is why even a very small rate of gene introgression from exotic or transgenic plantations into native stands could likely result in the long-term presence and eventual fixation of nonnative genes in the wild (Haygood et al. 2004). For this reason, it has been argued that the most important parameter for risk assessment of GM organisms should be the relative fitness of transgenes, rather than introgression rates *per se*, which might be assumed to be nonnull in practice (Lee and Natesan 2006; Chapman and Burke 2006). Yet, we lack information on lifetime fitness consequences of transgenes under contrasting ecological conditions, without which any prediction of the long-term implications of GM trees for the environment would be questionable (Farnum et al. 2007). As more powerful genomic analysis tools become available, future research should aim at dissecting the genetic architecture of ecologically important tree traits, testing for potential interactions with transgenic loci under relevant environmental conditions.

Assisted Migration and Adaptation to Climate Change

Many plants as well as forest tree species are declining toward extinction, which could be worsened by climate change (e.g., McLaughlin et al. 2002, 2007). However, even those species that are not endangered now would be at risk from climate change. Warming and severe droughts are projected to either significantly reduce or eliminate suitable habitat for most narrowly endemic taxa (Thomas et al. 2004; Hannah et al. 2005; Peterson et al. 2006), forcing species either to colonize new terrain to survive or to adapt to the new environment without moving. The capacity of trees to adapt to rapid climate change depends in part on the rate at which tree species are capable of extending their ranges. For instance, the ranges of many North American trees will have to expand at rates of 100–1,000 m/year to track the predicted climatic changes of this century (e.g., Davis and Zabinski 1992; Iverson and Prasad 2002; Iverson et al. 2004a). Recent models, analyses of paleoecological records, and interpretations of fossil pollen data for tree populations responding to postglacial warming suggest that tree populations are capable of rapid migration when climate warms (see McLachlan et al. 2005 for references). Fossil pollen is commonly interpreted as suggesting that the range of many temperate tree species expanded at rates of 100–1,000 m/year during the early Holocene, suggesting that temperate trees generally have the capacity to track future climate change through rapid migration. However, chloroplast DNA surveys demonstrated that postglacial range expansion in two eastern North American tree species, American beech (*Fagus grandifolia*) and red maple (*Acer rubrum*), were slower than that expected from pollen-based reconstructions and from patterns emerging from European molecular studies (McLachlan et al. 2005). The estimated rates of <100 m/year were consistent with model predictions based on life history and dispersal data for these species, and suggest that past migration rates were substantially slower than the rates that will be needed to track twenty-first-century warming. Iverson et al. (2004b), using a combination of two models, DISTRIB and SHIFT, to estimate potential migration of five tree species into suitable habitat due to climate change over the next 100 years, also found a serious lag between the potential movement of suitable habitat and the potential for the species to migrate into the new habitat. The simulation study to test whether Scots pine in northern Finland can change to the new predicted optimum through migration and local selection during the next 100 years also showed that genetic change will be slow and lag behind the moving optimum (Savolainen et al. 2004). Adaptation in species with fragmented populations and little migration could be even slower. However, humans could hasten the migration of certain tree species via artificial regeneration with suitable seed sources and assisted migration (Vitt et al. 2010), also called assisted colonization (Hunter 2007), managed relocation (Minteer and Collins 2010), or translocation (Shirey and Lamberti 2010), by physically moving the propagules, seeds, or seedlings, especially for certain rare species that are unable to move sufficiently through fragmented landscapes, or even for more common species that have lost many of their animal dispersers (Aitken et al. 2008; O'Neill et al. 2008; Vitt et al. 2010; Wang et al. 2010).

Foresters have been pioneers of assisted migration transferring forest tree reproductive material within and outside of the natural distribution area that has probably had a bigger impact on genetic diversity and its distribution than natural factors (Geburek and Müller 2005). The uncontrolled transfer of forest tree reproductive material may negatively affect the productivity due to poor adaptation of new provenances to local environment (Rehfeldt et al. 2002), which may be aggravated by climate change (Rehfeldt et al. 1999; Jump and Peñuelas 2005; Matyas 2010). Understanding of phenotypic plasticity and knowledge of adaptive range of tree populations are much needed for efficient use of assisted migration. The data collected in provenance trials are excellent to study adaptation and to assess the adaptive range of tree using genomic tools (Krutovsky and Neale 2005b; González-Martínez et al. 2006b). Provenance trials are reciprocal common garden tests where tree populations collected from different regions are tested for survival, growth, phenology, productivity, and other adaptive traits to study the environment effects (e.g., Matyas 1994). The data obtained in these tests can be used in assisted migration via defining seed transfer rules and seed zones taking into account potential climate change (Matyas 2005). The assisted migration certainly must be considered when it is the only management option to prevent climate-driven extinction. Mitigation of climate change and providing reserve population networks that foster connectivity and gene flow should remain a priority (e.g., Hannah et al. 2002; Smulders et al. 2009), but we have to recognize that even optimistic estimates of natural seed dispersal may be insufficient for tree species to keep pace with climate change. However, other natural mechanisms, such as phenotypic plasticity, epigenetic effects, the potential for rapid local adaptation upon strong selection on ample local genetic variation, and long-distance movement of preadapted gametes might attenuate the pessimistic and somewhat simplistic predictions based solely on the potential of forest trees for demographic migration (Davis and Shaw 2001; Aitken et al. 2008).

It is worth noting that existing forest tree inventory, provenance trials, geographic cultures, ecological plots, urban trees, arboretum, and botanical garden collections are invaluable material for experimental assisted migration (e.g., O'Neill et al. 2008; Rehfeldt and Jaquish 2010; Woodall et al. 2010; Wang et al. 2010). For instance, using significant genecological, phenotypic, morphological, physiological, and phenological data collected in multiple experiments and provenance tests in Douglas-fir (*Pseudotsuga menziesii* var. *menziesii*), the Taskforce on Adapting Forests to Climate Change (http://tafcc.forestry.oregonstate.edu; St. Clair and Howe 2009) developed the online Seedlot Selection Tool (SST; http://sst.forestry.oregonstate.edu/PNW/index.html) that is based on a GIS mapping program designed to help forest managers match seedlots with planting sites, based on climatic information. The SST can be used to map current climates, or future climates based on selected climate change scenarios. Although it is tailored for matching seedlots and planting sites, it can be used for Douglas-fir assisted migration by anyone interested in mapping present or future climates defined by temperature and precipitation.

Approximately 185,000 forest inventory and ecological plots from both USA and Canada were used to predict the contemporary distribution of western larch (*Larix occidentalis* Nutt.) from climate variables (Rehfeldt and Jaquish 2010). Genetic variation among 143 populations within western larch's natural distribution was predicted from multiple regression models, using variables describing the climate of the seed source as predictors and response data from two separate genetic tests: (1) 15-year height at a field site in British Columbia, Canada, and (2) two principal components of eight variables describing growth, disease tolerance, and phenology of 6-year-old trees in a test in Idaho, USA. Presence and absence of the species and genetic variation within the species were projected into future climates provided by three General Circulation Models and two scenarios. Although the projections described pronounced impacts on the species and its populations, concurrence among the six projections pinpointed areas where the probability would be high that the future climate would be suitable for western larch. Concurrence among projections also was used to locate those sources of seed that should be best attuned genetically to future climates. These data can be used for western larch assisted migration and outline a logical approach for developing management strategies for accommodating climate change while taking into account the variability imposed by the differences among climatic estimates.

Assisted migration is a controversial issue that places different conservation objectives at odds with one another. To help trees adapt to rapid climate change via assisted migration, we need to collect more data on contemporary and historic migration rates in tree populations that should help to develop new experimental and theoretical models and conceptual framework for biodiversity dynamics following environmental change. Such models should incorporate lags in extinction and immigration, which lead to extinction debt and immigration credit, respectively (e.g., Jackson and Sax 2010). Collectively, these models and concepts would enable a balanced consideration of changes in biodiversity following climate change, habitat fragmentation, and other forcing events. They also reveal transient phenomena, such as biodiversity surpluses and deficits, which have important ramifications for biological conservation and the preservation of ecosystem services. Predicting such transient dynamics poses a serious conservation challenge in a time of rapid environmental change. The major risk associated with assisted migration or colonization is introducing ecologically harmful varieties or species. Therefore, the current policies should be very carefully examined and accordingly revised, if needed, to provide the best scientifically based management options for deciding when and where to move a species to mitigate for climate change (e.g., Minteer and Collins 2010; Shirey and Lamberti 2010).

Tree Population Genomics and Gene Flow

Forest tree genomics is mostly associated now with new generation (NG) sequencing and genotyping technologies that allow massively parallel high-throughput sequencing and genotyping of numerous genes and entire genomes in multiple individuals and populations (Morozova and Marra 2008; Delsenya et al. 2010; Metzker 2010). These technologies are now used to study differential gene expression (transcriptome profiling), genome-wide variation, adaptive differentiation, and species divergence in forest tree populations at the molecular level (González-Martínez et al. 2006b; Neale and Kremer 2011). With NG technologies, a virtually unlimited number of molecular markers can be generated (Futschik and Schlötterer 2010; Geraldes et al. 2011; Helyar et al. 2011; Seeb et al. 2011), and differentiation and gene flow can be monitored now at the global genomic level by scanning genome-wide nucleotide variation within and between populations (Teeter et al. 2008; Tautz et al. 2010; Ekblom and Galindo 2011). Forest trees often comprise large populations interconnected by extensive gene flow. Such systems are very informative and convenient for dissecting evolutionary factors, such as selection, genetic drift, and gene flow using NG technologies. Knowledge of nucleotide variation that underlies adaptive phenotypic and physiological variation would help researchers understand how gene flow and tree migration affect local adaptation. Tree population genomics has the potential to improve studies of gene flow by facilitating the identification of adaptive and selectively neutral molecular variation and by improving inferences about population demography and evolutionary history, leading to a more precise estimation of important parameters such as population size, migration rates, and phylogenetic relationships (Luikart et al. 2003). Neutral genetic markers that are not affected by selection have a greater potential for investigating gene flow, migration, and dispersal (Holderegger et al. 2006). Tree genomics provide researchers with a practically unlimited source of such markers (Neale and Kremer 2011).

Conclusions

Gene flow is very intense in most forest trees with common long-distance dispersal via pollen that effectively counteracts genetic drift and prevents genetic differentiation for selectively neutral or weakly selected alleles. Gene flow is difficult to estimate, but new methods and efficient genetic markers make gene flow estimates more reliable. The advance of forest tree genomics provides practically unlimited genomic resources to study current gene flow and infer historic gene flow and to partition the contributions of selection, genetic drift, and gene flow into genetic differentiation and adaptation. The detrimental processes in current forest tree populations, such as intense habitat loss and fragmentation, might disturb gene flow for particularly sensitive species, promoting genetic drift, inbreeding and, as a result, loss of genetic variation. Given that this variation is likely required for adaptation to new environmental conditions and climate changes, research efforts are needed to decide on a case-by-case basis whether gene flow should be restored and facilitated via artificial regeneration with suitable seed sources and assisted migration. Without human help, current gene flow and tree migration could be too slow to promote adaptation to fast-changing climate for some populations and species.

References

Adams WT, Birkes DS (1991) Estimating mating patterns in forest tree populations. In: Fineschi S, Malvoti ME, Cannata F, Hattemer HH (eds) Biochemical markers in the population genetics of forest tree. SPB Academic Publishing, The Hague, pp 157–172

Adams WT, Burczyk J (2000) Magnitude and implications of gene flow in gene conservation reserves. In: Young A, Boshier D, Boyle T (eds) Forest conservation genetics: principles and practice. CSIRO Publishing, Collingwood, pp 215–224

Adams WT, Hipkins VD, Burczyk J, Randall WK (1997) Pollen contamination trends in a maturing Douglas-fir seed orchards. Can J For Res 27:131–134

Ahmed S, Compton SG, Butlin RK, Gilmartin PM (2009) Wind-borne insects mediate directional pollen transfer between desert fig trees 160 kilometers apart. Proc Natl Acad Sci USA 106:20342–20347

Ahuja MR (2009) Transgene stability and dispersal in forest trees. Trees 23:1125–1135

Ahuja MR (2011) Fate of transgenes in the forest tree genome. Tree Genet Genomes 7:221–230

Aitken SN, Yeaman S, Holliday JA, Wang T, Curtis-McLane S (2008) Adaptation, migration or extirpation: climate change outcomes for tree population. Evol Appl 1:95–111

Alleaume-Benharira M, Pen IR, Ronce O (2006) Geographical patterns of adaptation within a species' range: interactions between drift and gene flow. J Evol Biol 19:203–215

Arnold ML (2006) Evolution through genetic exchange. Oxford University Press, Oxford

Ashley MV (2010) Plant parentage, pollination, and dispersal: how DNA microsatellites have altered the landscape. Crit Rev Plant Sci 29:148–161

Austerlitz F, Smouse PE (2001) Two-generation analysis of pollen flow across a landscape. II. Relation between Φ_{FT}, pollen dispersal and interfemale distance. Genetics 157:851–857

Austerlitz F, Mariette S, Machon N, Gouyon P-H, Godelle B (2000) Effects of colonization processes on genetic diversity: differences between annual plants and tree species. Genetics 154:1309–1321

Austerlitz F, Dick CW, Dutech C, Klein EK, Oddou-Muratorio S, Smouse PE, Sork VL (2004) Using genetic markers to estimate the pollen dispersal curve. Mol Ecol 13:937–954

Ayele TB, Gailing O, Umer M, Finkeldey R (2009) Chloroplast DNA haplotype diversity and postglacial recolonization of *Hagenia abyssinica* (Bruce) JF Gmel. in Ethiopia. Plant Syst Evol 280:175–185

Bacilieri R, Ducousso A, Petit RJ, Kremer A (1996) Mating system and asymmetric hybridization in a mixed stand of European oaks. Evolution 50:900–908

Bacles CFE, Lowe AJ, Ennos RA (2006) Effective seed dispersal across a fragmented landscape. Science 311:628

Bagnoli F, Fady B, Fineschi S, Oddou-Muratorio S, Piotti A, Sebastiani F, Vendramin GG (2011) Neutral patterns of genetic variation and applications to conservation in conifer species. Chapter 4. In: Plomion C, Bousquet J, Kole C (eds) Genetics, genomics and breeding of conifers. CRC Press/Science Publishers, Inc., Enfield, pp 141–195

Bai WN, Zeng YF, Zhang DY (2007) Mating patterns and pollen dispersal in a heterodichogamous tree, *Juglans mandshurica* (Juglandaceae). New Phytol 176:699–707

Barton NH (1993) The probability of fixation of a favoured allele in a subdivided population. Genet Res 62:149–158

Beaulieu J, Perron M, Bousquet J (2004) Multivariate patterns of adaptive genetic variation and seed source transfer in black spruce. Can J For Res 34:531–545

Beerli P, Felsenstein J (2001) Maximum likelihood estimation of a migration matrix and effective population sizes in n subpopulations by using a coalescent approach. Proc Natl Acad Sci USA 98:4563–4568

Berry O, Tocher M, Sarre S (2004) Can assignment tests measure dispersal? Mol Ecol 13: 551–561

Bohrerova Z, Bohrer G, Cho KD, Bolch MA, Linden KG (2009) Determining the viability response of pine pollen to atmospheric conditions during long-distance dispersal. Ecol Appl 19:656–667

Bossuyt B (2007) Genetic rescue in an isolated metapopulation of a naturally fragmented plant species, *Parnassia palustris*. Conserv Biol 21:832–841

Broquet T, Petit E (2009) Molecular estimation of dispersal for ecology and population genetics. Annu Rev Ecol Evol Syst 20:193–216

Brunner AM, Li J, DiFazio SP, Shevchenko O, Montgomery BE, Mohamed R, Wei H, Ma C, Elias AA, Van Wormer K, Strauss SH (2007) Genetic containment of forest plantations. Tree Genet Genomes 3:75–100

Burczyk J, Chybicki IJ (2004) Cautions on direct gene flow estimation in plant populations. Evolution 58:956–963

Burczyk J, Adams WT, Shimizu JY (1996) Mating patterns and pollen dispersal in a natural knobcone pine (*Pinus attenuate* Lemmon) stand. Heredity 77:251–260

Burczyk J, Adams WT, Birkes DS, Chybicki IJ (2006) Using genetic markers to directly estimate gene flow and reproductive success parameters in plants based on naturally regenerated seedlings. Genetics 173:363–372

Burdon RD, Wilcox PL (2011) Integration of molecular markers in breeding. Chapter 7. In: Plomion C, Bousquet J, Kole C (eds) Genetics, genomics and breeding of conifers. CRC Press/ Science Publishers, Inc., Enfield, pp 276–322

Burger R, Gimelfarb A (2002) Fluctuating environments and the role of mutation in maintaining quantitative genetic variation. Genet Res Camb 80:31–46

Campbell RK (1979) Genecology of Douglas-fir in a watershed in the Oregon cascades. Ecology 60:1036–1050

Campbell ID, McDonald K, Flannigan MD, Kringayark J (1999) Long-distance transport of pollen into the Arctic. Nature 399:29–30

Cannon CH, Morley RJ, Bush ABG (2009) The current refugial rainforests of Sundaland are unrepresentative of their biogeographic past and highly vulnerable to disturbance. Proc Natl Acad Sci USA 106:11188–11193

Cao C-P, Gailing O, Siregar I, Siregar U, Finkeldey R (2009) Genetic variation in nine *Shorea* species (Dipterocarpaceae) in Indonesia revealed by AFLPs. Tree Genet Genomes 5:407–420

Cavers S, Navarro C, Lowe AJ (2003) Chloroplast DNA phylogeography reveals colonization history of a Neotropical tree, *Cedrela odorata* L., in Mesoamerica. Mol Ecol 12:1451–1460

Chagné D, Batley J, Edwards D, Forster JW (2007) Single nucleotide polymorphisms genotyping in plants. Chapter 5. In: Oraguzie N, Rikkerink E, Gardiner S, De Silva H (eds) Association mapping in plants. Springer, New York, pp 77–94

Chapman MA, Burke JM (2006) Letting the gene out of the bottle: the population genetics of genetically modified crops. New Phytol 170:429–443

Charlesworth D, Charlesworth B (1987) Inbreeding depression and its evolutionary consequences. Annu Rev Ecol Syst 18:237–268

Cheliak WM, Pitel JA (1984) Techniques for starch gel electrophoresis of enzymes from forest tree species. Report PI-X-42. Petawawa National Forest Institute, Ontario

Chybicki IJ, Burczyk J (2010a) NM+: software implementing parentage-based models for estimating gene dispersal and mating patterns in plants. Mol Ecol Res 10:1071–1075

Chybicki IJ, Burczyk J (2010b) Realized gene flow within mixed stands of *Quercus robur* L. and *Q. petraea* (Matt.) L. revealed at the stage of naturally established seedling. Mol Ecol 19:2137–2151

Chybicki I, Trojankiewicz M, Oleksa A, Dzialuk A, Burczyk J (2009) Isolation-by-distance within naturally established populations of European beech (*Fagus sylvatica*). Botany 87:791–798

Colinvaux PA, De Oliveira PE (2001) Amazon plant diversity and climate through the Cenozoic. Palaeogeogr Palaeoclimatol Palaeoecol 166:51–63

Compton SG, Ellwood MDF, Davis AJ, Welch K (2000) The flight heights of chalcid wasps (Hymenoptera, Chalcidoidea) in a lowland Bornean rain forest: fig wasps are the high fliers. Biotropica 32:515–522

Conkle MT, Hodgskinss P, Nunnally L, Hunter S (1982) Starch gel electrophoresis of conifer seeds: a laboratory manual. General technical report PSW-64. U.S. Forest Service, Washington, DC

Corner EJH (1954) The evolution of tropical trees. In: Huxley J, Hardy AC, Ford EC (eds) Evolution as a process. Allen & Unwin, London, pp 34–46

Craft KJ, Ashley MV (2010) Pollen-mediated gene flow in isolated and continuous stands of bur oak, *Quercus macrocarpa* (Fagaceae). Am J Bot 97:1999–2006

Craft KJ, Ashley MV, Koenig WD (2002) Limited hybridization between *Quercus lobata* and *Quercus douglasii* (Fagaceae) in a mixed stand in central coastal California. Am J Bot 89:1792–1798

Crawford TJ (1984) The estimation of neighbourhood parameters for plant populations. Heredity 52:273–283

Crispo E (2008) Modifying effects of phenotypic plasticity on interactions among natural selection, adaptation and gene flow. J Evol Biol 21:1460–1469

Curtu A-L, Gailing O, Finkeldey R (2009) Patterns of contemporary hybridization inferred from paternity analysis in a four-oak-species forest. BMC Evol Biol 9:284. doi:10.1186/1471-2148-9-284

Davis MB, Shaw R (2001) Range shifts and adaptive responses to Quaternary climate change. Science 292:673–679

Davis MB, Zabinski C (1992) Changes in geographic range resulting from greenhouse warming: effects on biodiversity in forests. In: Peters RL, Lovejoy TE (eds) Global warming and biodiversity. Yale University Press, New Haven, pp 297–308

De Carvalho D, Ingvarsson PK, Joseph J, Suter L, Sedivy C, Macaya-Sanz D, Cottrell J, Heinze B, Schanzer I, Lexer C (2010) Admixture facilitates adaptation from standing variation in the European aspen (*Populus tremula* L.), a widespread forest tree. Mol Ecol 19:1638–1650

Dean JFD (2006) Genomics resources for conifers. Chapter 4. In: Williams CG (ed) Landscapes, genomics and transgenic conifers, vol 9, Section II, Managing forest ecosystems. Springer, Dordrecht, pp 55–74

Delsenya M, Han B, Hsing YI (2010) High throughput DNA sequencing: the new sequencing revolution. Plant Sci 179:407–422

Derero A, Gailing O, Finkeldey R (2011) Maintenance of genetic diversity in *Cordia africana* Lam., a declining forest tree species in Ethiopia. Tree Genet Genomes 7(1):1–9. doi:10.1007/s11295-010-0310-1

Devlin B, Ellstrand NC (1990) The development and application of a refined method for estimating gene flow from angiosperm paternity analysis. Evolution 44:248–259

DeWoody J, Nason JD, Hipkins VD (2006) Mitigating scoring errors in microsatellite data from wild populations. Mol Ecol Notes 6:951–957

Dick CW, Etchelecu G, Austerlitz F (2003) Pollen dispersal of tropical trees (*Dinizia excelsa*: Fabaceae) by native insects and African honeybees in pristine and fragmented Amazonian rainforest. Mol Ecol 12:753–764

Dick CW, Jones FA, Hardy OJ, Petit R (2008a) Spatial scales of seed and pollen-mediated gene flow in tropical forest trees. Trop Plant Biol 1:20–33

Dick CW, Hardy OJ, Jones FA, Petit R (2008b) Spatial scales of pollen and seed-mediated gene flow in tropical rain forest trees. Springer, New York

DiFazio SP, Slavov G, Burczyk J, Leonardi S, Strauss SH (2004) Gene flow from tree plantations and implications for transgenic risk assessment. In: Walter C, Carson M (eds) Plantation forest biotechnology for the 21st century. Research Signpost, Trivandrum, pp 405–422

Di-Giovanni F, Kevan PG, Arnold J (1996) Lower planetary boundary layer profiles of atmospheric conifer pollen above a seed orchard in northern Ontario. Can For Ecol Manage 83:87–97

Dow BD, Ashley MV (1996) Microsatellite analysis of seed dispersal and parentage of saplings in bur oak, *Quercus macrocarpa*. Mol Ecol 5:120–132

Dow BD, Ashley MV (1998a) Factors influencing male mating success in bur oak, *Quercus macrocarpa*. Springer, Dordrecht

Dow BD, Ashley MV (1998b) High levels of gene flow in bur oak revealed by paternity analysis using microsatellites. J Hered 89:62–70

Dutech C, Sork VL, Irwin AJ, Smouse PE, Davis FW (2005) Gene flow and fine-scale genetic structure in a wind-pollinated tree species, *Quercus lobata* (Fagaceaee). Am J Bot 92:252–261

Eckenwalder JE (1984) Natural intersectional hybridization between North American species of *Populus* (Salicaceae) in sections Aigeiros and Tacamahaca. III. Paleobotany and evolution. Can J Bot 62:336–342

Eckert AJ, Pande B, Ersoz ES, Wright MH, Rashbrook VK, Nicolet CM, Neale DB (2009) High-throughput genotyping and mapping of single nucleotide polymorphisms in loblolly pine (*Pinus taeda* L.). Tree Genet Genomes 5:225–234

Edmands S (2007) Between a rock and a hard place: evaluating the relative risks of inbreeding and outbreeding for conservation and management. Mol Ecol 16:463–475

Ekblom R, Galindo J (2011) Applications of next generation sequencing in molecular ecology of non-model organisms. Heredity 107(1):1–15. doi:10.1038/hdy.2010.152

El-Kassaby YA, White EE (1985) Isozymes and forest trees: an annotated bibliography. Canadian Forest Service Information Report BC-X-267, p 79

Ellstrand NC (1992a) Gene flow among seed plant populations. New Forests 6:241–245

Ellstrand NC (1992b) Gene flow by pollen: implications for plant conservation genetics. Oikos 63:77–86

Ellstrand NC, Elam DR (1993) Population genetic consequences of small population size: implications for plant conservation. Annu Rev Ecol Syst 24:217–242

Epperson BK (2003) Geographical genetics. Princeton University Press, Princeton/Oxford

Ewens WJ (2004) Mathematical population genetics. I. Theoretical introduction, 2nd edn. Springer, New York

Excoffier L, Smouse PE, Quattro JM (1992) Analysis of molecular variance inferred from metric distances among DNA haplotypes: application to human mitochondrial DNA restriction data. Genetics 131:479–491

Farnum P, Lucier A, Meilan R (2007) Ecological and population genetics research imperatives for transgenic trees. Tree Genet Genomes 3:119–133

Faubet P, Waples RS, Gaggiotti OE (2007) Evaluating the performance of a multilocus Bayesian method for the estimation of migration rates. Mol Ecol 16:1149–1166

Fedorov AA (1966) The structure of the tropical rain forest and speciation in the humid tropics. J Ecol 54:1–11

Finkeldey R (1995) Homogeneity of pollen allele frequencies of single seed trees in *Picea abies* (L.) Karst. plantations. Heredity 74:451–463

Finkeldey R, Hattemer HH (2007) Tropical forest genetics. Springer, Berlin/Heidelberg/New York

Finkeldey R, Ziehe M (2004) Genetic implications of silvicultural regimes. For Ecol Manage 197:231–244

Finkeldey R, Leinemann L, Gailing O (2010) Molecular genetic tools to infer the origin of forest plants and wood. Appl Microbiol Biotechnol 85:1251–1258

Franklin IR (1980) Evolutionary change in small populations. In: Soulé ME, Wilcox BA (eds) Conservation biology: an evolutionary, ecological perspective. Sinauer, Sunderland, pp 135–150

Futschik A, Schlötterer C (2010) The next generation of molecular markers from massively parallel sequencing of pooled DNA samples. Genetics 186:207–218

Gailing O, Wachter H, Schmitt H-P, Curtu A-L, Finkeldey R (2007) Characterization of different provenances of Slavonian pedunculate oaks (*Quercus robur* L.) in Münsterland (Germany) with chloroplast DNA markers: PCR-RFLPs and chloroplast microsatellites. Allg Forst Jagdztg 178:85–90

Garcia-Ramos G, Kirkpatrick M (1997) Genetic models of adaptation and gene flow in peripheral populations. Evolution 51:21–28

Geburek T, Müller F (2005) How can silvicultural management contribute to genetic conservation? In: Geburek T, Turok J (eds) Conservation and management of forest genetic resources in Europe. Arbora Publishers, Zvolen, pp 651–669

Geng QF, Lian CL, Goto S, Tao JM, Kimura M, Islam MDS, Hogetsu T (2008) Mating system, pollen and propagule dispersal, and spatial genetic structure in a high-density population of the mangrove tree *Kandelia candel*. Mol Ecol 17:4724–4739

Geraldes A, Pang J, Thiessen N, Cezard T, Moore R, Zhao Y, Tam A, Wang S, Friedmann M, Birol I, Jones SJM, Cronk QCB, Douglas CJ (2011) SNP discovery in black cottonwood (*Populus trichocarpa*) by population transcriptome resequencing. Mol Ecol Resour 11(suppl 1):81–92

Gerard P, Klein EK, Austerlitz F, Fernandez-Manjarres J, Frascaria-Lacoste N (2006) Assortative mating and differential male mating success in an ash hybrid zone population. BMC Evol Biol 6:96. doi:10.1186/1471-2148-6-96

Gerber S, Mariette S, Streiff R, Bodénès C, Kremer A (2000) Comparison of microsatellites and amplified fragment length polymorphism markers for parentage analysis. Mol Ecol 9:1037–1048

Gillet EM (ed) (1999) Which DNA marker for which purpose? Final compendium of the research project "Development, optimisation and validation of molecular tools for assessment of biodiversity in forest trees" in the European Union DGXII biotechnology FW IV research programme molecular tools for biodiversity. http://webdoc.sub.gwdg.de/ebook/y/1999/whichmarker/index.htm

Goldstein DB, Pollock DD (1997) Launching microsatellites: A review of mutation processes and methods of phylogenetic inference. J Heredity 88:335–342

Gömöry D, Foffová E, Kmeť J, Longauer R, Romšáková I (2010) Norway spruce (*Picea abies* [L.] Karst.) provenance variation in autumn cold hardiness: adaptation or acclimation? Acta Biol 52(2):42–49

González-Martínez SC, Burczyk J, Nathan R, Nanos N, Gil L, Alía R (2006a) Effective gene dispersal and female reproductive success in Mediterranean maritime pine (*Pinus pinaster* Aiton). Mol Ecol 15:4577–4588

González-Martínez SC, Krutovsky KV, Neale DB (2006b) Forest tree population genomics and adaptive evolution. New Phytol 170(2):227–238

González-Martínez SC, Dillon S, Garnier-Géré P, Krutovsky KV, Alía R, Burgarella C, Eckert A, Garcia MR, Grivet D, Heuertz M, Jaramillo-Correa JP, Lascoux M, Neale DB, Savolainen O, Tsumura Y, Vendramin G (2011) Patterns of nucleotide diversity and association mapping. Chapter 6. In: Plomion C, Bousquet J, Kole C (eds) Genetics, genomics and breeding of conifers. CRC Press/Science Publishers, Inc., Enfield, pp 239–275

Goto S, Shimatani K, Yoshimaru H, Takahashi Y (2006) Fat-tailed gene flow in the dioecious canopy tree species *Fraxinus mandshurica* var. *japonica* revealed by microsatellites. Mol Ecol 15:2985–2996

Griffin AR, Burgess IP, Wolf L (1988) Patterns of natural and manipulated hybridization in the genus *Eucalyptus* L'Herit – a review. Aust J Bot 36:41–66

Grivet D, Robledo-Arnuncio JJ, Smouse PE, Sork VL (2009) Relative contribution of contemporary pollen and seed dispersal to the effective parental size of seedling population of California valley oak (*Quercus lobata*, Née). Mol Ecol 18:3967–3979

Groover AT (2005) What genes make a tree a tree? Trends Plant Sci 10:210–214

Guillot G, Estuop A, Mortier F, Cosson JF (2005) A spatial statistical model for landscape genetics. Genetics 170:1261–1280

Hadfield JD, Richardson DS, Burke T (2006) Towards unbiased parentage assignment: combining genetic, behavioural and spatial data in a Bayesian framework. Mol Ecol 15:3715–3730

Hall D, Luquez V, Garcia VM, St Onge KR, Jansson D, Ingvarsson PK (2007) Adaptive population differentiation in phenology across a latitudinal gradient in European Aspen (*Populus tremula*, L.): a comparison of neutral markers, candidate genes and phenotypic traits. Evolution 61:2849–2860

Hamilton MB (1999) Tropical tree gene flow and seed dispersal. Nature 401:129

Hamrick JL (2004) Response of forest trees to global environmental changes. For Ecol Manage 197:323–335

Hamrick JL, Nason JD (2000) Gene flow in forest trees. In: Young A, Boshier D, Boyle T (eds) Forest conservation genetics: principles and practice. CSIRO Publishing, Collingwood, pp 81–90

Hamrick JL, Godt MJW, Sherman-Broyles SL (1992) Factors influencing levels of genetic diversity in woody plant species. New Forests 6:95–124

Hannah L, Midgley GF, Millar D (2002) Climate change-integrated conservation strategies. Glob Ecol Biogeogr 11:485–495

Hannah L, Midgley G, Hughes G, Bomhard B (2005) The view from the cape. Extinction risk, protected areas, and climate change. Bioscience 55:231–242

Hardesty BD, Hubbell SP, Bermingham E (2006) Genetic evidence of frequent long-distance recruitment in a vertebrate-dispersed tree. Ecol Lett 9:516–525

Hardy OJ, Maggia L, Bandou E, Breyne P, Caron H, Chevallier M-H, Doligez A, Dutech C, Kremer A, Latouche-Hallé C, Troispoux V, Veron V, Degen B (2006) Fine-scale genetic structure and gene dispersal inferences in 10 Neotropical tree species. Mol Ecol 15:559–571

Haygood R, Ives AR, Andow DA (2004) Population genetics of transgene containment. Ecol Lett 7:213–220

Helyar SJ, Hemmer-Hansen J, Bekkevold D, Taylor MI, Ogden R, Limborg MT, Cariani A, Maes GE, Diopere E, Carvalho GR, Nielsen EE (2011) Application of SNPs for population genetics of nonmodel organisms: new opportunities and challenges. Mol Ecol Resour 11(suppl 1):123–136

Henry RJ (2008) Plant genotyping II: SNP technology. Oxford University Press, New York, p 285

Heuertz M, Vekemans X, Hausman J-F, Palada M, Hardy OJ (2003) Estimating seed versus pollen dispersal from spatial genetic structure in the common ash. Mol Ecol 12:2483–2495

Heuertz M, Teufel J, González-Martínez SC, Soto A, Fady B, Alía R, Vendramin GG (2010) Geography determines genetic relationships between species of mountain pine (*Pinus mugo* complex) in western Europe. J Biogeogr 37:541–556

Hewitt G (2000) The genetic legacy of the Quaternary ice ages. Nature 405:907–913

Hoffman JI, Amos W (2005) Microsatellite genotyping errors: detection approaches, common sources and consequences for paternal exclusion. Mol Ecol 14:599–612

Holderegger R, Wagner HH (2008) Landscape genetics. Bioscience 58:199–207

Holderegger R, Kamm U, Gugerli F (2006) Adaptive vs. neutral genetic diversity: implications for landscape genetics. Landsc Ecol 21:797–807

Holderegger R, Buehler D, Gugerli F, Manel S (2010) Landscape genetics of plants. Trends Plant Sci 15(12):675–683

Howe HF, Smallwood J (1982) Ecology of seed dispersal. Annu Rev Ecol Syst 13:201–228

Hunter ML Jr (2007) Climate change and moving species: furthering the debate on assisted colonization. Conserv Biol 21:1356–1358

Ingvarsson PK (2010) Nucleotide polymorphism, linkage disequilibrium and complex trait dissection in *Populus*. In: Jansson S, Bhalerao R, Groover AT (eds) Genetics and genomics of *Populus*. Springer, New York, pp 91–112

Iverson LR, Prasad AM (2002) Potential redistribution of tree species habitat under five climate change scenarios in the eastern US. For Ecol Manage 155:205–222

Iverson LR, Schwartz MW, Prasad AM (2004a) How fast and far might tree species migrate in the eastern United States due to climate change? Global Ecol Biogeogr 13:209–219

Iverson LR, Schwartz MW, Prasad AM (2004b) Potential colonization of newly available tree-species habitat under climate change: an analysis for five eastern US species. Landsc Ecol 19:787–799

Jackson ST, Sax DF (2010) Balancing biodiversity in a changing environment: extinction debt, immigration credit and species turnover. Trends Ecol Evol 25:153–160

Janzen DH (1979) How to be a fig. Annu Rev Ecol Syst 10:13–51

Janzen DH (1986) The future of tropical ecology. Annu Rev Ecol Syst 17:305–324

Jech K, Wheeler N (1984) Laboratory manual for horizontal starch gel electrophoresis. Weyerhaeuser technical report 050-3210/7, 61 pp

Jehle R, Wilson GA, Arntzen JW, Burke T (2005) Contemporary gene flow and the spatio-temporal genetic structure of subdivided newt populations (*Triturus cristatus*, *T. marmoratus*). J Evol Biol 18:619–628

Johnsen Ø, Daehlen OG, Østreng G, Skrøppa T (2005) Daylength and temperature during seed production interactively affect adaptive performance of *Picea abies* progenies. New Phytol 168:589–596

Johnsen Ø, Kvaalen H, Yakovlev I, Dæhlen OG, Fossdal CG, Skrøppa T (2009) An epigenetic memory from time of embryo development affects climatic adaptation in Norway spruce. In: Gusta LV, Wisniewski ME, Tanino KK (eds) Plant cold hardiness from the laboratory to the field. CABI, Oxfordshire, pp 99–107

Johnson KL, Raybould AF, Hudson MD, Poppy GM (2006) How does scientific risk assessment of GM crops fit within the wider risk analysis? Trends Plant Sci 12:1–5

Johnsson H (1945) Interspecific hybridization within the genus *Betula*. Hereditas 31:163–176

Jones TH, Vaillancourt RE, Potts BM (2007) Detection and visualization of spatial genetic structure in continuous *Eucalyptus globulus* forest. Mol Ecol 16:697–707

Jones B, Walsh D, Werner L, Fiumera A (2009) Using blocks of linked single nucleotide polymorphisms as highly polymorphic genetic markers for parentage analysis. Mol Ecol Resour 9:487–497

Jones AG, Small CM, Paczolt KA, Ratterman NL (2010) A practical guide to methods of parentage analysis. Mol Ecol Resour 10:6–30

Jordano P (2010) Pollen, seeds and genes: the movement ecology of plants. Heredity 105:329–330

Jump AS, Peñuelas J (2005) Running to stand still: adaptation and the response of plants to rapid climate change. Ecol Lett 8:1010–1020

Kalinowski ST, Taper ML, Marshall TC (2007) Revising how the computer program CERVUS accommodates genotyping error increases success in paternity assignment. Mol Ecol 16:1099–1106

Kamm U, Rotach P, Gugerli F, Siroky M, Edwards P, Holderegger R (2009) Frequent long-distance gene flow in a rare temperate forest tree (*Sorbus domestica*) at the landscape scale. Heredity 103:476–482

Katul G, Williams CG, Siqueira M, Poggi D, Porporato A, Mccarthy H, Oren R (2006) Dispersal of transgenic conifer pollen. Chapter 8. In: Williams CG (ed) Landscapes, genomics and transgenic conifers, vol 9, Section III, Managing forest ecosystems. Springer, Dordrecht, pp 121–146

Kaya N, Isik K, Adams WT (2006) Mating system and pollen contamination in a *Pinus brutia* seed orchard. New Forests 31(3):409–416

Kikuchi A, Watanabe K, Tanaka Y, Kamada H (2008) Recent progress on environmental biosafety assessment of genetically modified trees and floricultural plants in Japan. Plant Biotechnol 25:9–15

Klein EK, Lavigne C, Gouyon PH (2006) Mixing of propagules from discrete sources at long distance: comparing a dispersal tail to an exponential. BMC Ecol 6:3

Klein EK, Desassis N, Oddou-Muratorio S (2008) Pollen flow in the wildservice tree, *Sorbus torminalis* (L.) Crantz. IV. Whole inter-individual variance of male fecundity estimated jointly with dispersal kernel. Mol Ecol 17:3323–3336

Kole C (ed) (2007) Genome mapping and molecular breeding in plants, vol 7: Forest trees. Springer, Berlin/Heidelberg

Konnert M, Maurer W (1995) Isozymic investigation on Norway spruce (*Picea abies* L. Karst.) and European Silver fir (*Abies alba* Mill.): practical guide to separation methods and zymogram evaluation. German Federal-State Working Group "Conservation of Forest Gene Resources", 79 pp

Konuma A, Tsumura Y, Lee C-T, Lee SL, Okuda T (2000) Estimation of gene flow in the tropical-rainforest tree *Neobalanocarpus heimii* (Dipterocarpaceae), inferred from paternity analysis. Mol Ecol 9:1843–1852

Kramer AT, Ison JL, Ashley MV, Howe HF (2008) The paradox of forest fragmentation genetics. Conserv Biol 22:878–885

Krutovskii KV, Bergmann F (1995) Introgressive hybridization and phylogenetic relationships between Norway, *Picea abies* (L.) Karst., and Siberian, *P. obovata* Ledeb., spruce species studied by isozyme loci. Heredity 74:464–480

Krutovsky KV, Neale DB (2005a) Nucleotide diversity and linkage disequilibrium in coldhardiness- and wood quality-related candidate genes in Douglas fir. Genetics 171:2029–2041

Krutovsky KV, Neale DB (2005b) Forest genomics and new molecular genetic approaches to measuring and conserving adaptive genetic diversity in forest trees. In: Geburek T, Turok J (eds) Conservation and management of forest genetic resources in Europe. Arbora, Zvolen, pp 369–390

Kuparinen A (2006) Mechanistic models for wind dispersal. Trends Plant Sci 11:296–301

Kvaalen H, Johnson Ø (2008) Timing of bud set in *Picea abies* is regulated by a memory of temperature during zygotic and somatic embryogenesis. New Phytol 177:49–59

Lanner RM (1965) Needed: a new approach to the study of pollen dispersion. Silvae Genet 15:50–52

LeCorre V, Kremer A (2003) Genetic variability at neutral markers, quantitative trait loci and trait in a subdivided population under selection. Genetics 164:1205–1219

Ledig FT, Jacob-Cervantes V, Hodgskiss PD, Eguiluz-Piedra T (1997) Recent evolution and divergence among populations of a rare Mexican endemic, Chihuahua spruce, following Holocene climatic warming. Evolution 51:1815–1827

Lee D, Natesan E (2006) Evaluating genetic containment strategies for transgenic plants. Trends Biotechnol 24:109–114

Lenormand T (2002) Gene flow and the limits to natural selection. Trends Ecol Evol 17:183–189

Lepoittevin C, Frigerio J-M, Garnier-Géré P, Salin F, Cervera M-T, Vornam B, Harvengt L, Plomion C (2010) *In vitro vs in silico* detected SNPs for the development of a genotyping array: what can we learn from a non-model species? PLoS One 5(6):e11034. doi:10.1371/journal.pone.0011034

Lexer C, Fay M, Joseph J, Nica M-S, Heinze B (2005) Barrier to gene flow between two ecologically divergent *Populus* species, *P. alba* (white poplar) and *P. tremula* (European aspen): the role of ecology and life history in gene introgression. Mol Ecol 14:1045–1057

Lexer C, Kremer A, Petit RJ (2006) Comment: shared alleles in sympatric oaks: recurrent gene flow is a more parsimonious explanation than ancestral polymorphism. Mol Ecol 15:2007–2012

Lindgren D, Paule L, Xihuan S, Yazdani R, Segerström U, Wallin J-E, Lejdebro ML (1995) Can viable pollen carry Scots pine genes over long distances? Grana 34:64–69

Lopez S, Rousset F, Shaw FH, Shaw RG, Ronce O (2008) Migration load in plants: role of pollen and seed dispersal in heterogeneous landscapes. J Evol Biol 21:293–309

Lörz H, Wenzel G (eds) (2004) Biotechnology in agriculture and forestry: molecular marker systems, vol 55. Springer, Berlin/Heidelberg/New York

Lourmas M, Kjellberg F, Dessard H, Joly HI, Chevallier M (2007) Reduced density due to logging and its consequences on mating system and pollen flow in the African mahogany *Entandrophragma cylindricum*. Heredity 99:151–160

Luikart G, England PR, Tallmon D, Jordan S, Taberlet P (2003) The power and promise of population genomics: from genotyping to genome typing. Nat Rev Genet 4:981–994

Malécot G (1948) Les mathématiques de l'héredité. Masson, Paris
Manchenko G (2003) Handbook of detection of enzymes on electrophoretic gels, 2nd edn. CRC Press, New York, 554 p
Manel S, Gaggiotti OE, Waples RS (2005) Assignments methods: matching biological questions with appropriate techniques. Trends Ecol Evol 20:136–142
Manel S, Schwartz M, Luikart G, Taberlet P (2003) Landscape genetics: combining landscape ecology and population genetics. Trends Ecol Evol 18:189–197
Marshall TC, Slate J, Kruuk LEB, Pemberton JM (1998) Statistical confidence for likelihood-based paternity inference in natural populations. Mol Ecol 7:639–655
Matyas C (1994) Modelling climate change effects with provenance test data. Tree Physiol 14:797–804
Matyas C (2005) Expected climate instability and its consequences for conservation of forest genetic resources. In: Geburek T, Turok J (eds) Conservation and management of forest genetic resources in Europe. Arbora, Zvolen
Matyas C (2010) Forecasts needed for retreating forests. Nature 464:1271
McCauley DE (1995) The use of chloroplast DNA polymorphism in studies of gene flow in plants. Trends Ecol Evol 10:198–202
McKay HK, Latta RG (2002) Adaptive population divergence: markers, QTL and traits. Trends Ecol Evol 17:285–291
McLachlan JS, Clark JS, Manos PS (2005) Molecular indicators of tree migration capacity under rapid climate change. Ecology 86:2088–2098
McLachlan JS, Hellmann JJ, Schwartz MW (2007) A framework for debate of assisted migration in an era of climate change. Conserv Biol 21:297–302
McLaughlin JF, Hellmann JJ, Boggs CL, Ehrlich PR (2002) Climate change hastens population extinctions. Proc Natl Acad Sci USA 99:6070–6074
Metzker ML (2010) Sequencing technologies – the next generation. Nat Rev Genet 11(1):31–46
Mimura M, Aitken SN (2010) Local adaptation at the range peripheries of Sitka spruce. J Evol Biol 23:249–258
Mimura M, Barbour RC, Potts BM, Vaillancourt RE, Watanabe KN (2009) Comparison of contemporary mating patterns in continuous and fragmented *Eucalyptus globulus* native forests. Mol Ecol 18:4180–4192
Minteer BA, Collins JP (2010) Move it or lose it? The ecological ethics of relocating species under climate change. Ecol Appl 20:1801–1804
Mitton JB, Williams CG (2006) Gene flow in conifers. Chapter 9. In: Williams CG (ed) Landscapes, genomics and transgenic conifers, vol 9, Section III, Managing forest ecosystems. Springer, Dordrecht, pp 147–168
Moran EV, Clark JS (2011) Estimating seed and pollen movement in a monoecious plant: a hierarchical Bayesian approach integrating genetic and ecological data. Mol Ecol 20:1248–1262
Morgante M, Olivieri AM (1993) PCR-amplified microsatellites as markers in plant genetics. Plant J 3:175–182
Morozova O, Marra MA (2008) Applications of next-generation sequencing technologies in functional genomics. Genomics 92(5):255–264
Muir G, Schlötterer C (2005) Evidence for shared ancestral polymorphism rather than recurrent gene flow at microsatellite loci differentiating two hybridizing oaks (*Quercus* spp.). Mol Ecol 14:549–561
Muir G, Schlötterer C (2006) Comment: moving beyond single-locus studies to characterize hybridization between oaks (*Quercus* spp.). Mol Ecol 15:2301–2304
Müller-Starck G (1977) Cross-fertilization in a conifer stand inferred from enzyme gene-markers in seeds. Silvae Genet 26:223–226
Murawski D, Hamrick J (1991) The effect of the density of flowering individuals on the mating systems of nine tropical tree species. Heredity 67:167–174
Mutke S, Gordo J, Chambel MR, Prada MA, Álvarez D, Iglesias S, Gil L (2010) Phenotypic plasticity is stronger than adaptative differentiation among Mediterranean stone pine provenances. For Syst 19(3):354–366

Namkoong G, Koshy MP (2001) Application of genetic markers to forest tree species. Draft report to IPGRI of the project "Developing decision-making strategies on priorities for conservation and use of forest genetic resources", 26 p

Nason JD, Herre EA, Hamrick JL (1998) The breeding structure of a tropical keystone plant resource. Nature 391:685–687

Nathan R, Casagrandi R (2004) A simple mechanistic model of seed dispersal, predation and plant establishment: Janzen-Connell and beyond. J Ecol 92:733–746

Neale DB, Kremer A (2011) Forest tree genomics: growing resources and applications. Nat Rev Genet 12:111–122

Neigel JE (1997) A comparison of alternative strategies for estimating gene flow from genetic markers. Annu Rev Ecol Syst 28:105–128

Ng KKS, Lee SL, Koh CL (2004) Spatial structure and genetic diversity of two tropical tree species with contrasting breeding systems and different ploidy levels. Mol Ecol 13:657–669

Oddou-Muratorio S, Klein E (2008) Comparing direct vs. indirect estimates of gene flow within a population of a scattered tree species. Mol Ecol 17:2743–2754

Oddou-Muratorio S, Klein EK, Austerlitz F (2005) Pollen flow in the wildservice tree, *Sorbus torminalis* (L.) Crantz. II. Pollen dispersal and heterogeneity in mating success inferred from parent-offspring analysis. Mol Ecol 14:4441–4452

Oline DK, Mitton JB, Grant MC (2000) Population and subspecific genetic differentiation in the Foxtail pine (*Pinus balfouriana*). Evolution 54:1813–1819

O'Neill GA, Ukrainetz NK, Carlson MR, Cartwright CV, Jaquish BC, King JN, Krakowski J, Russell JH, Stoehr MU, Xie C, Yanchuk AD (2008) Assisted migration to address climate change in British Columbia: recommendations for interim seed transfer standards. Technical report 048. British Columbia Ministry of Forests and Range, Research Branch, Victoria. www.for.gov.bc.ca/hfd/pubs/Docs/Tr/Tr048.htm

Pavy N, Pelgas B, Beauseigle S, Blais S, Gagnon F, Gosselin I, Lamothe M, Isabel N, Bousquet J (2008) Enhancing genetic mapping of complex genomes through the design of highly-multiplexed SNP arrays: application to the large and unsequenced genomes of white spruce and black spruce. BMC Genomics 9:17

Penaloza-Ramirez JM, Gonzalez-Rodriguez A, Mendoza-Cuenca L, Caron H, Kremer A, Oyama K (2010) Interspecific gene flow in a multispecies oak hybrid zone in the Sierra Tarahumara of Mexico. Ann Bot Lond 105:389–399

Peterson AT, Sanchez-Cordero V, Martinez-Meyer E, Navarro-Siguenza AG (2006) Tracking population extirpations via melding ecological niche modeling with land-cover information. Ecol Model 195:229–236

Petit RJ, Hampe A (2006) Some evolutionary consequences of being a tree. Annu Rev Ecol Evol Syst 37:187–214

Petit R, Brewer S, Bordács S, Burg K, Cheddadi R, Coart E, Cottrell J, Csaikl U, van Dam B, Deans D, Espinel S, Fineschi S, Finkeldey R, Glaz I, Goicoechea PG, Jensen JS, König AO, Lowe AJ, Madsen SF, Mátyás G, Munro RC, Popescu F, Slade D, Tabbener H, de Vries SGM, Ziegenhagen B, de Beaulieu J-L, Kremer A (2002a) Identification of refugia and post-glacial colonisation routes of European white oaks based on chloroplast DNA and fossil pollen evidence. For Ecol Manage 156:49–74

Petit R, Csaikl U, Bordács S, Burg K, Coart E, Cottrell J, van Dam B, Deans D, Dumolin-Lapègue S, Fineschi S, Finkeldey R, Gillies A, Glaz I, Goicoechea PG, Jensen JS, König AO, Lowe AJ, Madsen SF, Mátyás G, Munro RC, Olalde M, Pemonge M-H, Popescu F, Slade D, Tabbener H, Taurchini D, de Vries SGM, Ziegenhagen B, Kremer A (2002b) Chloroplast DNA variation in European white oaks. Phylogeography and patterns of diversity based on data from over 2600 populations. For Ecol Manage 156:5–26

Petit R, Aguinagalde I, Beaulieu J, Bittkau C, Brewer S, Cheddadi R, Ennos R, Fineschi S, Grivet D, Lascoux M, Mohanty A, Müller-Starck G, Demesure-Musch B, Palmé A, Martín JP, Rendell S, Vendramin GG (2003) Glacial refugia: hotspots but not melting pots of genetic diversity. Science 300:1563–1565

Petit RJ, Bodencs C, Ducousso A, Roussel G, Kremer A (2004) Hybridization as a mechanism of invasion in oaks. New Phytol 161:151–164

Pijut PM, Woeste KE, Vengadesan G, Michler CH (2007) Technological advances in temperate hardwood tree improvement including breeding and molecular marker applications. In Vitro Cell Dev Plant 43:283–303

Pluess AR, Sork VL, Dolan B, Davis FW, Grivet D, Merg K, Papp J, Smouse PE (2009) Short distance pollen movement in a wind-pollinated tree, *Quercus lobata* (Fagaceae). For Ecol Manage 258:735–744

Potts BM, Barbour RC, Hingston AB, Vaillancourt RE (2003) Genetic pollution of native eucalyptus gene pools – identifying the risks. Aust J Bot 51:1–25

Pritchard JK, Stephens M, Donnelly P (2000) Inference of population structure using multilocus genotype data. Genetics 155:945–959

Pyhäjärvi T, Kujala ST, Savolainen O (2011) Revisiting protein heterozygosity in plants – nucleotide diversity in allozyme coding genes of conifer *Pinus sylvestris*. Tree Genet Genomes 7(2):385–397. doi:10.1007/s11295-010-0340-8

Rehfeldt GE, Jaquish BC (2010) Ecological impacts and management strategies for western larch in the face of climate-change. Mitig Adapt Strateg Global Change 15:283–306

Rehfeldt GE, Cheng CY, Spittlehouse DL, Hamilton DA (1999) Genetic responses to climate in *Pinus contorta*: niche breadth, climate change, and reforestation. Ecol Monogr 69:375–407

Rehfeldt GE, Tchebakova NM, Parfenova YI, Wykoff WR, Kuzmina NA, Milyutin LI (2002) Intraspecific responses to climate in *Pinus sylvestris*. Glob Chang Biol 8:912–929

Rieseberg LH, Raymond O, Rosenthal DM, Lai Z, Livingstone K, Nakazato T, Durphy JL, Schwarzbach AE, Donovan LA, Lexer C (2003) Major ecological transitions in annual sunflowers facilitated by hybridization. Science 301:1211–1216

Robledo-Arnuncio JJ (2011) Wind pollination over mesoscale distances: an investigation with Scots pine. New Phytol 190:222–233

Robledo-Arnuncio JJ, García C (2007) Estimation of the seed dispersal kernel from exact identification of source plants. Mol Ecol 16:5098–5109

Robledo-Arnuncio JJ, Gil L (2005) Patterns of pollen dispersal in a small population of *Pinus sylvestris* L. revealed by total exclusion paternity analysis. Heredity 94:13–22

Robledo-Arnuncio JJ, Rousset F (2010) Isolation by distance in a continuous population under stochastic demographic fluctuations. J Evol Biol 23:53–71

Robledo-Arnuncio JJ, Austerlitz F, Smouse PE (2006) A new method of estimating the pollen dispersal curve independently of effective density. Genetics 173:1033–1045

Robledo-Arnuncio JJ, González-Martínez SC, Smouse PE (2010) Theoretical and practical considerations of gene flow. In: El-Kassaby YA (ed) Forests and genetically modified trees. FAO, Rome, pp 147–162

Rousset F (2000) Genetic differentiation between individuals. J Evol Biol 13:58–62

Rushton BS (1993) Natural hybridization within the genus *Quercus* L. Ann Sci For 50(Supl 1): 73s–90s

Sambatti JBM, Rice KJ (2006) Local adaptation, patterns of selection, and gene flow in the Californian serpentine sunflower (*Helianthus exilis*). Evolution 60:696–710

Savolainen O, Bokma F, García-Gil R, Komulainen P, Repo T (2004) Genetic variation in cessation of growth and frost hardiness and consequences for adaptation of *Pinus sylvestris* to climatic changes. For Ecol Manage 197:79–89

Savolainen O, Pyhäjärvi T, Knürr T (2007) Gene flow and local adaptation in trees. Annu Rev Ecol Syst 38:595–619

Sax DF, Stachowicz JJ, Brown JH, Bruno JF, Dawson MN, Gaines SD, Grosberg RK, Hastings A, Holt RD, Mayfield MM, O'Connor MI, Rice WR (2007) Ecological and evolutionary insights from species invasions. Trends Ecol Evol 22:465–471

Schoen DL, Reichman JR, Ellstrand NC (2008) Transgene escape monitoring, population genetics, and the law. Bioscience 58:71–77

Seeb JE, Carvalho G, Hauser L, Naish K, Roberts S, Seeb LW (2011) Single-nucleotide polymorphism (SNP) discovery and applications of SNP genotyping in nonmodel organisms. Mol Ecol Resour 11(suppl 1):1–8

Semagn K, Bjørnstad Å, Ndjiondjop MN (2006) An overview of molecular marker methods for plants. Afr J Biotechnol 5:2540–2568
Shirey PD, Lamberti GA (2010) Assisted colonization under the U.S. Endangered Species Act. Conserv Lett 3:45–52
Skrøppa T, Tollefsrud MM, Sperisen C, Johnsen Ø (2009) Rapid change in adaptive performance from one generation to the next in *Picea abies* – Central European trees in a Nordic environment. Tree Genet Genomes 6:93–99
Slatkin M (1976) The rate of spread of an advantageous allele in a subdivided population. In: Karlin S, Nevo E (eds) Population genetics and ecology. Academic Press, New York, pp 767–780
Slatkin M (1981) Estimating levels of gene flow in natural populations. Genetics 99:323–333
Slatkin M (1985) Gene flow in natural populations. Annual Review of Ecology and Systematics 16:393–430
Slavov GT, Howe GT, Adams WT (2005a) Pollen contamination and mating patterns in a Douglas-fir seed orchard as measured by simple sequence repeat markers. Can J For Res 35(7): 1592–1603
Slavov GT, Howe GT, Gyaourova AV, Birkes DS, Adams WT (2005b) Estimating pollen flow using SSR markers and paternity exclusion: accounting for mistyping. Mol Ecol 14:3109–3121
Slavov GT, Leonardi S, Burczyk J, Adams WT, Strauss SH, DiFazio SP (2009) Extensive pollen flow in two ecologically contrasting populations of *Populus trichocarpa*. Mol Ecol 18:357–373
Smith DB, Adams WT (1983) Measuring pollen contamination in clonal seed orchards with the aid of genetic markers. In: Proceedings of the 17th southern forest tree improvement conference, Athens, Georgia, USA, 6–9 June 1983, pp 69–77. http://www.rngr.net/publications/tree-improvement-proceedings/sftic/1983
Smouse PE, Sork VL (2004) Measuring pollen flow in forest trees: an exposition of alternative approaches. For Ecol Manage 197:21–38
Smouse PE, Meagher T, Kobak C (1999) Parentage analysis in *Chamaelirium luteum* (L.) Gray (Liliaceae): why do some males have higher reproductive contributions? J Evol Biol 12:1069–1077
Smouse PE, Dyer RJ, Westfall RD, Sork VL (2001) Two-generation analysis of pollen flow across a landscape. I. Male gamete heterogeneity among females. Evolution 55:260–271
Smulders MJM, Cobben MMP, Arens P, Verboom J (2009) Landscape genetics of fragmented forests: anticipating climate change by facilitating migration. iForest 2:128–132
Sork VL, Campbell D, Dyer R, Fernandez J, Nason J, Petit R, Smouse P, Steinberg E (1998) Proceedings from a workshop on gene flow in fragmented, managed, and continuous populations. Research paper no. 3. National Center for Ecological Analysis and Synthesis, Santa Barbara. http://www.nceas.ucsb.edu/projects/2057
Sork VL, Nason J, Campbell DR, Fernandez J (1999) Landscape approaches to contemporary gene flow in plants. Trends Ecol Evol 14:219–224
St. Clair JB, Howe GT (2009) Genetic options for adapting forests to climate change. West Forester 54:9–11
Stacy EA, Hamrick JL, Nason JD, Hubbell SP, Foster RB, Condit R (1996) Pollen dispersal in low-density populations of three neotropical tree species. Am Nat 148:275–298
Stefenon VM, Gailing O, Finkeldey R (2008) The role of gene flow in shaping genetic structures of the subtropical conifer species *Araucaria angustifolia*. Plant Biol 10:356–364
Takahata N (1991) Genealogy of neutral genes and spreading of selected mutations in a geographically structured population. Genetics 129:585–595
Tallmon DA, Luikart G, Waples RS (2005) The alluring simplicity and complex reality of genetic rescue. Trends Ecol Evol 16:330–342
Tautz D, Ellegren H, Weigel D (2010) Next generation molecular ecology. Mol Ecol 19:1–3
Teeter KC, Payseur BA, Harris LW, Bakewell MA, Thibodeau LM, O'Brien JE et al (2008) Genome-wide patterns of gene flow across a house mouse hybrid zone. Genome Res 18:67–76
Templeton AR (2006) Population genetics and microevolutionary theory. Wiley, Hoboken

Thomas CD, Cameron A, Green RE, Bakkenes M, Beaumont LJ, Collingham YC, Erasmus BF, de Siqueira MF, Grainger A, Hannah L, Hughes L, Huntley B, Van Jaarsveld A, Midgley GF, Miles L, Ortega-Huerta MA, Peterson AT, Phillips OL, Williams SE (2004) Extinction risk from climate change. Nature 427:145–148

Van Frankenhuyzen K, Beardmore T (2004) Current status and environmental impact of transgenic forest trees. Can J For Res 34:1163–1180

van Kleunen M, Burczyk J (2008) Estimating selection on floral traits through male fertility in a large natural population of *Mimulus guttatus*. Evol Ecol 22:39–54

Varis S, Pakkanen A, Galofré A, Pulkkinen P (2009) The extent of south-north pollen transfer in Finnish Scots pine. Silva Fenn 43:717–726

Vekemans X, Hardy OJ (2004) New insights from fine-scale spatial genetic structure analyses in plant populations. Mol Ecol 13:921–935

Viera MF, de Carvalho-Okano RM (1996) Pollination biology of *Mabea fistulifera* (Euphorbiaceae) in Southeastern Brazil. Biotropica 28:61–68

Vitt P, Havens K, Kramer AT, Sollenberger D, Yates E (2010) Assisted migration of plants: changes in latitudes, changes in attitudes. Biol Conserv 143:18–27

Vos P, Hogers R, Bleeker M, Reijans M, van de Lee T, Hornes M, Frijters A, Pot J, Peleman J, Kuiper M, Zabeau M (1995) AFLP: a new technique for DNA fingerprinting. Nucleic Acids Res 23:4407–4414

Wang JL (2004) Sibship reconstruction from genetic data with typing errors. Genetics 166:1963–1979

Wang XR, Szmidt AE (2001) Molecular markers in population genetics of forest trees. Scand J For Res 16:199–220

Wang T, O'Neill GA, Aitken SN (2010) Integrating environmental and genetic effects to predict responses of tree populations to climate. Ecol Appl 20:153–163

Wheeler NC, Guries RP (1987) A quantitative measure of introgression between lodgepole and jack pines. Can J Bot 65:1876–1885

White TL, Adams WT, Neale DB (2007) Forest genetics. CABI Publishing, Oxfordshire

Whitlock MC (2003) Fixation probability and time in subdivided populations. Genetics 164:767–779

Whitlock MC, McCauley DE (1999) Indirect measures of gene flow and migration: $F_{ST} \neq 1/(4Nm+1)$. Heredity 82:117–125

Whittemore AT, Schaal BA (1991) Interspecific gene flow in sympatric oaks. Proc Natl Acad Sci USA 88:2540–2544

Willi Y, Buskirk JV (2005) Genomic compatibility occurs over a wide range of parental genetic similarity in an outcrossing plant. Proc R Soc B 272:1333–1338

Williams CG, LaDeau SL, Oren R, Katul GG (2006) Modeling seed dispersal distances: implications for transgenic *Pinus taeda*. Ecological Applications 16:117–124

Williams CG (2010) Long-distance pine pollen still germinates after mesoscale dispersal. American Journal of Botany 97:846–855

Wilson GA, Rannala B (2003) Bayesian inference of recent migration rates using multilocus genotypes. Genetics 163:1177–1191

Woodall CW, Nowak DJ, Liknes GC, Westfall JA (2010) Assessing the potential for urban trees to facilitate forest tree migration in the eastern United States. For Ecol Manage 259:1447–1454

Wright S (1931) Evolution in Mendelian populations. Genetics 16:97–159

Wright S (1943) Isolation by distance. Genetics 28:114–138

Wright S (1946) Isolation by distance under diverse systems of mating. Genetics 31:39–59

Yakovlev IA, Fossdal CG, Johnsen Ø (2010) MicroRNAs, the epigenetic memory and climatic adaptation in Norway spruce. New Phytol 187(4):1154–1169

Yeaman S, Jarvis A (2006) Regional heterogeneity and gene flow maintain variance in a quantitative trait within populations of lodgepole pine. Proc R Soc B 273:1587–1593

Ziegenhagen B, Fladung M (2004) DNA markers for identification and evaluation of genetic resources in forest trees: case studies in *Abies*, *Picea* and *Populus*. In: Lörz H, Wenzel G (eds) Biotechnology in agriculture and forestry: molecular marker systems, vol 55. Springer, Berlin/Heidelberg/New York, pp 413–429

Chapter 5
Genetic Transformation of Fruit Trees

Richard E. Litz and Guillermo Padilla

Abstract Since the twentieth century, classical breeding strategies have been used to produce improved fruit trees. However, breeders have to deal with many problems due to their long juvenile period, open pollination, self-incompatibility, heterozygosity and the time involved before selected phenotypes can be properly evaluated. Transformation with genes that mediate horticulturally important traits is a truly revolutionary approach for improving perennial species, as it implies that superior cultivars can be modified for a specific trait without otherwise altering the integrity of the clone. This review discusses the progress that has been made using transformation technologies to address important breeding objectives of perennial fruit species and also underlines the current limitations of this technology.

Keywords *Agrobacterium tumefaciens* • Breeding • Embryogenesis • Disease resistance • Fruit tree • Genetic transformation • Organogenesis • Pest resistance

Introduction

Fruit crop species, vegetatively propagated for many centuries, are ancient selections of perennial plants that bear superior quality fruits. Most of the fruit crop species are trees, although some fruit species are also herbaceous perennials, e.g., pineapple, strawberry, and banana (and plantain), and are also propagated vegetatively. Some herbaceous fruit species are not perennials and are seed-propagated, e.g., the melons

R.E. Litz, Ph.D. (✉)
Tropical Research & Education Center, University of Florida,
18905 SW 280 Street, Homestead, FL 33031-3314, USA
e-mail: relitz@ufl.edu

G. Padilla, Ph.D.
Agrobiología y Medio Ambiente, Instituto de Productos Naturales y
Agrobiología (IPNA-CSIC), Avda. Astrofísico Francisco Sánchez, 3,
38206 San Cristóbal de La Laguna, Santa Cruz de Tenerife, Spain

R.J. Schnell and P.M. Priyadarshan (eds.), *Genomics of Tree Crops*,
DOI 10.1007/978-1-4614-0920-5_5, © Springer Science+Business Media, LLC 2012

and papaya. Most fruit tree cultivars have been derived from seedlings that were the result of open and uncontrolled pollinations. Only since the twentieth century has classical breeding been utilized to produce improved fruit trees. The usual breeding strategy has involved screening the adult population of seedlings that have been derived from controlled crosses. This *modus operandi* is not particularly well suited for the breeding and improvement of perennial species due to their long juvenile period, open pollination, self-incompatibility, heterozygosity and the time involved before selected phenotypes can be properly evaluated. Backcrossing is used for introgression of traits that are mediated by one or very few genes. Inbreeding depression and self-incompatibility, however, are obstacles for gene introgression through backcrossing or for inbreeding for production of hybrids. Pseudo-backcrossing can be carried out with different elite cultivars, but it is very time-consuming (Gessler and Patocchi 2007). If a trait is expressed during the juvenile period, evaluation and selection can involve much less time; however, if the trait cannot be evaluated until flowering or fructification, the time frame is much longer. According to Flachowsky et al. (2009), development and subsequent release of a new apple cultivar requires 15–20 years using conventional breeding. In some cases, marker-assisted selection (MAS) could reduce the time invested, although the long juvenile period is unaffected so that several years are still required for evaluation of the progeny of controlled crosses at their mature phase and for recurrent crosses. At the time of writing, relatively few genes or markers that are associated with horticultural traits have been identified from fruit crop species.

Genomics must impact the efficiency of breeding of perennial fruit species in various ways, but within the terms of reference of *this chapter*, the identification of the function of horticulturally important genes will enable seedling populations to be screened for important traits and for genes to be transferred to cultivars that are deficient for this trait. Transformation with genes that mediate horticulturally important traits is a truly revolutionary approach for improving perennial species as it implies that superior cultivars can be modified for a specific trait without otherwise altering the integrity of the clone. Not all important horticultural traits are conferred by a single gene and can involve two or more genes, i.e., quantitative trait loci (QTLs) or polygenes. Genetic modification (GM) technologies are unlikely to impact these characters for several more years. It is probable that genetic improvement of perennial, vegetatively propagated fruit species will be impacted more from GM technologies than any other group of crop species. At the same time, genetic transformation using RNA interference (RNAi) technology will have an important impact on the study of plant organ development and will provide valuable insight into the evolution of form and function (Chanderbali et al. 2008).

Objectives

The breeding objectives for fruit trees and annual crops are similar: an ideotype for scion selections that includes such traits as (1) fruit quality and yield, (2) disease and pest resistance, (3) environmental stress tolerance, etc. However, certain breeding objectives are specific to fruit trees due to their physiology, phenology, and particular production practices. These include (1) short juvenile period, (2) regular yield, (3) tree architecture and size, and (4) rootstock/scion compatibility. Rootstock cultivar attributes include environmental stress tolerance, disease resistance, polyembryony, and tree size control.

This chapter discusses the progress that has been made using transformation technologies to address important breeding objectives of perennial fruit species and also underlines the current limitations of this technology. Genetic transformation bypasses sexual reproduction. As fruit varieties are vegetatively propagated, one stable transformation event is sufficient to produce a transgenic line that does not need to be fixed through a sexual cycle (Petri and Burgos 2005). The technology is based upon efficient cell culture techniques, molecular genetics, and successful gene transfer between distantly related species, i.e., genetic transformation, and within the species or between closely related sexually compatible species, i.e., cisgenics. Genetic transformation has significant limitations. At the time of writing, (1) the targeted trait must be mediated by a single gene and (2) cultivar improvement is dependent on an efficient regeneration protocol from cell cultures obtained from mature-phase explants. Many societies have rejected GM crops, and this has curtailed public and private funding of GM science and technology, particularly in the EU. In 2010, only GM papaya and plum have been approved for human consumption.

The Technologies

Regeneration of Perennial Fruit Crop Species

The transformation of fruit crop species has been based upon the theoretical totipotency of plant cells, i.e., the ability to regenerate plants from single cells. The two *de novo* regeneration pathways, organogenesis and somatic embryogenesis, have both been utilized. The embryogenic pathway occurs from cultures in which phase change from mature to embryonic has occurred, whereas the organogenic pathway occurs from tissue in which the phase change is incomplete or is intermediate between embryonic and juvenile or between juvenile and mature. As a result, there is incomplete expression of embryonic development genes in the organogenic pathway. A major impediment for using transformation or other GM technologies, e.g., *in vitro* mutagenesis, for improving perennial fruit crop species has been the difficulty of regenerating plants *de novo* from cells of woody plants that are in their mature phase. Overcoming this barrier is of critical importance since fruit trees are mature clones, often being several hundred years old. Many reports in the literature have described the transformation of morphogenic cultures that have been derived from either (zygotic) embryonic or juvenile tissues. Although these studies have scientific merit, for the most part they are less significant horticulturally.

The regeneration pathway that has been adapted for genetic transformations has varied with the species and is also influenced by genotype or cultivar, the source of the explant, and the degree of determination in the tissue (Litz and Gray 1995).

The regeneration pathway must be determined empirically, although certain types of explants have been favored: the nucellus, leaves in recent vegetative flushes of field-grown trees, and leaves of micropropagated plants. Regeneration must be optimized in order to achieve efficient rates of transformation. With somatic embryogenesis, either the explants or maintenance cultures, which consist of proembryonic cells and masses (PEMs), are the targets for transformation studies. Embryogenic cultures can be recovered from cells within certain explants that already have competence to develop as somatic embryos, e.g., the nucellus of many fruit tree species. Alternately, cells with competence to become somatic embryos are induced from explants in the presence of a strong auxin, usually 2,4-dichlorophenoxyacetic acid (2,4-D), picloram, dicamba, etc. In the presence of a strong auxin, proembryos are unable to develop as single somatic embryos, and cells in the protoderm proliferate, forming proembryonic cells and PEMs. Embryogenic cultures achieve maximum proliferation rates as suspension cultures.

In order to initiate somatic embryo development from proembryonic cells, embryogenic cultures are transferred to a medium which lacks the induction agent. Germination/conversion of somatic embryos normally can occur on somatic embryo development medium or on minimal medium. Unlike germinating somatic embryos, induction, maintenance, and somatic embryo development do not require light.

With the organogenesis regeneration pathway, a caulogenic or shoot-forming culture is induced from explants on a medium that contains cytokinin (and usually auxin). The caulogenic culture may be friable and easily subcultured or compact and difficult to subculture. Shoots that develop from adventitious meristems are proliferated on a medium supplemented with cytokinin. Rooting is induced on individual shoots in the presence of an auxin. With organogenic cultures, either the explant or caulogenic cultures have been utilized as targets for transformation studies.

Transformation

Most genetic transformations of fruit trees have been mediated by engineered disarmed strains of *Agrobacterium tumefaciens*, including GV3111SE, LBA4404, AT2260, C58, EHA101, EHA105, and others. Microprojectile bombardment has been less frequently used, but successful transformations of citrus (Yao et al. 1996), *Musa* (Becker 1999; Becker et al. 2000), passion fruit (Vieira et al. 2002), strawberry (Wang et al. 2004), and pineapple (Nan et al. 1996; Ko et al. 2006) have been reported; microprojectile bombardment has been most often reported for transient expression of reporter transgenes.

The vectors have been variable, although in recent years, there have been more reports that involve the pCAMBIA plasmids. The constructs have generally included the genes for hygromycin phosphotransferase (*hpt*), neophosphate transferase (*npt*II), and Basta® resistance (*bar*) as selectable markers and either the β-glucuronidase (*uidA*) or the green fluorescent protein (*gfp*) as reporter genes. The Cauliflower mosaic virus constitutive promoter CaMV 35S has most often been used, although other promoters including maize or rice ubiquitin (*ubi*) have been used with monocotyledonous species.

Transformation of Fruit Crops

Temperate Fruit

Apple: *Malus*

Controlled crosses between selected apple genotypes and the selection of full-sib families from these crosses have been the most common strategy for the development of new cultivars. However, very few cultivars have been developed by this method (Noiton and Shelbourne 1992; Gessler and Patocchi 2007; Pereira-Lorenzo et al. 2009). Transformation of apple was first reported by James et al. (1989) only 6 years after the first report of plant transformation (Herrera Estrella et al. 1983). Leaf disks of micropropagated plantlets have been targeted for transformations.

Disease Resistance

The major breeding objectives that have been targeted by genetic transformation include resistance to scab (*Venturia inaequalis*) and fire blight (*Erwinia amylovora*) diseases. Apple scab, caused by the ascomycete *Venturia inaequalis*, is the most damaging fungal disease in apple orchards. The earliest attempts to transform apple for enhanced resistance involved the pathogenesis-related chitinase gene from *Trichoderma*. Although transformants showed some resistance, there was a negative impact on plant growth (Bolar et al. 2000; Faize et al. 2003). Szankowski et al. (2003) introduced genes for antifungal proteins into ‘Elstar’ and ‘Holsteiner Cox’. The genes encoded stilbene synthase (*Vst1*) from grape (*Vitis vinifera* L.), which is involved in the biosynthesis of the phytoalexin resveratrol, and the gene for a polygalacturonase-inhibiting protein (*pgip*) from kiwi (*Actinidia deliciosa*). The integration and expression of the transgenes was demonstrated; however, the transgenic plants were not shown to have enhanced resistance to pathogenic fungi. Ruhmann et al. (2006) described resveratrol levels in transgenic apple fruits and their effect on other phenolic compounds.

Disease resistance can involve either a single transgene that expresses a protein that affects pathogen development or specific resistance genes from related resistant varieties. The latter strategy would be more acceptable to consumers in many countries and would involve the transfer of genes from the same species or from within the *Malus* genus, rather than genes from a microbe, an animal, or a sexually incompatible plant species. Functional resistance genes for the apple scab fungus are present in some wild *Malus* species. Eleven major resistance genes, or clusters of genes, have been mapped, and the genes of the *Vf* region from *M. floribunda* 821 have been used for transformation (Soriano et al. 2009). Transformation with these genes has conferred scab resistance to susceptible cultivars (Belfanti et al. 2004; Malnoy et al. 2008). Szankowski et al. (2009) introduced the *HcrVf2* gene of *M. floribunda* into the scab-susceptible 'Elstar' and 'Gala', and transgenic lines with scab resistance were regenerated. Moreover, since *M. floribunda* and *M. domestica* are sexually compatible, the transgenic lines are close to being cisgenic, except for the selection markers. At the time of writing, there has been only a single report of cisgenic apples (Krens et al. 2004).

Resistance to fire blight was initially addressed by inserting genes encoding antimicrobial proteins from different organisms: attacins and cecropins from insects (Norelli et al. 1994, 1999), hen egg white lysozymes (Norelli et al. 1999), and lysozymes from bacteriophages (Hanke et al. 1999). A different approach has involved the use of effector proteins from *Erwinia amylovora* in order to induce the systemic acquired resistance (SAR) response. Transgenic lines for the hairpin *N* gene, driven by an inducible promoter, were shown to have increased resistance (Aldwinckle et al. 2003; Gessler and Patocchi 2007). The silencing of apple proteins that interact with *dspE*, another pathogenicity effector from *E. amylovora*, also showed increased resistance in preliminary results (Borejsza-Wysocka et al. 2004). This work utilized a transgene from apple.

Flachowsky et al. (2008a) transformed 'Pinova' with a gene encoding for an extracellular polysaccharide (EPS)-depolymerase of the fire blight bacteriophage phi-Ea1h and evaluated its effects on the susceptibility to the disease. The regenerated transgenic plants were more resistant to fire blight than the control plants.

Overexpression of the apple *MdNPR1* gene, an *Arabidopsis NPR1* homolog that plays a key factor in the SAR response, reduced *E. amylovora* symptoms in transformed plants of 'Galaxy' and M26 compared with the non-transformed control plants. Some transgenic plants also showed some resistance to *Venturia inaequalis* and *Gymnosporangium juniperi-virginianae* (Malnoy et al. 2007). Overexpression of the resistance gene *mbr4* from *M. baccata*, driven by CaMV 35S, has resulted in increased resistance in 'Pinova' compared to its wild genotype (Flachowsky et al. 2008b).

Other Traits

Flachowsky et al. (2007) shortened the juvenile period of 'Pinova' using the *BpMADS4* gene from silver birch (*Betula pendula* Roth.). These transgenic trees could be utilized to shorten conventional breeding programs in which transgenes are outcrossed (Flachowsky et al. 2009). The juvenile period has also been reduced by silencing the *MdTFL1* gene, which is homologous to the *Arabidopsis Terminal Flower* 1 (*TFL1*) gene that controls inflorescence meristem identity (Kotoda et al. 2006). Espley et al. (2007) demonstrated that transformation with the transcription factor *MdMYB10* results in the accumulation of anthocyanin, which determines the red color of fruit. This may have important implications for the development of new cultivars. Teo et al. (2006) studied the relationship between sorbitol synthesis in leaves and glucose, fructose, starch, and malic acid accumulation in apple fruit by altering the levels of the gene for sorbitol-6-phosphate dehydrogenase.

Seedlessness occurs in obscure apple cultivars, and its introgression into commercial apple cultivars would require a long-term commitment to a breeding program. Yao et al. (2001) demonstrated that parthenocarpic apple production is caused by a mutation in a MADS-box gene that is related to floral organ identity. Transformation with this or another MADS-box gene using RNAi technology could be used to generate high-quality seedless apple cultivars.

Rootstocks

Apple rootstocks have been transformed in order to improve rooting ability and dwarfism. Most procedures involve *A. tumefaciens*-mediated transformation with the *rolABC* genes from *A. rhizogenes* (Holefors et al. 1998; Welander et al. 1998; Zhu et al. 2001a, b; Igarashi et al. 2002; Zhang et al. 2006). Transformation by *A. rhizogenes* has also been reported (Pawlicki-Jullian et al. 2002; Yamashita et al. 2004).

M26 rootstock was initially transformed with *rolA* and *rolB*. Holefors et al. (1998) used *A. tumefaciens* strain GV3101 with binary vector pMRK10 with *npt*II under the control of the CaMV 19S promoter and *rolA* with its own promoter to infect leaves of micropropagated shoots. Transformed plants had the characteristic *rolA* phenotype, i.e., wrinkled leaves and reduced growth. Welander et al. (1998) reported transformation with *A. tumefaciens* strain C58C1, harboring the binary vector pCMB-B:GUS containing *npt*II and the *nos* promoter and *gus* and *rolB* both under the control of the *rolB* promoter. Transformed shoots showed increased rooting and auxin sensitivity. Other rootstocks have been transformed with *rolA* and *rolB*; Zhu et al. (2001a) transformed A2 rootstock with *rolA*, obtaining two transgenic clones with reduced plant height and a shortened internode length compared with the control plants. Zhu et al. (2001b) reported transformation of rootstock M.9/29 with *rolB*, and transgenic shoots produced more roots than the controls with no change in their length or morphology. Marubakaidou (*M. prunifolia* Borkh. var. *ringo* Asami Mo 84-A) leaves from micropropagated shoots were transformed with *rolC* using *A. tumefaciens,* harboring plasmids pBIN19*rol*C1 and pBIN19*rol*C2, with *hpt* under the CaMV 35S promoter, *npt*II under the *nos* promoter, *uidA* under the CaMV 35S promoter, and *rolC* with its original promoter (Igarashi et al. 2002). Four different transformant phenotypes were obtained: (1) reduced height and short internodes, (2) reduced height with normal internodes, (3) normal height with shortened internodes, and (4) heights and internodes similar to the controls. Rooting ability of the transformant shoots was enhanced by *rolC* without auxin. Similar results were reported by Zhang et al. (2006) with transformation of Makino (*Malus micromalus*) with the *rolC* gene, although no differences within the transformed phenotypes were indicated.

Agrobacterium rhizogenes has been used to produce transgenic roots from which seven transgenic clones of Jork 9 rootstocks were obtained (Pawlicki-Jullian et al. 2002). All transgenic clones contained *rolB* and an increased ability to produce roots; five clones contained the *mas1* and the ORF 13 genes. Yamashita et al. (2004) infected stems of *M. prunifolia* Borkh. var. *ringo* Asami strain Nagano No. 1 with *A. rhizogenes,* and transgenic plants were obtained from hairy roots. A transgenic plant showed prolific branching, miniaturized leaves, and shortened internodes.

Zhu et al. (2008) transformed rootstock A2 with the *gai* gene from *Arabidopsis* that encodes a mutant protein which is expected to confer altered gibberellin responses; in greenhouse conditions, transgenic clones showed a reduction in stem length, internode length, and node number.

Pear: *Pyrus*

The first successful pear transformation was reported by Mourgues et al. (1996), who targeted wounded leaves of micropropagated ‘Conference’, ‘Doyenné du Comice’, and ‘Passe Crassane’ plants with *A. tumefaciens* strain EHA101 harboring the binary vector pFAJ3000 with the *npt*II and *uidA* genes.

Disease Resistance

Pear transformation has focused on resistance to fire blight caused by *E. amylovora.* Resistance genes for fire blight do occur in pear (Thompson et al. 1975), although they have not been cloned and this resource has remained unexploited. Reynoird et al. (1999) transformed ‘Passe Crassane’, which is highly susceptible to fire blight, with the *attacin E* gene that codes a lytic protein from the cecropia moth (*Hyalophora cecropia*) and obtained reduced symptoms in several regenerated lines. Malnoy et al. (2005a) reported the transformation of ‘Passe Crassane’ with a bovine lactoferrin gene, which codes for an antimicrobial protein. Transformants showed a significant reduction of susceptibility to fire blight and to two other bacterial pathogens: *Pseudomonas syringae* pv. *syringae* and *A. tumefaciens.* Other strategies have involved transformation with the EPS-depolymerase of bacteriophage *ØEa1h* (Malnoy et al. 2005b). Exopolysaccharide (EPS) is a major pathogenicity factor of *E. amylovora,* and the bacteria cannot colonize host plants without this factor. The authors observed a negative correlation between the degree of EPS-depolymerase expression and fire blight susceptibility of transgenic plants.

Other Traits

Polyamines, e.g., spermidine and spermine, have an important role in many regulatory functions and cellular processes related to defense against oxidative stresses (Liu et al. 2007). Transgenic ‘Ballad’ that overexpress spermidine synthase from apple showed higher tolerance of salinity, hyperosmosis, and aluminum stresses (He et al. 2008; Wen et al. 2008, 2009). Gao et al. (2007) transformed 'La France' with the sense and antisense of the apple gene ACC oxidase (ACO), which codes for an enzyme that mediates ethylene biosynthesis. Transgenic antisense lines were regenerated in which *in vitro* shoots showed reduced ethylene emission. Recovery of mature-phase transgenic plants in which ethylene biosynthesis has been suppressed during fruit maturation and ripening should result in fruit with improved shelf life.

Apricot, Peach, Plum, and Cherry: *Prunus*

Most reports involving *Prunus* genetic transformation have used seed-derived explants as targets. Only recently have transgenic plants derived from mature-phase leaf explants been reported, e.g., almond (*P. dulcis* Miller) (Ramesh et al. 2006), cherry (*P. cerasus* L. and *P. cerasus* × *P. canescens*) (Song and Sink 2006), and apricot (*P. armeniaca* L.) (Petri et al. 2008). Nonetheless, transformation that involves genes that confer an improved horticultural phenotype, i.e., disease resistance and tree architecture, has only been achieved with material from open-pollinated seeds. It is expected that in the next few years, these advances will be also be extended to clonal material.

Disease Resistance

Prunus transformation has focused on the recovery of plants with resistance to infection by Plum pox potyvirus (PPV). Plums and apricots have been transformed with the gene for the PPV coat protein (PPV-CP) in order to recovery regenerants with resistance to PPV. Cotyledons of openly pollinated immature embryos of 'Kecskemeter' apricot were cocultured with *A. tumefaciens* strain LBA4404, harboring pBinPPVm with the CaMV 35S promoter (Machado et al. 1992). Scorza et al. (1994) transformed openly pollinated immature embryos of 'Stanley' plum and breeding selections B69158 and B70146 using *A. tumefaciens* strains C58/Z707 and EHA101 with the binary plasmid pGA482GG containing *npt*II and *uidA* and subcloned with the PPV-CP cassette with the CaMV 35S promoter. Of the transgenic clones obtained, clone C5 contained a multicopy transgene insertion and produced a low level of CP mRNA and no detectable CP. Regenerants were resistant to PPV (Scorza et al. 2001). RNA-mediated silencing of PPV genes has been studied in apricot and peach genotypes (Damiano et al. 2007).

Transgenic lines of plum with the *Gastrodia* antifungal protein (GAFP-1) have been recovered. Hypocotyls from seeds of open-pollinated 'Stanley' were transformed with *A. tumefaciens* strain EHA101 harboring the binary vector pAVAT1 with the gene *gafp-1-vnf* under the control of the CaMV 35S promoter and *npt*II and *gus*. Regenerated plants showed some level of resistance to Phytophthora root rot, caused by the oomycete *Phytophthora cinnamomi*, and to the root-knot nematode (*Meloidogyne incognita*) (Nagel et al. 2008). Wild-type scions were grafted on two different transgenic lines, 4I and 4J; transcripts of *gafp-1* were detected in transgenic rootstock tissue and were not in the scion tissue (Nagel et al. 2010).

Other Traits

Different plum genotypes have been transformed with the peach ACC oxidase gene in the antisense under the control of the CaMV 35S promoter (Callahan and Scorza 2007). The authors observed delayed ethylene production and fruit softening for some of the regenerated transgenic lines. Transgenic peach lines expressing the *ipt* gene, a cytokinin biosynthesis gene that reduces apical dominance, have been produced (Hammerschlag and Smigocki 1998) in an attempt to manipulate tree architecture. Some regenerated transgenic lines showed shorter or greater branching than the control plants.

Plum is the only temperate fruit crop that has been granted regulatory approval for food and/or feed in the USA. The transgenic plum contains the PPV-CP for resistance to PPV (see above). According to the authors, it has not been commercially released, but it is available to breeders to improve other varieties (AGBIOS http://www.cera-gmc.org/?action=gm_crop_database).

Strawberry: *Fragaria sp.*

The modern cultivated strawberry (*Fragaria × ananassa* Duch.) is an interspecific hybrid between the wild octoploid species *F. chiloensis* L. and *F. virginiana* Duch. (Darrow 1966). The strawberry is self-compatible, and intraspecific crosses, i.e., recurrent mass selection, have been used to develop new cultivars. The species is susceptible to several diseases and environmental stresses. Genes for disease resistance are present in wild, related species, i.e., *F. vesca* and *Potentilla tucumanensis*, but crosses are difficult due to their lower ploidy (Arias et al. 2004).

Organogenesis has been achieved from leaves, runners, petioles, and peduncles of mature-phase plants (Foucault and Letouze 1987; Liu and Sanford 1988; Jones et al. 1988; Nehra et al. 1989). Somatic embryogenesis has been reported, albeit only from cotyledons (Wang et al. 1984). Genetic transformation has used organogenic cultures (Graham 2005) and has generally been mediated by *A. tumefaciens*, although biolistic transformation alone (Wang et al. 2004) or in combination with *A. tumefaciens* (de Mesa et al. 2000) and protoplast electroporation (Nyman and Wallin 1992) has also *been* reported. A study of the various factors involved in *Agrobacterium*-mediated transformation of strawberry has been reviewed by Husaini (2010).

Jelenkovic et al. (1986) first attempted strawberry transformation; however, a successful protocol for regeneration of transgenic plants with reporter and marker genes (*uidA* and/or *ntp*II) was described a few years later by James et al. (1990) and Nehra et al. (1990a, b). Since then, studies have targeted disease and pest resistance, abiotic stress resistance, and fruit quality improvement.

Disease Resistance

Chalavi et al. (2003) transformed 'Joliette' with *A. tumefaciens,* harboring the pathogenesis-related (PR) chitinase gene *pcht28* from *Lycopersicon chilense* under the control of the CaMV 35S promoter. Transgenic plants expressing the transgene had significantly greater resistance to wilt caused by *Verticillium dahliae* compared to the wild type. Vellicce et al. (2006) transformed 'Pájaro' with three different defense-related genes: *ch5B* (encoding for a PR chitinase from *Phaseolus vulgaris*) and *gnl2* and *ap24* (encoding for a PR glucanase and a thaumatin-like protein, respectively, from *Nicotiana tabacum*). Single expression constructs with *ch5B, gnl2,* and *ap24* were inserted into the binary vector pDE1001. Double-expression constructs with *ap24/ch5B, gnl2/ap24,* and *ch5B/gnl2* were also inserted into the same vector. All six constructs were independently transferred into *A. tumefaciens* strain LBA4404. Only transgenic plants expressing the *ch5B* gene showed resistance to gray mold disease caused by *Botrytis cinerea.* Transgenic lines of 'Firework' expressing an antifungial thaumatin-like protein from *Thaumatococcus daniellii* (*thau II*) were obtained by *Agrobacterium*-mediated transformation with *thau II,* driven by the CaMV 35S promoter and with *npt*II (Schestibratov and Dolgov 2005). Transgenic lines expressing *thau II* showed a significantly higher level of resistance to gray mold compared to the controls. Mercado et al. (2007) obtained 'Camarosa' transgenic plants with the *Trichoderma* genes encoding PR chitinase (*chit-42*) and PR β-1,3-glucanase proteins under the control of the CaMV 35S constitutive promoter. Some transgenic plants with significant chitinase activity showed significantly fewer anthracnose crown rot (*Colletotrichum acutatum*) lesions than the controls, although no correlation was found between chitinase activity and tolerance of the fungus. Some of the transgenic lines containing β-1,3-glucanase showed significantly fewer anthracnose crown rot lesions compared to the controls. The antimicrobial peptide-D gene (*apd*), driven by the CaMV 35S promoter, was introduced into 'Toyonoka'; integration of *apd* was demonstrated, and transgenic plants were acclimatized in the greenhouse for further studies (Qin and Zhang 2007; Qin et al. 2008).

Finstad and Martin (1995) transformed two strawberry cultivars with the coat protein (CP) gene of Strawberry mild yellow edge protexvirus. Integration and expression of the gene was demonstrated, although its effectiveness was not shown.

Insect Resistance

The cowpea (*Vigna unguiculata*) protease trypsin inhibitor gene (*CpTi*) was introduced into 'Melody', 'Rhapsody', and 'Symphony' (Graham et al. 1995), and transgenic lines showed a significant reduction in trypsin activity. Feeding bioassays performed with vine weevils (*Otiorhynchus sulcatus*) showed less root damage due to the weevil larvae on the transgenic lines than on the controls (Graham et al. 1997, 2002).

Other Traits

Strawberry fruit ripening results in rapid softening and reduced shelf life. Pectin solubilization is believed to play a key role in fruit softening; pectate lyase is implicated in ripening, and transformation with the antisense sequence of the pectate lyase gene (*FaplC)* has been utilized to block gene expression (Jiménez-Bermúdez et al. 2002; Santiago-Domenech et al. 2008). Although softening was reduced, transgenic lines also showed reduced fruit yield. Youssef et al. (2009) observed that sense transformation reduced softening by co-suppression and some transgenic lines showed a yield increase compared with the controls.

Although strawberry is a *non-climacteric fruit*, ripening does respond to low levels of ethylene. Reducing ethylene biosynthesis could delay the ripening of strawberries and increase their shelf life. Transgenic lines of 'Totem' with *S*-adenosylmethionine hydrolase (*SAMase*), the enzyme that degrades *S*-adenosylmethionine, a precursor of ethylene, were produced by Mathews et al. (1995); however, alteration of the phenotype was not reported. Mezzetti et al. (2004) observed that introgression of an auxin-synthesizing gene increased plant fecundity and fruit production. 'Alpina W. Original' and 'AN93.231.53' were transformed with the construct DefH9-iaaM, composed of the regulatory region of the *DefH9* gene from snapdragon (*Antirrhinum majus*), the *iaaM* gene from *Pseudomonas syringae* pv. *savastanoi,* and *npt*II. Transgenic plants had fruit yields that were significantly higher than the controls.

Wawrzynczak et al. (2005) studied the effect of increasing conjugated indol-3-acetic acid (IAA) and reducing free IAA on plant growth and development. The gene encoding the IAA-glucose synthase from maize (*iaglu*) was integrated in the pBIN19 binary plasmid with the *npt*II gene. *Agrobacterium tumefaciens* strain LBA4404 was used for transforming 'Kaster'. All transgenic lines showed a dwarfish genotype and more roots *in vitro* compared to the controls. No effects on fruit characteristics are mentioned.

Soluble sugar content is a major quality-determining factor in strawberry. ADP glucose pyrophosphorylase (ADPase) catalyzes the key step of starch biosynthesis. Park et al. (2006) suppressed the activity of ADPase by transformation with the antisense sequence of the coding gene (*FagpS*) under the control of the strawberry ascorbate peroxidase promoter (APX). 'Anther' transformation was mediated by *A. tumefaciens* strain LBA4404, harboring the modified binary vector plasmid pAPX-FagpS with APX::*FagpS* and *npt*II. Soluble sugar content of fruits increased in transgenic plants without changes in fruit weight and hardness. Sugaya et al. (2008) transformed 'Anther' with the gene encoding for miraculin, a taste-modifying protein from the miracle fruit (*Richadella dulcifica*). The transgene was inserted via *A. tumefaciens* strain GV2260 hosting the binary vectors 35S-MIR and El2-MIR, depending on the constitutive promoter used, with *npt*II. Transgenic plants were obtained with both constructs; however, the level of miraculin accumulation was not high enough for commercial production.

The osmotin protein is expressed in plants under different stress conditions, i.e., salinity, drought, or low temperatures, and it is believed to have antifungal activity.

Husaini and Abdin (2008) introduced the binary vector BinAR with the osmotin gene from *Nicotiana tabacum* under the control of 35S CaMV promoter and *npt*II in *A. tumefaciens* strain GV2260 and obtained transgenic 'Chandler' plants. Transgenic lines had enhanced levels of proline, total soluble protein and chlorophyll content under salt stress.

Transgenic lines of 'Calypso' were obtained with the *rolC* gene from *A. rhizogenes* in order to achieve better rooting of stem cuttings and early flowering (Landi et al. 2009). The *A. tumefaciens* LBA4004 strain containing the binary vector pBI121 with the *rolC* gene under the control of the natural promoter and *npt*II was used. In field tests, two transgenic lines showed a total yield significantly greater than the controls. Inoculations under greenhouse conditions showed that the transgenic lines had an increased tolerance to infection by *Phytophthora cactorum*.

Wang et al. (2004) reported strawberry transformation through particle bombardment with plasmid pNBY520 containing a regulated protein in response to stress, the embryogenesis abundant protein gene (LEA3) from barley (*Hordeum vulgare*). Transgenic lines showed greater salt tolerance than the controls. Another late embryogenesis abundant protein gene, *WCOR410* from wheat (*Triticum aestivum*), was used to transform 'Chambly' (Houde et al. 2004). Three transgenic lines were obtained which showed a 5°C improvement of freezing tolerance over the controls.

Grapevine: *Vitis* spp.

Both somatic embryogenesis and organogenesis from mature-phase explants of *Vitis vinifera* L. (grapevine) have been described, and enables the transformation of traditional cultivars. *Agrobacterium*-mediated transformation has been most commonly utilized, although there are reports involving microprojectile bombardment.

Disease Resistance

Agrobacterium-mediated transformation has been used to transform grapevine for resistance to a number of different diseases. Bornhoff et al. (2005) transformed leaf disks of 'Seyval blanc' with the PR chitinase gene and the gene for RIP (ribosome inactivating protein) to develop enhanced resistance to fungal diseases. Transgenic plants did not show any significant difference in resistance to downy mildew and powdery mildew compared to non-transformed controls. Grapevine fanleaf virus (GFLV), an important virus disease of grapevines, has been targeted by genetic transformation: expression of the coat protein gene of GFLV (GFLV-CP) and of the GFLV protein movement genes and their silencing by their antisense sequences in transgenic rootstocks and in scion cultivars (Mauro et al. 1995; Gambino et al. 2005; Valat et al. 2006; Jardak-Jamoussi et al. 2009).

Vitis genes *vvtl-1* and *eg-2* have been used to produce disease-resistant lines of 'Thompson Seedless', 'Merlot', and 'Shiraz' (Dhekney et al. 2009). Transgenic plants were tested for powdery mildew resistance, and fruit showed fewer symptoms than the controls. The genes were under the control of grapevine promoters and could enable the recovery of cisgenic grapevine plants (Gray et al. 2009). A co-transformation system for producing transgenic plants free of selectable markers has also been reported (Dutt et al. 2008).

Vidal et al. (2003) obtained transgenic grapevine with magainins, short peptides with broad-spectrum antimicrobial activity. Transgenic plants were tested for their resistance to infection by *Agrobacterium vitis* and *Uncinula necator*, with better results obtained for *Agrobacterium* (Vidal et al. 2006).

Other Traits

Cold-tolerant 'Centennial Seedless' plants that contain the transgene for cold-inducible transcription factor dehydration response element binding *DREB1b* from *A. thaliana* have been regenerated (Jin et al. 2009). Leaf disks from *in vitro* plantlets were cocultured with *A. tumefaciens* LBA4404 strain carrying the pBPDREB binary vector with *DEEB1b* gene under the control of the CaMV 35S promoter and with *npt*II. The presence of *DREB1b* transcripts was demonstrated in the transgenic lines, which showed enhanced tolerance of cold in comparison with non-transgenic controls. In order to improve the rooting of grape rootstocks, the *rolB* gene from *A. rhizogenes* was inserted into 'Richter 110' (*V. berlandieri* × *V. rupestris*) (Geier et al. 2008). Single nodes from 2-year-old greenhouse-grown plants, both transgenic and non-transgenic, were examined in aeroponic culture. Transgenic plants had almost twice as many roots as the control plants.

Kiwifruit: *Actinidia sp.*

Within the *Actinidia* genus, there are two species with horticultural importance: *A. chinensis* and *A. deliciosa*. Fruit of the former are almost hairless, with flesh that varies from pale green to bright yellow (Gutiérrez-Pesce and Rugini 2008) and a basic chromosome number of $n=x=29$. Cultivars are either diploid or tetraploid (Ferguson and Seal 2008). *Actinidia deliciosa* fruits have brown hairy skin and green flesh, and all cultivars are hexaploids. The species are closely related (Ferguson and Seal 2008; Gutiérrez-Pesce and Rugini 2008). All *Actinidia* species are dioecious, which is a breeding obstacle, and self-fertility is a goal of breeding programs (Messina et al. 1990; Shuxiang and Shufan 1995; Ferguson and Seal 2008). Breeding has focused on fruit flavor, size, flesh color, shelf life, time of harvest, and yield (Ferguson and Seal 2008).

Two methods have been used to genetically transform *Actinidia*: *Agrobacterium*-mediated transformation and direct gene transfer to protoplasts. Although direct DNA transfer to 'Hayward' protoplasts has been reported (Oliveira et al. 1991; Raquel and Oliveira 1996), most transformations have been *Agrobacterium*-mediated, e.g., *A. chinensis*, *A. deliciosa*, *A. arguta*, and *A. eriantha*. With the two latter species, the scope of the research focused on functional genomics (Wang et al. 2006; Han et al. 2010). The majority of studies have been performed *A. deliciosa* 'Hayward'. Most transformation protocols involve cocultivation of leaf explants with *A. tumefaciens* (Matsuta et al. 1993; Janssen and Gardner 1993). Uematsu et al. (1991) transformed hypocotyl cuttings from *in vitro*-germinated 'Hayward' seedlings with *A. tumefaciens* strain EHA101, harboring binary vectors pLAN411 and pLAN421 containing *ntp*II and *uidA,* with CaMV 35S and *nos* promoter for *npt*II and CaMV 35S promoter for *uidA*. Fraser et al. (1995) transformed plants of four different genotypes of *A. chinensis* with *A. tumefaciens* strains A281 and C58, harboring the pKIWI105 plasmid, with *npt*II driven by the nos promoter and *uidA* by the CaMV 35S promoter.

Relatively few horticultural traits have been addressed in kiwi fruit using genetic transformation, and these involve disease resistance and plant architecture.

Disease Resistance

Resistance against gray mold (*Botrytis cinerea*) has been approached by Nakamura et al. (1999), who cocultivated leaf disks and stem segments of 'Hayward' with *A. tumefaciens* strain LBA4404, harboring the binary vector pROKla-EG with PR β-1,3-endoglucanase from soybean under the control of the CaMV 35S promoter with *npt*II. Transformants inoculated with *B. cinerea* showed smaller necrotic lesions than the controls. The tobacco osmotin gene under the control of CaMV 35S promoter with *npt*II in plasmid pKYLX71 was introduced into 'Hayward' using *A. tumefaciens* strain LBA4404. There were different levels of resistance to *B. cinerea* in the transgenic plants (Rugini et al. 1999; Gutiérrez-Pesce and Rugini 2008). Kobayashi et al. (2000) transformed 'Hayward' with the stilbene synthase gene isolated from three different *Vitis* species (*V. vinifera, V. labrusca,* and *V. riparia*). Leaf disks and petioles were infected with *A. tumefaciens* strain LBA4404 containing three different plasmids, one for each gene under the control of the CaMV 35S promoter and *npt*II. Transgenic plants produced piceid instead of resveratrol, which showed no antifungal activity in inoculation tests with *B. cinerea*, although fruits with high amount of piceid may be beneficial for human health. Kobayashi et al. (1996) obtained transgenic plants with the chemically synthesized gene encoding the human epidermal growth factor (hEGF) under the control of the CaMV 35S promoter.

Other Traits

Leaf disks from a micropropagated male clone of 'Hayward' were infected with *A. tumefaciens* strain LBA4404 carrying the plasmid Bin19 with the *rol A, B,* and *C* genes from *A. rhizogenes* (Rugini et al. 1989, 1991). Transgenic plants maintained the hairy root phenotype and other alterations compared to the non-transformed control, i.e., lower number of flowers per plant, expanded root system, and drought tolerance (Gutiérrez-Pesce and Rugini 2008). *Actinidia deliciosa* hypocotyls were transformed with *A. rhizogenes* strain NIAES 1724, and two hairy root lines were obtained, from which plants with short internodes and small leaves were selected (Yazawa et al. 1995). Yamakawa and Chen (1996) transformed 'Hayward', 'Matsua', 'Abbott', and 'Bruno' with *A. rhizogenes* strain ArM 123.

Tropical and Subtropical Fruits

Mango: *Mangifera indica* L.

The procedure for transforming mango was described by Mathews et al. (1992, 1993), who utilized 'Hindi' and 'Keitt' embryogenic cultures that were derived from the nucellus and zygotic embryos, respectively. These studies utilized two different disarmed, engineered strains of *A. tumefaciens*, i.e., strain C58C1 containing binary vector pGV 3850::1103 with the selectable marker gene *npt*II and driven by the CaMV 35S promoter (Mathews et al. 1993) and strain A208 containing the pTiT37-SE::pMON9749, a cointegrate vector, with *npt*II and *uid*A both driven by the CaMV 35S (Mathews et al. 1992).

Only a single horticultural trait of mango, control of fruit ripening, has been targeted by genetic transformation (Cruz-Hernández et al. 1997). Genes that mediate ethylene biosynthesis in the antisense were utilized: alternative oxidase, ACC oxidase, and ACC synthase. ACC oxidase cDNA was from an *Arabidopsis thaliana* cDNA library; alternative oxidase cDNA was isolated from a ripe mango fruit mesocarp cDNA library; ACC synthase was from DNA isolated from mango leaves (Gómez-Lim 1993; Cruz-Hernández and Gómez-Lim 1995; Cruz-Hernández et al. 1997), and transferred to a pGEM7Zf(+) vector in the pBI121 binary vector and controlled by the CaMV 35S promoter. The vectors were introduced into *A. tumefaciens* LBA4404. Although transformations were confirmed, the altered fruit ripening phenotype was not reported.

Grapefruit, Limes, and Oranges: *Citrus*

Genetic transformation of *Citrus* has focused on disease resistance and improved fruit quality traits. *Agrobacterium*-mediated transformation has been most successfully exploited, and different strains have been effective, including strain C58C1, EHA101 and LBA4404 (Bond and Roose 1998), and EHA105 (Ghorbel et al. 2001). Plasmids have normally included *npt*II with a nopaline synthase promoter (nos) for selection and *uid*A driven by the CaMV 35S promoter. Transformation efficiency of citrus is strongly genotype dependent.

Transformation of citrus initially was achieved using protoplast technology (Kobayashi and Uchimiya 1989) with 'Trovita' using the pCT2T3 vector containing *npt*II. A similar procedure was used by Vardi et al. (1990), who transformed rough lemon (*C. jambhiri* Lush.) protoplasts, and plants were recovered using a selection medium containing paromomycin. Transgenic sweet orange was obtained by electroporation of embryogenic protoplasts (Niedz et al. 1995); Fleming et al. (2000) also transformed sweet orange protoplasts with a plasmid containing *gfp*. Yao et al. (1996) attempted particle bombardment to transform 'Page' tangelo with a plasmid DNA that contained *npt*II and *uid*A, but plants were not regenerated.

The first successful *A. tumefaciens*-mediated transformation of *Citrus* was reported by Hidaka et al. (1990) with embryogenic 'Trovita' and 'Washington navel' sweet orange using a construct that contained *npt*II and *hpt* genes driven by the CaMV 35S promoter. *Agrobacterium tumefaciens* EHA101 containing pMON9793 with *npt*II and *uidA* was used to transform organogenic 'Carrizo' citrange stem sections. Sweet and sour oranges were later transformed using EHA101 with pGA482GG containing the Citrus tristeza virus coat protein gene (CTV-CP) (Moore et al. 1992, 1993; Gutierrez et al. 1997), and plants were recovered.

Kaneyoshi et al. (1994) reported a similar procedure that resulted in highly efficient transformation of *Poncirus trifoliata* with a pBI vector containing *ntp*II and *uid*A. Luth and Moore (1999), Yang et al. (2000), and Peña et al. (1995a, b) adopted this procedure to ensure highly efficient transformation. Different strategies have been adopted to enhance the efficiency of recovery of transformants of recalcitrant genotypes. Peña et al. (1995a) micrografted transformed shoots onto 'Troyer' citrange to obtain 100% survival. Grafted plants were then grafted *ex vitro* on rough lemon for more rapid plant recovery. Peña et al. (1997) cocultivated stem pieces of Key lime with tomato nurse cultures and recovered transformants using EHA105 with p35SGUSINT. Cervera et al. (1998a), Ghorbel et al. (1999, 2000), Dominguez et al. (2000), and Yu et al. (2002) utilized a similar procedure to optimize transformation efficiency with different citrus genotypes, and Cervera et al. (1998a) varied the protocol, utilizing mature-phase sweet orange.

Disease and Pest Resistance

Mexican lime, 'Carrizo' citrange, 'Duncan' grapefruit, and sour orange have been transformed with the p25 major coat protein gene of CTV (Gutierrez et al. 1997; Dominguez et al. 2000; Ghorbel et al. 2000; Febres et al. 2003). Dominguez et al. (2002c) demonstrated that 10–33% of transgenic Mexican lime plants appeared to be immune to CTV, and the remaining plants showed delay in the expression of symptoms. Dominquez et al. (2002a, b) were unsuccessful in developing RNA-mediated resistance to CTV by transforming Mexican lime with an untranslatable version of the p25 gene. Yang et al. (2000) transformed 'Rio Red' grapefruit with an untranslatable version of the p25 gene and with the *Galanthus nivalis* agglutinin gene, a plant-derived insecticidal gene. The *P. trifoliata* CTV allelic region that confers resistance to CTV has been cloned (Deng et al. 2000, 2001; Yang et al. 2000), and two genes, CTVR.5 and CTVR.6, were introduced separately into 'Duncan' grapefruit. Preliminary data have not clearly demonstrated that the regenerated transformants were resistant to CTV, although it is possible that both genes may be necessary to confer resistance (Grosser et al. 2009).

'Duncan' grapefruit was transformed with the coat protein of Citrus psorosis virus; however, regenerants all displayed characteristic symptoms when leaves were inoculated with the pathogen (Grosser et al. 2009). Iwanami et al. (2004) transformed *P. trifoliata* with the coat protein gene from Citrus mosaic virus (CiMV) and identified one line that showed some resistance. 'Ruby Red' and 'Rio Red' grapefruit were transformed separately with the spinach defensin gene *SoD2* in a plasmid containing CaMV 35S and *uidA* and with the bacteriolytic bovine lysozyme gene (*BVLZ*) (Grosser et al. 2009). Preliminary data indicate that *SoD2* confers resistance to Asiatic citrus canker (Xac). A tomato pathogenesis-related protein P23 (PR-5) has been constitutively expressed in 'Pineapple' sweet orange, and a transformed line showed increased tolerance of *Phytophthora citrophthora* (Fagoaga et al. 2001).

Other Traits

Costa et al. (2002) transformed 'Duncan' grapefruit with three genes that are implicated in the carotenoid biosynthetic pathway: lycopene-β-cyclase, phytoene desaturase, and phytoene synthase. 'Carrizo' citrange and *P. trifoliata* were transformed with the gene ACC synthase (in antisense) from citrus (CS-ACS1) (Wong et al. 2001). The authors speculated that enhanced accumulation of ACC would confer greater cold tolerance. In order to stimulate the hydrolysis of the methyl ester of pectin, Guo et al. (2005) transformed 'Valencia' orange with a pectin methylesterase gene (*Cs-PME4*). Cervera et al. (2000) transformed 'Carrizo' citrange with *HAL2* from yeast, which is implicated in salt tolerance; however, regenerated plants did not display better tolerance than the controls. In contrast, Molinari et al. (2004) transformed 'Carrizo' citrange with a Δ^1-pyrroline-5-carboxylate synthetase mutant gene (*P5CS*) from *Vigna aconitifolia* to enhance the accumulation of proline and thereby enhance tolerance of drought stress. The transgenic plants were reported to be able to withstand water stress. 'Ponkan' and 'Valencia' oranges were transformed with the barnase ribonuclease gene from *Bacillus amyloliquefaciens* under the control of an anther tapetum-specific promoter pTA29 in order to produce pollen-sterile, seedless fruit (Li et al. 2002, 2003). Mexican limes were transformed in order to produce fruit with fewer seeds (Koltunow et al. 2000). The success of these transformation events has not been confirmed.

'Carrizo' citrange was transformed to express constitutively either the *Arabidopsis LEAFY* (*LFY*) or the *APETALA1* (*AP1*) genes for flower initiation (Peña et al. 2001). The regenerated plants showed an abnormal phenotype; plants expressing *AP1* were fertile and bore fruit in the first year. Perez-Molphe-Balch and Ochoa-Alejo (1998) inoculated Mexican lime stem segments with a wild-type *A. rhizogenes* strain containing pESC4 that contained *npt*II and *uid*A. The segments either produced shoots or roots from which shoots could be regenerated.

Avocado: *Persea americana* Mill.

The procedure for transforming embryogenic avocado cultures was initially described by Cruz Hernández et al. (1998), who described transformation with *A. tumefaciens* strain 9749 ASE2, which contained a cointegrate vector pMON9749 containing *npt*II and *uidA* both driven by the CaMV 35S promoter. Following elimination of *A. tumefaciens*, an initial selection in the presence of 50 mg L^{-1} kanamycin sulfate followed by a second, more intensive selection with 100–200 mg L^{-1} kanamycin sulfate was utilized. A similar strategy has been followed for other transformations of avocado.

Disease Resistance

Candidate genes for resistance to Phytophthora root rot (PRR) have not been cloned; however, 'Hass' has been transformed with the antifungal defensin gene *pdf1.2* from *A. thaliana* (Epple et al. 1997; Raharjo et al. 2008). The defensin gene was driven by the CaMV 35S promoter in binary vector pGPTV, which also contained *uidA* and *bar*. Transformation was mediated by *A. tumefaciens* EHA105. Plants do not appear to have resistance to PRR.

Embryogenic avocado cultures have also been transformed to address other breeding objectives, including seedlessness and resistance to Avocado sunblotch viroid disease (ASBVd) (Perea et al., unpublished data). The latter strategy involved transformation of an ASBVd-infected cultivar with the *pac1* ribonuclease gene in a construct that linked it with the transit peptide (TP) of the precursor to the small subunit of ribulose 1,5-biphosphate carboxylase from pea, *uidA*, and *npt*II and under the control of the CaMV 35S promoter. Since the viroid replicates in the chloroplasts, the TP targets the pac1 gene to the chloroplasts. Preliminary RT PCR data indicate that the ASBVd in half of the transformed lines was eliminated.

Other Traits

On-tree fruit storage and extended shelf life of Antillean and West Indian avocados are being addressed by controlling ethylene biosynthesis, using the bacteriophage gene S-adenosyl-L-methionine hydrolase (*SAMase*) under the control of an avocado fruit cellulase promoter (Raharjo et al. 2003). S-adenosyl-L-methionine (SAM) hydrolase is in the binary vector pAG4092 (Agritope 1999; Hughes et al. 1987) with the *npt*II gene under the control of the AGT01 promoter. The *SAMase* gene catalyzes conversion of SAM to methylthioadenosine (Good et al. 1994) and results in a depleted pool of ACC, the precursor of ethylene, thereby inhibiting ethylene biosynthesis (Good et al. 1994; Kramer et al. 1997). Transformation of 'Suardia' was mediated by *A. tumefaciens* EHA105. Plants have been under evaluation in a screenhouse for 5 years (Litz et al. 2007).

Banana and Plantain: *Musa* (Groups AAA, AAB, ABB)

Banana and plantain present unique problems for classical breeding. Both the dessert (AAA) and cooking bananas or plantains (AAB and ABB) are triploids and are sterile with the exception of a few genotypes. There has been considerable interest in the application of biotechnologies such as mutation breeding and genetic transformation in order to target breeding objectives of specific cultivars. Most transformation reports have focused on disease resistance, primarily the real threats of Panama disease and Black Sigatoka.

Sagi et al. (1994) reported the transformation of *Musa* protoplasts by means of electroporation; however, this procedure has been superseded by microprojectile bombardment and *Agrobacterium*-mediated transformation of embryogenic cultures. Becker et al. (2000) transformed embryogenic cultures of 'Grand Nain' (AAA), and Becker (1999) and Sagi et al. (1995) transformed scalped shoot cultures of 'Bluggoe' (ABB) using microprojectile bombardment. Acetosyringone-activated *Agrobacterium* (LBA4404) has been used to transform meristems of *in vitro* plantlets (May et al. 1995), although this procedure has not been favored due to problems associated with recovery of solid transformants. Ganapathi et al. (2001) transformed 'Rasthali' embryogenic suspension (ABB) cultures with *A. tumefaciens* EHA105 strain and regenerated plants efficiently.

Constructs have included one of the following genes for selection: neomycin phosphotransferase (*npt*II) (May et al. 1995; Becker 1999), hygromycin B phosphotransferase (*hpt*) (Sagi et al. 1995), and acetolactate synthase (*als*) (Ganapathi et al. 2001). The reporter genes have included *uid*A (Sagi et al. 1995; May et al. 1995) and *gfp* (Dugdale et al. 1998, 2000; Becker 1999). Promoters have included maize polyubiquitin gene (ubi-1) and CaMV 35S.

Disease Resistance

At least two groups have attempted to address the problem of Black Sigatoka disease (*Mycosphaerella fijiensis*) of banana and plantain and have adopted slightly different strategies. Gomez Kosky et al. (http://www.redbio.org/rdominicana/redbio2004rd/Memoria_REDBIO_2004/Talleres-PDF/t19-PDF/t19-05.pdf) transformed 'Grand Nain' (AAA) and 'Navolean' with different combinations of PR genes β-1,6-glucanase and chitinase in a construct with *bar* and a ubiquitin promoter. There was no significant difference between the infection of control and transgenic plants. The other strategy involved the transformation of 'Gros Michel' (AAA) using either rice chitinase *rcc2* or *rcg3* in a construct with the CaMV 35S promoter and *hpt* (http://www.biovisioneastafrica.com/publications/Experience_GM_Banana.pdf). In field trials, the transformed plants were not protected against infection by *M. fijiensis*.

'Rasthali' banana (AAB) has been transformed with a synthetic analogue (*MSI-99*) of the gene encoding the antimicrobial peptide magainin derived from the skin of the African clawed frog *Xenopus laevis* (Chakrabarti et al. 2003). The vectors contained *npt*II; *MSI-99* was driven either by the ubiquitin 3 (*ubq*-3) promoter or the pea vicilin secretory signal to target the peptide into the extracellular space. Transformation was mediated by *A. tumefaciens.* Regenerated bananas showed resistance to *Fusarium oxysporum* f. sp. cubense, the cause of Panama disease, and *Mycosphaerella musicola.*

Pest Resistance

Atkinson et al. (2004) have transformed ‘Grand Nain’ banana with a rice cystatin as a means for conferring resistance to the nematode *Radopholus similis.* The protocol involved a plasmid construct containing the rice cystatin *OcIΔD86* under the control of the maize ubiquitin promoter UBA-1. Regenerated lines were assessed for resistance 8 weeks after challenge with the nematode, and 16 regenerated lines were putative positives.

Other Traits

Kumar et al. (2005) reported the transformation of embryogenic cultures of ‘Rasthali’ (AAB) with the “s” gene of the hepatitis B surface antigen (HBsAg).

Pineapple: *Ananas comosus* L. (Merr.)

A single cultivar, ‘Smooth Cayenne’ accounts for most of the world production of pineapple, and its improvement has resisted conventional breeding approaches. According to Botella et al. (2000), Graham et al. (2000), and Rohrbach et al. (2000), several traits, including nematode resistance, resistance to fungal diseases (Espinosa et al. 2002) and Pineapple mealybug wilt virus, control of flowering and fruit ripening, and blackheart resistance, have been targeted using transformation.

Microprojectile bombardment (Nan et al. 1996; Ko et al. 2006; Sripaoraya et al. 2006a, b) and *Agrobacterium*-mediated transformation (Firoozabady and Gutterson 1998; Isidron et al. 1998; Graham et al. 2000; Espinosa et al. 2002; Yabor et al. 2006; Firoozabady et al. 2006; Trusov and Botella 2006) have both been utilized, and organogenic and embryogenic cultures and leaf base explants have been successfully targeted. *Agrobacterium tumefaciens* strains LBA4404, AT2260, AGLO and C58 were used by Espinosa et al. (2002) Graham et al. (2000), and Firoozabady et al. (2006), respectively. The vectors have included *npt*II, *bar*, *hph,* and *surB* as selectable marker genes; reporter genes have included *uidA* and *gfp*; the promoters have included ubi-1, CaMV 35S, SCSV4, Smas, and p45.

In order to address blackheart resistance, pineapple has been transformed with the pineapple polyphenol oxidase gene (PINPPO1) in sense, antisense, and hairpin constructs (Stewart et al. 2001; Ko et al. 2006). Espinosa et al. (2002) transformed pineapple with two pathogenesis-related genes, chitinase (*chi*) and *ap24* (for fungal tolerance), in order to control infection by *Phytophthora nicotianae* var. *parasitica.* Sripaoraya et al. (2006a, b) and Yabor et al. (2006) assessed transgenic plants under field conditions. Transgenic plants, with the exception of the appearance of somaclonal variants, were comparable to the controls.

Papaya: *Carica papaya* L.

Unlike the fruit crop species that have been addressed in this chapter, the papaya is not a perennial plant species. Generally seed-propagated, the papaya has a relatively short juvenile period, and conventional breeding has had considerable impact on its improvement. Commercial papaya production is based upon dioecious cultivars in the subtropics and hermaphroditic cultivars in the tropics. The latter group includes the 'Solo' type of papayas, which are highly inbred and highly susceptible to the disease caused by Papaya ringspot virus (PRSV). Because of their homozygosity, the 'Solo' papayas have mostly been targeted for genetic transformation; resistance to PRSV has been the major focus, although other traits are also being addressed, including resistance to fungal and oomycete diseases, soil and climate stress tolerance, improved shelf life, etc.

Papaya has been transformed by means of biolistics and *Agrobacterium*. Various constructs have been utilized as vectors and have included the *npt*II, *hpt,* and *bar* as selective marker genes, the *gfp* and *uidA* as reporter marker genes, the CaMV 35S and ubi3 ubiquitin as promoters, and various genes of interest: *CSb*, citrate synthase (de la Fuente et al. 1997), *DmAMP1*, a dahlia defensin (Zhu et al. 2007), *MSCH*, chitinase from *Manduca sexta* (McCafferty et al. 2006), *pmi*, phosphomannose isomerase (Zhu et al. 2005), *VST1*, stilbene synthase (Zhu et al. 2004), and the C-repeat binding factor (*CBF*) (Dhekney et al. 2007). The efficiency of transforming embryogenic cultures has been optimized: wounding of embryogenic suspensions (Cheng et al. 1996; Yeh et al. 1998) and somatic embryos (Ying et al. 1999), and use of biolistics (Cai et al. 1999).

Disease Resistance

Following the earliest transformation (Fitch et al. 1990), 'Sunset' was transformed with the PRSV coat protein gene of a mild cross protecting strain, HA5-1 (Ling et al. 1991; Fitch et al. 1990), and a few lines were resistant to PRSV (Fitch 2002; Tennant et al. 1994; Lius et al. 1997). An R_0 female, 55-1, containing a single transgene insertion, was backcrossed with non-transgenic 'Sunset', and homozygous R_2 plants were identified by *gus* expression in R_3 seeds. Homozygous R_4 plants ('SunUp') were crossed with non-transgenic 'Kapoho', yielding 'Rainbow'. Although 'SunUp' was resistant to the Hawaiian strain of PRSV, it was highly susceptible to Thailand and Taiwan strains, which are divergent from the Hawaiian strain (Tennant 1996; Tennant et al. 2001); however, line 63-1, which contained two copies of the PRSV CP gene, was resistant to all strains screened (Tennant et al. 2001). Bau et al. (2003) demonstrated that some transgenic Taiwanese selections are resistant to several strains of PRSV.

'SunUp' and 'Rainbow' are the first genetically engineered fruits to be deregulated and licensed for commercial production; however, consumer resistance in Japan, the EU, and elsewhere continues to impede the acceptance of transgenic papayas. Programs that focus on transgenic PRSV resistance exist in several countries and have, for the most part, relied on vector constructs that contain the CP gene of the local strain(s) of PRSV.

Resistance to another virus, Papaya leaf distortion mosaic virus (PLDMV), has been a priority in Taiwan (Lai et al. 2002), where PRSV transformants were highly susceptible to PLDMV (Yeh and Bau 2001). Yeh et al. (1998) transformed papaya with PRSV and PLDMV CP genes, and regenerants were resistant to both PRSV and PLDMV.

Zhu et al. (1999, 2001c, d, 2002, 2003a, b) transformed papaya with different PR genes, including rice, *Manduca* and *Streptomyces albidoflavus* chitinases, defensin gene *DmAMP1* from dahlia, and grapevine stilbene synthase (*VST*1), a PR protein controller gene. *Phytophthora palmivora* growth was inhibited *in vitro* by papaya tissue containing rice chitinase and survived root drench bioassays (Zhu et al. 1999, 2002). Transgenic papaya regenerants with *DmAMP1* defensin were resistant to *P. palmivora* in root drench and leaf bioassays (Zhu et al. 2007). According to Zhu et al. (2004), regenerated papaya plants containing the *VST*1 stilbene synthase gene were resistant to *P. palmivora*.

Pest Resistance

The *Manduca sexta* chitinase gene was introduced into papaya to control infestations of the carmine spider mite (*Tetranychus cinnabarinus*) (McCafferty et al. 2003, 2006), and the regenerated transformants showed enhanced resistance. McCafferty et al. (2008) transformed 'Kapoho' with the lectin gene (*gna*) from the snowdrop *Galanthus nivalis* and also reported enhanced resistance to the carmine spider mite.

Other Traits

In order to extend fruit shelf life, Neupane (1997) and Neupane et al. (1997) transformed papaya with ACC and ACC oxidase (ACO) in the antisense; however, delayed ripening and extended shelf life were not observed (Fitch 2005). Botella (2002) filed an application for a field trial of 'Solo' papayas that had been transformed with sense and antisense ACC oxidase (cloned from papaya) and the *etr1-1* ethylene receptor gene from *Arabidopsis*. Magdalita et al. (2002) also reported that transformation of papaya with the ACC synthase gene (in antisense) has been a priority on the Philippines.

Cabrera-Ponce et al. (1995) described the production of herbicide-tolerant (Basta®) papaya following transformation with phosphinothricin acetyl transferase (*pat*). Papaya was transformed with citrate synthase in order to confer tolerance of aluminum in tropical soils (de la Fuente et al. 1997). This study was repeated with tobacco, however, and the putative advantages could not be confirmed (Delhaize et al. 2001). Dhekney et al. (2007) transformed papaya with a C-repeat binding factor gene (*CBF*) from *Arabidopsis* in an attempt to confer cold tolerance (although *CBF* is present in cold-tolerant mountain papaya *Vasconcellea cundinamarcensis*, it is absent from tropical papaya). Enhanced cold tolerance could not be confirmed in transgenic papaya.

Passion Fruit: *Passiflora* spp.

Genetic transformation of passion fruit has been a component of strategies to address disease problems which include passion fruit woodiness caused by Cowpea aphid-borne mosaic virus (CABMV) and Passion fruit woodiness virus (PWV), Fusarium wilt caused by *Fusarium oxysporum* f. sp. *passiflorae,* and bacterial blight caused by *Xanthomonas axonopodis* pv. *passiflorae*. Manders et al. (1994) first reported *A. tumefaciens*-mediated genetic transformation of passion fruit with the cointegrate pMON200 vector containing the *nptII* gene and regenerated 3 transformed shoots via organogenesis from leaf explants. *Agrobacterium* strain LBA4404 was used. Transient gene expression mediated by *A. tumefaciens* has also been reported (Cancino et al. 1998; Silva 1998; Hall et al. 2000).

Disease Resistance

Alfenas et al. (2005) recovered transgenic plants containing the CABMV coat protein (CP) gene from hypocotyl segments. They used the pBI121 binary vector that included two-thirds of the *Nib* gene and one-third of the CABMV CP gene, both driven by the CaMV 35S promoter, and *npt*II driven by the *nos* promoter. Transformation was mediated by *A. tumefaciens* strain LBA4404. Plants were resistant to the Mg-Avr strain only. R_1 progeny that resulted from self-pollinated R_0, i.e., homozygous for the resistant gene, was resistant to three strains of CABMV. Trevisan et al. (2006) transformed passion fruit with the full-length CP of CABMV cloned in pCAMBIA 2300, and transgenic regenerants were resistant to three strains of CABMV.

Biolistics was used by Takahashi (2002) and Monteiro (2005) to transform passion fruit with the attA gene, which encodes a bacteriocidal protein, in order to confer resistance to bacterial blight.

Conclusions

The major fruit crop species of the world have been genetically transformed, although a few transformations have involved non-clonal materials, e.g., plum and passion fruit. Transformations of papaya, citrus, and plum have seen great progress.

To date, most transformations have been concerned with enhancing resistance to diseases and extending the shelf life of fruit. The absence of candidate genes for transforming these plants for other traits is a major bottleneck for addressing other breeding objectives at this time. Advances in genomics during the next few years will therefore have a major impact on this field. While it is difficult to determine changes in public acceptance of transgenic fruit in the future, the advancement of the alternative, cisgenics, may significantly affect public opinion.

References

Agritope (1999) Research and development-Agritope, Inc. http://www.agritope.com/rd_main.html

Aldwinckle HS, Borejsza Wysocka EE, Malnoy M, Brown SK, Norelli JL, Beer SV, Meng X, He SY, Jin QL (2003) Development of fire blight resistant apple cultivars by genetic engineering. Acta Hortic 622:105–111

Alfenas PF, Braz ASK, Torres LB, Santana EN, Nascimiento AVS, Carvalho MG, Otoni WC, Zerbini FM (2005) Transgenic passionfruit expressing RNA derived from Cowpea aphid-borne mosaic virus is resistant to passionfruit woodiness disease. Fitopatol Bras 30:33–38

Arias M, Camadro E, Diaz-Ricci J, Castagnaro A (2004) Breeding barriers between the cultivated strawberry, *Fragaria × ananassa*, and related wild germplasm. Euphytica 136:139–150

Atkinson HJ, Grimwood S, Johnston K, Green J (2004) Prototype demonstration of transgenic resistance to the nematode *Radopholus similis* conferred on banana by a cystatin. Transgenic Res 13:135–142

Bau HJ, Cheng YH, Yu TA, Yang JS, Yeh SD (2003) Broad-spectrum resistance to different geographic strains of *Papaya ringspot virus* in coat protein gene transgenic papaya. Phytopathology 93:112–120

Becker DK (1999) The transformation of banana with potential virus resistance genes. Ph.D. thesis, Queensland University of Technology, Brisbane

Becker DK, Dugdale B, Smith MK, Harding RM, Dale JL (2000) Genetic transformation of Cavendish banana (Musa spp. AAA group) cv. 'Grand Nain' via microprojectile bombardment. Plant Cell Rep 19:229–234

Belfanti E, Silfverberg-Dilworth E, Tartarini S, Patocchi A, Barbieri M, Zhu J, Vinatzer BA, Gianfranceschi L, Gessler C, Sansavini S (2004) The *HcrVf2* gene from a wild apple confers scab resistance to a transgenic cultivated variety. Proc Natl Acad Sci USA 101:886–890

Bolar JP, Norelli JL, Wong K-W, Harman GE, Aldwinckle HS (2000) Expression of endochitinase from *Trichoderma harzianum* in transgenic apple increases resistance to apple scab and reduces vigor. Phytopathology 90:72–77

Bond JE, Roose ML (1998) *Agrobacterium*-mediated transformation of the commercially important citrus cultivar Washington navel orange. Plant Cell Rep 18:229–234

Borejsza-Wysocka EE, Malnoy M, Meng X, Bonasera JM, Nissinen RM, Kim JF, Beer SV, Aldwinckle HS (2004) Silencing of apple proteins that interact with *dspE*, a pathogenicity effector from *Erwinia amylovora*, as a strategy to increase resistance to fire blight. Acta Hortic 663:469–473

Bornhoff B-A, Harst M, Zyprian E, Töpfer R (2005) Transgenic plants of *Vitis vinifera* cv. Seyval blanc. Plant Cell Rep 24:433–438

Botella J (2002) Application for licence for intentional release of a GMO into the environment: Application No. DIR 026/2002. www.health.gov.au/internet/ogtr/publishing.nsf/Content/…4/…/dir026.rtf

Botella JR, Cavallaro AS, Cazzonelli CI (2000) Towards the production of transgenic pineapple to control flowering and ripening. Acta Hortic 529:115–122

Cabrera-Ponce JL, Vegas-Garcia A, Herrera-Estrella L (1995) Herbicide resistant transgenic papaya plants produced by an efficient particle bombardment transformation method. Plant Cell Rep 15:1–7

Cai W, Gonsalves C, Tennant P, Fermin G, Souza M, Sarindu N, Jan F, Zhu H, Gonsalves D (1999) A protocol for efficient transformation and regeneration of *Carica papaya* L. In Vitro Cell Dev Biol-Plant 35:61–69

Callahan A, Scorza R (2007) Effects of a peach antisense acc oxidase gene on plum fruit quality. Acta Hortic 738:567–573

Cancino GO, Davey MR, Lowe KC, Power JB (1998) Shoot regeneration from leaf and root explants of *Passiflora mollissima*. J Exp Bot 49(Suppl):65 (Abstract)

Cervera M, Juarez J, Navarro A, Pina JA, Duran-Vila N, Navarro L, Peña L (1998a) Genetic transformation and regeneration of mature tissues of woody fruit plants bypassing the juvenile stage. Transgenic Res 7:51–59

Cervera M, Pina JA, Juarez J, Navarro L, Peña L (2000) A broad exploration of a transgenic population of citrus: stability of gene expression and phenotype. Theor Appl Genet 100:670–677

Chakrabarti A, Ganapathi TR, Mukherjee PK, Bapat VA (2003) MSI-99, a magainin analogue, imparts enhanced disease resistance in transgenic tobacco and banana. Planta 216:587–596

Chalavi V, Tabaeizadeh Z, Thibodeau P (2003) Enhanced resistance to *Verticillium dahliae* in transgenic strawberry plants expressing a *Lycopersicon chilense* chitinase gene. J Am Soc Hort Sci 128:747–753

Chanderbali A, Albert V, Ashworth V, Clegg M, Litz RE, Soltis D, Soltis P (2008) *Persea americana* (avocado): bringing ancient flowers to fruit in the genomics era. Bioessays 30:386–396

Cheng YH, Yang JS, Yeh SD (1996) Efficient transformation of papaya by coat protein gene of papaya ringspot virus mediated by *Agrobacterium* following liquid-phase wounding of embryogenic tissues with carborundum. Plant Cell Rep 16:127–132

Costa MGC, Otoni WC, Moore GA (2002) An evaluation of factors affecting the efficiency of Agrobacterium-mediated transformation of *Citrus paradisi* (Macf.) and production of transgenic plants containing biosynthetic genes. Plant Cell Rep 21:365–373

Cruz-Hernández A, Gómez-Lim MA (1995) Alternative oxidase from mango (*Mangifera indica* L.) is differentially regulated during fruit ripening. Planta 197:569–576

Cruz-Hernández A, Gómez-Lim MA, Litz RE (1997) Transformation of mango somatic embryos. Acta Hortic 455:292–298

Cruz-Hernández A, Witjaksono, Litz RE, Gómez-Lim M (1998) *Agrobacterium tumefaciens*-mediated transformation of embryogenic avocado cultures and regeneration of somatic embryos. Plant Cell Rep 17:497–503

Damiano C, Monticelli S, Gentile A, Di Nicola-Negri E, Ilardi V (2007) Efficiency of *Prunus* transformation for PPV-resistance by gene silencing and PPV coat protein gene strategies. Acta Hortic 764:63–70

Darrow GM (1966) The strawberry: history, breeding and physiology. Holt, Rinehart and Winston, New York

de la Fuente JM, Ramirez-Rodriguez V, Cabrera-Ponce JL, Herrera-Estrella L (1997) Aluminum tolerance in transgenic plants by alteration of citrate synthesis. Science 276:1566–1568

de Mesa MC, Jimenez-Bermudez S, Pliego-Alfaro F, Quesada MA, Mercado JA (2000) *Agrobacterium* cells as microprojectile coating: a novel approach to enhance stable transformation rates in strawberry. Aust J Plant Physiol 27:1093–1100

Delhaize E, Hebb DM, Ryan PR (2001) Expression of a *Pseudomonas aeruginosa* citrate synthase gene in tobacco is not associated with either enhanced citrate accumulation of efflux. Plant Physiol 125:2059–2067

Deng Z, Huang S, Ling P, Chen C, Yu C, Weber CA, Moore GA, Gmitter FG Jr (2000) Cloning and characterization of NBS-LRR class resistance-gene candidate sequences in citrus. Theor Appl Genet 101:814–822

Deng Z, Huang S, Ling P, Yu C, Weber CA (2001) Cloning and characterization of NBS-LRR class resistance gene candidate sequences in citrus. Theor Appl Genet 101:814–822

Dhekney SA, Litz RE, Moraga DA, Yadav AK (2007) Potential for introducing cold tolerance into papaya by transformation with C-repeat binding factor (Cbf) genes. In Vitro Cell Dev Biol-Plant 43:195–202

Dhekney SA, Li ZT, Zimmerman TW, Gray DJ (2009) Using endogenous *Vitis* genes to produce disease resistant transgenic grapevines. In Vitro Cell Dev Biol-Plant 45:S38–S38

Dominguez A, Guerri J, Cambra M, Navarro L, Moreno P, Peña L (2000) Efficient production of transgenic citrus plants expressing the coat protein gene of citrus tristeza virus. Plant Cell Rep 19:427–433

Dominguez A, Fagoaga C, Navarro L, Moreno P, Pena L (2002a) Regeneration of transgenic citrus plants under non selective conditions results in high frequency recovery of plants with silenced transgenes. Mol Genet Genom 267:544–556

Dominguez A, Fagoaga C, Navarro L, Moreno P, Pena L (2002b) Constitutive expression of untranslatable versions of the p25 coat protein gene in Mexican lime (*Citrus aurantifolia* (Christm.) Swing.) transgenic plants does not confer resistance to Citrus tristeza virus (CTV). In: Duran-Vila N, Milne RG, Da Graca JV (eds) Proceedings of the 15th conference of the international organization of citrus virologists. University of California, Riverside, pp 341–344

Dominguez A, Guerri J, Cambra M, Navarro L, Moreno P, Pena L (2002c) Pathogen-derived resistance to Citrus tristeza virus (CTV) in transgenic Mexican lime (*Citrus aurantifolia* (Christm.) Swing.) plants expressing its p25 coat protein gene. Mol Breed 10:1–10

Dugdale B, Beetham PR, Becker DK, Harding RM, Dale JL (1998) Promoter activity associated with the intergenic regions of banana bunchy top virus DNA-1 to -6 in transgenic tobacco and banana cells. J Gen Virol 79:2301–2311

Dugdale B, Becker DK, Beetham PR, Harding RM, Dale JL (2000) Promoters derived from banana bunchy top virus DNA-1 to -5 direct vascular-associated expression in transgenic banana (*Musa* spp.). Plant Cell Rep 19:810–814

Dutt M, Li ZT, Dhekney SA, Gray DJ (2008) A co-transformation system to produce transgenic grapevines free of marker genes. Plant Sci 175:423–430

Epple P, Apel K, Bohlmann H (1997) ESTs reveal a multigene family for plant defensins in *Arabidopsis thaliana*. FEBS Lett 400:168–172

Espinosa P, Lorenzo JC, Iglesias A, Yabor L, Menendez E, Borroto J, Hernandez L, Arencibia AD (2002) Production of pineapple transgenic plants assisted by temporary immersion bioreactors. Plant Cell Rep 21:136–140

Espley RV, Hellens RP, Putterill J, Stevenson DE, Kutty-Amma S, Allan AC (2007) Red colouration in apple fruit is due to the activity of the MYB transcription factor, MdMYB10. Plant J 49:414–427

Fagoaga C, Rodrigo I, Conejero V, Hinarejos C, Tuset JJ, Arnau J, Pina JA, Navarro L, Peña L (2001) Increased tolerance to Phytophthora citrophthora in transgenic orange plants constitutive expressing a tomato pathogenesis related protein PR-5. Mol Breed 7:175–185

Faize M, Malnoy M, Dupuis F, Chevalier M, Parisi L, Chevreau E (2003) Chitinases of *Trichoderma atroviride* induce scab resistance and some metabolic changes in two cultivars of apple. Phytopathology 93:1496–1504

Febres VJ, Niblett CL, Lee RF, Moore GA (2003) Characterization of grapefruit plants (*Citrus paradisi* Macf.) transformed with citrus tristeza closterovirus genes. Plant Cell Rep 21:421–428

Ferguson AR, Seal AG (2008) Kiwifruit. In: Hancock JF (ed) Temperate fruit crop breeding. Springer, New York, pp 235–263

Finstad K, Martin RR (1995) Transformation of strawberry for virus resistance. Acta Hortic 385:86–90

Firoozabady E, Gutterson N (1998) Genetically transformed pineapple plants and methods for their production. International Patent Application No. PCT/US98?03681

Firoozabady E, Heckert M, Gutterson N (2006) Transformation and regeneration of pineapple. Plant Cell Tissue Organ Cult 84:1–16

Fitch MMM (2002) Transgenic papayas in Hawaii-a useful tool for new cultivar development and clonal propagation. In: Khachatourians T, McHughen A, Scorza R, Nip WK (eds) Transgenic plants and crops. Marcel Dekker Publishers, New York, pp 437–448

Fitch MMM (2005) Carica papaya Papaya. In: Litz RE (ed) Biotechnology of fruit and nut crops. CABI, Wallingford Oxon, pp 174–207

Fitch MMM, Manshardt RM, Gonsalves D, Slightom JL, Sanford JC (1990) Stable transformation of papaya via microprojectile bombardment. Plant Cell Rep 9:189–194

Flachowsky H, Peil A, Sopanen T, Elo A, Hanke V (2007) Overexpression of *BpMADS4* from silver birch (*Betula pendula* Roth.) induces early flowering in apple (*Malus domestica* Borkh.). Plant Breed 126:137–145

Flachowsky H, Peil A, Rollins J, Hanke M-V, Richter K, Lee DH (2008a) Improved fire blight resistance in transgenic apple lines by constitutive overexpression of the *mbr4* gene of *Malus baccata*. Acta Hortic 793:287–291

Flachowsky H, Richter K, Kim WS, Geider K, Hanke MV (2008b) Transgenic expression of a viral EPS-depolymerase is potentially useful to induce fire blight resistance in apple. Ann Appl Biol 153:345–355

Flachowsky H, Hanke MV, Peil A, Strauss SH, Fladung M (2009) A review on transgenic approaches to accelerate breeding of woody plants. Plant Breed 128:217–226

Fleming GH, Olivares-Fuster O, Fatta Del-Bosco S, Grosser JW (2000) An alternative method for the genetic transformation of sweet orange. In Vitro Cell Dev Biol-Plant 36:450–455

Foucault C, Letouze R (1987) *In vitro* regeneration from excised petiole segments and from flower buds. Biol Plant 29:409–414

Fraser LG, Kent J, Harvey CF (1995) Transformation studies of *Actinidia chinensis* Planch. N Z J Crop Hort Sci 23:407–413

Gambino G, Gribaudo I, Leopold S, Schartl A, Laimer M (2005) Molecular characterization of grapevine plants transformed with GFLV resistance genes: I. Plant Cell Rep 24:655–662

Ganapathi TR, Higgs NS, Balint-Kurti PJ, Arntzen CJ, May GD, Van Eck JM (2001) *Agrobacterium*-mediated transformation of embryogenic cell suspensions of the banana cultivar Rasthali (AAB). Plant Cell Rep 20:157–162

Gao M, Matsuta N, Murayama H, Toyomasu T, Mitsuhashi W, Dandekar AM, Tao R, Nishimura K (2007) Gene expression and ethylene production in transgenic pear (*Pyrus communis* cv. 'La France') with sense or antisense cDNA encoding ACC oxidase. Plant Sci 173:32–42

Geier T, Eimert K, Scherer R, Nickel C (2008) Production and rooting behaviour of *rolB*-transgenic plants of grape rootstock 'Richter 110' (*Vitis berlandieri* × *V. rupestris*). Plant Cell Tissue Organ Cult 94:269–280

Gessler C, Patocchi A (2007) Recombinant DNA technology in apple. Adv Biochem Eng/Biotechnol 107:113–132

Ghorbel R, Juarez J, Navarro L, Peña L (1999) Green fluorescent protein as a screenable marker to increase the efficiency of generating transgenic woody fruit plants. Theor Appl Genet 99:350–358

Ghorbel R, Dominguez A, Navarro L, Peña L (2000) High efficiency genetic transformation of sour orange (*Citrus aurantium*) and production of transgenic trees containing the coat protein gene of citrus tristeza virus. Tree Physiol 20:1183–1189

Ghorbel R, La-Malfa S, Lopez MM, Petit A, Navarro L, Peña L (2001) Additional copies of virG from pTiBo542 provide a super-transformation ability to *Agrobacterium tumefaciens* in citrus. Physiol Mol Plant Pathol 58:103–110

Gómez-Lim MA (1993) Mango fruit ripening: physiology and molecular biology. Acta Hortic 341:484–499

Good X, Kellogg JA, Wagoner W, Langhoff D, Matsumura W, Bestwick RK (1994) Reduced ethylene synthesis by transgenic tomatoes expressing S-adenosylmethionine hydrolase. Plant Mol Biol 26:781–790

Graham J (2005) *Fragaria* strawberry. In: Litz RE (ed) Biotechnology of fruit and nut crops. CABI, Wallingford Oxon, pp 456–474

Graham J, McNicol RJ, Greig K (1995) Towards genetic based insect resistance in strawberry using the cowpea trypsin inhibitor gene. Ann Appl Biol 127:163–173

Graham J, Gordon SC, McNicol RJ (1997) The effect of the CpTi gene in strawberry against attack by vine weevil (*Otiorhynchus sulcatus* f. Coleoptera: Curculionidae). Ann Appl Biol 131:133–139

Graham M, Ko L, Hardy V, Robinson S, Sawyer B, O'Hare T, Jobin M, Dahler J, Underhill S, Smith M (2000) The development of blackheart resistant pineapples through genetic engineering. Acta Hortic 529:133–136

Graham J, Gordon SC, Smith K, McNicol RJ, McNicol JW (2002) The effect of the Cowpea trypsin inhibitor in strawberry on damage by vine weevil under field conditions. J Hort Sci Biotechnol 77:33–40

Gray DJ, Li ZT, Dhekney SA, Hopkins DL, Zimmerman TW (2009) Green genetic engineering technology: rearrangement of endogenous functional genetic elements to create improved grapevines. In Vitro Cell Dev Biol-Plant 45:S39–S39

Grosser JW, Gmitter FG Jr, Orbovic V, Moore GA, Graham JH, Soneji J, Gonzalez-Ramos J, Mirkov TE, Kayim M (2009) Grapefruit. In: Hall TC, Kole C (eds) Transgenic tropical and subtropical fruits and nuts. Wiley Blackwell, Chichester, pp 63–76

Guo WW, Duan YX, Olivares-Fuster O, Wu ZC, Arias CR, Burns JK, Grosser JW (2005) Protoplast transformation and regeneration of transgenic Valencia sweet orange plants containing a juice quality-related pectin methylesterase gene. Plant Cell Rep 24:482–486

Gutierrez-E MA, Luth D, Moore GA (1997) Factors affecting the *Agrobacterium*-mediated transformation in *Citrus* and production of sour orange (*Citrus aurantium* L.) plants expressing the coat protein gene of citrus tristeza virus. Plant Cell Rep 18:271–278

Gutiérrez-Pesce P, Rugini E (2008) Kiwifruit. In: Kole C, Hall TH (eds) Transgenic tropical and subtropical fruits and nuts. Wiley Blackwell, Chichester, pp 185–211

Hall RM, Drew RA, Higgins CM, Dietzgen RG (2000) Efficient organogenesis of an Australian passionfruit hybrid (*Passiflora edulis* x *Passiflora edulis* var. *flavicarpa*) suitable for gene delivery. Aust J Bot 48:673–680

Hammerschlag FA, Smigocki AC (1998) Growth and in vitro propagation of peach plants transformed with the shooty mutant strain of *Agrobacterium tumefaciens*. HortScience 33:897–899

Han ML, Gleave AP, Wang TC (2010) Efficient transformation of *Actinidia arguta* by reducing the strength of basal salts in the medium to alleviate callus browning. Plant Biotechnol Rep 4:129–138

Hanke V, Düring K, Norelli JL, Aldwinckle HS (1999) Transformation of apple cultivars with t4-lysozyme-gene to increase fire blight resistance. Acta Hortic 489:253–256

He LX, Ban Y, Inoue H, Matsuda N, Liu JH, Moriguchi T (2008) Enhancement of spermidine content and antioxidant capacity in transgenic pear shoots overexpressing apple spermidine synthase in response to salinity and hyperosmosis. Phytochemistry 69:2133–2141

Herrera Estrella L, Depicker A, Van Montagu M, Schell J (1983) Expression of chimaeric genes transferred into plant cells using a Ti-plasmid-derived vector. Nature 303:209–213

Hidaka T, Omura M, Ugaki M, Kato A, Oshima M, Motoyoshi F (1990) *Agrobacterium*-mediated transformation and regeneration of *Citrus* spp. from suspension cells. Jpn J Breed 40:199–207

Holefors A, Xue ZT, Welander M (1998) Transformation of the apple rootstock M26 with the *rolA* gene and its influence on growth. Plant Sci 136:69–78

Houde M, Dallaire S, N'Dong D, Sarhan F (2004) Overexpression of the acidic dehydrin WCOR410 improves freezing tolerance in transgenic strawberry leaves. Plant Biotechnol J 2:381–387

Hughes JA, Brown LR, Ferro AJ (1987) Nucleotide sequence and analysis of the coliphage T3S-adenosylmethionine hydrolase gene and its surrounding ribonuclease III processing sites. Nucleic Acid Res 15:717–729

Husaini AM (2010) Pre- and post-agroinfection strategies for efficient leaf disk transformation and regeneration of transgenic strawberry plants. Plant Cell Rep 29:97–110

Husaini AM, Abdin MZ (2008) Development of transgenic strawberry (*Fragaria × ananassa* Duch.) plants tolerant to salt stress. Plant Sci 174:446–455

Igarashi M, Ogasawara H, Hatsuyama Y, Saito A, Suzuki M (2002) Introduction of *rolC* into Marubakaidou (*Malus prunifolia* Borkh. Var. *ringo* Asami Mo 84-A) apple rootstock via *Agrobacterium tumefaciens*. Plant Sci 163:463–473

Isidron PM, Benega GR, Cisneros PA, Arias VE, Lorenzo FJC, Espinosa AP, Borroto NCG (1998) Application of biotechnological and traditional methods in Cuban pineapple breeding program. In: Proceedings of the third international pineapple symposium, Pattaya, p 45 (Abstract)

Iwanami T, Shimizu T, Ito T, Hirabayashi T (2004) Tolerance to citrus mosaic virus in transgenic trifoliate orange lines harbouring capsid polyprotein gene. Plant Dis 88:865–868

James DJ, Passey AJ, Barbara DJ, Bevan MW (1989) Genetic transformation of apple (*Malus pumila* Mill.) using a disarmed Ti-binary vector. Plant Cell Rep 7:658–661

James DJ, Passey AJ, Barbara DJ (1990) *Agrobacterium* mediated transformation of the cultivated strawberry (*Fragaria × ananassa* Duch.) using disarmed binary vectors. Plant Sci 69:79–94

Janssen B-J, Gardner RC (1993) The use of transient GUS expression to develop an *Agrobacterium*-mediated gene transfer system for kiwifruit. Plant Cell Rep 13:28–31

Jardak-Jamoussi R, Winterhagen P, Bouamama B, Dubois C, Mliki A, Wetzel T, Ghorbel A, Reustle GM (2009) Development and evaluation of a GFLV inverted repeat construct for genetic transformation of grapevine. Plant Cell Tissue Organ Cult 97:187–196

Jelenkovic G, Chin CK, Billings S (1986) Transformation studies of *Fragaria × ananassa* by Ti-plasmids of *Agrobacterium tumefaciens*. HortScience 21:695 (Abstract)

Jiménez-Bermúdez S, Redondo-Nevado J, Munoz-Blanco J, Caballero JL, Lopez-Aranda JM, Valpuesta V, Pliego-Alfaro F, Quesada MA, Mercado JA (2002) Manipulation of strawberry fruit softening by antisense expression of a Pectate lyase gene. Plant Physiol 128:751–759

Jin WM, Dong J, Hu YL, Lin ZP, Xu XF, Han ZH (2009) Improved cold-resistant performance in transgenic grape (*Vitis vinifera* L.) overexpressing cold-inducible transcription Factors *AtDREB1b*. HortScience 44:35–39

Jones OP, Waller BJ, Beech MG (1988) The production of strawberry plants from callus cultures. Plant Cell Tissue Organ Cult 12:235–241

Kaneyoshi J, Kobayashi S, Nakamura Y, Shigemoto N, Doi Y (1994) A simple and efficient gene transfer system of trifoliate orange (*Poncirus trifoliata* Raf.). Plant Cell Rep 13:541–545

Ko HL, Campbell PR, Jobin-Decor MP, Eccleston KL, Graham MW, Smith MK (2006) The introduction of transgenes to control blackheart in pineapple (*Ananas comosus* L.) cv. Smooth Cayenne by microparticle bombardment. Euphytica 150:387–395

Kobayashi S, Uchimiya H (1989) Expression and integration of a foreign gene in orange (*Citrus sinensis* Osb.) protoplasts by direct DNA transfer. Jpn J Genet 64:91–97

Kobayashi S, Nakamura Y, Kaneyoshi J, Higo H, Higo K (1996) Transformation of kiwifruit (*Actinidia chinensis*) and trifoliate orange (*Poncirus trifoliata*) with a synthetic gene encoding the human epidermal growth factor (hEGF). J Jpn Soc Hort Sci 64:763–769

Kobayashi S, Ding CK, Nakamura Y, Nakajima I, Matsumoto R (2000) Kiwifruits (*Actinidia deliciosa*) transformed with a *Vitis* stilbene synthase gene produce piceid (resveratrol-glucoside). Plant Cell Rep 19:904–910

Koltunow AM, Brennan P, Protopsaltis S, Nito N (2000) Regeneration of West Indian limes (*Citrus aurantifolia*) containing genes for decreased seed set. Acta Hortic 535:81–91

Kotoda N, Iwanami H, Takahashi S, Abe K (2006) Antisense expression of MdTFL1, a TFL1-like gene, reduces the juvenile phase in apple. J Am Soc Hort Sci 131:74–81

Kramer MG, Kellogg J, Wagoner W, Matsumura W, Good X, Peters S, Clough G, Bestwick RK (1997) Reduced ethylene synthesis and ripening control in tomatoes expressing S-adenosylmethionine hydrolase. In: Kannelis AK, Chang C, Kende H, Grierson D (eds) Biology and biotechnology of the plant hormone ethylene. Kluwer Academic Publishers, Dordrecht, pp 307–319

Krens FA, Pelgrom KTB, Schaart JG, den Nijs APM, Rouwendal GJA (2004) Clean vector technology for marker-free transgenic fruit crops. Acta Hortic 663:431–435

Kumar BBS, Ganapathi tr, Revathi CJ, Srinvas L, Bapat VA (2005) Expression of hepatitis B surface antigen in transgenic banana plants. Planta 222:484–493
Lai C, Yu T, Yeh S, Yang J (2002) Non-cryogenic, long-term preservation of in vitro multiple bud clusters of papaya. International Association of Plant Tissue Culture and Biotechnology, Orlando, p 104A (Abstracts)
Landi L, Capocasa F, Costantini E, Mezzetti B (2009) *ROLC* strawberry plant adaptability, productivity, and tolerance to soil-borne disease and mycorrhizal interactions. Transgenic Res 18:933–942
Li DD, Shi W, Deng XX (2002) Agrobacterium-mediated transformation of embryogenic calluses of Ponkan mandarin and the regeneration of plants containing the chimeric ribonuclease gene. Plant Cell Rep 21:153–156
Li DD, Shi W, Deng XX (2003) Factors influencing Agrobacterium-mediated embryogenic callus transformation of Valencia sweet orange (*Citrus sinensis*) containing the pTA29-barnase gene. Tree Physiol 23:1209–1215
Ling K, Namba S, Consalves C, Slightom JL, Gonsalves D (1991) Protection against detrimental effects of potyvirus infection in transgenic tobacco plants expressing the papaya ringspot virus coat protein gene. Biotechnology 9:752–758
Litz RE, Gray DJ (1995) Somatic embryogenesis for agriculture improvement. World J Microbiol Biotechnol 11:416–425
Litz RE, Efendi D, Raharjo SHT, Padilla G, Moon PA (2007) Genetic transformation of avocado with S-adenosylmethionine hydrolase (samase) and evaluation of transformants after three years. In: Proceedings VI world Avocado congress CD, 1c-24, 10 p
Liu ZR, Sanford JC (1988) Plant regeneration by organogenesis from strawberry leaf and runner tissue. HortScience 23:1057–1059
Liu JH, Kitashiba H, Wang J, Ban Y, Moriguchi T (2007) Polyamines and their ability to provide environmental stress tolerance to plants. Plant Biotechnol 24:117–126
Lius S, Manshardt RM, Fitch MMM, Slightom JL, Sanford JC, Gonsalves G (1997) Pathogen-derived resistance provides papaya with effective protection against papaya ringspot virus. Mol Breed 3:161–168
Luth D, Moore G (1999) Transgenic grapefruit plants obtained by *Agrobacterium tumefaciens*-mediated transformation. Plant Cell Tissue Organ Cult 57:219–222
Machado MLD, Machado AD, Hanzer V, Weiss H, Regner F, Steinkellner H, Mattanovich D, Plail R, Knapp E, Kalthoff B, Katinger H (1992) Regeneration of transgenic plant of *Prunus armenaica* containing the coat protein gene of plum pox virus. Plant Cell Rep 11:25–29
Magdalita PM, Laurena AC, Yabut-Perez BM, Mendoza EMT, Villegas VN, Botella JR (2002) Progress in the development of transgenic papaya: Transformation of Solo papaya using ACC synthase antisense construct. Acta Hortic 575:171–176
Malnoy M, Faize M, Venisse JS, Geider K, Chevreau E (2005a) Expression of viral EPS-depolymerase reduces fire blight susceptibility in transgenic pear. Plant Cell Rep 23:632–638
Malnoy M, Venisse JS, Brisset MN, Chevreau E (2005b) Expression of bovine lactoferrin cDNA confers resistance to *Erwinia amylovora* in transgenic pear. Mol Breed 12:231–244
Malnoy M, Jin Q, Borejsza-Wysocka EE, He SY, Aldwinckle HS (2007) Overexpression of the apple *MpNPR1* gene confers increased disease resistance in *Malus × domestica*. Mol Plant Microbe Interact 20:1568–1580
Malnoy M, Xu M, Borejsza-Wysocka E, Korban SS, Aldwinckle HS (2008) Two receptor-like genes, Vfa1 and Vfa2, confer resistance to the fungal pathogen *Venturia inaequalis* inciting apple scab disease. Mol Plant Microbe Interact 21:448–458
Manders G, Otoni WC, d'Utra Vaz FB, Davey MR, Power JB (1994) Transformation of passion fruit (*Passiflora edulis* f. *flavicarpa* Degener) using *Agrobacterium tumefaciens*. Plant Cell Rep 13:234–238
Mathews H, Litz RE, Wilde DH, Merkel S, Wetzstein HY (1992) Stable integration and expression of β-glucuronidase and NPT II genes in mango somatic embryos. In Vitro Cell Dev Biol-Plant 28:172–178

Mathews H, Litz RE, Wilde HD, Wetzstein HY (1993) Genetic transformation of mango. Acta Hortic 341:93–97

Mathews H, Wagoner W, Kellogg J, Bestwick R (1995) Genetic transformation of strawberry: stable integration of a gene to control biosynthesis of ethylene. In Vitro Cell Dev Biol-Plant 31:36–43

Matsuta N, Iketani H, Hayashi T (1993) Transformation in grape and kiwifruit. In: Hayashi T, Omura M, Scott NS (eds) Techniques on gene diagnosis and breeding in fruit trees. Fruit Tree Research Station, Tsukuba, pp 184–192

Mauro MC, Toutain S, Walter B, Pinck L, Otten L, Coutos-Thevenot P, Deloire A, Barbier P (1995) High efficiency regeneration of grapevine plants transformed with the GFLV coat protein gene. Plant Sci 112:97–106

May GD, Afza R, Mason HS, Wiecko A, Novak FJ, Arntzen CJ (1995) Generation of transgenic banana (*Musa acuminata*) plants via *Agrobacterium*-mediated transformation. Biotechnology 13:486–492

McCafferty HR, Zhu J, Moore P (2003) Towards improved insect resistance in papaya. Plant Biol 2003:58 (Abstract)

McCafferty HR, Moore P, Zhu YJ (2006) Improved *Carica papaya* tolerance to carmine spider mite by the expression of *Manduca sexta* chitinase transgene. Transgen Res 15:337–347

McCafferty HR, Moore P, Zhu YJ (2008) Papaya transformed with the *Galanthus nivalis* GNA gene produces a biologically active lectin with spider mite control activity. Plant Sci 175:385–393

Mercado JA, Martin-Pizarro C, Pascual L, de los Santos B, Romero F, Quesada MA, Pliego-Alfaro F, Galvez J, Rey M, de la Vina G, Llobell A, Yubero-Serrano EM, Munoz-Blanco J, Caballero JL (2007) Evaluation of tolerance of *Colletotrichum acutatum* in strawberry plants transformed with Trichoderma-derived genes. Acta Hortic 738:383–388

Messina R, Vischi M, Marchetti S, Testolin R, Milani N (1990) Observations on subdioeciousness and fertilisation in a kiwifruit breeding program. Acta Hortic 282:377–386

Mezzetti B, Costantini E, Chionchetti F, Landi L, Pandolfini T, Spena A (2004) Genetic transformation in strawberry and raspberry for improving plant productivity and fruit quality. Acta Hortic 649:107–110

Molinari HBC, Marur CJ, Bespalhok JCF, Kobayashi AK, Pereira LFP, Vieira LGE (2004) Osmotic adjustment in transgenic citrus rootstock Carrizo citrange (*Citrus sinensis* Osb. *Poncirus trifoliata* L. Raf.) overproducing proline. Plant Sci 167:1375–1381

Monteiro M (2005) Genetic transformation of yellow passion fruit to confer resistance to Xanthomonas axonopodis pv. Passiflorae (in Portuguese). Ph.D. dissertation, Escola Superior de Agricultura "Luis de Queiroz", Universidade de Sao Paulo, Piracicaba, SP, Brazil

Moore GA, Jacono CC, Neidigh JL, Lawrence SD, Cline K (1992) *Agrobacterium*-mediated transformation of citrus stem sections and regeneration of transgenic plants. Plant Cell Rep 11:238–242

Moore GA, Jacono CC, Neidigh JL, Lawrence SD, Cline K (1993) Transformation in citrus. In: Bajaj YPS (ed) Biotechnology in agriculture and forestry, vol 23: Plant protoplasts and genetic engineering IV. Springer, Berlin, pp 194–208

Mourgues F, Chevreau E, Lambert C, de Bondt A (1996) Efficient Agrobacterium-mediated transformation and recovery of transgenic plants from pear (*Pyrus communis* L.). Plant Cell Rep 16:245–249

Nagel AK, Schnabel G, Petri C, Scorza R (2008) Generation and characterization of transgenic plum lines expressing the *Gastrodia* antifungal protein. HortScience 43:1514–1521

Nagel AK, Kalariya H, Schnabel G (2010) The *Gastrodia* antifungal protein (GAFP-1) and its transcript are absent from scions of chimeric-grafted plum. HortScience 45:188–192

Nakamura Y, Sawada H, Kobayashi S, Nakajima I, Yoshikawa M (1999) Expression of soybean beta-1,3-endoglucanase cDNA and effect on disease tolerance in kiwifruit plants. Plant Cell Rep 18:527–532

Nan G-L, Nagai C, Moore PH, Sun SSM, Sipes BS, Paull RE, Rohrbach K (1996) Tissue culture and genetic transformation studies on pineapple. In Vitro Cell Dev Biol-Plant 32:102A (Abstract)

Nehra NS, Stushnoff C, Kartha KK (1989) Direct shoot regeneration from strawberry leaf disks. J Am Soc Hort Sci 114:1014–1018

Nehra NS, Chibbar RN, Kartha KK, Datla RSS, Crosby WL, Stushnoff C (1990a) Genetic transformation of strawberry by *Agrobacterium tumefaciens* using a leaf disk regeneration system. Plant Cell Rep 9:293–298

Nehra NS, Chibbar RN, Kartha KK, Datla RSS, Crosby WL, Stushnoff C (1990b) *Agrobacterium*-mediated transformation of strawberry calli and recovery of transgenic plants. Plant Cell Rep 9:10–13

Neupane KR (1997) Genetic engineering of papaya (*Carica papaya* L.) for modified ethylene biosynthesis. Unpublished Ph.D. dissertation, University of Hawaii, Honolulu, Hawaii, 173 p

Neupane KR, Mukatira UT, Kato C, Stiles JI (1997) Cloning and characterization of fruit-expressed ACC synthase and ACC oxidase from papaya. International symposium on biotechnology of tropical and subtropical species, Brisbane, Australia, 29 Sept–3 Oct 1997, p 73 (Abstract)

Niedz RP, Sussman MR, Satterlee JS (1995) Green fluorescent protein: an in vivo reporter of plant gene expression. Plant Cell Rep 14:403–406

Noiton D, Shelbourne CJA (1992) Quantitative genetics in an apple breeding strategy. Euphytica 60:213–219

Norelli JL, Aldwinckle HS, Destefano-Beltran L, Jaynes JM (1994) Transgenic 'Malling 26' apple expressing the attacin E gene has increased resistance to *Erwinia amylovora*. Euphytica 77:123–128

Norelli JL, Borejsza-Wysocka E, Momol MT, Mills JZ, Grethel A, Aldwinckle HS, Ko K, Brown SK, Bauer DW, Beer SV, Abdul-Kader AM, Hanke V (1999) Genetic transformation for fire blight resistance in apple. Acta Hortic 489:295–296

Nyman M, Wallin A (1992) Improved culture techniques for strawberry (*Fragaria* × *ananassa* Duch.) protoplasts and the determination of DNA content in protoplast derived plants. Plant Cell Tissue Organ Cult 30:127–133

Oliveira MM, Barroso J, Pais MS (1991) Direct gene transfer into *Actinidia deliciosa* protoplasts: analysis of transient expression of the *CAT* gene using TLC autoradiography and a GC-MS-based method. Plant Mol Biol 17:235–242

Park JI, Lee YK, Chung WI, Lee IH, Choi JH, Lee WM, Ezura H, Lee SP, Kim IJ (2006) Modification of sugar composition in strawberry fruit by antisense suppression of an ADP-glucose pyrophosphorylase. Mol Breed 17:269–279

Pawlicki-Jullian N, Sedira M, Welander M (2002) The use of *Agrobacterium rhizogenes* transformed roots to obtain transgenic shoots of the apple rootstock Jork 9. Plant Cell Tissue Organ Cult 70:163–171

Peña L, Cervera M, Juarez J, Ortega C, Pina JA, Duran-Vila N, Navarro L (1995a) High-efficiency *Agrobacterium*-mediated transformation and regeneration of citrus. Plant Sci 104:183–191

Peña L, Cevera M, Juarez J, Navarro A, Pina JA, Duran-Vila N, Navarro L (1995b) *Agrobacterium*-mediated transformation of sweet orange and regeneration of transgenic plants. Plant Cell Rep 14:616–619

Peña L, Cevera M, Juarez J, Navarro A, Pina JA, Navarro L (1997) Genetic transformation of lime (*Citrus aurantifolia* Swing.): factors affecting transformation and regeneration. Plant Cell Rep 17:731–737

Peña L, Martin-Trillo M, Juarez J, Pina JA, Navarro L, Martinez-Zapater JM (2001) Constitutive expression of Arabidopsis *LEAFY* and *APETALA1* genes in citrus reduces their generation time. Nat Biotechnol 19:263–267

Pereira-Lorenzo S, Ramos-Cabrer AM, Fischer M (2009) In: Jain SM, Priyadarshan PM (eds) Breeding plantation tree crops: temperate species. Springer, New York, pp 33–81

Perez-Molphe-Balch E, Ochoa-Alejo N (1998) Regeneration of transgenic plants of Mexican lime from *Agrobacterium rhizogenes*-transformed tissues. Plant Cell Rep 17:591–596

Petri C, Burgos L (2005) Transformation of fruit trees. Useful breeding tool or continued future prospect? Transgenic Res 14:15–26

Petri C, Wang H, Alburquerque N, Faize M, Burgos L (2008) *Agrobacterium*-mediated transformation of apricot (*Prunus armeniaca* L.) leaf explants. Plant Cell Rep 27:1317–1324

Qin Y, Zhang S (2007) Factors influencing the efficiency of *Agrobacterium*-mediated transformation in strawberry cultivar Toyonoka. J Nucl Agric Sci 21:461–465 (in Chinese with English abstract)
Qin Y, Teixeira da Silva JA, Zhang L, Zhang S (2008) Transgenic strawberry: state of the art for improved traits. Biotechnol Adv 26:219–232
Raharjo S, Witjaksono, Efendi D, Gomez-Lim MA, Suarez I, Litz RE (2003) Genetic transformation of avocado. In: Proceedings V Congreso Mundial de Aguacate, vol II, pp115–118
Raharjo SHT, Witjaksono, Padilla G, Litz RE (2008) Recovery of avocado (*Persea americana* Mill.) plants transformed with the antifungal plant defensin gene *pdf1.2*. In Vitro Cell Dev Biol-Plant 44:254–262
Ramesh SA, Kaiser BN, Franks T, Collins G, Sedgley M (2006) Improved methods in *Agrobacterium*-mediated transformation of almond using positive (mannose/pmi) or negative (kanamycin resistance) selection-based protocols. Plant Cell Rep 25:821–828
Raquel MH, Oliveira MM (1996) Kiwifruit leaf competent for plant regeneration and direct DNA transfer. Plant Sci 121:107–114
Reynoird JP, Mourgues F, Norelli J, Aldwinckle HS, Brisset MN, Chevreau E (1999) First evidence for improved resistance to fire blight in transgenic pear expressing the *attacin E* gene from *Hyalophora cecropia*. Plant Sci 149:23–31
Rohrbach KG, Christopher D, Hu J, Paull R, Sipes B, Nagai C, Moore P, McPherson M, Atkinson H, Levesley A, Oda C, Fleisch H, McLean M (2000) Management of a multiple goal pineapple genetic engineering program. Acta Hortic 529:111–113
Rugini E, Pellegrineschi A, Jacobini A, Mariotti D, Mencuccini M (1989) Trasformazione genetica dell'Actinidia (Actinidia deliciosa A. Chev.) mediante geni "Root inducing" di Agrobacterium rhizogenes. In: Proceedings of the Società Italiana di Genetica Agraria, Alghero, Italy, 23–26 Oct, p 147
Rugini E, Pellegrineschi A, Mencuccini M, Mariotti D (1991) Increase of rooting ability in the woody species kiwi (*Actinidia deliciosa* A. Chev.) by transformation with *Agrobacterium rhizogenes rol* genes. Plant Cell Rep 10:291–295
Rugini E, Muganu M, Pilotti M, Balestra GM, Varvaro L, Magro P, Bressan R, Taratufolo C (1999) Genetic stability, transgene hereditability and agronomic evaluation of transgenic kiwi (*Actinidia deliciosa* A. Chev.) plants for *rolABC, rolB* and osmotin gene. In: Fourth international symposium on kiwifruit, Santiago de Chile, 11–14 Feb 1999, pp 26 (Abstract)
Ruhmann S, Treutter D, Fritsche S, Briviba K, Szankowski I (2006) Piceid (resveratrol glucoside) synthesis in stilbene synthase transgenic apple fruit. J Agric Food Chem 54:4633–4640
Sagi L, Remy S, Panis B, Swennen R, Volckaert G (1994) Transient gene expression in electroporated banana (*Musa* spp, cv Bluggoe, ABB group) protoplasts isolated from regenerable embryogenetic cell suspensions. Plant Cell Rep 13:262–266
Sagi L, Panis B, Remy S, Schoofs H, Desmet K, Swennen R, Cammue BPA (1995) Genetic transformation of banana and plantain (*Musa* spp) via particle bombardment. Biotechnology 13:481–485
Santiago-Domenech N, Jimenez-Bemudez S, Matas AJ, Rose JKC, Munoz-Blanco J, Mercado JA, Quesada MA (2008) Antisense inhibition of a pectate lyase gene supports a role for pectin depolymerization in strawberry fruit softening. J Exp Bot 59:2769–2779
Schestibratov KA, Dolgov SV (2005) Transgenic strawberry plants expressing a thaumatin II gene demonstrate enhanced resistance to *Botrytis cinerea*. Sci Hortic 106:177–189
Scorza R, Ravelonandro M, Callahan AM, Cordts JM, Funchs M, Dunez J, Gonsalves D (1994) Transgenic plum (*Prunus domestica* L.) express the plum pox virus coat protein gene. Plant Cell Rep 14:18–22
Scorza R, Callahan A, Levy L, Damsteegt V, Webb K, Ravelonandro M (2001) Post-transcriptional gene silencing in plum pox virus resistant transgenic European plum containing the plum pox potyvirus coat protein gene. Transgenic Res 10:201–209
Shuxiang T, Shufan J (1995) Detection and observation on the bud mutation of *Actinidia chinensis* Planch. Acta Hortic 403:71–73
Silva MB (1998) *Agrobacterium tumefaciens*-mediated transformation of the yellow passionfruit (*Passiflora edulis f. flavicarpa Degener)*. M.Sc. thesis, Universidade Federal de Viçosa, Viçosa, MG, Brazil

Song GQ, Sink KC (2006) Transformation of Montmorency sour cherry (*Prunus cerasus* L.) and Gisela 6 (*P. cerasus* × *P. canescens*) cherry rootstock mediated by *Agrobacterium tumefaciens*. Plant Cell Rep 25:117–123

Soriano JM, Joshi SG, van Kaauwen M, Noordijk Y, Groenwold R, Henken B, van de Weg WE, Schouten HJ (2009) Identification and mapping of the novel apple scab resistance gene *Vd3*. Tree Genet Genom 5:475–482

Sripaoraya S, Keawsompong S, Insupa P, Davey MR, Power JB, Srinives P (2006a) Evaluation of transgene stability, gene expression and herbicide tolerance of genetically modified pineapple under field conditions. Acta Hortic 702:37–40

Sripaoraya S, Keawsompong S, Insupa P, Davey MR, Power JB, Srinives P (2006b) Genetically manipulated pineapple: transgene stability, gene expression and herbicide tolerance under field conditions. Plant Breed 125:411–413

Stewart RJ, Sawyer BJB, Bucheli CS, Robinson SP (2001) Polyphenol oxidase is induced by chilling and wounding in pineapple. Aust J Plant Physiol 28:181–191

Sugaya T, Yano M, Sun H-J, Hirai T, Ezura H (2008) Transgenic strawberry expressing the taste-modifying protein miraculin. Plant Biotechnol 25:329–333

Szankowski I, Briviba K, Fleschhut J, Schonherr J, Jacobsen HJ, Kiesecher H (2003) Transformation of apple (*Malus domestica* Borkh.) with the stilbene synthase gene from grapevine (*Vitis vinifera* L.) and a PGIP gene from kiwi (*Actinidia deliciosa*). Plant Cell Rep 22:141–149

Szankowski I, Waidmann S, Degenhardt J, Patocchi A, Paris R, Silfverberg-Dilworth E, Broggini G, Gessler C (2009) Highly scab-resistant transgenic apple lines achieved by introgression of *HcrVf2* controlled by different native promoter lengths. Tree Genet Genom 5:349–358

Takahashi EHK (2002) Transference of the attacin A gene to yellow passion fruit (*Passiflora edulis* Sims. F. *flavicarpa* Deg.) via biolistics (in Portuguese). M.Sc. thesis, Escola Superior de Agricultura "Luis de Queiroz", Universidade de Sao Paulo, Piracicaba, SP, Brazil

Tennant PF (1996) Evaluation of the resistance of coat protein transgenic papaya against papaya ringspot potyvirus isolates and development of transgenic papaya for Jamaica. Unpublished Ph.D. dissertation, Cornell University, Ithaca, NY, 318 pp

Tennant PF, Gonsalves C, Ling KS, Fitch M, Manshardt R, Slightom J, Gonsalves D (1994) Differential protection against papaya ringspot virus isolates in coat protein gene transgenic papaya and classically cross-protected papaya. Phytopathology 84:1359–1366

Tennant PF, Fermin G, Fitch MM, Manshardt RM, Slightom JL, Gonsalves D (2001) Papaya ringspot virus resistance of transgenic Rainbow and SunUp is affected by gene dosage, plant development and coat protein homology. Eur J Plant Pathol 107:645–653

Teo G, Suzuki Y, Uratsu SL, Lampinen B, Ormonde N, Hu WK, DeJong TM, Dandekar AM (2006) Silencing leaf sorbitol synthesis alters long-distance partitioning and apple fruit quality. Proc Natl Acad Sci USA 103:18842–18847

Thompson JM, Zimmerman RH, Vanderzwet T (1975) Inheritance of fire blight resistance in Pyrus.1. Dominant gene, SE, causing sensitivity. J Hered 66:259–264

Trevisan F, Mendes BMJ, Maciel SC, Vieira MLC, Meletti LMM, Rezende JAM (2006) Resistance to passionfruit woodiness virus in transgenic passionflower expressing the virus coat protein gene. Plant Dis 90:1026–1030

Trusov Y, Botella JR (2006) Delayed flowering in pineapples (*Ananas comosus* (L.) Merr.) caused by co-suppression of the ACACS2 gene. Acta Hortic 702:29–36

Uematsu C, Murase M, Ichikawa H, Imamura J (1991) *Agrobacterium*-mediated transformation and regeneration of kiwi fruit. Plant Cell Rep 10:286–290

Valat L, Fuchs M, Burrus M (2006) Transgenic grapevine rootstock clones expressing the coat protein or movement protein genes of Grapevine fanleaf virus: characterization and reaction to virus infection upon protoplast electroporation. Plant Sci 170:739–747

Vardi A, Bleichman S, Aviv D (1990) Genetic transformation of *citrus* protoplasts and regeneration of transgenic plants. Plant Sci 69:199–206

Vellicce GR, Ricci JCD, Hernández L, Castagnaro AP (2006) Enhanced resistance to *Botrytis cinerea* mediated by the transgenic expression of the chitinase gene *ch5B* in strawberry. Transgenic Res 15:57–68

Vidal JR, Kikkert JR, Wallace PG, Reisch BI (2003) High efficiency biolistic co-transformation and regeneration of 'Chardonnay' (*Vitis vinifera* L.) containing *npt-II* and antimicrobial peptide genes. Plant Cell Rep 22:252–260

Vidal JR, Kikkert JR, Malnoy MA, Wallace PG, Barnard J, Reisch BI (2006) Evaluation of transgenic 'Chardonnay' (*Vitis vinifera*) containing magainin genes for resistance to crown gall and powdery mildew. Transgenic Res 15:69–82

Vieira MLC, Takahashi EK, Falco MC, Vieira LG, Pereira LFP (2002) Direct gene transfer to passion fruit (*Passiflora edulis*) with the *attacin A* gene. In: Proceedings of the 10th international association for plant tissue culture and biotechnology congress, Orlando, p 103-A (Abstract)

Wang DY, Wergin WP, Zimmerman RH (1984) Somatic embryogenesis and plant regeneration from immature embryos of strawberry. HortScience 19:71–72

Wang JL, Ge HB, Peng SQ, Zhang HM, Chen PL, Xu JR (2004) Transformation of strawberry (*Fragaria ananassa* Duch.) with late embryogenesis abundant protein gene. J Hort Sci Biotechnol 79:735–738

Wang TC, Ran YD, Atkinson RG, Gleave AP, Cohen D (2006) Transformation of *Actinidia eriantha*: a potential species for functional genomics studies in *Actinidia*. Plant Cell Rep 25:425–431

Wawrzynczak D, Michalczuk L, Sowik W (2005) Modification in indole-3-acetic acid metabolism, growth and development of strawberry through transformation with maize 1AA-glucose synthase gene (*iaglu*). Acta Physiol Plant 27:19–27

Welander M, Pawlicki N, Holefors A, Wilson F (1998) Genetic transformation of the apple rootstock M26 with the *rolB* gene and its influence on rooting. J Plant Physiol 153:371–380

Wen XP, Pang XM, Matsuda N, Kita M, Inoue H, Hao YJ, Honda C, Moriguchi T (2008) Overexpression of the apple spermidine synthase gene in pear confers multiple abiotic stress tolerance by altering polyamine titers. Transgenic Res 17:251–263

Wen XP, Ban Y, Inoue H, Matsuda N, Moriguchi T (2009) Aluminum tolerance in a spermidine synthase-overexpressing transgenic European pear is correlated with the enhanced level of spermidine via alleviating oxidative status. Environ Exp Bot 66:471–478

Wong WS, Li GG, Ning W, Xu ZF, Hsaio WLW, Zhang LY, Li N (2001) Repression of chilling-induced ACC accumulation in transgenic citrus by overproduction of antisense 1-aminocyclopropane-1-carboxylate synthase RNA. Plant Sci 161:969–977

Yabor L, Arzola M, Aaragon C, Hernandez M, Arencibia A, Lorenzo JC (2006) Biochemical side effects of genetic transformation of pineapple. Plant Cell Tissue Organ Cult 86:63–67

Yamakawa Y, Chen L-H (1996) *Agrobacterium rhizogenes* -mediated transformation of kiwifruit (*Actinidia deliciosa*) by direct formation of adventitious buds. J Jpn Soc Hort Sci 64: 741–747

Yamashita H, Daimon H, Akasaka-Kennedy Y, Masuda T (2004) Plant regeneration from hairy roots of apple rootstock, *Malus prunifolia* Borkh. var. *ringo* Asami, strain Nagano no. 1, transformed by *Agrobacterium rhizogenes*. J Jpn Soc Hortic Sci 73:505–510

Yang ZN, Ingelbrecht IL, Louzada E, Skaria M, Mirkov TE (2000) *Agrobacterium*-mediated transformation of the commercially important grapefruit cultivar Rio Red (*Citrus paradisi* Macf.). Plant Cell Rep 19:1203–1211

Yao J-L, Wu J-H, Gleave AP, Morris BAM (1996) Transformation of citrus embryogenic cells using particle bombardment and production of transgenic embryos. Plant Sci 113:175–183

Yao J-L, Dong YH, Morris B (2001) Parthenocarpic apple fruit production conferred by transposon insertion mutations in a MADS-box transcription factor. Proc Natl Acad Sci USA 98: 1306–1311

Yazawa M, Suginuma C, Ichikawa K, Kamada H, Akihama T (1995) Regeneration of transgenic plants from hairy root of kiwi fruit (*Actinidia deliciosa*) induced by *Agrobacterium rhizogenes*. Breed Sci 45:241–244

Yeh S-D, Bau H-J (2001) Characterization of a distinct pathotype of leaf-distortion mosaic virus capable of breaking the transgenic resistance of papaya conferred by the coat protein gene of Papaya ringspot virus. In: The second international symposium on biotechnology of tropical and subtropical species, Taipei, Taiwan, 5–9 Nov 2001, p 21 (Abstract)

Yeh S-D, Bau H-J, Cheng Y, Yu T, Yang J (1998) Greenhouse and field evaluations of coat-protein transgenic papaya resistant to papaya ringspot virus. Acta Hortic 461:321–328

Ying ZT, Yu X, Davis MJ (1999) New method for obtaining transgenic papaya plants by *Agrobacterium*-mediated transformation of somatic embryos. Proc Florida State Hortic Soc 112:201–205

Youssef SM, Jimenez-Bermudez S, Bellido ML, Martin-Pizarro C, Barcelo M, Abdal-Aziz SA, Caballero JL, Lopez-Aranda JM, Pliego-Alfaro F, Munoz J, Quesada MA, Mercado JA (2009) Fruit yield and quality of strawberry plants transformed with a fruit specific strawberry pectate lyase gene. Sci Hortic 119:120–125

Yu C, Huang S, Chen C, Deng Z, Ling P, Gmitter FG (2002) Factors affecting *Agrobacterium*-mediated transformation of sweet orange and citrange. Plant Cell Tissue Organ Cult 71:147–155

Zhang Z, Sun AJ, Cong Y, Sheng BC, Yao QH, Cheng ZM (2006) *Agrobacterium*-mediated transformation of the apple rootstock *Malus micromalus* Makino with the *rolC* gene. In Vitro Cell Dev Biol-Plant 42:491–497

Zhu YJ, Fitch MMM, Moore PH (1999) Improving fungal disease resistance in papaya. In Vitro Cell Dev Biol-Plant 35:63-A (Abstract)

Zhu LH, Ahlman A, Li XY, Welander M (2001a) Integration of the *rolA* gene in the genome of the vigorous apple rootstock A2 reduced plant height and shortened internodes. J Hortic Sci Biotechnol 76:758–763

Zhu LH, Holefors A, Ahlman A, Xue Z-T, Welander M (2001b) Transformation of the apple rootstock M.9/29 with the *rol B* gene and its influence on rooting and growth. Plant Sci 160:433–439

Zhu YJ, Tang CS, Fitch M, Moore P (2001c) Transformation with a pathogen inducible stilbene synthase gene for increased fungal resistance in papaya. In Vitro Cell Dev Biol-Plant 37(26-A) (Abstract)

Zhu YJ, Tang CS, Fitch M, Moore P (2001d) Increased fungal resistance in papaya by transformation of a pathogen-inducible stilbene synthase gene. In: Second international symposium on biotechnology of tropical and subtropical species, Taipei, Taiwan, 5–9 Nov 2001, p 22 (Abstract)

Zhu YJ, Tang CS, Ferreira S, Fitch M, Moore P (2002) Genetic improvement of *Carica papaya* for disease resistance. In: 10th international association of plant tissue culture and biotechnology congress, plant biotechnology 2002 and beyond, Orlando, FL, 23–28 June 2002, Addendum booklet, p 2 (Abstract)

Zhu YJ, Tang CS, Ferreira S, Fitch M, Moore P (2003a) Genetic improvement of *Carica papaya* for disease resistance. Proc Int Congr Plant Pathol 2(P12.3):169 (Abstract)

Zhu YJ, Tang CS, Moore P (2003b) Metabolic engineering of phytoalexins from the *Vitaceae* improves antifungal activity in a tropical plant. In Vitro Cell Dev Biol-Plant 39:24-A (Abstract)

Zhu YJ, Agbayani R, Jackson MC, Tang CS, Moore P (2004) Expression of the grapevine stilbene synthase gene VST1 in papaya provides increased resistance against diseases caused by *Phytophthora palmivora*. Planta 220:241–250

Zhu YJ, Agbayani R, McCafferty H, Albert H, Moore P (2005) Effective selection of transgenic papaya plants with the PMI/Man selection system. Plant Cell Rep 24:426–432

Zhu YJ, Agbayani R, Moore P (2007) Ectopic expression of Dahlia merckii defensin DmAMP1 improves papaya resistance to *Phytophthora palmivora* by reducing pathogen vigor. Planta 226:87–97

Zhu LH, Li XY, Welander M (2008) Overexpression of the *Arabidopsis gai* gene in apple significantly reduces plant size. Plant Cell Rep 27:289–296

Chapter 6
Genomics of Temperate Fruit Trees

María José Aranzana, Iban Eduardo, Santiago Vilanova, Carlos Romero, and Ana Montserrat Martín-Hernández

Abstract Temperate fruit species are those adapted to climates in the middle latitudes and according to their fruits are usually classified as fruit and nut trees, vine, and berries. Most of the best-known temperate tree crops are members of the Rosaceae family, including pome fruits (e.g., apple, pear, loquat, and quince) and stone fruits (e.g., apricot, cherry, peach, and plum). During the last decades, many efforts have been made in the field of genetics and genomics of these species. Such efforts have generated important tools like molecular markers, genetic and physical maps, SNP arrays and microarrays, among others, ready to be applied in variability studies, in breeding programs, and in general in the better understanding of the genetics behind agricultural important traits in these species. Here we present a brief summary of genomic studies on the most important temperate fruit trees grown for edible use.

Keywords Temperate fruit trees • Rosaceae • Pome fruits • Stone fruits • Genetic variability • Fruit trees self-incompatibility • Marker assisted selection • Fruit trees breeding

---

M.J. Aranzana, Ph.D. (✉) • I. Eduardo, Ph.D. • A.M. Martín-Hernández
Department of Plant Genetics, IRTA – Centre for Research in Agricultural Genomics, CSIC-IRTA-UAB-UB, Campus UAB. Bellaterra (Cerdanyola del Vallès), 08193 Barcelona, Spain
e-mail: mariajose.aranzana@irta.cat

S. Vilanova
Instituto Universitario de Conservación y Mejora de la Agrodiversidad Valenciana, Universidad Politécnica de Valencia, Valencia 46022, Spain

C. Romero
Citricultura y Producción Vegetal, Instituto Valenciano de Investigaciones Agrarias, Moncada, Valencia 46113, Spain

R.J. Schnell and P.M. Priyadarshan (eds.), *Genomics of Tree Crops*,
DOI 10.1007/978-1-4614-0920-5_6,

Introduction

Temperate fruit species are those adapted to climates in the middle latitudes, requiring a chilling period to flower and complete their life cycle. Temperate fruit crops are usually classified as fruit and nut trees, vines, and berries. Here we present a brief summary of genomic studies on the most important temperate fruit trees grown for edible use: peach, cherry, plum, apricot, apple, pear, loquat, quince, and olive.

Most of the world's best-known temperate tree crops are members of the Rosaceae family, classified as pome and stone depending on their fruit morphology. In pome fruits (e.g., apple, pear, loquat, and quince), the succulent edible part is the mesocarp, formed by the expanded ovary and containing seeds. In stone fruits (e.g., apricot, cherry, peach, and plum), the fleshy part surrounds a shell of hardened endocarp with a single seed inside.

Temperate fruits are rich sources of vitamins, fibers, minerals, and other compounds associated with a beneficial effect on health. They are consumed fresh or dried; processed in juices, jams, and sauces; used for cakes; or, as with olives, pressed for oil. In addition, their juice can be fermented to obtain alcoholic beverages such as apple cider or brandy (e.g., from apple and plum).

Apple

Malus × *domestica* Borkh. (*M. pumila*). There are many species of *Malus*, but the domesticated types appear to be derived from a native species from the Kazakhstan area, *M. sieversii*. Apples are grown in all temperate and subtropical countries, with minor production at high altitudes in tropical countries. One of the main advantages of apple is its adaptability to a wide range of climate and soil conditions. With 4.8 MHa and a production of 69 Mt in 2008, apple is the fourth major fruit crop grown worldwide (FAOSTAT 2010). The four most important producers are China, the USA, Iran, and Turkey, accounting for more than 50% of world production. More than 10,000 apple cultivars have been described, with a tremendous genetic variability, although only 10–20 are grown on a commercial scale, with 'Red Delicious' and 'Golden Delicious' the most commonly cultivated varieties in the world. The genome of the cultivated apple is about 750 Mb in size, with 17 chromosome pairs. Most of the cultivars are diploid.

Pear

There are three economically important pear species: *P. communis* L. (European pear), *P. pyrifolia* (Burm) Nakai (Japanese pear or Nashi), and *P. ussuriensis* Maxim (Chinese pear), eaten when crisp. Pears are closely related to apples and have similar uses, but they are less popular, with an annual world production of less than 21 Mt in 2008, covering 1.73 MHa (FAOSTAT 2010). China, Argentina, Italy, and the USA are the principal producers, accounting for 65% of world production. Most of the *Pyrus* species are diploid $2n = 2x = 34$.

Loquat

Eryobotria japonica (Thunb.) Lindl. is native from China, introduced into Japan and more recently to the Mediterranean regions, where it is acquiring some importance. China, Japan, Pakistan, and Israel are the main producers, with Spain by far the main exporter (Caballero and Fernández 2004).

Quince

Although two closely related genera, *Cydonia and Chaenomeles*, are referred to as quince, *Cydonia oblonga* Mill. is the one commonly used for fruit production. Most varieties are too astringent and sour to be eaten raw; hence, they are usually cooked to make jam, jelly, and marmalade. Some quinces are also used as rootstocks. Annual quince world production was around 0.5 Mt in 2008 (FAOSTAT 2010).

Peach

Prunus persica (L.) Batsch. is native to China, and from there spread worldwide, becoming adapted to local conditions. In the middle of the nineteenth century, a few Chinese peach varieties of good quality were introduced into North American and intensively used in breeding programs. Most of the current American and European commercial varieties are their descendants, resulting in a reduction of genetic variability. Despite this scarce genetic variability, new varieties arise every year. Peach cultivars have a wide variety of fruit morphology, in most cases caused by single gene mutations: the smooth skin of nectarines, the flat shape of fruits, or the yellow or white color of the fruit flesh. Fruits are eaten raw and are processed for canning or used in jams. In 2008, world peach production reached 18 Mt covering a surface of 1.6 MHa (FAOSTAT 2010). China, Italy, Spain, and the USA are the principal producers, accounting for almost 67% of world production. Peaches are self-compatible. The cultivated peach is a diploid species ($2n=2x=16$).

Cherries

Cherries belong to more than 50 species of the *Prunus* genus. The edible ones are included in the species *avium* and *cerasus*. *P. avium* L. is a diploid species ($2n=2x=16$) native to the Caucasus and includes sweet cherry trees, cultivated for human consumption, and wild cherry trees, for their timber. *P. cerasus* L., a tetraploid ($2n=4x=32$), is thought to be the result of a natural cross between wild *P. avium* L. and *P. fruticosa* Pall (Horvath et al. 2008). Sour cherries are principally used in pies and pastries. In 2008, world annual production was about 1.9 Mt of sweet cherries and 1.15 Mt of sour cherries, covering 0.4 and 0.2 MHa, respectively (FAOSTAT 2010). The main production areas for sweet cherry are Turkey, the USA, Iran, and Italy; whereas Russia, Turkey, Ukraine, and the USA are, in this order, the principal producers of sour cherry.

Plum

Plums belong to species of European, Asian, and American origin. European plums belong to *Prunus domestica* L. (European plum) and *P. insititia* L. (damson plum), both hexaploid ($2n=6x=48$). Asian plums are commonly included in *P. salicina* Lindl. (Japanese plum) and *P. simonii* Carrière (apricot plum) species. American plums arose from numerous species such as *P. americana* Marshall, *P. munsoniana* W. Wight & Hedrick, and *P. angustifolia* Marshall, but none are widely cultivated. *P. domestica* and *P. salicina* are the major plum species in cultivation today, with cultivation exceeding 2.4 MHa and more than 10 Mt, worldwide, in 2008 (FAOSTAT 2010). China, Serbia, the USA, and Romania are the main producers. Plums are consumed raw or dried as well as processed into jelly and jam.

Apricot

Apricots arose from the section Armeniaca within the genus *Prunus*. Although the section Armeniaca comprises five separate species, the common apricot is *P. armeniaca* L. China and Central Asia are the two primary centers of apricot origin. It has been suggested that some cultivars originated directly from the primary centers, while others may have arisen from the hybridization of genotypes from Near Eastern countries. Genetic variability in plum is much higher than that observed in peach commercial varieties (Mnejja et al. 2005), more so in central Asia and the Near East than in occidental varieties (Romero et al. 2003; Pedryc et al. 2009). Apricots are diploid ($2n=2x=16$).

Olives

Olive (*Olea europaea* L.) is one of the most important fruit crops in the Mediterranean area. It appears to be native to the eastern Mediterranean basin, northern Iran, and south of the Caspian Sea and is one of the first domesticated trees, probably grown in the Near East about 5,500–5,700 years ago (Zohary and Hopf 1994). Olive is cultivated mainly for oil but also as table olives. In 2008, the world production of olives reached 18 Mt, covering almost 11 MHa, with an oil production of 2.9 Mt (FAOSTAT 2010). It is a diploid species ($2n=46$), and more than 2,000 cultivars have been described, with great genetic variability. The existing cultivars are highly variable in fruit oil content, fruit size, canopy shape, and adaptation to local environmental conditions (Bartolini et al. 1998).

Here we aim to give an overview of the genetics and genomics of temperate fruit trees. Most of the fruit species included here belong to the Rosaceae family. A complete and extensive review on Rosaceae genetics and genomics can be found in Folta and Gardiner (2009).

Genetic Diversity in Temperate Fruit Trees

The need to characterize and preserve the genetic variability of temperate fruit tree species is crucial for future breeding projects. Although traditionally cultivars have been, and are still, morphologically characterized, molecular techniques have been proven to be more efficient in distinguishing among accessions, establishing genetic similarities and tracking the processes of domestication. Different techniques have been broadly applied in temperate fruit crops.

The first type of molecular markers used to study diversity in plants and particularly in temperate fruit trees were isozymes. Despite the limitations of the technique (i.e., low number of markers and differential expression due to environmental factors), they provided a useful tool for cultivar identification and important information about variability levels within and among species. For example, isozyme markers were used to detect the low level of variability present in peach in comparison with other *Prunus* species (Messeguer et al. 1987). These markers were soon replaced by other, more informative molecular markers with greater rates of polymorphism, including restriction fragment length polymorphisms (RFLPs), random amplified polymorphic DNA (RAPDs) and amplified fragment length polymorphisms (AFLPs). For example, these marker types have been used to hypothesize about separated domestication events in apple (Ishikawa et al. 1992) and to detect bottleneck events in the recent history of apricot. However, the discovery of microsatellites or simple sequence repeats (SSRs), genetic regions with highly polymorphic tandem repeats of di- to tetranucleotide sequence motifs and broad distribution in the genome whose variation can be captured by the PCR technique, have subsequently replaced these molecular markers. Since the late 1990s, SSRs have been the marker of choice for diversity studies in temperate fruit trees.

SSRs were first used in apple for cultivar identification by Guilford et al. (1997). One of the drawbacks of these markers is the need for previous sequence information. To obtain this sequence information, the authors first developed a library enriched for (GA) repeats from a 'Royal Gala' apple cultivar. By sequencing some of the obtained fragments, they developed 28 SSRs which were used for cultivar fingerprinting. Since then, numerous SSRs have been developed and used in apple for cultivar identification as well as for diversity analysis (Gianfranceschi et al. 1998; Hokanson et al. 1998, 2001; Liebhard et al. 2002; Oraguzie et al. 2005; Kitahara et al. 2005; Guarino et al. 2006; Ramos-Cabrer et al. 2007; Pereira-Lorenzo et al. 2008; Garkava-Gustavsson et al. 2008; Cavanna et al. 2008; Gharghani et al. 2009; Volk et al. 2009; Richards et al. 2009).

A major advantage of SSR markers is their transferability between closely related species. Apple-derived SSRs combined with some designed from *Pyrus* SSR libraries have been used to characterize wild and cultivated *Pyrus* species (Yamamoto et al. 2001a, 2002a; Kimura et al. 2002; Ghosh et al. 2006; Volk et al. 2006; Bao et al. 2007; Katayama et al. 2007; Brini et al. 2008), quince varieties (Yamamoto et al. 2004a; Dumanoglu et al. 2009), and loquat (Soriano et al. 2005; Watanabe et al. 2008). More recently, Gisbert et al. (2009a, b) developed and characterized the first 21 loquat polymorphic microsatellite loci, from an enriched genomic library, to study the genetic relationships among 83 loquat accessions belonging to the European loquat germplasm collection, from different countries.

Similarly, a large number of SSRs have been developed from genomic libraries and EST sequences of *Prunus* species, including *persica* (Cipriani et al. 1999; Sosinski et al. 2000; Testolin et al. 2000; Aranzana et al. 2002; Dirlewanger et al. 2002; Yamamoto et al. 2002b), *avium* (Cantini et al. 2001; Clarke and Tobutt 2003), *cerasus* (Downey and Iezzoni 2000; Cantini et al. 2001), *salicina* (Mnejja et al. 2004), *domestica* (Decroocq et al. 2004), and *armeniaca* (Lopes et al. 2002; Messina et al. 2004; Hagen et al. 2004; Vilanova et al. 2006b). Due to the high transferability of SSRs between closely related species, these markers have been used to study and compare different species within a genus (Dirlewanger et al. 2002; Wünsch and Hormaza 2002; Mnejja et al. 2005; Wünsch 2009). Transferability within a family has also been addressed by Gasic et al. (2009), who amplified apple-derived SSRs in other Rosaceae species, including the temperate tree species pear, peach, apricot, European and Japanese plums, and sweet and sour cherries. Transferability ranged from 59% in pear to 25% in apricot. Among the three Rosaceae subfamilies, Maloideae, Rosoideae, and Prunoideae, transferability was considered high at 59%, 53%, and 56%, respectively. Mnejja et al. (2010) also tackled this question and studied the transferability of *Prunus* SSRs across rosaceous crops. Of the *Prunus* SSRs studied, 37.9% of the SSR primers amplified fragments in all non-*Prunus* rosaceous species tested, with 44.8% amplification in apple and 38.6% in pear. The authors also assessed the levels of polymorphism detected by the markers, finding that *Prunus* SSRs were more polymorphic in *Prunus* (mean heterozygosity 63.9%) than in non-*Prunus* species (16.3%).

In olive, SSRs derived from enriched genomic libraries have been developed (Rallo et al. 2000; Cipriani et al. 2002; Carriero et al. 2002) and used to study genetic diversity and relationships among olive cultivars (Sefc et al. 2000; Rallo et al. 2003; Belaj et al. 2003, 2007; Lopes et al. 2004; Gil et al. 2006; Breton et al. 2008).

Recent advances in sequencing technologies have resulted in the massive identification of single nucleotide polymorphisms (SNPs). They are broadly distributed throughout the genome and are reported to be the major contributors to genetic variation. Although thousands of SNP markers are widely used in animal and human genome analysis, their use in plants is still at an early stage. For example, SNPs have already been used for variability analysis in Japanese apricot (Fang et al. 2006) and olive (Reale et al. 2006; Consolandi et al. 2007, 2008; Muleo et al. 2009). Moreover, the recent availability of expressed sequence tag (EST) datasets has made possible the discovery of putative SNPs *in silico*, using bioinformatics tools that permit the selection of polymorphisms which occur multiple times within a set of aligned sequences. Databases like ESTree (http://www.itb.cnr.it/estree/) (Lazzari et al. 2005) for peach and the genomic database for Rosaceae (GDR, http://www.rosaceae.org/) now contain thousands of SNPs useful for diversity analysis as well as for mapping. Chagne et al. (2008) developed a set of SNP markers from the 350,000 EST sequences of apple accessions.

Initially SNPs were tested in cultivars using low-throughput analysis methods which do not permit massive SNP genotyping. Nowadays, the large amount of SNPs developed thanks to next-generation sequencing technologies demands the development of high-throughput SNP genotyping technologies. Recently Micheletti et al. (2010) genotyped 200 apple varieties with several hundred SNPs, originally discovered within the FEM-IASMA 'Golden Delicious' genome sequencing project, using the Illumina SNPlex Technology. They found that 41% of the 'Golden Delicious' SNPs were transferable to cultivated apple accessions, whereas a much lower rate was observed with most of the wild species.

In general, diversity studies with molecular markers have been oriented to characterize and measure the genetic variability in the cultivated germplasm as well as to find new sources of genetic variability. Wild relatives or closely related species are promising donors of such new variability. This germplasm can be screened for a clearly defined character that can be introduced into elite material through crosses. However, this approach has only been feasible, so far, to introduce characters or traits controlled by one or a few genes due to the requirement for three or more generations of backcrossing to remove genetic drag. Despite the long generation time of fruit tree species, interspecific crosses have been widely used especially in plum and for rootstock development (reviewed in Layne and Sherman 1986). However, they have been used more moderately in cultivar development.

Currently, plant breeders face the challenges of increasing crop productivity and the uncertain problems that may arise from climatic change which will require the wise use of plant genetic resources. Although national collections preserve and characterize germplasm, in most cases the genetic information is poor and difficult to access. Current efforts are focused on managing such resources. A collaborative program among European countries, the European Cooperative programme Plant Genetic Resources (ECPGR), was created in 1980 to facilitate the long-term *in situ* and *ex situ* conservation and utilization of plant genetic resources in Europe. The ECPGR aims to strengthen links between all plant genetic resources programs in Europe and promote the integration of countries that are not members of the ECPGR. The ECPGR fruit network includes temperate fruit species of *Malus*, *Pyrus,* and *Prunus*. The information on these accessions is centralized in the ECPGR European Central Crop Databases (ECCDBs).

Molecular Markers and Linkage Maps

One of the main goals in plant genetics is to identify genes affecting important agronomic traits such as pest resistance, fruit quality, and growth habit. Although some economically important traits are controlled by major genes, most variation is explained by quantitative trait loci (QTLs), making the study more complex.

Gene mapping has this as its goal. Although gene mapping has been used since the early 1990s, it was not developed in fruit trees until the late twentieth century. Initially, the number of available markers was small, so the principal barrier to gene mapping was the large genomic distance between the gene and marker. Since the explosion in molecular genetics techniques, it has been possible to cover the genome with a considerable amount of informative markers.

Linkage maps in temperate fruit trees were first developed in apple (Hemmat et al. 1994; Maliepaard et al. 1998), peach (Chaparro et al. 1994), sweet cherry (Stockinger et al. 1996), and sour cherry (Wang et al. 1998). In pear, the first genetic map was published in 2001 (Iketani et al. 2001), whereas the first map in olive was published in 2003 (de la Rosa et al. 2003). Initially, they were mainly constructed using RAPD molecular markers because these were easy to develop and cost-efficient. However, the main drawback of RAPDs is that they are dominant, not very reproducible, and not transferable to other progenies. Later on, several maps were constructed using high-quality transferable molecular markers such as RFLPs or SSRs (reviewed in Arús et al. 2006). Today's sequencing methods, mainly next-generation technologies, provide an enormous amount of SNP markers. These are particularly appropriate for map construction given that SNPs are the most abundant form of variation found in the genomes and suitable for multiplexing, reducing the cost per data point. Indeed, one of the main goals in Rosaceae research nowadays is SNP discovery and the development of a high-throughput genotyping platform based on SNP markers that can be used in several temperate fruit tree species to speed up genetic map construction, synteny studies, and MAS, among others.

Some of the developed maps are considered as reference maps of their respective species or genera. This is the case of 'Fiesta' × 'Discovery' in apple (Maliepaard et al. 1998; Liebhard et al. 2003), 'Texas' × 'Earlygold' in *Prunus* (Joobeur et al. 1998; Aranzana et al. 2003), and more recently the maps developed in European and Japanese pears (Yamamoto et al. 2007, 2009). Highly saturated reference maps are an important genetic tool because (1) they are usually built in highly polymorphic populations with a high probability of finding polymorphisms, (2) they provide a common terminology for linkage groups, (3) they allow selection of markers in specific genomic regions that facilitate the construction of framework maps in other populations or map comparison with other species or genera, (4) they allow the localization of major genes and QTLs in the same map, and (5) bin mapping strategy can be used to increase the mapping efficiency using a limited number of plants losing a minimum of resolution (Howad et al. 2005). The bin mapping strategy has been recently applied in peach and apple to map a high number of markers, including 51 BAC-end sequences (Han et al. 2009) and 93 SNPs (Celton et al. 2009) in apple and 613 RosCOS markers (Cabrera et al. 2009) and 133 candidate genes (Ogundiwin et al. 2009) in peach.

Linkage maps of several temperate fruit tree species, together with other important Rosaceae crops, such as raspberry, strawberry, and rose, can be viewed and compared at the GDR (http://www.rosaceae.org/).

A particularity of temperate fruit species such as apple, pear, apricot, sweet or sour cherries, and olive is the obligatory outbreeding reproduction system in most of their varieties, such that F_1 populations have to be used. Peach is an exception, being self-compatible so F_2 populations can be generated. Genetic maps from F_1 populations are widely constructed through the "pseudotest cross" strategy (Ritter et al. 1990). Briefly, this strategy consists of the development of one genetic map for each parent, using the markers segregating in both parents as bridges to align both maps.

In 1997, Tanksley and McCouch pointed out the potential role of genome mapping for efficiently managing the genetic diversity of wild relatives, especially if combined with the advanced backcross strategy. The procedure, although time consuming, is conceptually easy and was proposed by Wehrhahn and Allard (1965). Basically an elite variety is crossed with an exotic one and the hybrid backcrossed with the same elite material. This produces a collection of plants with a small number of introgressions from the exotic in the elite genetic background. Combining this approach with the use of genetic markers, the number, extension, and position of the introgressions can be detected. Furthermore, molecular markers may make it possible to decrease the number of generations needed to obtain a small number of introgressions and recover the recurrent parent as fast as possible. This strategy has been widely used in herbaceous crops, and it is starting to be used in fruit trees like peach, where the development of a near-isogenic line (NIL) collection of almond chromosome fragments in the peach genome is in process (Arús et al. 2010).

Important Agronomic Traits and Molecular Markers Available

In temperate fruit tree species belonging to the Rosaceae family, such as apple, pear, and stone fruits, the most important traits for breeding programs embrace pest and disease resistance, fruit quality (including texture, size, color, and flavor), productivity, plant architecture, abiotic stress, self-incompatibility, blooming time, and chilling requirement.

These fruit tree species are adapted to seasonal changes and have the ability to halt meristem activity in the fall and recover it in the spring. The chilling requirement refers to the minimum period of cold weather that the plant needs to recover meristem activity and flower. Heat requirement is another factor involved in meristem activity but is not as well understood or quantified as chilling requirement. Both traits, especially the former, are very important because they determine blooming time. Early blooming in regions with cold winters can produce major losses due to frost damage. Furthermore, reducing chilling requirement can increase the production of temperate fruit trees in subtropical regions. A clear example of this is the *Evergrowing* peach mutant, a nondormant genotype from southern Mexico that fails to cease growth and enter dormancy under dormancy-inducing conditions (Jimenez et al. 2010).

Some of the earlier named traits are controlled totally or in a high proportion by a single gene, although most are controlled by several genes or quantitative trait loci (QTLs). Moreover, some traits may behave as quantitative in some populations and as qualitative in others.

One of the crucial aspects in breeding strategies is phenotyping. Whereas some traits are very easy to score, others require expensive and advanced techniques and very specialized labor. For example, fruit quality can include very different traits including sensory aspects such as flavor and texture, nutritional aspects such as vitamin and antioxidant contents, processing aspects such as ease of skin and seed removal or juiciness, and postharvest aspects such as shelf life and chilling injury. Most of the studies in fruit quality traditionally evaluate the so-called standard quality which includes fruit size, fruit weight, fruit shape, soluble solid content, acidity and pH, skin and flesh colors, and flesh texture, based on the penetrometer. Studies with other fruit quality traits such as individual chemical compound content have become more usual recently, thanks to advances in metabolomic studies. Fruit acidity can easily be scored as pH or titratable acidity of the fruit juice or using complex chromatographic protocols to identify and quantify the individual acids. The phenotyping of some traits are even starting to be revised. For example, fruit firmness has been traditionally phenotyped using the penetrometer, but this does not take into account the speed of softening; international fruit consortiums are starting to talk about standardizing phenotypic methods (http://www.rosaceae.org/node/227) to allow a more reliable comparison of the results from different studies.

Here we present a summary of the different traits for which linked markers have been identified using bulk segregant analysis (BSA), gene mapping, or QTL analysis in different species.

In apple, the best studied traits are pest and disease resistance, specifically the fungal diseases apple scab (*Venturia inaequalis*) and powdery mildew (*Podosphaera leucotricha*) and the bacterial fire blight (*Erwinia amylovora*). Molecular markers that are tightly linked to major scab resistance genes have been identified and are being used for pyramiding in the same background (Bus et al. 2009). In addition, novel resistance genes have been recently investigated to cover new durable resistances (Bus et al. 2010), and molecular markers closely linked to five different sources of powdery mildew resistance (Gardiner et al. 2007), to two different kinds of aphids, to the woolly apple aphid (*Eriosoma lanigerum* Hausn.) (Bus et al. 2008) and to the leaf-curling aphid (*Dysaphis devecta* Wlk.), have been described (Stoeckli et al. 2008).

Other apple traits for which linked molecular markers have been identified include: self-incompatibility and acidity (Maliepaard et al. 1998), the presence of root suckers (Weeden et al. 1994), fruit skin color (Cheng et al. 1996), flesh color (Chagne et al. 2007), fruit acidity (Maliepaard et al. 1998), columnar habit (Tian et al. 2005), fruit allergens (Gao et al. 2005), fruit texture (Harada et al. 2000; Costa et al. 2008), vegetative budbreak (van Dyk et al. 2010), and aroma and volatile organic compound (VOC) content (Dunemann et al. 2009; Rowan et al. 2009a, b).

In pear, the major diseases are fire blight (*Erwinia amylovora*), pear scab (*Venturia naschicola* and *V. pyrina*), and black spot (*Alternaria alternata*). QTL analysis of fire blight revealed the quantitative nature of the trait (Dondini et al. 2004; Bokszczanin et al. 2009). In Japanese pear, molecular markers linked to the major resistance gene *Vnk* to *V. naschicola* have recently been identified (Terakami et al. 2006), and in European pear, two major QTLs for *V. pyrina* (Pierantoni et al. 2007) and another for *Dysaphys pyri* resistance (Evans et al. 2008). Molecular markers for self-incompatibility (Takasaki et al. 2004), fruit texture, and skin color have also been described (Inoue et al. 2006; Dondini et al. 2008).

Peach is one of the best genetically characterized fruit trees in the *Prunus* reference map; 28 major genes affecting important agronomic traits have been mapped: peach flesh color, petal color, plum pox virus resistance, nematode resistance (B), broomy plant habit (Br), flower shape, flesh color around the stone, anther color, polycarpel, blooming date, flesh adherence to stone, nonacid fruit, fruit skin pubescence, leaf shape, dwarf plants, male sterility, fruit skin color, leaf color, fruit shape, self-incompatibility, nematode resistance from myrobalan plum, leaf gland shape, and resistance to powdery mildew. QTL analysis in peach has also been carried out, using intra- and interspecific populations segregating for fruit quality (Abbott et al. 1998; Dirlewanger et al. 1999; Quarta et al. 2000; Etienne et al. 2002; Quilot et al. 2004; Ogundiwin et al. 2009), chilling requirement and blooming date (Fan et al. 2010), and tree short life syndrome (Blenda et al. 2007).

The major pest and diseases in peach are brown rot (*Monilinia* sp.), powdery mildew (*Sphaerotheca pannosa*, *Podosphaera clandestina*), leaf curl (*Taphrina deformans*), plum pox virus (PPV) (Decroocq et al. 2005; Rubio et al. 2010), and green aphids (*Myzus persicae*) which are the PPV vector.

Molecular markers have been identified for acidity (Boudehri et al. 2009), sweetness (Dirlewanger et al. 1999; Etienne et al. 2002), skin pubescence (Bliss et al. 2002; Dirlewanger et al. 1999), flesh texture (Peace et al. 2005), flesh color (Bliss et al. 2002), fruit shape (Dirlewanger et al. 1999, 2007), peach short life syndrome (Blenda et al. 2006), and aroma and VOC content (Eduardo I. unpublished). A European project (ISAFRUIT) has constructed genetic maps in two peach and two apricot populations for fruit quality QTL analysis (Illa et al. 2009).

The main breeding objective in apricot is to obtain varieties resistant to sharka, caused by the plum pox virus (PPV). An SSR linked to a major QTL for resistance to sharka (Lambert et al. 2007; Lalli et al. 2008; Soriano et al. 2008; Marandel et al. 2009) was first identified and then confirmed in another population (Pilarova et al. 2010), and even the origin of the resistance was located with molecular markers (Zhebentyayeva et al. 2008a). QTLs for chilling requirement and budbreak have been also described (Olukolu et al. 2009). Other important breeding objectives are climatic adaptation, fruit quality, introduction of self-compatible cultivars, chilling requirement, and resistance to diseases such as apricot chlorotic leaf roll, bacterial canker, and Monilinia (Hormaza et al. 2007).

No mapping information is yet available for plum (Esmenjaud and Dirlewanger 2007). The most studied trait in plum is resistance to PPV and to root-knot nematodes (RKN) in myrobalan plum and in interspecific plum hybrids used for rootstock development. Resistance to RKN, together with chlorosis and drought adaptation, water logging tolerance, graft compatibility, and rooting ability are the main objectives of *Prunus* rootstock breeding programs.

In cherry, molecular markers for self-incompatibility (Dirlewanger et al. 2004a), bloom and maturity date, fruit weight, and soluble solid content (SSC) have been described (Wang et al. 1998; Canli 2004; Zhang et al. 2009), self-incompatibility being the only trait with Mendelian inheritance.

In loquat, a RAPD marker for canker disease resistance was identified using the BSA strategy (Fukuda et al. 2005). In olive, important traits like early bearing, resistance to pests and to abiotic stresses such as frost and drought, limited alternate bearing, suitability to intensive culture and to mechanical harvesting, as well as high-quality production, in terms of both organoleptic characteristics of fruits and oils and high content of substances good for human health, are gradually being included in breeding programs (Fabbri et al. 2009).

Self-Incompatibility in Temperate Fruit Trees

Gametophytic self-incompatibility (GSI) is a widespread mechanism in flowering plants often controlled by a single multi-allelic locus, the *S*-locus, which prevents inbreeding and promotes outcrossing (de Nettancourt 2001). Among families including temperate fruit trees, GSI has been reported in Oleaceae and Rosaceae (Igic and Kohn 2001). Studies on the Oleaceae GSI are scarce, but there have been major advances in the last few years. However, there is still no consistent evidence of the nature of proteins controlling the mechanism (Serrano et al. 2010). In contrast, during the last two decades, significant progress has been made in understanding the molecular basis of GSI in Rosaceae (Yamane and Tao 2009).

In Rosaceae, as in Solanaceae and Scrophulariaceae, the *S*-locus is considered to contain at least two linked genes encoding pistil and pollen proteins involved in the specific recognition mechanism. The female *S*-determinant was first identified in *Nicotiana* as a glycoprotein with ribonuclease activity (*S*-RNases) in the pistils (McClure et al. 1989). *S*-RNases have also been found in *Pyrus* (Sassa et al. 1992), *Prunus* (Boskovic and Tobutt 1996), and *Antirrhinum* (Xue et al. 1996). *S*-RNases were shown to be essential for rejection of haploid pollen when the *S*-allele of the pollen matches either of the diploid pistil *S*-alleles (Huang et al. 1994; Xue et al. 1996; Sassa et al. 1997). The male *S*-determinant remained elusive but was then identified as an F-box pollen-expressed gene (named *SLF* or *SFB*) through the genomic analysis of the *S*-locus region of *Antirrhinum hispanicum* (Lai et al. 2002). Similar approaches then led to the identification of *S*-linked F-box genes in *Prunus* (Entani et al. 2003; Ushijima et al. 2003) and *Petunia* (Wang et al. 2004; Sijacic et al. 2004).

F-box proteins are involved in the ubiquitin/26S proteasome proteolytic pathway (Deshaies 1999), suggesting that the function of *S*-linked F-box genes may be to inactivate nonself *S*-RNases by proteolytic degradation once inside the pollen tubes (Ushijima et al. 2003; Ikeda et al. 2004). This hypothesis is supported by the interaction between the $AhSLF_2$ F-box pollen protein and self/cross *S*-RNases observed in *Antirrhinum* (Qiao et al. 2004). What is generally known as the GSI inhibitor model proposes that all *S*-RNases, regardless of their *S*-haplotype, enter the pollen tubes (Luu et al. 2000) and are all degraded except the self *S*-RNase that degrades the pollen RNA. However, the mechanism specifically protecting the self *S*-RNase is still unknown.

In contrast to this model, Goldraij et al. (2006) recently showed that *S*-RNases enter the pollen tubes sequestered in vacuoles. Moreover, they showed that self *S*-RNases remain stable in compatible and incompatible crosses, while the HT-B protein (McClure et al. 1999) is degraded in compatible crosses. These authors point out that the pollen endomembrane system has a key role in the GSI model and suggest that compartmentalization rather than *S*-RNase degradation is the basis to escape rejection in compatible pollinations. While all these findings have thrown light on the underlying GSI mechanism, they have raised new questions still to be explored.

Rosaceae is comprised of four different subfamilies: Prunoideae (Amygdaloideae), Maloideae, Rosoideae and Spiraeoideae, but GSI has only been studied at the molecular level in the first two. As mentioned above, in Prunoideae, and more specifically in *Prunus*, genomic approaches aimed at identifying the pollen-*S* gene were initiated at the beginning of the 2000s decade. In almond, a 70-Kb region of the *S*-locus (S_C specific haplotype) was sequenced completely using the shotgun strategy on three genomic clones (Ushijima et al. 2003). The analysis of this sequence allowed the identification of 12 ORFs including one with homology to almond *S*-RNases and two homologous to F-box proteins (ORF1 and ORF3). Finally, only the ORF3 has been shown to fulfill all the expected features for the pollen *S*-gene (tight linkage to the *S-RNase*, high level of *S*-haplotype-specific sequence polymorphism, and pollen-specific expression). Similarly, nucleotide sequence analyses of cosmid clones spanning the S_1 and S_7 specific haplotypes in Japanese apricot (*Prunus mume* Sieb. et Zucc.) showed the presence of the *S-RNase* and at least four independent F-box genes (Entani et al. 2003). Again, only one of these F-box genes fulfilled the conditions for the pollen *S*-determinant. These analyses were also useful to understand the role of the *S*-locus genomic structure in the GSI model. Distances between *S*-genes were found to be highly variable among the *S*-haplotypes, while comparative analysis revealed that the region around *S-RNases* exhibits extreme sequence diversity among *S*-haplotypes while the gene organization outside this region was highly collinear. Both features seem to contribute to the heteromorphism of the region, ensuring the tight association of *S-RNase* and *SFB* genes and avoiding recombination events that would compromise the GSI system. Subsequent studies confirmed these observations in *Prunus* species such as apricot (Romero et al. 2004), sweet cherry (Ikeda et al. 2005), and peach (Tao et al. 2007).

Attempts to isolate the pollen *S*-gene in Maloideae through the homology-based approach with *Prunus SFB* have been unsuccessful, probably due to sequence divergence between both subfamilies. This lack of results prompted Sassa et al. (2007) to sequence a 317-Kb region of the apple (*Malus* × *domestica* Borkh.) S^9-haplotype. Two closely related *F-box* genes were identified and named *SFBB* (*S-locus F-box brothers*). These authors also found two additional *SFBBs* by analyzing BAC clones containing the apple S^3-haplotype and detected three *SFBB* genes in the Japanese pear (*Pyrus pyrifolia* Nakai) S^4- and S^5-haplotypes. The *SFBB* genes are specifically expressed in pollen and show *S*-haplotype-specific polymorphism. However, in contrast with *SFB/SLF* genes found in *Prunus,* they were not single-copy but multiple, and therefore, it is still unclear whether *SFBB* is the pollen *S*-gene in Maloideae.

Regarding the *S*-locus, this is not the only difference found between Prunoideae and Maloideae, since *Prunus S-RNases* have been found to contain two introns while *Malus* and *Pyrus S-RNases* contain only one (Igic and Kohn 2001).

Besides the structural analysis of the *S*-locus, spontaneous and induced self-compatible mutants have been used extensively for its functional analysis. Stylar part mutations in the *S*-locus were first reported in *Lycopersicon peruvianum* (Royo et al. 1994) and *Pyrus serotina* (Sassa et al. 1997), revealing that the RNase activity of *S*-RNases is needed to inhibit pollen growth. *S*-RNase mutations found in *Prunus* have since confirmed earlier findings. In *Prunus cerasus*, a *Mu*-like element (2,600 pb) insertion upstream of the S_{6m}-RNase reduces its expression (Yamane et al. 2003). A similar mutation has been found in the *Prunus salicina* S^{e}-RNase, with its reduced expression leading to insufficient accumulation in the pistil, breaking the rejection mechanism (Watari et al. 2007). The *Prunus persica* S^{2m}-RNase has reduced stability as a consequence of the cysteine residue being replaced by a tyrosine in the C5 domain (Tao et al. 2007). Similarly, genomic analysis of self-compatible pollen-part mutants (PPM) in *Prunus* has found defective *SFB* genes in most cases, providing additional evidence for *SFB* as the pollen *S*-gene.

PPMs have been identified and analyzed in *Prunus avium* (Ushijima et al. 2004; Sonneveld et al. 2005; Marchese et al. 2007), *Prunus armeniaca* (Vilanova et al. 2006b), *Prunus cerasus* (Hauck et al. 2006), *Prunus mume* (Ushijima et al. 2004), and *Prunus persica* (Tao et al. 2007). The pollen function breakdown has usually been associated with insertions and deletions of different sizes in the *SFB* coding region, leading to a frameshift with a premature stop codon and the production of nonfunctional truncated proteins (Yamane and Tao 2009). In contrast to these findings, mutant analyses in Solanaceae suggest that the loss of function of the pollen *S*-gene would be lethal (Golz et al. 2001). In fact, most PPMs reported in this family are consistent with the competitive interaction phenomenon in which *S*-heteroallelic pollen containing two different *S*-alleles fails to function in self-incompatibility (Golz et al. 2001). However, this does not seem to be the case in *Prunus*, where it has been shown that *S*-heteroallelic pollen from the tetraploid acid cherry (*Prunus cerasus* L.) is self-incompatible (Hauck et al. 2006). Interestingly, competitive interaction has also been reported in Maloideae (Lewis and Modlibowska 1942). This observation, along with the structural differences described above, prompted some authors to speculate that the *Prunus* GSI mechanism may have differences compared to the mechanisms in Solanaceae and Maloideae (Tao et al. 2007). Further studies are needed to understand the reasons behind the differences and similarities of GSI systems.

In addition to the *S*-locus products, other pistil factors essential for the GSI, which do not contribute to *S* specificity, have already been found in Solanaceae but not yet in Rosaceae, such as HT-B and the 120-KDa glycoprotein (McClure et al. 2000). Similarly, breakdown of self-incompatibility has also been associated with mutations affecting GSI pollen modifier factors in Solanaceae (McClure et al. 2000) as well as in *Prunus avium* (Wünsch and Hormaza 2004), and *Prunus armeniaca* (Vilanova et al. 2006b). Other GSI-related proteins have also been recently identified in Solanaceae, such as thioredoxin h (Juárez-Díaz et al. 2006), NaStEP

(Busot et al. 2008), and NaPCCP (Lee et al. 2009). Most of these factors are currently being characterized and may provide crucial data to improve our understanding of the GSI molecular mechanism.

Beyond the huge recent interest in the GSI genetic, molecular, and evolutionary bases, their important applications in plant production and breeding should also be highlighted. In Rosaceae, breeders have been interested in the GSI trait since the beginning of the twentieth century, when the first intercompatible cultivars were identified in sweet cherry by means of cross-pollinations (Crane and Brown 1937). In the last decade, molecular genotyping, mainly based on *S-RNase* sequence polymorphisms, has progressively replaced controlled pollination, pollen tube growth tests, and the enzymatic assays used to determine *S*-genotypes (Burgos et al. 1998). Molecular techniques have accelerated the identification of new *S*-alleles in all Rosaceae species of agronomic interest, and currently more than 100 rosaceous *S-RNases* alleles from *Malus*, *Pyrus,* and *Prunus* genera can be found in the GenBank (Yamane and Tao 2009). Other rosaceous species such as loquat have only been analyzed very recently, but information on *S-RNase* alleles is already available (Carrera et al. 2009). The known number of intercompatibility groups has doubled in sweet cherry, Japanese plum, apricot, and apple, facilitating the election of suitable combinations for orchards (Goldway et al. 2007; Yamane and Tao 2009). The *S*-genotype of the pollinator trees may in fact significantly influence yields and fruit quality of the crops. Furthermore, *S*-genotyping is useful in designing crosses and selecting self-compatible hybrids produced in the Rosaceae breeding programs (Goldway et al. 2007).

It can be concluded that structural and functional genomics approaches have contributed decisively to the current knowledge on the GSI molecular basis in Rosaceae. There is new data complementing those from Scrophulariaceae and Solanaceae, and molecular *S*-genotyping has become a common tool for breeding and production purposes. Despite this progress, much work still remains to be done to complete our understanding of the GSI.

Synteny and Comparative Genomics

Originally, the term synteny designated the presence of several loci on the same chromosome. Today the concept of synteny has been expanded to address questions of homeology (McCouch 2001), i.e., the position of a linear series of markers in two or more species. Macrosynteny refers to the preservation of synteny in large portions of a chromosome and microsynteny in small genomic regions. Synteny analysis and comparative mapping can provide useful information on the evolutionary relationships between different species and be used to further saturate a specific genomic region or to predict the position of a gene based on its position in other species.

Molecular markers or genome sequences transferable between species are essential for synteny analysis. Several studies (some described in section "Genetic

Diversity in Temperate Fruit Trees") have evaluated the transferability of markers between temperate fruit trees (Yamamoto et al. 2001b, 2004b; Dirlewanger et al. 2002; Decroocq et al. 2003, 2004; Mnejja et al. 2004, 2010; Dondini et al. 2007; Sargent et al. 2007, 2009; Vendramin et al. 2007; Gasic et al. 2009). A very useful tool for comparative mapping in Rosaceae has recently been published by Cabrera et al. (2009). It consists of a very large, Rosaceae, conserved orthologous set (COS) also called RosCOS markers. This is a set of genes that have been highly conserved in sequence and copy number throughout Rosaceae evolution. Based on the analysis of the available Rosaceae ESTs, mainly from peach, apple, and strawberry, 857 RosCOS were selected for primer design that resulted in successful amplification of more than 50% of the RosCOS in each genus. RosCOS markers have been mapped on the TxE *Prunus* reference map and used in comparison mapping between *Rosaceae*, *Arabidopsis,* and *Populus* (Cabrera et al. 2009).

The use of transferable markers between *Prunus* species has revealed a complete collinearity of their genomes, and therefore, *Prunus* genera can be considered as a unique genetic entity (Arús et al. 2006). The only exceptions were found in the inversion of several contiguous markers in different maps, probably due to errors in the mapping process, and in a reciprocal translocation found in an interspecific cross of peach and almond (Jauregui et al. 2001). This translocation was cytogenetically confirmed.

Apple and pear genomes have also been compared using a small number of SSRs (Yamamoto et al. 2004b). All pear linkage groups were successfully aligned to the apple consensus map by at least one apple SSR, suggesting that positions and linkages of SSR loci are well conserved between pear and apple. Similar conclusions were drawn by Celton et al. (2009) with a higher number of markers. Both genera, *Prunus* and *Malus*, have also been compared using 30 common loci between the *Prunus* TxE and *apple* PxF maps (Dirlewanger et al. 2004a). These preliminary results indicate that there is a high level of synteny between the *Prunus* and *Malus* genomes, with each chromosome of *Prunus* corresponding to two of *Malus*.

The *Prunus* reference map has also been compared with other model species such as *Arabidopsis*, *Medicago,* and *Populus* for which full genome sequences are available. Initially, 177 DNA probes were used to compare the TxE *Prunus* map with the *Arabidopsis* genome (Dirlewanger et al. 2004a). These probes corresponded to 227 loci in *Prunus* and 703 in *Arabidopsis,* allowing the detection of 37 syntenic regions that covered 23% of the *Prunus* map and 17% of the *Arabidopsis* genome. Jung et al. (2006), using 475 peach ESTs anchored to *Prunus* genetic maps and 1,097 peach ESTs anchored to BAC contigs, detected microsynteny between the five *Arabidopsis* chromosomes and seven of the eight linkage groups of the *Prunus* genome. Later, Jung et al. (2009) found, with the previously described RosCOS markers, 6 large, 21 medium, and 20 small syntenic blocks between *Prunus* and *Populus*, suggesting that Rosaceae gene order could be predicted based on their order in poplar. They concluded that syntenic blocks between *Prunus* and *Arabidopsis* are smaller than those between *Prunus* and *Populus*.

Physical Maps (BAC Libraries) and Genome Sequencing Projects

After linkage maps, physical maps are the major components of structural genomics and allow genetic maps and genome sequences to be bridged. Physical maps consist of sequences, generally BACs or ESTs, anchored to genetic linkage maps. In the last few years, physical maps have been generated for apple (Han et al. 2007) and peach (Zhebentyayeva et al. 2008b).

In apple, the physical map was constructed using the agarose gel-based restriction fingerprinting method and FPC software. A total of 2,702 contigs were assembled from 82,503 BAC clones. Three BAC libraries were used, two from the cultivated apple, *M.* × *domestica* cv. 'GoldRush', and the other from the wild crabapple species, *M. floribunda* 821. Each library represented about 5–7x haploid genome equivalents. Nearly all BAC clones used for physical map construction were from the two 'GoldRush' libraries, except for a few BACs from the *M. floribunda* 821 library. The physical length of the physical map was estimated at 927 Mb and represented 10.5x haploid genome equivalents. This physical map is currently being integrated with the 'Royal Gala' × 'A689/24' genetic map, bin mapping SNP identified in the BAC-end sequences (Han et al. 2009).

In peach, the physical map was constructed using high-information content fingerprinting (HICF) and FPC software. In total, 2,138 contigs containing 15,655 clones were assembled. For this, two BAC libraries were used, 'Nemared', containing 44,160 BAC clones with an average insert size of 70 kb representing an 8.8-fold genome coverage (Georgi et al. 2002), and a second library based on a haploid strain (Plov 2-1N) derived from the peach rootstock Lovell, containing 34,560 clones with an average insert size of 80 kb, providing a 9.2-fold genome coverage (L. Georgi unpublished). The total physical length of all contigs was estimated to be 303 Mb, 104.5% of the peach genome. The framework physical map is anchored on the *Prunus* genetic reference map and integrated with the peach transcriptome map. The physical length of anchored contigs has been estimated to be 45.0 Mb, 15.5% of the genome.

Other BAC libraries have also been constructed in apricot (Vilanova et al. 2003) and in myrobolan plum (Claverie et al. 2004), using cultivars resistant to plum pox virus and to root-knot nematodes, respectively.

Currently, apple and peach physical maps are being used to anchor the genome sequences that are under construction (Sosinski et al. 2009). In peach, to avoid problems derived from the heterozygotic nature of the species, a double haploid line called 'Lovell', also used for the development of the physical map, has been selected for sequencing by the combination of the whole genome shotgun (WGS) strategy with the BAC by BAC strategy. The first draft of the peach genome sequence has been recently released (1st April 2010, www.rosaceae.org). In apple, two independent initiatives have been coordinated as an international consortium. The varieties used are 'Golden Delicious' and a double haploid derived from 'Golden Delicious'. In this case, appropriate strategies for assembling the heterozygous genome have to be applied, as has been done in grape (Velasco et al. 2007). A difficulty in assembling the apple genome is that apple is an old tetraploid species that now behaves as a diploid, so many duplicated regions are expected. A high-quality draft of the apple genome sequence was published in August 2010 (Velasco et al. 2010).

Gene Cloning: Positional Cloning and Candidate Gene Approach

Despite the usefulness of having molecular markers closely linked to loci responsible for phenotypic variations, the final aim is to generate markers on top of the polymorphisms responsible for these changes, to avoid recombination events between the markers and the loci. There are two main genomic approaches to identify such polymorphisms, the positional cloning strategy and the candidate gene approach. Positional cloning consists of narrowing down the region where the responsible locus is located till a relatively small piece of sequence is identified. Once there, sequence polymorphism can be identified and functionally validated. The candidate gene approach consists of mapping candidate genes putatively involved in the phenotypic variation of the trait and check if they colocate together. Functional validation is also required. Potential candidate genes could be genes involved in a specific metabolic pathway, orthologous genes that are involved in the same trait in other species, genes identified by comparative mapping as being close to the loci involved on the trait in other species, and genes identified in transcript profiling experiments. Both strategies have been used in temperate tree species, and several examples are available for both strategies.

One of the first examples of positional cloning in a temperate tree species was the apple scab resistance gene *Vf*, originating in the wild species *Malus floribunda* 821. This gene was introgressed through crosses in several apple varieties. To isolate the *Vf* gene, a bacterial artificial chromosome (BAC) library from *M. floribunda* 821 was constructed and *Vf*-linked SCAR markers were used to screen it (Xu et al. 2001). A BAC contig, of five overlapping BAC clones, was constructed spanning 290 kb of the *Vf* region. Four homologous receptor-like resistance genes co-segregated with the *Vf* locus in a segregant apple mapping population. These four *Vf* homologous genes were designated as *Vfa1*, *Vfa2*, *Vfa3*, and *Vfa4*. Deduced amino acid sequences revealed that each of the *Vf* gene homologues contained both an extracellular leucine-rich repeat (LRR) domain and a transmembrane (TM) domain. For functional analysis, transformed apple lines carrying *Vfa1*, *Vfa2,* and *Vfa4* genes were developed. It was shown that *Vfa3* was truncated and therefore was not used. Transformed lines expressing *Vfa4* were found to be susceptible to apple scab, whereas those expressing either *Vfa1* or *Vfa2* exhibited partial resistance to apple scab (Malnoy et al. 2008b). Transgenic plants under their own promoter control have also been constructed for *Vfa2*. Promoter length was found to influence both the constitutive transcription levels of *HcrVf2* in transgenic lines and the resistance level. Highly scab-resistant 'Elstar' and 'Gala' plants were obtained, proving that the *HcrVf2* gene controlled by its native promoter is effective in conferring resistance to apple scab, as with *Vf* introgressed in apple cultivars through classical breeding (Szankowski et al. 2009b).

Some other examples of positional cloning in temperate fruit tree species are the identification of the Sd-1 locus in apple that confers resistance to the rosy leaf-curling aphid (Cevik and King 2002), the *Evergrowing* mutation in peach (Bielenberg et al. 2008), and the cloning of the *D* locus, also in peach (Boudehri et al. 2009). The positional cloning approach will be greatly simplified once genome sequences become fully available.

An example to illustrate the candidate gene approach is the association between the transcription factor MdMYB10 and red flesh color in apple. In 2007, Espley et al. showed a clear correlation between the expression of the MYB10 transcription factor and red color levels in apple. MYB10 was then mapped, and it was shown that the MdMYB10 gene co-segregated with the Rni locus responsible for apple red color (Chagne et al. 2007). Finally, it has recently been shown that expression of MYB10 in transient assays in *Nicotiana benthamiana* can induce anthocyanin biosynthesis and that the R2R3 MYB10 transcription factor may also be responsible for fruit color in other fruit species (Lin-Wang et al. 2010).

Some other examples of candidate genes in fruit tree species include the MdMADS2.1 transcription factor associated with apple fruit flesh firmness (Cevik et al. 2010); the Md-Exp7 expansin associated with fruit softening, also in apple (Costa et al. 2008); the PpLDOX leucoanthocyanidin dioxygenase gene associated with cold storage-induced browning in peach (Ogundiwin et al. 2008); and the endoPG gene associated with freestone fruit, melting flesh, and mealiness in peach (Peace et al. 2005). Systematic approaches have also been used for *Prunus* resistance genes (Lalli et al. 2005) and fruit quality traits (Ogundiwin et al. 2009). Most of these examples show some co-segregation between the candidate gene and the phenotypic variation, but functional validation has not yet been done.

Marker Assisted Breeding

The main applications of molecular markers in temperate fruit tree breeding programs are: (1) the examination of genetic diversity and the use of genetic distance to choose parental crosses, (2) marker assisted selection (MAS), (3) monitoring of introgression of novel variability, and (4) genetic profiling for variety protection. All these applications have been included in the marker assisted breeding (MAB) concept.

Recently, Byrne (2007) conducted a survey of marker application in applied breeding programs of fruit and ornamental perennial plants in more than 100 species. A similar approach was followed by Bassil and Lewers (2009) to assess marker use in Rosaceae breeding programs. They sent a questionnaire to Rosaceae breeders worldwide which was only answered by 23 of them, more than half being strawberry and raspberry breeders. Therefore, we must admit that very little reliable information about the use of molecular markers in temperate tree breeding programs exists. Byrne (2007) concluded that less than 50% of fruit and ornamental perennial plant breeding programs used molecular markers, the main reasons being the lack of available markers and the costs. Another reason can be that important traits are often quantitatively inherited and the uncertainty of some QTLs could imply that phenotypic selection is more efficient than MAS (Moreau et al. 2004). In fact, although molecular markers linked to QTLs have been identified in temperate fruit tree species, they have not been yet used in MAS (Bassil and Lewers 2009). Furthermore, genotype per environment and genetic background effects are not very well understood as most traits have only been mapped in one or a few populations. Another problem is that in most cases the genetic markers available are still far away from the gene of interest. This is even worse in species such as peach where only a very few number of markers segregate in breeding progenies.

Most of the examples of MAS found in the bibliography belong to apple and peach because they are the most-studied temperate tree species due to their economical importance. In olive, although it is the sixth most important oil crop, almost no olive sequence information was available in the literature till very recently (Alagna et al. 2009) and identification of molecular markers linked to important traits is still under development. In apple, breeders use MAS for resistance to apple scab, powdery mildew, and two types of aphids (Gardiner et al. 2007; Bus et al. 2008; Stoeckli et al. 2008); for columnar growth habit (Moriya et al. 2009) and for some fruit quality traits such as the skin color defined by the Rf locus (Cheng et al. 1996); and fruit firmness (Costa et al. 2008). In peach, MAS is possible for a few Mendelian traits that allow selection for specific fruit types, including fruit shape, flesh color, fruit texture, and pubescence (Dirlewanger et al. 2004a). Also, the Mi and Ma genes that confer resistance to root-knot nematodes are being used in *Prunus* rootstock breeding programs (Dirlewanger et al. 2004b). In apricot, plum, and sweet cherries, MAS has been used for self-incompatibility assessment (reviewed in Yamane and Tao 2009), and, in apricot, an SSR closely linked to a major QTL for plum pox virus resistance is being used (Soriano et al. 2008).

Functional Genomics

Functional genomics aims to determine the function of genes and how they interact, aided by the use of high-throughput technologies which are capable of assaying many functions or relationships simultaneously. Several genomic techniques have been developed for this purpose, including expressed sequence tags (ESTs), generation and annotation and microarray, and reverse transcriptase PCR (RT-PCR) analysis. These technologies have been already designed and broadly applied in animals and model plants and are now under development for their use in temperate fruit tree species. Several cDNA libraries and EST datasets have been developed in key species like apple, pear, peach, and olive, and transcriptomics have been used for gene discovery. Moreover, whole genome sequencing projects, such as the ones undertaken in apple and peach, support the development of new tools.

Proteomics and metabolomics address the study of the entire protein and metabolite array in organisms. These are developing as important functional genomics tools; however, their application in molecular biology, and especially in studies of temperate fruit trees, is very recent, and they are still under development (Aggarwal and Lee 2003; Bino et al. 2004).

EST Collections

ESTs are DNA sequences derived from cDNA libraries and therefore are transcribed sequences of the genes expressed in a specific tissue at a specific stage of development or subject to specific conditions. In species where genome sequences are largely unknown, ESTs generated in sequencing projects, annotated and deposited in publicly available databases, are a valuable tool to discover and clone important genes. The recent use of novel high-throughput sequencing technologies in cDNA libraries is providing large amounts of EST data, allowing for massive and efficient gene discovery and quantitative analysis of gene expression patterns in different tissues and development stages. Moreover, they are a source of molecular markers like SSRs and SNPs, facilitating candidate gene mapping.

Several EST sequencing projects have been conducted in temperate fruit tree crops. As of March 2010, the NCBI database (http://www.ncbi.nlm.nih.gov/) contained around 336,000 ESTs derived from more than 115 cDNA *Malus* libraries, about 325,000 of them from *M.* × *domestica* commercial varieties and rootstocks and some derived from fruit tissues, providing information on genes involved in fruit development and ripening (Sung et al. 1998; Newcomb et al. 2006). Park et al. (2006) analyzed approximately 200,000 EST sequences available in databases to identify apple genes highly expressed in fruit and/or regulated during fruit growth and development. These ESTs resulted from 70 cDNA libraries obtained from at least nine cultivars and with diverse transcriptional profiles from a variety of organs, fruit parts, developmental stages, and responses to biotic and abiotic stresses. The EST sequences assembled in approximately 44,000 unique sequences, identifying apple genes that may participate in generating flavor and aroma components of mature fruits. Other tissues sampled in apple include roots, shoots, buds, flowers, and leaves that may be subjected to external conditions such as pathogen infection or abiotic stress. For example, Wisniewski et al. (2008) developed nine cDNA libraries from 'Royal Gala' trees, non-stressed (control) and stressed by exposure to low temperatures or water deficit. Over 22,600 clones were clustered and annotated, identifying genes up- and downregulated in response to stress. Degenhardt et al. (2005) also studied genes differentially expressed in apple cultivars when infected by apple scab by constructing two cDNA libraries using the suppression subtractive hybridization (SSH) method.

The NCBI contains close to 100,000 ESTs for *Prunus*, most of them from peach and apricot (79,500 and 15,000, respectively). In peach, most research has focused on finding genes expressed in the fruit or at development, ripening, and postharvest stages. Horn et al. (2005) developed an EST collection from a cDNA library using fruit mesocarp tissue. Almost 10,000 EST sequences clustered in 3,842 unigenes. The analysis of gene expression associated with chilling injury in peach fruit has been addressed in the ChillPeach project (Granell et al. 2006). Within the framework of this initiative, Ogundiwin et al. (2008) created mesocarp tissue cDNA libraries from two full-sib progeny contrasting for chilling injury. A total of 7,862 high-quality ESTs (of 4,468 unigenes) were deposited in the ChillPeach database, and 4,261 of the unigenes obtained were printed in a microarray. Some of the cold-responsive genes from the microarray analysis of cold-treated versus untreated peach mesocarp tissues were mapped in the intraspecific peach map Pop-DG and in the *Prunus* reference map TxE to construct a fruit quality gene map. Vizoso et al. (2009) also looked for candidate genes expressed during ripening and in response to postharvest cold storage by creating a collection of 50,625 ESTs assembled in 10,830 unigenes. Tittarelli et al. (2009) conducted digital expression analyses on this EST dataset, revealing 164 cold-induced peach genes, several of which have similarities to genes associated with cold acclimation and cold stress responses.

Besides the interest on the genes involved in fruit processes, some research has addressed traits such as bud dormancy in peach. Leida et al. (2010) constructed four suppression subtractive hybridization (SSH) libraries from dormant buds, dormancy-released buds, and two cultivars with different chilling requirement, sampled after dormancy release. Hybridization of microarrays with 2,500 picked clones, with positive clones sequenced, revealed genes coding for MAD-box transcription factors previously found in the *Evergrowing* mutant of peach (Bielenberg et al. 2008) as well as others related to bud dormancy.

In apricot, Grimplet et al. (2005) constructed three cDNA libraries from fruit mesocarp tissue at three different stages of development, obtaining 13,000 ESTs assembled in 5,219 unigenes.

About 4,800 olive ESTs from fruit and leaf cDNAs are also available in the NCBI database. Most olive cDNA libraries aim to identify ESTs involved in fruit development, principally those related with phenolic and lipid metabolism. Galla et al. (2009) created four SSH libraries from fruit of the 'Leccino' olive cultivar sampled at three different stages (initial fruit set, completed pit hardening, and veraison) to identify differentially expressed genes involved in fruit development. A total of 1,132 clones were sequenced and annotated. Those showing differentially expressed unique sequences were further investigated by real-time PCR. Alagna et al. (2009) constructed four cDNA libraries starting from fruit tissues, at different developmental stages, of two olive cultivars contrasting for high phenolic content, 'Carotina' and 'Tendellone'. All clones were sequenced by 454 pyrosequencing, giving a total of 261,485 sequence reads assembling into 22,904 clusters and 75,570 singleton ESTs.

Recent EST sequencing initiatives in other temperate fruit tree species have generated around 1,400 *Pyrus* ESTs, most of them (46%) from *P. pyrifolia* (Nashi), and a few ESTs of loquat and quince.

Microarrays

Microarray techniques are able to characterize transcription profiles of collections of many genes simultaneously. Early microarrays often used cDNA probes.

For example, Fonseca et al. (2004) constructed a high-density cDNA microarray to study molecular events associated with fruit development and climacteric ripening in pear. In apple, Lee et al. (2007) also used this technology to study genes involved in early fruit development.

Nowadays, most available sequencing data has fostered the design of arrays based on oligonucleotide probes. Major EST sequencing projects, such as the ones previously referred to, usually include the transcriptomic study of the sequences, while others use the large amount of EST data available in databases to design microarrays. For example, Schaffer et al. (2007) used a 15,720-oligonucleotide apple microarray to study the expression pattern of 179 candidate genes found in databases and putatively involved in the ripening process of climacteric fruits, identifying 17 candidate genes likely to be ethylene control points for aroma production in apple. Janssen et al. (2008) studied genes expressed in apple fruit development by printing a microarray from apple ESTs including approximately 13,000 genes. In total 1,955 showed significant changes in expression during development. Functional analysis gave genes associated with different fruit development stages. The microarray expression was validated by quantitative reverse transcriptase PCR (qRT-PCR), showing that 74% of the genes had the same pattern of expression in microarray as in the qRT-PCR experiment.

In addition, some *Prunus* microarrays have been printed from publicly available ESTs, mostly to study genes expressed in processes involved in fruit development. Trainotii et al. (2006) studied genes expressed in the transition from the preclimateric to the postclimateric phase in peach fruits, using the first microarray available in peach (μPEACH1.0) with 4,800 oligonucleotide probes. Dardick et al. (2010) studied the process of stone formation in peach fruits and printed an array with 4,806 peach oligonucleotides (70 mer). They showed that lignin deposition in peach initiates near the blossom end within the endocarp layer and also revealed a previously unknown coordination of competing lignin and flavonoid biosynthetic pathway during early fruit development.

Proteomics

Proteomics is the study of the whole set of proteins encoded by the genome at a certain time and under a particular set of conditions. A typical proteomic workflow includes several steps: protein extraction, separation, quantification, identification, and subsequent data analysis.

Protein separation is usually carried out using two-dimensional electrophoresis (2-DE) (O'Farrell et al. 1977), two-dimensional in-gel electrophoresis (DIGE) (Alban et al. 2003), or liquid chromatography-mass spectrometry (LC-MS) (Roe and Griffin 2006). Mass spectrometry is the technology most widely used for protein identification, with an ion source and one or more mass analyzers. To ionize the peptides, matrix-assisted laser desorption ionization (MALDI) (Karas and Hillenkamp 1988) and electrospray ionization (ESI) (Fenn et al. 1989) are the techniques of choice. To analyze the mass, a time-of-flight (TOF) or quadrupole (*Q*) analyzer is usually used.

Proteomics is in the early stages for temperate tree fruits, with only a few publications where it has been used to study different disease. Heyens et al. (2006) used 2-DE to study the differential expression of proteins in apple following inoculation with *Erwinia amylovora*. Seventy-three protein spots were picked and identified by mass spectrometry, and the 2-D gels were analyzed to identify differentially expressed proteins. This method was also applied to study core breakdown disorder in 'Conference' pears, by Pedreschi et al. (2007). Statistical approaches revealed interesting differentially expressed proteins in healthy and disordered pears. Obenland et al. (2008) identified proteins related to the development of mealiness in peaches, also using 2-DE combined with MS. Five proteins were identified that differed significantly in amount in peaches that had mealy flesh and those that remained juicy.

Allergens are very important in tree fruits, and their presence has been confirmed both in pollen and in fruits. Different authors have used a proteomics approach. Reuter et al. (2005) identified a new isoform of Pru av 1, the major cherry allergen, by a combination of molecular biology and proteomic tools. Herndl et al. (2007) identified several apple allergens by 2-D electrophoresis and IgE reactivity. The major allergen of the Oleaceae family is Ole e 1. Napoli et al. (2008) applied a proteomics approach to detect Ole e 1 isoallergens, using a preliminary chemical fractionation procedure followed by MALDI MS and MS/MS measurements. Castro et al. (2010) used 2-DE coupled to MS analysis to examine the molecular variability of the Ole e 1 allergen in three olive cultivars. They observed that 'Picual' and 'Arbequina' cultivars had the highest and lowest degree of Ole e 1 polymorphism, respectively.

Almost all fruits are stored at some point, and different treatments are applied to improve the quality and longer shelf life of the product. Exposure of plants to low, nonfreezing temperatures results in genetically programmed changes in the physiology and biochemistry that are critical for low-temperature survival of plants. Renaut et al. (2008) conducted a quantitative assessment of changes in the peach bark proteome during cold acclimation using DIGE technology. Later, Lara et al. (2009) have studied the effects of heat treatment on peach fruit quality using a proteomic approach, characterizing the differential proteome of heated fruit, revealing that heat-induced chilling injury tolerance may be acquired by the activation of different molecular mechanisms. Zhang and Tian (2009) examined proteome patterns in peach fruit stored at different low temperatures. Their results led to the understanding of the mechanisms by which peach fruit have a higher chilling tolerance when stored at 0°C than that at 5°C.

Pedreschi et al. (2009) used the DIGE approach to study the protein expression profiles of 'Conference' pear tissue exposed to extreme gas conditions (anoxia and air). Changes in expression, up to fourfold, were identified in proteins involved in respiration, protein synthesis, and defense mechanisms. Pedreschi et al. (2009) analyzed the physiology of pears subjected to four controlled atmosphere conditions by means of the proteomics approach. The results show that impaired respiration is highly related to protein synthesis alterations and activation of defense mechanisms. Peach fruit undergoes a rapid softening process that involves a number of metabolic changes. Nilo et al. (2010) used a 2-D DIGE approach to screen for differentially accumulated proteins in peach fruit during normal softening as well as under conditions that lead to fruit chilling injury. They showed that a significant proportion of the proteins identified had not previously been associated with softening, cold storage, or chilling injury-altered fruit.

Proteomics has also been used to study developmental process. Grimplet et al. (2005) used transcriptomics and proteomics tools to study fruit ripening in apricot. Wang et al. (2007) characterized seed storage proteins and their synthesis during seed development in olive using SDS-PAGE.

Fruit senescence has been reported to be an oxidative phenomenon, but the detailed mechanisms by which reactive oxygen species (ROS) regulates this process remain largely unknown. Qin et al. (2009a, b) analyzed mitochondrial proteome variations on exposure to high oxygen levels (100%), which induces oxidative stress and accelerates fruit senescence. These data suggest that ROS may regulate fruit senescence by changing the expression profiles of specific mitochondrial proteins and impairing their biological function.

Guarino et al. (2007) analyzed the protein profile from fruit tissues of three accessions of the apple cultivar Annurca, a regional variety from southern Italy, which is known for crispness, excellent taste, and long shelf life. Forty-four protein spots were identified and related to important physiological processes including energy production, ripening, and the stress response. Chan et al. (2008) reported a comparative analysis of sweet cherry fruit proteomes induced by salicylic acid (SA) at different maturity stages. Based on proteomics analysis, after SA treatment, 13 proteins were identified at an earlier, and 28 at a later, stage of maturity. The findings indicated that younger sweet cherry fruits had stronger resistance against pathogen invasion after SA treatment.

Metabolomics

Metabolomics has been carried out since the mid-1970s but has only become a standard laboratory technique in the past decade. This new approach can be described as the study of a complete collection of metabolites present in a cell or tissue under a particular set of conditions, generating a biochemical profile. This profile describes the biochemistry defining different physiological states, including disease states, that a cell or tissue may adapt.

Two techniques dominate metabolite profiling strategies: mass spectrometry (MS) and nuclear magnetic resonance (NMR), gas chromatography-mass spectrometry (GC-MS), gas chromatography time-of-flight mass spectrometry (GC-TOF-MS), and liquid chromatography-mass spectrometry (LC-MS) are currently the standard mass spectrometry methods for metabolite analyses. In addition, capillary electrophoresis-mass spectrometry (CE-MS) and Fourier transform-ion cyclotron resonance-mass spectrometry (FT-ICR-MS) systems have been also applied.

In fruit trees, metabolomics has been used mainly for identification of volatile components. Takeoka et al. (1992) identified a total of 72 components, evaluating the volatiles of Asian pear by the GC-MS technique. Results highlight that (E)-2-

hexenal and bornyl acetate are probably major contributors to the persimmon aroma. Guichard and Souty (1988) isolated 82 volatile components analyzing fresh apricots by GC-MS. Volatile compounds from 'La France' pears, responsible for their smell, were examined by Oshita et al. (2000) after three different storage treatments. The odors were classified into three classes, depending on their physiological states, through a distinctive odor pattern formed by 32 outputs. Campeol et al. (2001) were able to distinguish three olive cultivars by examining their volatile fraction composition. Vikram et al. (2004a, b) evaluated the volatile metabolite profiles of apple fruits to detect fungal diseases. For this propose, they inoculated the McIntosh, Cortland, and Empire cultivars with four different fungi and subsequently profiled using GC-MS. A number of specific compounds that could be used for the early detection of apple diseases in storage were found.

Guillot et al. (2006) characterized the aroma profile of six apricot cultivars in order to obtain objective data concerning the aromatic quality. They found 23 common volatile compounds, responsible for aromatic notes involved in apricot aroma, which could be utilized to discriminate apricot varieties. Chen et al. (2006) investigated the changes in the volatile compounds and the chemical and physical properties of Yali pear (*Pyrus bertschneideri* Reld) during storage by different techniques. Chen et al. (2005) studied the aroma constituents present in apricot fruit during three different developmental stages. They concluded that aromatic constituents behaved differently during the fruit developmental period. Solis-Solis et al. (2007) recognized and classified eight apricot varieties by means of the aroma compounds. Wang et al. (2009) evaluated the volatile characteristics of 50 peaches and nectarines of different germplasm origins. They found that volatile composition was relatively consistent and their contents depended on genotypic background and germplasm origin. All the peaches and nectarines could be classified into four groups by principal component analysis of the volatiles

Takahashi et al. (2000) evaluated the volatile constituents of fresh loquat, its canned product immediately after canning, and the stored products. They were able to identify 78 compounds by a capillary column GC-MS of which 15 and 19 significantly contributed to the aroma of fresh and canned loquat, respectively. Lu and Liu (2008) analyzed the chemical components of the volatile oil in processed loquat by GC-MS, concluding that it was a good method for further evaluating loquat quality.

Rudell et al. (2008) used global metabolic profiling of 'Granny Smith' apple peel for evaluating metabolomic alterations resulting from prestorage UV-white light irradiation. The profile revealed changes in the metabolome provoked by UV-white light irradiation and cold storage. Rudell et al. (2009) and Rudell and Mattheis (2009) used the same approach in order to study scald development in apple. Analysis of the peel metabolome revealed associations between scald status, light treatment duration, and individual metabolites from multiple pathways. To study core breakdown disorder in pear, Pedreschi et al. (2009) compared the metabolic profiling of brown and sound tissue, demonstrating that brown tissue was clearly characterized by a distinctive metabolic pattern.

Peschel et al. (2007) studied the composition of wax and cutin from developing sweet cherry fruit by GC-MS. There were no qualitative and only minor quantitative differences in wax and cutin composition between cultivars at maturity, indicating that deposition of some constituents of wax and cutin ceased during early fruit development.

Füzfai et al. (2004) developed a method to simultaneously identify and quantify sugar, sugar alcohol, and carboxylic acid contents of sour cherry, apple, and ber (*Ziziphus mauritiana* L.) fruits. Bureau et al. (2009) developed a fast and accurate Fourier transform mid-infrared spectroscopy method for simultaneously determining sugar and organic acid contents in apricot fruit slurries.

Genetic Engineering of Temperate Fruit Trees

The limitation of introducing new traits through conventional plant breeding in trees has always been the long generation time needed. One possible shortcut can be the introduction of such characters via plant transformation. In fact, genetic engineering is a powerful tool for plant improvement as genes from virtually any organism can be inserted into a plant, singly or in combination. This allows for targeted improvement of elite cultivars as well as the study of gene function.

Here we summarize several achievements reported in the last few years on gene transfer, overexpression, and RNA silencing of genes controlling characters of agricultural importance.

Gene Transfer and Regeneration

Numerous species of agronomic interest have been genetically engineered worldwide, including temperate fruit trees, where the application of biotechnological methods can contribute to overcome some of the handicaps for conventional breeding, such as the long juvenile period, the reproductive barriers, and the high degree of heterozygosity (Petri and Burgos 2005). In addition, it has several inherent advantages. For instance, once a transformant is fixed and determined as a valuable new line, it can be vegetatively propagated, unlimitedly (Petri and Burgos 2005). However, in spite of the potential benefits, to date the GM Crops Database (http://www.cera-gmc.org) only contains one temperate fruit tree, a plum resistant to PPV (plum pox virus) (Scorza et al. 1994) that can be found in the Canadian and United States markets. This situation is due not only to the general public concern about consumption of transgenic food products but also to the particular limitations of transformation and regeneration protocols in woody fruit species.

Disarmed strains of *Agrobacterium tumefaciens* (LBA4404, EHA105, etc.) have become the most popular means of introducing foreign DNA into plant cells for most plant species including fruit trees (Petri and Burgos 2005). Biolistics (otherwise known as particle bombardment) has also been used as transformation method. Significant effort has been made to study the factors determining the efficiency of transformation techniques in different species: the virulence of the *Agrobacterium* strains (Cervera et al. 1998), the influence of phenolic compounds (James et al. 1993), and the length of coculture (Ainsley et al. 2002), among others. Furthermore, in gene transfer, it is crucial to develop regeneration procedures which permit the production of buds, shoots, and entire plants from transformed tissues.

Regeneration can be carried out by two commonly recognized ways: adventitious organogenesis and somatic embryogenesis. The most suitable method should be selected for each particular case, and this is especially true in fruit trees where there are often difficulties with plant regeneration. Characteristically, gene transfer procedures are strongly dependent on the specific genotype, and this is probably one of the major constraints. Usually only a small set of cultivars can be reliably transformed within each species, and most of the time they have no commercial quality. The development of genotype-independent procedures is therefore a challenge for genetic transformation in fruit trees (Petri and Burgos 2005). Some promising strategies have been proposed to tackle it, the transformation of meristematic cells with high regeneration potential (Mezzeti et al. 2002) or the transformation with regeneration-promoting genes (i.e., *ipt* gene from *A. tumefaciens* coding an isopentenyl transferase involved in cytokinin synthesis) (Zuo et al. 2002; López-Noguera et al. 2009). An additional advantage of this latter system is that it avoids the use of selecting genes conferring resistance to antibiotics or herbicides (i.e., *NPT*II, *bar*, etc.) that are not accepted by consumers and remain restricted by European laws. Alternative marker genes, such as the *manA* gene encoding a phosphomannose isomerase (Degenhardt et al. 2006), are being used to transform fruit trees in order to avoid the use of antibiotics. Lastly, cisgenesis should be emphasized as a global alternative to transgenic approaches. Currently, this concept is being developed in apple (Joshi et al. 2009).

Commercial transgenic fruits are exceptional, but many fruit tree species have already been successfully transformed with diverse objectives (Table 6.1). Apple is probably the most commonly transformed temperate fruit tree species, probably due to its economic importance. The first apple transgenic plants, expressing the selectable gene *bar*, were obtained by De Bondt et al. (1996). Since then, a few genotypes have been transformed, mainly expressing resistance to major apple pathogens such as fire blight (*Erwinia amylovora*) (Ko et al. 2000) or scab (*Venturia inaequalis*) (Bolar et al. 2000; Faize et al. 2003, 2004; Belfanti et al. 2004). However, other interesting approaches have been developed, for instance, Welander et al. (1998) introgressed the *rolB* gene from *Agrobacterium rhizogenes*, obtaining shorter internodes and improved rooting, and Broothaerts et al. (2004) silenced the expression of an *S-RNase*, achieving self-compatible apples. More recently, transgenic apples have been used to produce a vaccine against the human respiratory syncytial virus (Lau and Korban 2010). In *Pyrus*, there are several examples of genetic modifications aimed at conferring resistance to fire blight (Reynoird et al. 1999; Puterka et al. 2002; Malnoy et al. 2005a, b), to improve rooting (Bell et al. 1999; Zhu et al. 2003) or to bring forward flowering (Matsuda et al. 2009).

In general terms, transformation approaches in *Prunus* have been less successful than in *Malus* or *Pyrus*. Nevertheless, there are significant achievements that should be highlighted. As reported above, transgenic plums expressing the PPV coat

Table 6.1 Some of the most outstanding genetic modification approaches in fruit trees

Species	Cultivar	Genes	Phenotype	References
Malus × *domestica* (apple)	'Jonagold'	*bar*	Resistance to Basta	De Bondt et al. (1996)
	'Jonagold'	*Rs-AFP2* *AMP1*	Antimicrobial and antifungal activity	De Bondt et al. (1998)
	'M.26'	*rol B*	Improved rooting	Welander et al. (1998)
	–	Endochitinase	Scab resistance	Bolar et al. (2000)
	'Galaxy' 'M.26'	*attE*	Resistance to fire blight	Ko et al. (2000)
	'Galaxy'	*ech42* and *nag70*	Scab resistance	Faize et al. (2003)
	'Elstar' 'Holsteiner Cox'	*Vst1* *PGIP*	Resistance to fungal diseases	Szankowski et al. (2003)
	'Ariane'	*pinB*	Scab resistance	Faize et al. (2004)
	'Gala'	HcrVf2	Scab resistance	Belfanti et al. (2004)
	'Elstar'	S_3-*RNase*	Self-fertility	Broothaerts et al. (2004)
	'Holsteiner Cox'	*GUS* *manA*	Phosphomannose isomerase/ mannose selection	Degenhardt et al. (2006)
	'Orin'	*MdTFL-1 anti sense*	Reduction of juvenile phase	Kotoda et al. (2006)
	'Royal Gala'	*Na-PI*	Light-brown apple moth resistance	Maheswaran et al. (2007)
	'Royal Gala'	*RSV-F*	Expression of antigenic protein	Lau and Korban (2010)
	'Galaxy'	*HIMP Inverted Repeat (IR)*	Reduced susceptibility to *Erwinia amylovora*	Malnoy et al. (2008a)
		Mal d 1 IR	Reduced allergenicity	Gilissen et al. (2005)
	'Greensleeves'	*ACC synthase ACC oxidase* co-suppression	Ethylene biosynthesis reduction	Dankedar et al. (2004)
	'Greensleeves'	*Sorbitol-6-phosphate dehydrogenase*	Altered sorbitol distribution	Teo et al. (2006)

(continued)

Table 6.1 (continued)

Species	Cultivar	Genes	Phenotype	References
		Anthocyanin synthase IR	Decrease in flavonoids and related polyphenols	Szankowski et al. (2009a)
	'A2'and 'Gravenstein and McIntosh'	*A. thaliana gai* co-suppression	Gibberellins metabolism with reduced growth	Zhu et al. (2008)
		GA oxidase sense and antisense	Gibberellins biosynthesis with reduced growth	Bulley et al. (2005)
	'Galaxy'and 'M26'	*MpNPR1*	Increased resistance to *Erwinia amylovora*	Malnoy et al. (2007)
	'Pinova'	*BPMADS4*	Early flowering	Flachowsky et al. (2007)
	'Pinova'	*A. thaliana LEAFY*	Reduced growth	Flachowsky et al. (2010)
	Holsteiner Cox	Maize *Lc* transcription factor	Increased biosynthesis of anthocyanins and specific flavonoids	Li et al. (2007)
Prunus armeniaca (apricot)	'Kecskemeter' seeds	CP-PPV	Sharka resistance	Laimer da Câmara Machado et al. (1992)
	'Helena'	gfp	GFP expression	Petri et al. (2008)
Prunus avium (cherry)	'Mazzard F12/1'	Ri-1855-T-DNA	Increased rooting ability, various hairy root phenotypes	Gutiérrez-Pesce et al. (1998)
P. avium × *pseudocerasus* (cherry rootstock)	'Colt'			
	'Colt'	Phytochrome A	Resistance to bacterial canker	Cirvilleri et al. (2008)
Prunus cerasus (sour cherry)	'Montmorency'	NPTII-GUS	Resistance to kanamycin and GUS expression	Song and Sink (2006)
P. cerasus × *canescens*(cherry rootstock)	'Gisela 6'			
P. dawykensis (cherry rootstock)	'Damil'	Ri-T-DNA *bar*	Improved rooting, hairy roots. Basta resistance	Druart et al. (1998)
P. domestica (plum)	'Bluebyrd' C5 clone	*CP-PPV*	Sharka resistance	Scorza et al. (1994)
	'Stanley' seeds	*hpt*	Resistance to hygromycin	Tian et al. (2009)
	'Bluebyrd'	CP-PPV (IR)	Sharka resistance	Hily et al. (2007)

P. incisa × serrula	'Inmil'	T-DNA *(ipt) bar*	Improved rooting	Druart et al. (1998)
			Basta resistance	
P. persica (peach)	'Redhaven'	T-DNA *(ipt)*	More branching	Hammerschlag and Smigocki (1998)
			Reduced rooting	
	–	*gfp*	GFP expression	Pérez-Clemente et al. (2004)
Pyrus communis (pear)	'Beurre Bosc'	*rolC*	Increased rooting	Bell et al. (1999)
	'Passe Crassane'	*attE*	Resistance to fire blight	Reynoird et al. (1999)
	'Bartlett'	*D5C1*	Resistance to fire blight	Puterka et al. (2002)
	'BP10030'	*rolB*	Increased rooting	Zhu et al. (2003)
	'Passe Crassane'	*Harpin N*	Resistance to fire blight	Malnoy et al. (2005a)
	'Passe Crassane'	EPS depolymerase	Decreased fire blight susceptibility	Malnoy et al. (2005b)
	'La France'	*ACC oxidase sense and antisense*	Ethylene biosynthesis alteration	Gao et al. (2007)
	'Ballad'	*MdSPDS1*	Abiotic stress tolerance	Wen et al. (2008)
	'La France' 'Ballad'	*CiFT*	Early flowering	Matsuda et al. (2009)

Adapted and updated from Petri and Burgos (2005)

protein, and resistant to PPV, are already commercially available (Scorza et al. 1994), and attempts to incorporate this resistance into elite plum cultivars by cross-pollination have been initiated (Ravelonandro et al. 2002). PPV is the causal agent of sharka, probably the major disease in *Prunus* affecting plums, peaches, and apricots, making PPV resistance a priority in most breeding programs. Unfortunately, the success in plum still has not been reproduced in other *Prunus* species. In apricot, known to be highly recalcitrant to gene transfer, Laimer da Câmara Machado et al. (1992) reported the achievement of PPV-resistant plants, but further testing revealed their chimerical nature (Escalettes et al. 1994). In fact, different studies on factors affecting apricot transformation and regeneration have been published in the last 20 years, but, only recently, transgenic plants expressing the *gfp* gene were finally obtained by Petri et al. (2008). Similarly, in peach, the most important *Prunus* species in economic terms, only two successful approaches have been reported evaluating marker (*ipt*) (Hammerschlag and Smigocki 1998) and reporter (*gfp*) genes (Pérez-Clemente et al. 2004). These and other examples of transgenic *Prunus* are shown in Table 6.1.

In the light of these data, it can be stated that gene transfer in temperate fruit trees has progressed enormously over the last two decades. However, advances are needed in the near future, to face consumer requirements and law restrictions on one side and to overcome technical difficulties on the other.

Silencing and Gene Overexpression in Temperate Fruit Trees

Some of the achievements cited above and in Table 6.1 have been reached through RNA silencing or gene overexpression. Here we describe these experiments in more detail.

RNA Silencing of Endogenous Genes

Resistance to viruses through silencing has been achieved mainly in *Prunus* against plum pox virus (PPV). PPV produces one of the major diseases in the *Prunus* species. Few highly resistant cultivars have been developed through conventional plant breeding, hence the interest in developing pathogen-derived resistance through overexpression of the viral coat protein (CP), which has been reported to confer resistance against the corresponding virus (Abel et al. 1986). In plum, one of the CP transformed lines (line C5) had no detectable expression of the transgene and was highly resistant to PPV (Scorza et al. 1994). This line harbored several copies of the transgene, which resulted in silencing instead of overexpression. This was the first posttranscriptional gene silencing (PTGS)-based virus resistance reported in a temperate woody perennial and remained virus free for 4 years in a field exposed to natural aphid vectors (Hily et al. 2004). This line showed neither differences in the visiting aphid populations nor alterations in the dynamics of PPV populations, and there was no evidence of recombination between the CP transgene and the CP gene of the virus over a period of 8 years (Capote et al. 2008). The silencing C5 clone produced 25–26-nt-length siRNAs and methylation of the transgene, both traits associated with PPV resistance, whereas the susceptible clones only showed 21–22-nt siRNAs (Kundu et al. 2008).

Silencing in the C5 clone was obtained randomly, by aberrant multiple integrations of the transgene. In a search for a more efficient system, inverted repeats of PPV sequences under either a 35S or a phloem promoter were transformed into *N. benthamiana*, where more than 90% and 80% T0 transgenic plants, respectively, were resistant to PPV (Di Nicola-Negri et al. 2005; Pandolfini et al. 2003). Likely to be a good system to transfer to stone fruit plants, it was subsequently addressed by Hily et al. (2007), who used a PPV CP hairpin construct for plum transformation. The transformed lines showed similar levels of siRNA accumulation to the *P. domestica* C5 line described previously (Hily et al. 2005). The resistance to PPV of these transformants still remains to be tested.

A similar strategy has been followed against the bacteria *Erwinia amylovora*, the causal agent of fire blight in apple. The HIPM apple protein interacts with the bacterial protein HrpN, which plays an important role in pathogenicity, since it is involved in eliciting the hypersensitive response. A hairpin construct of the HIPM gene was used for silencing the gene through apple transformation, producing some lines with reduced susceptibility to *E. amylovora*, which correlated with a reduced expression of the HIPM gene and impairment of the interaction with the bacterial protein (Malnoy et al. 2008a).

Apart from resistance-related genes, silencing of endogenes in temperate fruit trees has largely addressed issues important in the fruit quality in these species or mainly occurring in these plants, such as self-incompatibility. Most of the rosaceae trees are self-incompatible, unable to produce self-fertilizing pollen. This character is controlled by a single gene (*S*-locus) encoding an RNAse that specifically degrades the self-pollen (see section "Silencing and Gene Overexpression in Temperate Fruit Trees" of this chapter for more details). Transgenic apple trees harboring extra copies of the *S*-gene under the CaMV35S promoter suffered co-suppression, resulting in silencing of the endogenous *S*-gene. Expression of the *S*-gene in the pistil was inhibited, producing normal self-pollen tube growth with the plants able to self-pollinate, unlike the control non-silenced plants (Broothaerts et al. 2004).

Among characters involving fruit quality, delaying fruit ripening is an important goal. It has been addressed by silencing an ACC oxidase in apple (Dandekar et al. 2004) and pear (Gao et al. 2007) or an ACC synthase in apple. The silencing of an ACC synthase or an ACC oxidase gene, both involved in ethylene biosynthesis and ripening, in apple, produced firmer fruits that also had longer shelf life than controls. Synthesis of volatile esters was also suppressed, but not their aldehyde and alcohol precursors (Dandekar et al. 2004). In a genomic approach, Schaffer et al. (2007) showed that, by co-suppression of an ACC oxidase, the production of volatile aroma compounds was downregulated and microarray experiments indicated that the first, or preferably last, enzyme of each biosynthetic aroma-producing pathway was the key step regulated by ethylene.

Manipulation of taste and sugar content was achieved by silencing of apple sorbitol-6-phosphate dehydrogenase, the key enzyme for sorbitol production. The experiment revealed differences in the pathways controlling sorbitol transport and degradation, indicating a role for sorbitol in the carbon metabolism and sugar distribution in the fruit, with higher accumulation of sucrose and lower accumulation of other sugars, such as fructose and starch. Interestingly, no overall differences in fruit firmness were observed (Teo et al. 2006).

Fruit color is another important trait required by consumers. Silencing of anthocyanin synthase in red-leaved apple produced a decrease in flavonoids and related polyphenols. In addition, there was a major increase in necrotic leaf lesions and hence viability (Szankowski et al. 2009a). This study, together with those of overexpression of the Myb factor Lc (see below) (Li et al. 2007; Flachowsky et al. 2010b), suggests the essential function of anthocyanins in apple. There is also the need to decrease allergenicity of apples, in relation to the health of the consumer. This trait is related to the reaction of IgE with the product of the Mal d 1 gene in areas were birch pollen is endemic. Transgenic apple trees produced by RNAi using a hairpin of the Mal d 1 gene showed reduced expression of the gene and significantly reduced in vivo allergenicity compared to wild-type apples (Gilissen et al. 2005). The plants remained silenced and hypoallergenic after grafting and the phenotype has been stable over more than 3 years, throughout all developmental stages (Krath et al. 2009).

Plant architecture is another important trait for breeders. Smaller apple plants require reduced applications of chemical retardants. This phenotype has been successfully controlled by manipulating, among others, the gibberellin (GA) biosynthesis or sensitivity. Overexpression of the Arabidopsis gai (gibberellic acid insensitive) gene, in apple cultivars A2 and Gravenstein and McIntosh, resulted in co-suppression and silencing of the apple Gai gene, with phenotypes having reduced growth, shorter stems and internode length, and impaired rooting ability (Zhu et al. 2008). Likewise, silencing via sense and antisense overexpression of a GA20 oxidase, an enzyme that participates in GA biosynthesis, also produced reduced growth, and the phenotype remained after grafting the transgenic scions onto invigorating rootstocks (Bulley et al. 2005). Shortening of the juvenile period is another desirable trait. Antisense silencing of MdTFL (a transcription factor that delays flowering) in Orin apple plants promoted early flowering, producing transgenic plants flowering three times earlier than non-transgenic plants (Kotoda et al. 2006).

Until now all the silencing technologies used in temperate trees have been based on stable transformation of the required construct. Transient systems based on virus-induced gene silencing (VIGS) still have not been set up in these species. VIGS technology would be useful since it would provide results on gene function analysis in a shorter time, since the silencing phenotype would be seen during the infection time of the virus, within a few weeks. Possible candidate viruses are PPV or, more promising, apple latent spherical virus (ALSV), a VIGS vector that has already proved to be useful for silencing in a broad range of herbaceous species (Igarashi et al. 2009).

Overexpression of Heterologous Genes

Stable overexpression of heterologous genes in temperate fruit trees has been achieved mainly in apple and pear. *Prunus* species remain difficult to transform despite some success in the last few years (Maghuly et al. 2007; Urtubia et al. 2008; López-Noguera et al. 2009).

In apple and pear, there are recent reports on expression of heterologous proteins, mainly related to flavor and taste, defense, and plant morphology. For more detailed information on previous achievements, see Malnoy and Aldwinckle (2009).

Overexpression of defense-related genes has provided a means of introducing new sources of resistance in apple and pear cultivars. NPR1 is a gene involved in mounting systemic acquired resistance against pathogens. Overexpression of the apple homologue, MpNPR1, in apple cultivars Galaxy and M26 resulted in activation of PR genes (pathogenesis-related genes) and subsequent increased resistance to a virulent strain of the bacteria *Erwinia amylovora*. Some of the transgenic lines also showed resistance to the two fungal pathogens *Venturia inaequalis*, the causal agent of apple scab, and *Gymnosporangium juniperi-virginianae*, the causal agent of apple rust (Malnoy et al. 2007). The exopolysaccharide (EPS) of the capsule of *Erwinia amylovora* is the main pathogenicity factor. In pear, cultivar Passe Crassane, overexpression of an EPS depolymerase gene (Malnoy et al. 2005b) produced plants with decreased fire blight susceptibility, since the transgene degrades the EPS of the bacterial capsules once the bacteria enter the cell, thus reducing the infection. Overexpression of proteins from pathogen origin has also given promising results. The viral protein F from human respiratory syncytial virus (RSV) was successfully expressed at high levels in *Malus* x *domestica Borkh.* cv. Royal Gala, producing the first report on the possible use of transgenic apple for developing a plant-based vaccine for RSV (Lau and Korban 2010).

Plant architecture and flowering time of apple trees have also been altered by overexpression of genes, not only by silencing. Early flowering has been achieved by overexpression of the BPMADS4 gene from silver birch, resulting in flowering after 3–4 months in the greenhouse with viable pollen and fruits with viable seeds (Flachowsky et al. 2007). Overexpression of the LEAFY gene from *A. thaliana* produced reduced growth, with shorter internodes and columnar phenotype, but still had not shown early flowering after 2 years, which would be a desired phenotype to shorten the generation time in trees (Flachowsky et al. 2010a).

Fruit quality plays an important role from the consumer's point of view. Overexpression of the maize Lc transcription factor in apple, cultivar Holsteiner Cox, produced increased biosynthesis of anthocyanins and specific flavonoids (Li et al. 2007). These compounds enhance fruit quality, not only in apple fruit color and flavor, but also because their antioxidant properties offer potential improvement of consumer's health. Additionally, these plants showed increased resistance against apple scab and fire blight, although there was also altered growth tropism and size and the leaves often developed necrotic lesions, hypersensitive response-like (Flachowsky et al. 2010b).

Plant survival under abiotic stress has been addressed by overexpression of the apple spermidine synthase gene in European pear (*Pyrus communis* L. 'Ballad'), producing a line (line 32) with increased tolerance to salt, osmotic, and heavy metal stresses (Wen et al. 2008). This effect could be of interest if maintained in field conditions since it would allow these plants to live in polluted environments. Additionally, line 32 had increased antioxidant enzyme activities (He et al. 2008), which are responsible for the heavy metal tolerance (Wen et al. 2010).

References

Abbott AG, Rajapakse S, Sosinski B, Lu ZX, Sossey-Alaoui K, Gannavarapu M, Reighard G, Ballard RE, Baird WV, Scorza R, Callahan A (1998) Construction of saturated linkage maps of peach crosses segregating for characters controlling fruit quality, tree architecture and pest resistance. Fourth Int Peach Symp 1–2:41–49

Abel PP, Nelson RS, Hoffmann DB, Rogers SG, Fraley RT, Beachy RN (1986) Delay of disease development in transgenic plants that express the tobacco mosaic virus coat protein gene. Science 232:738–743

Aggarwal K, Lee KH (2003) Functional genomics and proteomics as a foundation for systems biology. Brief Funct Genomic Proteomic 2:175–184

Ainsley PJ, Collins GG, Sedgley M (2002) Factors affecting Agrobacterium-mediated gene transfer and the selection of transgenic calli in paper shell almond (*Prunus dulcis* Mill.). J Hortic Sci Biotechnol 76:522–528

Alagna F, D'Agostino N, Torchia L, Servili M, Rao R, Pietrella M, Giuliano G, Chiusano ML, Baldoni L, Perrotta G (2009) Comparative 454 pyrosequencing of transcripts from two olive genotypes during fruit development. BMC Genomics 10:399

Alban A, David SO, Bjorkesten L, Andersson C, Sloge E, Lewis S, Currie I (2003) A novel experimental design for comparative two-dimensional gel analysis: two-dimensional difference gel electrophoresis incorporating a pooled internal standard. Proteomics 3:36–44

Aranzana MJ, Garcia-Mas J, Carbo J, Arus P (2002) Development and variability analysis of microsatellite markers in peach. Plant Breed 121:184–184

Aranzana MJ, Pineda A, Cosson P, Dirlewanger E, Ascasibar J, Cipriani G, Ryder CD, Testolin R, Abbott A, King GJ, Iezzoni AF, Arus P (2003) A set of simple-sequence repeat (SSR) markers covering the Prunus genome. Theor Appl Genet 106:819–825

Arús PYT, Dirlewanger E, Abbott AG (2006) Synteny in the Rosaceae. Plant Breed Rev 27:175–211

Arús P, Picañol R, Howad W (2010) Almond (*Prunus dulcis*) as a source of genetic variability for peach (*Prunus persica*). In: Plant and animal genome XVIII conference, San Diego, CA, USA

Bao L, Chen KS, Zhang D, Cao YF, Yamamoto T, Teng YW (2007) Genetic diversity and similarity of pear (Pyrus L.) cultivars native to East Asia revealed by SSR (simple sequence repeat) markers. Genet Resour Crop Evol 54:959–971

Bartolini G, Prevost G, Messeri C, Carignani G (1998) Olive germplasm: cultivars and world-wide collections. FAO Library, Rome

Bassil N, Lewers K (2009) Genomics opportunities, new crops and new products. In: Genetics and genomics of Rosaceae. Springer, New York, pp 55–70

Belaj A, Satovic Z, Cipriani G, Baldoni L, Testolin R, Rallo L, Trujillo I (2003) Comparative study of the discriminating capacity of RAPD, AFLP and SSR markers and of their effectiveness in establishing genetic relationships in olive. Theor Appl Genet 107:736–744

Belaj A, Munoz-Diez C, Baldoni L, Porceddu A, Barranco D, Satovic Z (2007) Genetic diversity and population structure of wild olives from the north-western Mediterranean assessed by SSR markers. Ann Bot 100:449–458

Belfanti E, Silfverberg-Dilworth E, Tartarini S, Patocchi A, Barbieri M, Zhu J, Vinatzer BA, Gianfranceschi L, Gessler C, Sansavini S (2004) The HcrVf2 gene from a wild apple confers scab resistance to a transgenic cultivated variety. Proc Natl Acad Sci USA 101:886–890

Bell RL, Scorza R, Srinivasan C, Webb K (1999) Transformation of 'Beurre Bosc' pear with the rolC gene. J Am Soc Hortic Sci 124:570–574

Bielenberg DG, Wang Y, Li ZG, Zhebentyayeva T, Fan SH, Reighard GL, Scorza R, Abbott AG (2008) Sequencing and annotation of the evergrowing locus in peach [*Prunus persica* (L.) Batsch] reveals a cluster of six MADS-box transcription factors as candidate genes for regulation of terminal bud formation. Tree Genet Genomes 4:495–507

Bino RJ, Hall RD, Fiehn O, Kopka J, Saito K, Draper J, Nikolau BJ, Mendes P, Roessner-Tunali U, Beale MH, Trethewey RN, Lange BM, Wurtele ES, Sumner LW (2004) Potential of metabolomics as a functional genomics tool. Trends Plant Sci 9:418–425

Blenda AV, Wechter WP, Reighard GL, Baird WV, Abbott AG (2006) Development and characterization of diagnostic AFLP markers in *Prunus persica* for its response to peach tree short life syndrome. J Hortic Sci Biotechnol 81:281–288

Blenda AV, Verde I, Georgi LL, Reighard GL, Forrest SD, Munoz-Torres M, Baird WV, Abbott AG (2007) Construction of a genetic linkage map and identification of molecular markers in peach rootstocks for response to peach tree short life syndrome. Tree Genet Genomes 3:341–350

Bliss FA, Arulsekar S, Foolad MR, Becerra V, Gillen AM, Warburton ML, Dandekar AM, Kocsisne GM, Mydin KK (2002) An expanded genetic linkage map of Prunus based on an interspecific cross between almond and peach. Genome 45:520–529

Bokszczanin K, Dondini L, Przybyla AA (2009) First report on the presence of fire blight resistance in linkage group 11 of *Pyrus ussuriensis* Maxim. J Appl Genet 50:99–103

Bolar JP, Norelli JL, Wong KW, Hayes CK, Harman GE, Aldwinckle HS (2000) Expression of endochitinase from *Trichoderma harzianum* in transgenic apple increases resistance to apple scab and reduces vigor. Phytopathology 90:72–77

Boskovic R, Tobutt KR (1996) Correlation of stylar ribonuclease zymograms with incompatibility alleles in sweet cherry. Euphytica 90:245–250

Boudehri K, Bendahmane A, Cardinet G, Troadec C, Moing A, Dirlewanger E (2009) Phenotypic and fine genetic characterization of the D locus controlling fruit acidity in peach. BMC Plant Biol 9:59

Breton C, Pinatel C, Medail F, Bonhomme F, Berville A (2008) Comparison between classical and Bayesian methods to investigate the history of olive cultivars using SSR-polymorphisms. Plant Sci 175:524–532

Brini W, Mars M, Hormaza JI (2008) Genetic diversity in local Tunisian pears (*Pyrus communis* L.) studied with SSR markers. Sci Hortic 115:337–341

Broothaerts W, Keulemans J, Van Nerum I (2004) Self-fertile apple resulting from S-RNase gene silencing. Plant Cell Rep 22:497–501

Bulley SM, Wilson FM, Hedden P, Phillips AL, Croker SJ, James DJ (2005) Modification of gibberellin biosynthesis in the grafted apple scion allows control of tree height independent of the rootstock. Plant Biotechnol J 3:215–223

Bureau S, Ruiz D, Reich M, Gouble B, Bertrand D, Audergon JM, Jean-Marc RCMGC (2009) Application of ATR-FTIR for a rapid and simultaneous determination of sugars and organic acids in apricot fruit. Food Chem 115:1133–1140

Burgos L, Pérez-Tornero O, Ballester J, Olmos E (1998) Detection and inheritance of stylar ribonucleases associated with incompatibility alleles in apricot. Sex Plant Reprod 11:153–158

Bus VGM, Chagne D, Bassett HCM, Bowatte D, Calenge F, Celton JM, Durel CE, Malone MT, Patocchi A, Ranatunga AC, Rikkerink EHA, Tustin DS, Zhou J, Gardiner SE (2008) Genome mapping of three major resistance genes to woolly apple aphid (*Eriosoma lanigerum* Hausm.). Tree Genet Genomes 4:223–236

Bus VGM, Esmenjaud D, Buck E, Laurens F (2009) Application of genetic markers in rosaceous crops. In: Genetics and Genomics of Rosaceae. Springer, New York. pp. 563–600

Bus V, Bassett H, Bowatte D, Chagné D, Ranatunga C, Ulluwishewa D, Wiedow C, Gardiner S (2010) Genome mapping of an apple scab, a powdery mildew and a woolly apple aphid resistance gene from open-pollinated Mildew Immune Selection. Tree Genet Genomes 6:477–487

Busot GY, McClure B, Ibarra-Sánchez CP, Jiménez-Durán K, Vázquez-Santana S, Cruz-García F (2008) Pollination in *Nicotiana alata* stimulates synthesis and transfer to the stigmatic surface of NaStEP, a vacuolar Kunitz proteinase inhibitor homologue. J Exp Bot 59:3187–3201

Byrne DH (2007) Molecular marker use in perennial plant breeding. In: Proceedings of the Ivth international symposium on rose research and cultivation, pp 163–167

Caballero P, Fernández MA (2004) Loquat, production and market. Options Méditerrannéennes 58:11–20

Cabrera A, Kozik A, Howad W, Arus P, Iezzoni AF, van der Knaap E (2009) Development and bin mapping of a Rosaceae Conserved Ortholog Set (COS) of markers. BMC Genomics 10:562

Campeol E, Flamini G, Chericoni S, Catalano S, Cremonini R (2001) Volatile compounds from three cultivars of *Olea europaea* from Italy. J Agric Food Chem 49:5409–5411

Canli FA (2004) Development of a second generation genetic linkage map for sour cherry using SSR markers. Pak J Biol Sci 7:1676–1683

Cantini C, Iezzoni AF, Lamboy WF, Boritzki M, Struss D (2001) DNA fingerprinting of tetraploid cherry germplasm using simple sequence repeats. J Am Soc Hortic Sci 126:205–209

Capote N, Perez-Panades J, Monzo C, Carbonell E, Urbaneja A, Scorza R, Ravelonandro M, Cambra M (2008) Assessment of the diversity and dynamics of Plum pox virus and aphid populations in transgenic European plums under Mediterranean conditions. Transgenic Res 17:367–377

Carrera L, Sanzol J, Herrero M, Hormaza JI (2009) Genomic characterization of self-incompatibility ribonucleases (S-RNases) in loquat (*Eriobotrya japonica* Lindl.) (Rosaceae, Pyrinae). Mol Breed 23:539–551

Carriero F, Fontanazza G, Cellini F, Giorio G (2002) Identification of simple sequence repeats (SSRs) in olive (*Olea europaea* L.). Theor Appl Genet 104:301–307

Castro AJ, Bednarczyk A, Schaeffer-Reiss C, Rodriguez-Garcia MI, Van Dorsselaer A, Alche JD (2010) Screening of Ole e 1 polymorphism among olive cultivars by peptide mapping and N-glycopeptide analysis. Proteomics 10:953–962

Cavanna M, Marinoni DT, Bounous G, Botta R (2008) Genetic diversity in ancient apple germplasm from northwest Italy. J Hortic Sci Biotechnol 83:549–554

Celton JM, Tustin DS, Chagne D, Gardiner SE (2009) Construction of a dense genetic linkage map for apple rootstocks using SSRs developed from Malus ESTs and Pyrus genomic sequences. Tree Genet Genomes 5:93–107

Cervera M, López MM, Navarro L, Peña L (1998) Virulence and supervirulence of *Agrobacterium tumefaciens* in woody fruit plants. Physiol Mol Plant Pathol 52:67–78

Cevik V, King GJ (2002) Resolving the aphid resistance locus Sd-1 on a BAC contig within a sub-telomeric region of Malus linkage group 7. Genome 45:939–945

Cevik V, Ryder C, Popovich A, Manning K, King G, Seymour G (2010) A FRUITFULL-like gene is associated with genetic variation for fruit flesh firmness in apple (*Malus domestica* Borkh.). Tree Genet Genomes 6:271–279

Chagne D, Carlisle CM, Blond C, Volz RK, Whitworth CJ, Oraguzie NC, Crowhurst RN, Allan AC, Espley RV, Hellens RP, Gardiner SE (2007) Mapping a candidate gene (MdMYB10) for red flesh and foliage color in apple. BMC Genomics 8:212

Chagne D, Gasic K, Crowhurst RN, Han Y, Bassett HC, Bowatte DR, Lawrence TJ, Rikkerink EHA, Gardiner SE, Korban SS (2008) Development of a set of SNP markers present in expressed genes of the apple. Genomics 92:353–358

Chan ZL, Wang Q, Xu XB, Meng XH, Qin GZ, Li BQ, Tian SP (2008) Functions of defense-related proteins and dehydrogenases in resistance response induced by salicylic acid in sweet cherry fruits at different maturity stages. Proteomics 8:4791–4807

Chaparro JX, Werner DJ, Omalley D, Sederoff RR (1994) Targeted mapping and linkage analysis of morphological isozyme, and RAPD markers in peach. Theor Appl Genet 87:805–815

Chen M, Chen X, Zhou J, Liu Y, Ci Z, Wu Y (2005) Changes of aroma constituents in apricot during fruit development. Sci Agric Sin 38:1244–1299

Chen JL, Yan S, Feng Z, Xiao L, Hu HS (2006) Changes in the volatile compounds and chemical and physical properties of Yali pear (*Pyrus bertschneideri* Reld) during storage. Food Chem 97:248–255

Cheng FS, Weeden NF, Brown SK (1996) Identification of co-dominant RAPD markers tightly linked to fruit skin color in apple. Theor Appl Genet 93:222–227

Cipriani G, Lot G, Huang WG, Marrazzo MT, Peterlunger E, Testolin R (1999) AC/GT and AG/CT microsatellite repeats in peach *Prunus persica* (L) Batsch: isolation, characterisation and cross-species amplification in Prunus. Theor Appl Genet 99:65–72

Cipriani G, Marrazzo MT, Marconi R, Cimato A, Testolin R (2002) Microsatellite markers isolated in olive (*Olea europaea* L.) are suitable for individual fingerprinting and reveal polymorphism within ancient cultivars. Theor Appl Genet 104:223–228

Cirvilleri G, Spina S, Iacona C, Catara A, Muleod R (2008) Study of rhizosphere and phyllosphere bacterial community and resistance to bacterial canker in genetically engineered phytochrome A cherry plants. J Plant Physiol 165:1107–1119

Clarke JB, Tobutt KR (2003) Development and characterization of polymorphic microsatellites from *Prunus avium* 'Napoleon'. Mol Ecol Notes 3:578–580

Claverie M, Dirlewanger E, Cosson P, Bosselut N, Lecouls AC, Voisin R, Kleinhentz M, Lafargue B, Caboche M, Chalhoub B, Esmenjaud D (2004) High-resolution mapping and chromosome landing at the root-knot nematode resistance locus Ma from Myrobalan plum using a large-insert BAC DNA library. Theor Appl Genet 109:1318–1327

Consolandi C, Palmieri L, Doveri S, Maestri E, Marmiroli N, Reale S, Lee D, Baldoni L, Tosti N, Severgnini M, De Bellis G, Castiglioni B (2007) Olive variety identification by ligation detection reaction in a universal array format. J Biotechnol 129:565–574

Consolandi C, Palmieri L, Severgnini M, Maestri E, Marmiroli N, Agrimonti C, Baldoni L, Donini P, De Bellis G, Castiglioni B (2008) A procedure for olive oil traceability and authenticity: DNA extraction, multiplex PCR and LDR-universal array analysis. Eur Food Res Technol 227:1429–1438

Costa F, Van de Weg WE, Stella S, Dondini L, Pratesi D, Musacchi S, Sansavini S (2008) Map position and functional allelic diversity of Md-Exp7, a new putative expansin gene associated with fruit softening in apple (*Malus* × *domestica* Borkh.) and pear (*Pyrus communis*). Tree Genet Genomes 4:575–586

Crane MB, Brown AG (1937) Incompatibility and sterility in the sweet cherry, *Prunus avium* L. J Pomol Hortic Sci 15:86–116

da Câmara L, Machado M, da Câmara MA, Hanzer V, Weiss H, Regner F, Steinkeliner H, Mattanovich D, Plail R, Knapp E, Kaltho B, Katinger HWD (1992) Regeneration of transgenic plants of *Prunus armeniaca* containing the coat protein gene of Plum pox virus. Plant Cell Rep 11:25–29

Dandekar AM, Teo G, Defilippi BG, Uratsu SL, Passey AJ, Kader AA, Stow JR, Colgan RJ, James DJ (2004) Effect of down-regulation of ethylene biosynthesis on fruit flavor complex in apple fruit. Transgenic Res 13:373–384

Dardick CD, Callahan AM, Chiozzotto R, Schaffer RJ, Piagnani MC, Scorza R (2010) Stone formation in peach fruit exhibits spatial coordination of the lignin and flavonoid pathways and similarity to Arabidopsis dehiscence. BMC Biol 8:13

de Bondt A, Eggermont K, Penninckx I, Goderis I, Broekaert WF (1996) Agrobacterium-mediated transformation of apple (*Malus* × *domestica* Borkh): an assessment of factors affecting regeneration of transgenic plants. Plant Cell Rep 15:549–554

de Bondt A, Zaman S, Broekaert WF, Cammue B, Keulemans J (1998) Genetic transformation of apple (*Malus pumila* Mill.) for increased fungal resistance: in vitro antifungal activity in protein extracts of transgenic apple expressing RS-AFP2 or ACE-AMP1. Acta Hortic 484:565–570

de La Rosa R, Angiolillo A, Guerrero C, Pellegrini M, Rallo L, Besnard G, Berville A, Martin A, Baldoni L (2003) A first linkage map of olive (*Olea europaea* L.) cultivars using RAPD, AFLP, RFLP and SSR markers. Theor Appl Genet 106:1273–1282

de Nettancourt D (2001) Incompatibility and incongruity in wild and cultivated plants. Springer, Berlin, pp 1–215

Decroocq V, Fave MG, Hagen L, Bordenave L, Decroocq S (2003) Development and transferability of apricot and grape EST microsatellite markers across taxa. Theor Appl Genet 106:912–922

Decroocq V, Hagen LS, Fave MG, Eyquard JP, Pierronnet A (2004) Microsatellite markers in the hexaploid *Prunus domestica* species and parentage lineage of three European plum cultivars using nuclear and chloroplast simple-sequence repeats. Mol Breed 13:135–142

Decroocq V, Foulongne M, Lambert P, Le Gall O, Mantin C, Pascal T, Schurdi-Levraud V, Kervella J (2005) Analogues of virus resistance genes map to QTLs for resistance to sharka disease in *Prunus davidiana*. Mol Genet Genomics 272:680–689

Degenhardt J, Al-Masri AN, Kurkcuoglu S, Szankowski I, Gau AE (2005) Characterization by suppression subtractive hybridization of transcripts that are differentially expressed in leaves of apple scab-resistant and susceptible cultivars of *Malus domestica*. Mol Genet Genomics 273:326–335

Degenhardt J, Poppe A, Montag J, Szankowski I (2006) The use of phosphomannose-isomerase/mannose selection system to recover transgenic apple plants. Plant Cell Rep 25:1149–1156

Deshaies RJ (1999) SCF and Cullin/Ring H2-based ubiquitin ligases. Annu Rev Cell Dev Biol 15:435–467

Di Nicola-Negri E, Brunetti A, Tavazza M, Ilardi V (2005) Hairpin RNA-mediated silencing of Plum pox virus P1 and HC-Pro genes for efficient and predictable resistance to the virus. Transgenic Res 14:989–994

Dirlewanger E, Moing A, Rothan C, Svanella L, Pronier V, Guye A, Plomion C, Monet R (1999) Mapping QTLs controlling fruit quality in peach (*Prunus persica* (L.) Batsch). Theor Appl Genet 98:18–31

Dirlewanger E, Cosson P, Tavaud M, Aranzana MJ, Poizat C, Zanetto A, Arus P, Laigret F (2002) Development of microsatellite markers in peach *Prunus persica* (L.) Batsch and their use in genetic diversity analysis in peach and sweet cherry (*Prunus avium* L.). Theor Appl Genet 105:127–138

Dirlewanger E, Cosson P, Howad W, Capdeville G, Bosselut N, Claverie M, Voisin R, Poizat C, Lafargue B, Baron O, Laigret F, Kleinhentz M, Arus P, Esmenjaud D (2004a) Microsatellite genetic linkage maps of myrobalan plum and an almond-peach hybrid – location of root-knot nematode resistance genes. Theor Appl Genet 109:827–838

Dirlewanger E, Graziano E, Joobeur T, Garriga-Caldere F, Cosson P, Howad W, Arus P (2004b) Comparative mapping and marker-assisted selection in Rosaceae fruit crops. Proc Natl Acad Sci USA 101:9891–9896

Dirlewanger E, Cosson P, Boudehri K, Renaud C, Capdeville G, Tauzin Y, Laigret F, Moing A (2007) Development of a second-generation genetic linkage map for peach [*Prunus persica* (L.) Batsch] and characterization of morphological traits affecting flower and fruit. Tree Genet Genomes 3:1–13

Dondini L, Pierantoni L, Gaiotti F, Chiodini R, Tartarini S, Bazzi C, Sansavini S (2004) Identifying QTLs for fire-blight resistance via a European pear (*Pyrus communis* L.) genetic linkage map. Mol Breed 14:407–418

Dondini L, Lain O, Geuna F, Banfi R, Gaiotti F, Tartarini S, Bassi D, Testolin R (2007) Development of a new SSR-based linkage map in apricot and analysis of synteny with existing Prunus maps. Tree Genet Genomes 3:239–249

Dondini L, Pierantoni L, Ancarani V, D'Angelo M, Cho KH, Shin IS, Musacchi S, Kang SJ, Sansavini S (2008) The inheritance of the red color character in European pear (*Pyrus communis*) and its map position in the mutated cultivar 'Max Red Bartlett'. Plant Breed 127:524–526

Downey SL, Iezzoni AF (2000) Polymorphic DNA markers in black cherry (*Prunus serotina*) are identified using sequences from sweet cherry, peach, and sour cherry. J Am Soc Hortic Sci 125:76–80

Druart P, Delporte F, Brazda M, Ugarte-Ballon C, da Câmara MA, da Câmara L, Machado M, Jacquemin J, Watillon B (1998) Genetic transformation of cherry trees. Acta Hortic 468:71–76

Dumanoglu H, Gunes NT, Aygun A, San B, Akpinar AE, Bakir M (2009) Analysis of clonal variations in cultivated quince (*Cydonia oblonga* 'Kalecik') based on fruit characteristics and SSR markers. N Z J Crop Hortic Sci 37:113–120

Dunemann F, Ulrich D, Boudichevskaia A, Grafe C, Weber WE (2009) QTL mapping of aroma compounds analyzed by headspace solid-phase microextraction gas chromatography in the apple progeny 'Discovery' × 'Prima'. Mol Breed 23:501–521

Entani T, Iwano M, Shiba H, Che FS, Isogai A, Takayama S (2003) Comparative analysis of the self-incompatibility (S-) locus region of *Prunus mume*: identification of a pollen-expressed F-box gene with allelic diversity. Genes Cells 8:203–213

Escalettes V, Dahuron F, Ravelonandro M, Dosba F (1994) Utilisation de la transgénose pour l'obtention de pruniers et d'abricotiers exprimant le gène de la protéine capside du Plum pox potyvirus. Bull OEPP/EPPO Bull 24:705–711

Esmenjaud D, Dirlewanger E (2007) Plum. In: Kole C (ed) Genome mapping and molecular breeding in plants, vol 4: Fruits and nuts. Springer, Berlin, pp 119–135

Espley RV, Hellens RP, Putterill J, Stevenson DE, Kutty-Amma S, Allan AC (2007) Red colouration in apple fruit is due to the activity of the MYB transcription factor, MdMYB10. Plant J 49:414–427

Etienne C, Rothan C, Moing A, Plomion C, Bodenes C, Svanella-Dumas L, Cosson P, Pronier V, Monet R, Dirlewanger E (2002) Candidate genes and QTLs for sugar and organic acid content in peach [*Prunus persica* (L.) Batsch]. Theor Appl Genet 105:145–159

Evans KM, Govan CL, Fernandez-Fernandez F (2008) A new gene for resistance to *Dysaphis pyri* in pear and identification of flanking microsatellite markers. Genome 51:1026–1031

Fabbri A, Lambardi M, Ozden-Tokatli Y (2009) Olive breeding. In: Priyadarshan PM, Jain M (eds) Breeding plantation tree crops: tropical species. Springer, New York, pp 423–465

Faize M, Malnoy M, Dupuis F, Chevalier M, Parisi L, Chevreau E (2003) Chitinases of *Trichoderma atroviride* induce scab resistance and some metabolic changes in two cultivars of apple. Phytopathology 93:1496–1504

Faize M, Sourice S, Dupuis F, Parisi L, Gautier MF, Chevreau E (2004) Expression of wheat puroindoline-b reduces scab susceptibility in transgenic apple (*Malus* × *domestica* Borkh.). Plant Sci 167:347–354

Fan S, Bielenberg DG, Zhebentyayeva TN, Reighard GL, Okie WR, Holland D, Abbott AG (2010) Mapping quantitative trait loci associated with chilling requirement, heat requirement and bloom date in peach (*Prunus persica*). New Phytol 185:917–930

Fang JG, Twito T, Zhang Z, Chao CCT (2006) Genetic relationships among fruiting-mei (*Prunus mume* Sieb. et Zucc.) cultivars evaluated with AFLP and SNP markers. Genome 49:1256–1264

FAOSTAT (2010) ProdStat: http://faostat.fao.org/site/339/default.aspx. Last time accessed Mar 2010

Fenn JB, Mann M, Meng CK, Wong SF, Whitehouse CM (1989) Electrospray ionization for mass-spectrometry of large biomolecules. Science 246:64–71

Flachowsky H, Peil A, Sopanen T, Elo A, Hanke V (2007) Overexpression of BpMADS4 from silver birch (*Betula pendula* Roth.) induces early-flowering in apple (*Malus* × *domestica* Borkh.). Plant Breed 126:137–145

Flachowsky H, Hättasch C, Höfer M, Peil A, Hanke MV (2010a) Overexpression of LEAFY in apple leads to a columnar phenotype with shorter internodes. Planta 231:251–263

Flachowsky H, Szankowski I, Fischer TC, Richter K, Peil A, Hofer M, Dorschel C, Schmoock S, Gau AE, Halbwirth H, Hanke MV (2010b) Transgenic apple plants overexpressing the Lc gene of maize show an altered growth habit and increased resistance to apple scab and fire blight. Planta 231:623–635

Folta KM, Gardiner SE (2009) Genetics and genomics of Rosaceae. Springer, New York, pp 636

Fonseca S, Hackler L, Zvara A, Ferreira S, Balde A, Dudits D, Pals MS, Puskas LG (2004) Monitoring gene expression along pear fruit development, ripening and senescence using cDNA microarrays. Plant Sci 167:457–469

Fukuda S, Hiehata N, Yamamoto T, Terai O, Nesumi H (2005) Development of DNA markers linked to the resistance gene (Pse a) to loquat canker. J Jpn Soc Hortic Sci 74:345–349

Fuzfai Z, Katona ZF, Kovacs E, Molnar-Perl I (2004) Simultaneous identification and quantification of the sugar, sugar alcohol, and carboxylic acid contents of sour cherry, apple, and ber fruits, as

their trimethylsilyl derivatives, by gas chromatography-mass spectrometry. J Agric Food Chem 52:7444–7452
Galla G, Barcaccia G, Ramina A, Collani S, Alagna F, Baldoni L, Cultrera N, Martinelli F, Sebastiani L, Tonutti P (2009) Computational annotation of genes differentially expressed along olive fruit development. BMC Plant Biol 9:128
Gao ZS, van de Weg WE, Schaart JG, van Arkel G, Breiteneder H, Hoffmann-Sommergruber K, Gilissen L (2005) Genomic characterization and linkage mapping of the apple allergen genes Mal d 2 (thaumatin-like protein) and Mal d 4 (profilin). Theor Appl Genet 111:1087–1097
Gao M, Matsuta N, Murayama H, Toyomasu T, Mitsuhashi W, Dandekar AM, Tao R, Nishimura K (2007) Gene expression and ethylene production in transgenic pear (*Pyrus communis* cv. 'La France') with sense or antisense cDNA encoding ACC oxidase. Plant Sci 173:32–42
Gardiner SE, Bus VGM, Rusholme RL, Chagné D, Rikkerink EHA (2007) Apple. In: Kole C (ed) Genome mapping and molecular breeding in plants, vol 4: Fruits and nuts. Springer, Berlin, pp 1–62
Garkava-Gustavsson L, Brantestam AK, Sehic J, Nybom H (2008) Molecular characterisation of indigenous Swedish apple cultivars based on SSR and S-allele analysis. Hereditas 145:99–112
Gasic K, Han YP, Kertbundit S, Shulaev V, Iezzoni AF, Stover EW, Bell RL, Wisniewski ME, Korban SS (2009) Characteristics and transferability of new apple EST-derived SSRs to other Rosaceae species. Mol Breed 23:397–411
Georgi LL, Wang Y, Yvergniaux D, Ormsbee T, Inigo M, Reighard G, Abbott AG (2002) Construction of a BAC library and its application to the identification of simple sequence repeats in peach [*Prunus persica* (L.) Batsch]. Theor Appl Genet 105:1151–1158
Gharghani A, Zamani Z, Talaie A, Oraguzie NC, Fatahi R, Hajnajari H, Wiedow C, Gardiner SE (2009) Genetic identity and relationships of Iranian apple (*Malus* × *domestica* Borkh.) cultivars and landraces, wild *Malus* species and representative old apple cultivars based on simple sequence repeat (SSR) marker analysis. Genet Resour Crop Evol 56:829–842
Ghosh AK, Lukens LN, Hunter DM, Strommer JN (2006) European and Asian pears: simple sequence repeat-polyacrylamide gel electrophoresis-based analysis of commercially important North American cultivars. HortScience 41:304–309
Gianfranceschi L, Seglias N, Tarchini R, Komjanc M, Gessler C (1998) Simple sequence repeats for the genetic analysis of apple. Theor Appl Genet 96:1069–1076
Gil FS, Busconi M, Machado AD, Fogher C (2006) Development and characterization of microsatellite loci from *Olea europaea*. Mol Ecol Notes 6:1275–1277
Gilissen L, Bolhaar STH, Matos CI, Rouwendal GJA, Boone MJ, Krens FA, Zuidmeer L, van Leeuwen A, Akkerdaas J, Hoffmann-Sommergruber K, Knulst AC, Bosch D, van de Weg WE, van Ree R (2005) Silencing the major apple allergen Mal d 1 by using the RNA interference approach. J Allergy Clin Immunol 115:364–369
Gisbert AD, Lopez-Capuz I, Soriano JM, Llacer G, Romero C, Badenes ML (2009a) Development of microsatellite markers from loquat, *Eriobotrya japonica* (Thunb.) Lindl. Mol Ecol Resour 9:803–805
Gisbert AD, Romero C, Martinez-Calvo J, Leida C, Llacer G, Badenes ML (2009b) Genetic diversity evaluation of a loquat (*Eriobotrya japonica* (Thunb) Lindl) germplasm collection by SSRs and S-allele fragments. Euphytica 168:121–134
Goldraij A, Kondo K, Lee CB, Hancock CN, Sivaguru M, Vazquez-Santana S, Kim S, Phillips TE, Cruz-García F, McClure B (2006) Compartmentalization of S-RNase and HT-B degradation in self-incompatible Nicotiana. Nature 439:805–810
Goldway M, Sapir G, Stern RA (2007) Molecular basis and horticultural application of the gametophytic self-incompatibility system in Rosaceous tree fruit. Plant Breed Rev 28:215–237
Golz JF, Oh HY, Su V, Kusaba M, Newbigin E (2001) Genetic analysis of Nicotiana pollen-part mutants is consistent with the presence of an S-ribonuclease inhibitor at the S locus. Proc Natl Acad Sci USA 98:15372–15376
Granell A, Crisosto CH, Martí-Ibáñez C, Gradziel TM, Froment J, Peace C (2006) "ChillPeach" a functional database to understand peach chilling injury. In: Third international Rosaceae genomics conference, Napier. http://mainlabclemsonedu/gdr/comunity/conferences/RG3_abstractspdf

Grimplet J, Romieu C, Audergon J-M, Marty I, Albagnac G, Lambert P, Bouchet J-P, Terrier N (2005) Transcriptomic study of apricot fruit (*Prunus armeniaca*) ripening among 13006 expressed sequence tags. Physiol Plant 125:281–292

Guarino C, Santoro S, De Simone L, Lain O, Cipriani G, Testolin R (2006) Genetic diversity in a collection of ancient cultivars of apple (*Malus* × *domestica* Borkh.) as revealed by SSR-based fingerprinting. J Hortic Sci Biotechnol 81:39–44

Guarino C, Arena S, De Simone L, D'Ambrosio C, Santoro S, Rocco M, Scaloni A, Marra M (2007) Proteomic analysis of the major soluble components in Annurca apple flesh. Mol Nutr Food Res 51:255–262

Guichard E, Souty M (1988) Comparison of the relative quantities of aroma compounds found in fresh apricot (*Prunus armeniaca*) from 6 different varieties. Z Lebensm Unters Forsch 186:301–307

Guilford P, Prakash S, Zhu JM, Rikkerink E, Gardiner S, Bassett H, Forster R (1997) Microsatellites in *Malus* × *domestica* (apple): abundance, polymorphism and cultivar identification. Theor Appl Genet 94:249–254

Guillot S, Peytavi L, Bureau S, Boulanger R, Lepoutre JP, Crouzet J, Schorr-Galindo S (2006) Aroma characterization of various apricot varieties using headspace-solid phase microextraction combined with gas chromatography-mass spectrometry and gas chromatography-olfactometry. Food Chem 96:147–155

Gutiérrez-Pesce P, Taylor K, Muleo R, Rugini E (1998) Somatic embryogenesis and shoot regeneration from transgenic roots of the cherry rootstock Colt (*Prunus avium* × *P. pseudocerasus*) mediated by pRi 1855T-DNA of *Agrobacterium rhizogenes*. Plant Cell Rep 17:574–580

Hagen LS, Chaib J, Fady B, Decroocq V, Bouchet JP, Lambert P, Audergon JM (2004) Genomic and cDNA microsatellites from apricot (*Prunus armeniaca* L.). Mol Ecol Notes 4:742–745

Hammerschlag FA, Smigocki AC (1998) Growth and in vitro propagation of peach plants transformed with the shooty mutant strain of *Agrobacterium tumefaciens*. HortScience 33:897–899

Han Y, Gasic K, Marron B, Beever JE, Korban SS (2007) A BAC-based physical map of the apple genome. Genomics 89:630–637

Han YP, Chagne D, Gasic K, Rikkerink EHA, Beever JE, Gardiner SE, Korban SS (2009) BAC-end sequence-based SNPs and Bin mapping for rapid integration of physical and genetic maps in apple. Genomics 93:282–288

Harada T, Sunako T, Wakasa Y, Soejima J, Satoh T, Niizeki M (2000) An allele of the 1-aminocyclopropane-1-carboxylate synthase gene (Md-ACS1) accounts for the low level of ethylene production in climacteric fruits of some apple cultivars. Theor Appl Genet 101:742–746

Hauck NR, Yamane H, Tao R, Iezzoni AF (2006) Accumulation of non-functional S-haplotypes results in the breakdown of gametophytic self-incompatibility in tetraploid Prunus. Genetics 172:1191–1198

He LX, Ban Y, Inoue H, Matsuda N, Liu JH, Moriguchi T (2008) Enhancement of spermidine content and antioxidant capacity in transgenic pear shoots overexpressing apple spermidine synthase in response to salinity and hyperosmosis. Phytochemistry 69:2133–2141

Hemmat M, Weeden NF, Manganaris AG, Lawson DM (1994) Molecular marker linkage map for apple. J Hered 85:4–11

Herndl A, Marzban G, Kolarich D, Hahn R, Boscia D, Hemmer W, Maghuly F, Stoyanova E, Katinger H, Laimer M (2007) Mapping of *Malus domestica* allergens by 2-D electrophoresis and IgE-reactivity. Electrophoresis 28:437–448

Heyens K, Valcke R, Dumont D, Robben J, Noben JP (2006) Differential expression of proteins in apple following inoculation with *Erwinia amylovoria*. Acta Hort(ISHS) 704:489–494

Hily JM, Scorza R, Malinowski T, Zawadzka B, Ravelonandro M (2004) Stability of gene silencing-based resistance to Plum pox virus in transgenic plum (*Prunus domestica* L.) under field conditions. Transgenic Res 13:427–436

Hily JM, Scorza R, Webb K, Ravelonandro M (2005) Accumulation of the long class of siRNA is associated with resistance to Plum pox virus in a transgenic woody perennial plum tree. Mol Plant Microbe Interact 18:794–799

Hily JM, Ravelonandro M, Damsteegt V, Bassett C, Petri C, Liu Z, Scorza R (2007) Plum poxvirus coat protein gene intron-hairpin-RNA (ihpRNA) constructs provide resistance to Plum pox virus in *Nicotiana benthamiana* Domin. and plum (*Prunus domestica* L.). J Am Soc Hortic Sci 132:850–858

Hokanson SC, Szewc-McFadden AK, Lamboy WF, McFerson JR (1998) Microsatellite (SSR) markers reveal genetic identities, genetic diversity and relationships in a *Malus* × *domestica* Borkh. core subset collection. Theor Appl Genet 97:671–683

Hokanson SC, Lamboy WF, Szewc-McFadden AK, McFerson JR (2001) Microsatellite (SSR) variation in a collection of Malus (apple) species and hybrids. Euphytica 118:281–294

Hormaza JI, Yamane H, Rodrigo J (2007) Apricot. In: Genome Mapping and Molecular Breeding in plants (Fruits and Nuts). Springer-Verlag, Berlin pp 171–187

Horn R, Lecouls AC, Callahan A, Dandekar A, Garay L, McCord P, Howad W, Chan H, Verde I, Main D, Jung S, Georgi L, Forrest S, Mook J, Zhebentyayeva T, Yu YS, Kim HR, Jesudurai C, Sosinski B, Arus P, Baird V, Parfitt D, Reighard G, Scorza R, Tomkins J, Wing R, Abbott AG (2005) Candidate gene database and transcript map for peach, a model species for fruit trees. Theor Appl Genet 110:1419–1428

Horvath A, Zanetto A, Tavaud M, Christmann H, Laigret F (2008) Origin of sour cherry (*Prunus cerasus* L.) genomes. In: Proceedings of the Vth international cherry symposium, vol 1–2, pp 131–136

Howad W, Yamamoto T, Dirlewanger E, Testolin R, Cosson P, Cipriani G, Monforte AJ, Georgi L, Abbott AG, Arus P (2005) Mapping with a few plants: using selective mapping for microsatellite saturation of the Prunus reference map. Genetics 171:1305–1309

Huang S, Lee HS, Karunanandaa B, Kao TH (1994) Ribonuclease activity of *Petunia inflata* S proteins is essential for rejection of self-pollen. Plant Cell 6:1021–1028

Igarashi A, Yamagata K, Sugai T, Takahashi Y, Sugawara E, Tamura A, Yaegashi H, Yamagishi N, Takahashi T, Isogai M, Takahashi H, Yoshikawa N (2009) Apple latent spherical virus vectors for reliable and effective virus-induced gene silencing among a broad range of plants including tobacco, tomato, *Arabidopsis thaliana*, cucurbits, and legumes. Virology 386:407–416

Igic B, Kohn JR (2001) Evolutionary relationships among self-incompatibility RNases. Proc Natl Acad Sci USA 98:13167–13171

Ikeda K, Igic B, Ushijima K, Yamane H, Hauck NR, Nakano R, Sassa H, Iezzoni AF, Kohn JR, Tao R (2004) Primary structural features of the S haplotype-specific F-box protein, SFB, in Prunus. Sex Plant Reprod 16:235–243

Ikeda K, Ushijima K, Yamane H, Tao R, Hauck NR, Sebolt AM, Iezzoni AF (2005) Linkage and physical distances between the S-haplotype S-RNase and SFB genes in sweet cherry. Sex Plant Reprod 17:289–296

Iketani H, Abe K, Yamamoto T, Kotobuki K, Sato Y, Saito T, Terai O, Matsuta N, Hayashi T (2001) Mapping of disease-related genes in Japanese pear using a molecular linkage map with RAPD markers. Breed Sci 51:179–184

Illa E, Lambert P, Quilot B, Audergon JM, Dirlewanger E, Howad W, Dondini L, Tartarini S, Lain O, Testolin R, Bassi D, Arus P (2009) Linkage map saturation, construction, and comparison in four populations of Prunus. J Hortic Sci Biotechnol 84:168–175

Inoue E, Kasumi M, Sakuma F, Anzai H, Amano K, Hara H (2006) Identification of RAPD marker linked to fruit skin color in Japanese pear (*Pyrus pyrifolia* Nakai). Sci Hortic 107:254–258

Ishikawa S, Kato S, Imakawa S, Mikami T, Shimamoto Y (1992) Organelle DNA polymorphism in apple cultivars and rootstocks. Theor Appl Genet 83(8):963–967

James DJ, Uratsu SL, Cheng J, Negri P, Viss P, Dandekar AM (1993) Acetosyringone and osmoprotectans like betaine or proline synergistically enhance Agrobacterium-mediated transformation of apple. Plant Cell Rep 12:559–563

Janssen BJ, Thodey K, Schaffer RJ, Alba R, Balakrishnan L, Bishop R, Bowen JH, Crowhurst RN, Gleave AP, Ledger S, McArtney S, Pichler FB, Snowden KC, Ward S (2008) Global gene expression analysis of apple fruit development from the floral bud to ripe fruit. BMC Plant Biol 8:16

Jauregui B, de Vicente MC, Messeguer R, Felipe A, Bonnet A, Salesses G, Arus P (2001) A reciprocal translocation between 'Garfi' almond and 'Nemared' peach. Theor Appl Genet 102:1169–1176

Jimenez S, Li ZG, Reighard GL, Bielenberg DG (2010) Identification of genes associated with growth cessation and bud dormancy entrance using a dormancy-incapable tree mutant. BMC Plant Biol 10:25

Joobeur T, Viruel MA, de Vicente MC, Jauregui B, Ballester J, Dettori MT, Verde I, Truco MJ, Messeguer R, Batlle I, Quarta R, Dirlewanger E, Arus P (1998) Construction of a saturated linkage map for Prunus using an almond x peach F-2 progeny. Theor Appl Genet 97:1034–1041

Joshi S, Soriano JM, Schaart J, Broggini GAL, Szankowski I, Jacobsen E, Krens F, Schouten H (2009) Approaches for development of cisgenic apples. Transgenic Plant J 3:40–46

Juárez-Díaz JA, McClure B, Vázquez-Santana S, Guevara-García A, León-Mejía P, Márquez-Guzmán J, Cruz–García F (2006) A novel thioredoxin h is secreted in *Nicotiana alata* and reduces S-RNases in vitro. J Biol Chem 281:3418–3424

Jung S, Main D, Staton M, Cho IH, Zhebentyayeva T, Arus P, Abbott A (2006) Synteny conservation between the Prunus genome and both the present and ancestral Arabidopsis genomes. BMC Genomics 7:8

Jung S, Jiwan D, Cho IH, Lee TI, Abbott A, Sosinski B, Main D (2009) Synteny of Prunus and other model plant species. BMC Genomics 10:76

Karas M, Hillenkamp F (1988) Laser desorption ionization of proteins with molecular masses exceeding 10000 daltons. Anal Chem 60:2299–2301

Katayama H, Adachi S, Yamamoto T, Uematsu C (2007) A wide range of genetic diversity in pear (*Pyrus ussuriensis* var. aromatica) genetic resources from Iwate, Japan revealed by SSR and chloroplast DNA markers. Genet Resour Crop Evol 54:1573–1585

Kimura T, Shi YZ, Shoda M, Kotobuki K, Matsuta N, Hayashi T, Ban Y, Yamamoto T (2002) Identification of Asian pear varieties by SSR analysis. Breed Sci 52:115–121

Kitahara K, Matsumoto S, Yamamoto T, Soejima J, Kimura T, Komatsu H, Abe K (2005) Molecular characterization of apple cultivars in Japan by S-RNase analysis and SSR markers. J Am Soc Hortic Sci 130:885–892

Ko K, Norelli JL, Reynoird JP, Boresjza-Wysocka E, Brown SK, Aldwinckle HS (2000) Effect of untranslated leader sequence of AMV RNA 4 and signal peptide of pathogenesis-related protein 1b on attacin gene expression, and resistance to fire blight in transgenic apple. Biotechnol Lett 22:373–381

Kotoda N, Iwanami H, Takahashi S, Abe K (2006) Antisense expression of MdTFL1, a TFL1-like gene, reduces the juvenile phase in apple. J Am Soc Hortic Sci 131:74–81

Krath BN, Eriksen FD, Pedersen BH, Gilissen L, Van de Weg WE, Dragsted LO (2009) Development of hypo-allergenic apples: silencing of the major allergen Mal d 1 gene in 'Elstar' apple and the effect of grafting. J Hortic Sci Biotechnol 84:52–57

Kundu JK, Briard P, Hily JM, Ravelonandro M, Scorza R (2008) Role of the 25–26 nt siRNA in the resistance of transgenic *Prunus domestica* graft inoculated with Plum pox virus. Virus Genes 36:215–220

Lai Z, Ma W, Han B, Liang L, Zhang Y, Hong G, Xue Y (2002) An F-box gene linked to the self-incompatibility (S) locus of Antirrhinum is expressed specifically in pollen and tapetum. Plant Mol Biol 50:29–42

Lalli DA, Decroocq V, Blenda AV, Schurdi-Levraud V, Garay L, Le Gall O, Damsteegt V, Reighard GL, Abbott AG (2005) Identification and mapping of resistance gene analogs (RGAs) in Prunus: a resistance map for Prunus. Theor Appl Genet 111:1504–1513

Lalli DA, Abbott AG, Zhebentyayeva TN, Badenes ML, Damsteegt V, Polak J, Krska B, Salava J (2008) A genetic linkage map for an apricot (*Prunus armeniaca* L.) BC1 population mapping Plum pox virus resistance. Tree Genet Genomes 4:481–493

Lambert P, Dicenta F, Rubio M, Audergon JM (2007) QTL analysis of resistance to sharka disease in the apricot (*Prunus armeniaca* L.) 'Polonais' × 'Stark Early Orange' F1 progeny. Tree Genet Genomes 3:299–309

Lara MV, Borsani J, Budde CO, Lauxmann MA, Lombardo VA, Murray R, Andreo CS, Drincovich MF (2009) Biochemical and proteomic analysis of 'Dixiland' peach fruit (*Prunus persica*) upon heat treatment. J Exp Bot 60:4315–4333

Lau JM, Korban SS (2010) Transgenic apple expressing an antigenic protein of the human respiratory syncytial virus. J Plant Physiol. doi:10.1016/j.jplph.2010.02.003
Layne REC, Sherman WB (1986) Interspecific hybridization of Prunus. HortScience 21:48–51
Lazzari B, Caprera A, Vecchietti A, Stella A, Milanesi L, Pozzi C (2005) ESTree db: a tool for peach functional genomics. BMC Bioinformatics 6:S16
Lee YP, Yu GH, Seo YS, Han SE, Choi YO, Kim D, Mok IG, Kim WT, Sung SK (2007) Microarray analysis of apple gene expression engaged in early fruit development. Plant Cell Rep 26:917–926
Lee CB, Kim S, McClure B (2009) A pollen protein, NaPCCP, that binds pistil arabinogalactan proteins also binds phosphatidylinositol 3-phosphate and associates with the pollen tube endomembrane system. Plant Physiol 149:791–802
Leida C, Martí GJT, Agustí M, Llácer G, Badenes ML, Ríos G (2010) Identification of genes associated with bud dormancy-release in *Prunus persica* by suppression subtractive hybridization. Tree Physiol 30:655–666
Lewis D, Modlibowska I (1942) Genetical studies in pears IV. Pollen-tube growth and incompatibility. J Genet 43:211–222
Li HH, Flachowsky H, Fischer TC, Hanke MV, Forkmann G, Treutter D, Schwab W, Hoffmann T, Szankowski I (2007) Maize Lc transcription factor enhances biosynthesis of anthocyanins, distinct proanthocyanidins and phenylpropanoids in apple (*Malus domestica* Borkh.). Planta 226:1243–1254
Liebhard R, Gianfranceschi L, Koller B, Ryder CD, Tarchini R, Van de Weg E, Gessler C (2002) Development and characterisation of 140 new microsatellites in apple (*Malus* × *domestica* Borkh.). Mol Breed 10:217–241
Liebhard R, Koller B, Gianfranceschi L, Gessler C (2003) Creating a saturated reference map for the apple (*Malus* × *domestica* Borkh.) genome. Theor Appl Genet 106:1497–1508
Lin-Wang K, Bolitho K, Grafton K, Kortstee A, Karunairetnam S, McGhie T, Espley R, Hellens R, Allan A (2010) An R2R3 MYB transcription factor associated with regulation of the anthocyanin biosynthetic pathway in Rosaceae. BMC Plant Biol 10:50
Lopes MS, Sefc KM, Laimer M, Machado AD (2002) Identification of microsatellite loci in apricot. Mol Ecol Notes 2:24–26
Lopes MS, Mendonca D, Sefc KM, Gil FS, Machado AD (2004) Genetic evidence of intra-cultivar variability within Iberian olive cultivars. HortScience 39:1562–1565
López-Noguera S, Petri C, Burgos L (2009) Combining a regeneration-promoting ipt gene and site-specific recombination allows a more efficient apricot transformation and the elimination of marker genes. Plant Cell Rep 28:1781–1790
Lu X-E, Liu Y-Q (2008) Analysis of volatile oil in pre and post processed pieces of *Eriobotrya japonica* by GC-MS. Zhong Yao Cai 31:1625–1626
Luu DT, Qin XK, Morse D, Cappadocia M (2000) S-RNase uptake by compatible pollen tubes in GSI. Nature 407:649–651
Maghuly F, Machado AD, Leopold S, Khan MA, Katinger H, Laimer M (2007) Long-term stability of marker gene expression in *Prunus subhirtella*: a model fruit tree species. J Biotechnol 127:310–321
Maheswaran G, Pridmore L, Franz P, Anderson MA (2007) A proteinase inhibitor from *Nicotiana alata* inhibits the normal development of light-brown apple moth, *Epiphyas postvittana* in transgenic apple plants. Plant Cell Rep 26:773–782
Maliepaard C, Alston FH, van Arkel G, Brown LM, Chevreau E, Dunemann F, Evans KM, Gardiner S, Guilford P, van Heusden AW, Janse J, Laurens F, Lynn JR, Manganaris AG, den Nijs APM, Periam N, Rikkerink E, Roche P, Ryder C, Sansavini S, Schmidt H, Tartarini S, Verhaegh JJ, Vrielink-van Ginkel M, King GJ (1998) Aligning male and female linkage maps of apple (*Malus pumila* Mill.) using multi-allelic markers. Theor Appl Genet 97:60–73
Malnoy M, Aldwinckle HS (2009) Apple transformation and translational genomics. In: Genetics and genomics of Rosaceae. Springer, New York, pp 143–162
Malnoy M, Venisse JS, Chevreau E (2005a) Expression of a bacterial effector, harpin N, causes increased resistance to fire blight in *Pyrus communis*. Tree Genet Genomes 1:41–49
Malnoy M, Faize M, Venisse JS, Geider K, Chevreau E (2005b) Expression of viral EPS-depolymerase reduces fire blight susceptibility in transgenic pear. Plant Cell Rep 23:632–638

Malnoy M, Jin Q, Borejsza-Wysocka EE, He Y, Aldwinckle HS (2007) Overexpression of the apple MpNPR1 gene confers increased disease resistance in *Malus × domestica*. Mol Plant Microbe Interact 20:568–1580

Malnoy M, Borejsza-Wysocka EE, Pascal-Omenaca L, Aldwinckle HS, Oh CS, Beer SV (2008a) Silencing of HIPM, the apple protein that interacts with HrpN Of *Erwinia amylovora*. Acta Hortic 793:261–264

Malnoy M, Xu M, Borejsza-Wysocka E, Korban SS, Aldwinckle HS (2008b) Two receptor-like genes, Vfa1 and Vfa2, confer resistance to the fungal pathogen *Venturia inaequalis* inciting apple scab disease. Mol Plant Microbe Interact 21:448–458

Marandel G, Pascal T, Candresse T, Decroocq V (2009) Quantitative resistance to Plum pox virus in *Prunus davidiana* P1908 linked to components of the eukaryotic translation initiation complex. Plant Pathol 58:425–435

Marchese A, Boskovic R, Caruso T, Raimondo A, Cutuli M, Tobutt KR (2007) A new self-incompatibility haplotype in sweet cherry 'Kronio', S5′ attributable to a pollen-part mutation in the SFB gene. J Exp Bot 58:4347–4356

Matsuda N, Ikeda K, Kurosaka M, Takashina T, Isuzugawa K, Endo T, Omura M (2009) Early flowering phenotype in transgenic pears (*Pyrus communis* L.) expressing the CiFT gene. J Jpn Soc Hortic Sci 78:410–416

McClure BA, Haring V, Ebert PR, Anderson MA, Simpson RJ, Sakiyama F, Clarke AE (1989) Style self-incompatibility gene products of *Nicotiana alata* are ribonucleases. Nature 342:955–957

McClure B, Mou B, Canevascini S, Bernatzky R (1999) A small asparagine-rich protein required for S-allele-specific pollen rejection in Nicotiana. Proc Natl Acad Sci USA 96:13548–13553

McClure BA, Cruz-García F, Beecher B, Sulaman W (2000) Factors affecting inter- and intra-specific pollen rejection in Nicotiana. Ann Bot 85:113–123

McCouch SR (2001) Genomics and synteny. Plant Physiol 125:152–155

Messeguer R, Arus P, Carrera M (1987) Identification of peach cultivars with pollen isozymes. Sci Hortic 31:107–117

Messina R, Lain O, Marrazzo MT, Cipriani G, Testolin R (2004) New set of microsatellite loci isolated in apricot. Mol Ecol Notes 4:432–434

Mezzetti B, Pandolfini T, Navacchi O, Landi L (2002) Genetic transformation of *Vitis vinifera* via organogenesis. BMC Biotechnol 2:18

Micheletti D, Troggio M, Baldi P, Costa F, Malnoy M, Magnano P, Velasco R, Salvi S (2010) LD estimation, analyses of diversity and domestication in apple. In: Second international symposium on genomics of plant genetic resources, Bologna, April 2010

Mnejja M, Garcia-Mas M, Howad W, Badenes ML, Arus P (2004) Simple-sequence repeat (SSR) markers of Japanese plum (*Prunus salicina* Lindl.) are highly polymorphic and transferable to peach and almond. Mol Ecol Notes 4:163–166

Mnejja M, Garcia-Mas J, Howad W, Arus P (2005) Development and transportability across Prunus species of 42 polymorphic almond microsatellites. Mol Ecol Notes 5:531–535

Mnejja M, Garcia-Mas J, Audergon J-M, Arús P (2010) Prunus microsatellite marker transferability across rosaceous crops. Tree Genet Genomes. doi: 10.1007/s11295-010-0284-z

Moreau L, Charcosset A, Gallais A (2004) Use of trial clustering to study QTL x environment effects for grain yield and related traits in maize. Theor Appl Genet 110:92–105

Moriya S, Iwanami H, Kotoda N, Takahashi S, Yamamoto T, Abe K (2009) Development of a marker-assisted selection system for columnar growth habit in apple breeding. J Jpn Soc Hortic Sci 78:279–287

Muleo R, Colao MC, Miano D, Cirilli M, Intrieri MC, Baldoni L, Rugini E (2009) Mutation scanning and genotyping by high-resolution DNA melting analysis in olive germplasm. Genome 52:252–260

Napoli A, Aiello D, Di Donna L, Moschidis P, Sindona G (2008) Vegetable proteomics: the detection of ole e 1 isoallergens by peptide matching of MALDI MS/MS spectra of underivatized and dansylated glycopeptides. J Proteome Res 7:2723–2732

Newcomb RD, Crowhurst RN, Gleave AP, Rikkerink EHA, Allan AC, Beuning LL, Bowen JH, Gera E, Jamieson KR, Janssen BJ, Laing WA, McArtney S, Nain B, Ross GS, Snowden KC,

Souleyre EJF, Walton EF, Yauk Y-K (2006) Analyses of expressed sequence tags from apple. Plant Physiol 141:147–166
Nilo R, Saffie C, Lilley K, Baeza-Yates R, Cambiazo V, Campos-Vargas R, Gonzalez M, Meisel LA, Retamales J, Silva H, Orellana A (2010) Proteomic analysis of peach fruit mesocarp softening and chilling injury using difference gel electrophoresis (DIGE). BMC Genomics 11:43
O'farrell PZ, Goodman HM, Ofarrell PH (1977) High-resolution 2-dimensional electrophoresis of basic as well as acidic proteins. Cell 12:1133–1141
Obenland DM, Vensel WH, Hurkman WJ (2008) Alterations in protein expression associated with the development of mealiness in peaches. J Hortic Sci Biotechnol 83:85–93
Ogundiwin EA, Marti C, Forment J, Pons C, Granell A, Gradziel TM, Peace CP, Crisosto CH (2008) Development of ChillPeach genomic tools and identification of cold-responsive genes in peach fruit. Plant Mol Biol 68:379–397
Ogundiwin EA, Peace CP, Gradziel TM, Parfitt DE, Bliss FA, Crisosto CH (2009) A fruit quality gene map of Prunus. BMC Genomics 10:587
Olukolu BA, Trainin T, Fan SH, Kole C, Bielenberg DG, Reighard GL, Abbott AG, Holland D (2009) Genetic linkage mapping for molecular dissection of chilling requirement and budbreak in apricot (*Prunus armeniaca* L.). Genome 52:819–828
Oraguzie NC, Yamamoto T, Soejima J, Suzuki T, De Silva HN (2005) DNA fingerprinting of apple (*Malus* spp.) rootstocks using simple sequence repeats. Plant Breed 124:197–202
Oshita S, Shima K, Haruta T, Seo Y, Kawagoe Y, Nakayama S, Takahara H (2000) Discrimination of odors emanating from 'La France' pear by semi-conducting polymer sensors. Comput Electron Agric 26:209–216
Pandolfini T, Molesini B, Avesani L, Spena A, Polverari A (2003) Expression of self-complementary hairpin RNA under the control of the rolC promoter confers systemic disease resistance to Plum pox virus without preventing local infection. BMC Biotechnol 3:7
Park S, Sugimoto N, Larson MD, Beaudry R, van Nocker S (2006) Identification of genes with potential roles in apple fruit development and biochemistry through large-scale statistical analysis of expressed sequence tags. Plant Physiol 141:811–824
Peace CP, Crisosto CH, Gradziel TM (2005) Endopolygalacturonase: a candidate gene for freestone and melting flesh in peach. Mol Breed 16:21–31
Pedreschi R, Vanstreels E, Carpentier S, Hertog M, Lammertyn J, Robben J, Noben JP, Swennen R, Vanderleyden J, Nicolai BM (2007) Proteomic analysis of core breakdown disorder in conference pears (*Pyrus communis* L.). Proteomics 7:2083–2099
Pedreschi R, Hertog M, Robben J, Lilley KS, Karp NA, Baggerman G, Vanderleyden J, Nicolai B (2009) Gel-based proteomics approach to the study of metabolic changes in pear tissue during storage. J Agric Food Chem 57:6997–7004
Pedryc A, Ruthner S, Hermán R, Krska B, Hegedus A, Halász J (2009) Genetic diversity of apricot revealed by a set of SSR markers from linkage group G1. Sci Hortic 121:19–26
Pereira-Lorenzo S, Ramos-Cabrer AM, Gonzalez-Diaz AJ, Diaz-Hernandez MB (2008) Genetic assessment of local apple cultivars from La Palma, Spain, using simple sequence repeats (SSRs). Sci Hortic 117:160–166
Pérez-Clemente RM, Pérez-Sanjuán A, García-Férriz L, Beltrán JP, Cañas LA (2004) Transgenic peach plants (*Prunus persica* L.) produced by genetic transformation of embryo sections using the green fluorescent protein (GFP) as an in vivo marker. Mol Breed 14:419–427
Peschel S, Franke R, Schreiber L, Knoche M (2007) Composition of the cuticle of developing sweet cherry fruit. Phytochemistry 68:1017–1025
Petri C, Burgos L (2005) Transformation of fruit trees. Useful breeding tool or continued future prospect? Transgenic Res 14:15–26
Petri C, Wang H, Alburquerque N, Faize M, Burgos L (2008) Agrobacterium-mediated transformation of apricot (*Prunus armeniaca* L.) leaf explants. Plant Cell Rep 27:1317–1324
Pierantoni L, Dondini L, Cho KH, Shin IS, Gennari F, Chiodini R, Tartarini S, Kang SJ, Sansavini S (2007) Pear scab resistance QTLs via a European pear (*Pyrus communis*) linkage map. Tree Genet Genomes 3:311–317

Pilarova P, Marandel G, Decroocq V, Salava J, Krska B, Abbott AG (2010) Quantitative trait analysis of resistance to Plum pox virus in the apricot F1 progeny "Harlayne" × "Vestar". Tree Genet Genomes 6:467–475

Puterka GJ, Bocchetti C, Dang P, Bell RL, Scorza R (2002) Pear transformed with a lytic peptide gene for disease control affects nontarget organism, Pear Psylla (Homoptera: Psyllidae). J Econ Entomol 95:797–802

Qiao H, Wang H, Zhao L, Zhou J, Huang J, Zhang Y, Xue Y (2004) The F-box protein AhSLF-S2 physically interacts with S-RNases that may be inhibited by the ubiquitin/26S proteasome pathway of protein degradation during compatible pollination in Antirrhinum. Plant Cell 16:571–581

Qin GZ, Meng XH, Wang Q, Tian SP (2009a) Oxidative damage of mitochondrial proteins contributes to fruit senescence: a redox proteomics analysis. J Proteome Res 8:2449–2462

Qin GZ, Wang Q, Liu J, Li BQ, Tian SP (2009b) Proteomic analysis of changes in mitochondrial protein expression during fruit senescence. Proteomics 9:4241–4253

Quarta R, Dettori MT, Sartori A, Verde I (2000) Genetic linkage map and QTL analysis in peach. In: Proceedings of the Xxv international horticultural congress, vol 11, pp 233–241

Quilot B, Wu BH, Kervella J, Genard M, Foulongne M, Moreau K (2004) QTL analysis of quality traits in an advanced backcross between *Prunus persica* cultivars and the wild relative species P-davidiana. Theor Appl Genet 109:884–897

Rallo P, Dorado G, Martin A (2000) Development of simple sequence repeats (SSRs) in olive tree (*Olea europaea* L.). Theor Appl Genet 101:984–989

Rallo P, Tenzer I, Gessler C, Baldoni L, Dorado G, Martin A (2003) Transferability of olive microsatellite loci across the genus Olea. Theor Appl Genet 107:940–946

Ramos-Cabrer AM, Diaz-Hernandez MB, Pereira-Lorenzo S (2007) Morphology and microsatellites in Spanish apple collections. J Hortic Sci Biotechnol 82:257–265

Ravelonandro M, Briard P, Monsion M, Scorza R, Renaud R (2002) Stable transfer of the Plum Pox Virus (PPV) capsid transgene to seedlings of two French cultivars 'Prunier d'Ente 303' and 'Quetsche 2906', and preliminary results of PPV challenge assays. Acta Hortic 577:91–96

Reale S, Doveri S, Diaz A, Angiolillo A, Lucentini L, Pilla F, Martin A, Donini P, Lee D (2006) SNP-based markers for discriminating olive (*Olea europaea* L.) cultivars. Genome 49:1193–1205

Renaut J, Hausman JF, Bassett C, Artlip T, Cauchie HM, Witters E, Wisniewski M (2008) Quantitative proteomic analysis of short photoperiod and low-temperature responses in bark tissues of peach (*Prunus persica L. Batsch*). Tree Genetics & Genomes 4:589–600

Reuter A, Fortunato D, Garoffo LP, Napolitano L, Scheurer S, Giuffrida MG, Vieths S, Conti AD (2005) Novel isoforms of Pru av 1 with diverging immunoglobulin E binding properties identified by a synergistic combination of molecular biology and proteomics. Proteomics 5:282–289

Reynoird JP, Mourgues F, Norelli JL, Aldwinckle HS, Brisset MN, Chevreau E (1999) First evidence for improved resistance to fire blight in transgenic pear expressing the attacin E gene from *Hyalophora cecropia*. Plant Sci 149:23–31

Richards CM, Volk GM, Reilley AA, Henk AD, Lockwood DR, Reeves PA, Forsline PL (2009) Genetic diversity and population structure in *Malus sieversii*, a wild progenitor species of domesticated apple. Tree Genet Genomes 5:339–347

Ritter E, Gebhardt C, Salamini F (1990) Estimation of recombination frequencies and construction of RFLP linkage maps in plants from crosses between heterozygous parents. Genetics 125:645–654

Roe MR, Griffin TJ (2006) Gel-free mass spectrometry-based high throughput proteomics: tools for studying biological response of proteins and proteomes. Proteomics 6:4678–4687

Romero C, Pedryc A, Muñoz V, Llácer G, Badenes ML (2003) Genetic diversity of different apricot geographical groups determined by SSR markers. Genome 46:244–252

Romero C, Vilanova S, Burgos L, Martínez-Calvo J, Vicente M, Llácer G, Badenes ML (2004) Analysis of the S-locus structure in *Prunus armeniaca* L. Identification of S-haplotype specific S-RNase and F-box genes. Plant Mol Biol 56:145–157

Rowan DD, Hunt MB, Alspach PA, Whitworth CJ, Oraguzie NC (2009a) Heritability and genetic and phenotypic correlations of apple (*Malus × domestica*) fruit volatiles in a genetically diverse breeding population. J Agric Food Chem 57:7944–7952

Rowan DD, Hunt MB, Dimouro A, Alspach PA, Weskett R, Volz RK, Gardiner SE, Chagne D (2009b) Profiling fruit volatiles in the progeny of a 'Royal Gala' × 'Granny Smith' apple (*Malus × domestica*) cross. J Agric Food Chem 57:7953–7961

Royo J, Kuntz C, Kowyama Y, Anderson M, Clarke AE (1994) Loss of a histidine residue at the active site of S-locus ribonuclease is associated with self-compatibility in *Lycopersicon peruvianum*. Proc Natl Acad Sci USA 91:6511–6514

Rubio M, Pascal T, Bachellez A, Lambert P (2010) Quantitative trait loci analysis of Plum pox virus resistance in *Prunus davidiana* P1908: new insights on the organization of genomic resistance regions. Tree Genet Genomes 6:291–304

Rudell DR, Mattheis JR (2009) Superficial scald development and related metabolism is modified by postharvest light irradiation. Postharvest Biol Technol 51:174–182

Rudell DR, Mattheis JP, Curry FA (2008) Prestorage ultraviolet-white light irradiation alters apple peel metabolome. J Agric Food Chem 56:1138–1147

Rudell DR, Mattheis JP, Hertog M (2009) Metabolomic change precedes apple superficial scald symptoms. J Agric Food Chem 57:8459–8466

Sargent DJ, Rys A, Nier S, Simpson DW, Tobutt KR (2007) The development and mapping of functional markers in Fragaria and their transferability and potential for mapping in other genera. Theor Appl Genet 114:373–384

Sargent DJ, Marchese A, Simpson DW, Howad W, Fernandez-Fernandez F, Monfort A, Arus P, Evans KM, Tobutt KR (2009) Development of "universal" gene-specific markers from *Malus* spp. cDNA sequences, their mapping and use in synteny studies within Rosaceae. Tree Genet Genomes 5:133–145

Sassa H, Hirano H, Ikehashi H (1992) Self-incompatibility-related RNases in styles of Japanese pear (*Pyrus serotina* Redh.). Plant Cell Physiol 33:811–814

Sassa H, Hirano H, Nishio T, Koba T (1997) Style-specific self-compatible mutation caused by deletion of the S-RNase gene in Japanese pear (*Pyrus serotina*). Plant J 12:223–227

Sassa H, Kakui H, Miyamoto M, Suzuki Y, Hanada T, Ushijima K, Kusaba M, Hirano H, Koba T (2007) S locus F-box brothers: multiple and pollen-specific F-box genes with S haplotype-specific polymorphisms in apple and Japanese pear. Genetics 175:1869–1881

Schaffer RJ, Friel EN, Souleyre EJF, Bolitho K, Thodey K, Ledger S, Bowen JH, Ma JH, Nain B, Cohen D, Gleave AP, Crowhurst RN, Janssen BJ, Yao JL, Newcomb RD (2007) A genomics approach reveals that aroma production in apple is controlled by ethylene predominantly at the final step in each biosynthetic pathway([w]). Plant Physiol 144:1899–1912

Scorza R, Ravelonandro M, Callahan AM, Cordts JM, Fuchs M, Dunez J, Gonsalves D (1994) Transgenic plums (*Prunus domestica* L) express the Plum pox virus coat protein gene. Plant Cell Rep 14:18–22

Sefc KM, Lopes S, Mendonca D, Dos Santos MR, Machado MLD, Machado AD (2000) Identification of microsatellite loci in olive (*Olea europaea*) and their characterization in Italian and Iberian olive trees. Mol Ecol 9:1171–1173

Serrano I, Pelliccione S, Olmedilla A (2010) Programmed-cell-death hallmarks in incompatible pollen and papillar stigma cells of *Olea europaea* L. under free pollination. Plant Cell Rep 29(6):561–572. doi:10.1007/s00299-010-0845-5

Sijacic P, Wang X, Skirpan AL, Wang Y, Dowd PE, McCubbin AG, Huang S, T-h K (2004) Identification of the pollen determinant of S-RNase-mediated self-incompatibility. Nature 429:302–305

Solis-Solis HM, Calderon-Santoyo M, Gutierrez-Martinez P, Chorr-Galindo SS, Ragazzo-Sancheza JA (2007) Discrimination of eight varieties of apricot (*Prunus armeniaca*) by electronic nose, LLE and SPME using GC-MS and multivariate analysis. Sens Actuat B-Chem 125:415–421

Song GQ, Sink KC (2006) Transformation of Montmorency sour cherry (*Prunus cerasus* L.) and Gisela 6 (*P. cerasus × P. canescens*) cherry rootstock mediated by Agrobacterium tumefaciens. Plant Cell Rep 25:117–123

Sonneveld T, Tobbutt KR, Vaughan SP, Robbins TP (2005) Loss of pollen-S function in two self-compatible selections of *Prunus avium* is associated with deletion/mutation of an S haplotype-specific F-box gene. Plant Cell 17:37–51

Soriano JM, Romero C, Vilanova S, Llacer G, Badenes ML (2005) Genetic diversity of loquat germplasm (*Eriobotrya japonica* (Thunb) Lindl) assessed by SSR markers. Genome 48:108–114

Soriano JM, Vera-Ruiz EM, Vilanova S, Martinez-Calvo J, Llacer G, Badenes ML, Romero C (2008) Identification and mapping of a locus conferring Plum pox virus resistance in two apricot-improved linkage maps. Tree Genet Genomes 4:391–402

Sosinski B, Gannavarapu M, Hager LD, Beck LE, King GJ, Ryder CD, Rajapakse S, Baird WV, Ballard RE, Abbott AG (2000) Characterization of microsatellite markers in peach *Prunus persica* (L.) Batsch. Theor Appl Genet 101:421–428

Sosinski B, Shulaev V, Dhingra A, Kalyanaraman A, Bumgarner R, Rokhsar D, Verde I, Velasco R, Abbott AG (2009) Rosaceaous genome sequencing: perspectives and progress. In: Genetics and genomics of Rosaceae. Springer, New York, pp 601–615

Stockinger EJ, Mulinix CA, Long CM, Brettin TS, Iezzoni AF (1996) A linkage map of sweet cherry based on RAPD analysis of a microspore-derived callus culture population. J Hered 87:214–218

Stoeckli S, Mody K, Gessler C, Patocchi A, Jermini M, Dorn S (2008) QTL analysis for aphid resistance and growth traits in apple. Tree Genet Genomes 4:833–847

Sung SK, Jeong DH, Nam JM, Kim SH, Kim SR, An G (1998) Expressed sequence tags of fruits, peels, and carpels and analysis of mRNA expression levels of the tagged cDNAs of fruits from the Fuji apple. Mol Cells 8:565–577

Szankowski I, Briviba K, Fleschhut J, Schönherr J, Jacobsen HJ, Kiesecker H (2003) Transformation of apple (*Malus domestica* Borkh.) with stilbene synthase gene from grapevine (*Vitis vinifera* L.) and PGIP gene from kiwi (*Actinidia deliciosa*). Plant Cell Rep 22:141–149

Szankowski I, Flachowsky H, Li HH, Halbwirth H, Treutter D, Regos I, Hanke MV, Stich K, Fischer TC (2009a) Shift in polyphenol profile and sublethal phenotype caused by silencing of anthocyanidin synthase in apple (*Malus* sp.). Planta 229:681–692

Szankowski I, Waidmann S, Degenhardt J, Patocchi A, Paris R, Silfverberg-Dilworth E, Broggini G, Gessler C (2009b) Highly scab-resistant transgenic apple lines achieved by introgression of HcrVf2 controlled by different native promoter lengths. Tree Genet Genomes 5:349–358

Takahashi H, Sumitani H, Inada Y, Mori D, Nakano Y (2000) Potent aroma volatiles in fresh loquat and its canned product. J Jpn Soc Food Sci Technol-Nippon Shokuhin Kagaku Kogaku Kaishi 47:302–310

Takasaki T, Okada K, Castillo C, Moriya Y, Saito T, Sawamura Y, Norioka N, Norioka S, Nakanishi T (2004) Sequence of the S-9-RNase cDNA and PCR-RFLP system for discriminating S-1- to S-9-allele in Japanese pear. Euphytica 135:157–167

Takeoka GR, Buttery RG, Flath RA (1992) Volatile constituents of Asian pear (*Pyrus serotina*). J Agric Food Chem 40:1925–1929

Tanksley SD, McCouch SR (1997) Seed banks and molecular maps: unlocking genetic potential from the wild. Science 277:1063–1066

Tao R, Watari A, Hanada T, Habu T, Yaegaki H, Yamaguchi M, Yamane H (2007) Self-compatible peach (*Prunus persica*) has mutant versions of the S haplotypes found in self-incompatible Prunus species. Plant Mol Biol 63:109–123

Teo G, Suziki Y, Uratsu SL, Lampinen B, Ormonde N, Hu WK, DeJong TM, Dandekar AM (2006) Silencing leaf sorbitol synthesis alters long-distance partitioning and apple fruit quality. Proc Natl Acad Sci USA 103:18842–18847

Terakami S, Shoda M, Adachi Y, Gonai T, Kasumi M, Sawamura Y, Iketani H, Kotobuki K, Patocchi A, Gessler C, Hayashi T, Yamamoto T (2006) Genetic mapping of the pear scab resistance gene Vnk of Japanese pear cultivar Kinchaku. Theor Appl Genet 113:743–752

Testolin R, Marrazzo T, Cipriani G, Quarta R, Verde I, Dettori MT, Pancaldi M, Sansavini S (2000) Microsatellite DNA in peach (*Prunus persica* L. Batsch) and its use in fingerprinting and testing the genetic origin of cultivars. Genome 43:512–520

Tian YK, Wang CH, Zhang JS, James C, Dai HY (2005) Mapping Co, a gene controlling the columnar phenotype of apple, with molecular markers. Euphytica 145:181–188

Tian L, Canli FA, Wang X, Sibbald S (2009) Genetic transformation of *Prunus domestica* L. using the hpt gene coding for hygromycin resistance as the selectable marker. Sci Hortic 119:339–343

Tittarelli A, Santiago M, Morales A, Meisel LA, Silva H (2009) Isolation and functional characterization of cold-regulated promoters, by digitally identifying peach fruit cold-induced genes from a large EST dataset. BMC Plant Biol 9:121

Trainotti L, Bonghi C, Ziliotto F, Zanin D, Rasori A, Casadoro G, Ramina A, Tonutti P (2006) The use of microarray mu PEACH1.0 to investigate transcriptome changes during transition from pre-climacteric to climacteric phase in peach fruit. Plant Sci 170:606–613

Urtubia C, Devia J, Castro A, Zamora P, Aguirre C, Tapia E, Barba P, Dell'Orto P, Moynihan MR, Petri C, Scorza R, Prieto H (2008) Agrobacterium-mediated genetic transformation of *Prunus salicina*. Plant Cell Rep 27:1333–1340

Ushijima K, Sassa H, Dandekar AM, Gradziel TM, Tao R (2003) Structural and transcriptional analysis of the self-incompatibility locus of almond: identification of a pollen-expressed F-box gene with haplotype-specific polymorphism. Plant Cell 15:771–781

Ushijima K, Yamane H, Watari A, Kakehi E, Ikeda K, Hauck NR, Iezzoni AF, Tao R (2004) The S haplotype-specific F-box protein gene, SFB, is defective in self-compatible haplotypes of *Prunus avium* and *P. mume*. Plant J 39:573–586

van Dyk MM, Soeker MK, Labuschagne IF, Rees DJG (2010) Identification of a major QTL for time of initial vegetative budbreak in apple (*Malus* × *domestica* Borkh.). Tree Genet Genomes 6:489–502

Velasco R, Zharkikh A, Troggio M, Cartwright DA, Cestaro A, Pruss D, Pindo M, FitzGerald LM, Vezzulli S, Reid J, Malacarne G, Iliev D, Coppola G, Wardell B, Micheletti D, Macalma T, Facci M, Mitchell JT, Perazzolli M, Eldredge G, Gatto P, Oyzerski R, Moretto M, Gutin N, Stefanini M, Chen Y, Segala C, Davenport C, Dematte L, Mraz A, Battilana J, Stormo K, Costa F, Tao QZ, Si-Ammour A, Harkins T, Lackey A, Perbost C, Taillon B, Stella A, Solovyev V, Fawcett JA, Sterck L, Vandepoele K, Grando SM, Toppo S, Moser C, Lanchbury J, Bogden R, Skolnick M, Sgaramella V, Bhatnagar SK, Fontana P, Gutin A, Van de Peer Y, Salamini F, Viola R (2007) A high quality draft consensus sequence of the genome of a heterozygous grapevine variety. PLoS One 2(12):e1326. doi: 10.1371/journal.pone.0001326

Velasco R, Zharkikh A, Affourtit J, Dhingra A, Cestaro A, Kalyanaraman A, Fontana P, Bhatnagar SK, Troggio M, Pruss D, Salvi S, Pindo M, Baldi P, Castelletti S, Cavaiuolo M, Coppola G, Costa F, Cova V, Dal Ri A, Goremykin V, Komjanc M, Longhi S, Magnago P, Malacarne G, Malnoy M, Micheletti D, Moretto M, Perazzolli M, Si-Ammour A, Vezzulli S, Zini E, Eldredge G, Fitzgerald LM, Gutin N, Lanchbury J, Macalma T, Mitchell JT, Reid J, Wardell B, Kodira C, Chen Z, Desany B, Niazi F, Palmer M, Koepke T, Jiwan D, Schaeffer S, Krishnan V, Wu C, Chu VT, King ST, Vick J, Tao Q, Mraz A, Stormo A, Stormo K, Bogden R, Ederle D, Stella A, Vecchietti A, Kater MM, Masiero S, Lasserre P, Lespinasse Y, Allan AC, Bus V, Chagne D, Crowhurst RN, Gleave AP, Lavezzo E, Fawcett JA, Proost S, Rouze P, Sterck L, Toppo S, Lazzari B, Hellens RP, Durel C-E, Gutin A, Bumgarner RE, Gardiner SE, Skolnick M, Egholm M, Van de Peer Y, Salamini F, Viola R (2010) The genome of the domesticated apple (Malus [times] domestica Borkh.). Nat Genet 42:833–839

Vendramin E, Dettori MT, Giovinazzi J, Micali S, Quarta R, Verde I (2007) A set of EST-SSRs isolated from peach fruit transcriptome and their transportability across Prunus species. Mol Ecol Notes 7:307–310

Vikram A, Prithiviraj B, Hamzehzarghani H, Kushalappa A (2004a) Volatile metabolite profiling to discriminate diseases of McIntosh apple inoculated with fungal pathogens. J Sci Food Agric 84:1333–1340

Vikram A, Prithiviraj B, Kushalappa AC (2004b) Use of volatile metabolite profiles to discriminate fungal diseases of Cortland and empire apples. J Plant Pathol 86:215–225

Vilanova S, Romero C, Abernathy D, Abbott AG, Burgos L, Llacer G, Badenes ML (2003) Construction and application of a bacterial artificial chromosome (BAC) library of *Prunus armeniaca* L. for the identification of clones linked to the self-incompatibility locus. Mol Genet Genomics 269:685–691

Vilanova S, Badenes ML, Burgos L, Martínez-Calvo J, Llácer G, Romero C (2006a) Self-compatibility of two apricot selections is associated with two pollen-part mutations of different nature. Plant Physiol 142:629–641

Vilanova S, Soriano JM, Lalli DA, Romero C, Abbott AG, Llacer G, Badenes ML (2006b) Development of SSR markers located in the G1 linkage group of apricot (*Prunus armeniaca* L.) using a bacterial artificial chromosome library. Mol Ecol Notes 6:789–791

Vizoso P, Meisel LA, Tittarelli A, Latorre M, Saba J, Caroca R, Maldonado J, Cambiazo V, Campos-Vargas R, Gonzalez M, Orellana A, Silva H (2009) Comparative EST transcript profiling of peach fruits under different post-harvest conditions reveals candidate genes associated with peach fruit quality. BMC Genomics 10:421

Volk GM, Richards CM, Henk AD, Reilley AA, Bassil NV, Postman JD (2006) Diversity of wild *Pyrus communis* based on microsatellite analyses. J Am Soc Hortic Sci 131:408–417

Volk GM, Richards CM, Henk AD, Reilley A, Miller DD, Forsline PL (2009) Novel diversity identified in a wild apple population from the Kyrgyz Republic. HortScience 44:516–518

Wang D, Karle R, Brettin TS, Iezzoni AF (1998) Genetic linkage map in sour cherry using RFLP markers. Theor Appl Genet 97:1217–1224

Wang Y, Tsukamoto T, Yi KW, Huang S, McCubbin AG, Kao TH (2004) Chromosome walking in the Petunia inflata self-incompatibility (S-) locus and gene identification in an 881-kb contig containing S2-RNase. Plant Mol Biol 54:727–742

Wang W, Alche JD, Rodriguez-Garcia MI (2007) Characterization of olive seed storage proteins. Acta Physiol Plantarum 29:439–444

Wang YJ, Yang CX, Li SH, Yang L, Wang YN, Zhao JB, Jiang Q (2009) Volatile characteristics of 50 peaches and nectarines evaluated by HP-SPME with GC-MS. Food Chem 116:356–364

Watanabe M, Yamamoto T, Ohara M, Nishitani C, Yahata S (2008) Cultivar differentiation identified by SSR markers and the application for polyploid loquat plants. J Jpn Soc Hortic Sci 77:388–394

Watari A, Hanada T, Yamane H, Esumi T, Tao R, Yaegaki H, Yamaguchi M, Beppu K, Kataoka I (2007) A low transcriptional level of Se-RNase in the Se-haplotype confers self-compatibility in Japanese plum. J Am Soc Hortic Sci 132:396–406

Weeden NF, Hemmat M, Lawson DM, Lodhi M, Bell RL, Manganaris AG, Reisch BI, Brown SK, Ye GN (1994) Development and application of molecular marker linkage maps in woody fruit crops. Euphytica 77:71–75

Wehrhahn C, Allard RW (1965) Detection and measurement of effects of individual genes involved in inheritance of a quantitative character in wheat. Genetics 51:109

Welander M, Pawlicki N, Holefors A, Wilson F (1998) Genetic transformation of the apple rootstock M26 with the RolB gene and its influence on rooting. J Plant Physiol 153:371–380

Wen XP, Pang XM, Matsuda N, Kita M, Inoue H, Hao YJ, Honda C, Moriguchi T (2008) Over-expression of the apple spermidine synthase gene in pear confers multiple abiotic stress tolerance by altering polyamine titers. Transgenic Res 17:251–263

Wen XP, Ban YSK, Inoue H, Matsuda N, Moriguchi T (2010) Spermidine levels are implicated in heavy metal tolerance in a spermidine synthase overexpressing transgenic European pear by exerting antioxidant activities. Transgenic Res 19:91–103

Wisniewski M, Bassett C, Norelli J, Macarisin D, Artlip T, Gasic K, Korban S (2008) Expressed sequence tag analysis of the response of apple (*Malus* × *domestica* 'Royal Gala') to low temperature and water deficit. Physiol Plant 133:298–317

Wünsch A (2009) Cross-transferable polymorphic SSR loci in Prunus species. Sci Hortic 120:348–352

Wünsch A, Hormaza JI (2002) Molecular characterisation of sweet cherry (*Prunus avium* L.) genotypes using peach *Prunus persica* (L.) Batsch SSR sequences. Heredity 89:56–63

Wünsch A, Hormaza JI (2004) Genetic and molecular analysis in Cristobalina sweet cherry, a spontaneous self-compatible mutant. Sex Plant Reprod 17:203–210

Xu ML, Song JQ, Cheng ZK, Jiang JM, Korban SS (2001) A bacterial artificial chromosome (BAC) library of *Malus floribunda* 821 and contig construction for positional cloning of the apple scab resistance gene Vf. Genome 44:1104–1113

Xue Y, Carpenter R, Dickinson HG, Coen ES (1996) Origin of allelic diversity in Antirrhinum S locus RNases. Plant Cell 8:805–814
Yamamoto T, Kimura T, Sawamura Y, Kotobuki K, Ban Y, Hayashi T, Matsuta N (2001a) SSRs isolated from apple can identify polymorphism and genetic diversity in pear. Theor Appl Genet 102:865–870
Yamamoto T, Shimada T, Imai T, Yaegaki H, Haji T, Matsuta N, Yamaguchi M, Hayashi T (2001b) Characterization of morphological traits based on a genetic linkage map in peach. Breed Sci 51:271–278
Yamamoto T, Kimura T, Sawamura Y, Manabe T, Kotobuki K, Hayashi T, Ban Y, Matsuta N (2002a) Simple sequence repeats for genetic analysis in pear. Euphytica 124:129–137
Yamamoto T, Mochida K, Imai T, Shi YZ, Ogiwara I, Hayashi T (2002b) Microsatellite markers in peach [*Prunus persica* (L.) Batsch] derived from an enriched genomic and cDNA libraries. Mol Ecol Notes 2:298–301
Yamamoto T, Kimura T, Soejima J, Sanada T, Ban Y, Hayashi T (2004a) Identification of quince varieties using SSR markers developed from pear and apple. Breed Sci 54:239–244
Yamamoto T, Kirnura T, Saito T, Kotobuki K, Matsuta N, Liebhard R, Gessler C, van de Weg WE, Hayashi T (2004b) Genetic linkage maps of Japanese and European pears aligned to the apple consensus map. In: Proceedings of the XIth eucarpia symposium on fruit breeding and genetics, vols 1 and 2, pp 51–56
Yamamoto T, Kimura T, Terakami S, Nishitani C, Sawamura Y, Saito T, Kotobuki K, Hayashi T (2007) Integrated reference genetic linkage maps of pear based on SSR and AFLP markers. Breed Sci 57:321–329
Yamamoto T, Terakami S, Kimura T, Sawamura Y, Takada N, Hirabayashi T, Imai T, Nishitani C (2009) Reference genetic linkage maps of European and Japanese pears. Acta Hortic 814: 599–602
Yamane H, Tao R (2009) Molecular basis of self-(in)compatibility and current status of S-genotyping in Rosaceous fruit trees. J Jpn Soc Hortic Sci 78:137–157
Yamane H, Ikeda K, Hauck NR, Iezzoni F, Tao R (2003) Self-incompatibility (S) locus region of the mutated S6-haplotype of sour cherry (*Prunus cerasus*) contains a functional pollen S allele and a non-functional pistil S allele. J Exp Bot 54:2431–2437
Zhang CF, Tian SP (2009) Crucial contribution of membrane lipids' unsaturation to acquisition of chilling-tolerance in peach fruit stored at 0 degrees C. Food Chem 115:405–411
Zhang GR, Sebolt AM, Sooriyapathirana SS, Wang DC, Bink M, Olmstead JW, Iezzoni AF (2009) Fruit size QTL analysis of an F-1 population derived from a cross between a domesticated sweet cherry cultivar and a wild forest sweet cherry. Tree Genet Genomes 6:25–36
Zhebentyayeva TN, Reighard GL, Lalli D, Gorina VM, Krska B, Abbott AG (2008a) Origin of resistance to Plum pox virus in Apricot: what new AFLP and targeted SSR data analyses tell. Tree Genet Genomes 4:403–417
Zhebentyayeva TN, Swire-Clark G, Georgi LL, Garay L, Jung S, Forrest S, Blenda AV, Blackmon B, Mook J, Horn R, Howad W, Arus P, Main D, Tomkins JP, Sosinski B, Baird WV, Reighard GL, Abbott AG (2008b) A framework physical map for peach, a model Rosaceae species. Tree Genet Genomes 4:745–756
Zhu LH, Li XY, Ahlman A, Welander M (2003) The rooting ability of the dwarfing pear rootstock BP10030 (*Pyrus communis*) was significantly increased by introduction of the rolB gene. Plant Sci 165:829–835
Zhu LH, Li XY, Welander M (2008) Overexpression of the Arabidopsis gai gene in apple significantly reduces plant size. Plant Cell Rep 27:289–296
Zohary D, Hopf M (1994) Domestication of plants in the old world, 2nd edn. Clarendon, Oxford
Zuo J, Niu QW, Ikeda Y, Chua NH (2002) Marker-free transformation: increasing transformation frequency by the use of regeneration-promoting genes. Curr Opin Biotechnol 13:173–180

Chapter 7
Genomics of Tropical Fruit Tree Crops

Renée S. Arias, James W. Borrone, Cecile L. Tondo, David N. Kuhn, Brian M. Irish, and Raymond J. Schnell

Abstract The genetic improvement of tropical fruit trees is limited when compared to progress achieved in temperate fruit trees and annual crops. Tropical fruit tree breeding programs require significant resources to develop new cultivars that are adapted to modern shipping and storage requirements. The use of molecular markers in tropical fruit tree breeding is greatly assisting in solving a number of difficult challenges for breeders such as the development of complex family structures for recombination mapping and for recurrent selection. A review of the literature on molecular markers development and new techniques for increasing single-nucleotide polymorphic markers is discussed. The development of marker-assisted breeding for these tropical tree crops is also discussed.

Keywords SNP • Microsatellites • SSR • Molecular markers • Mango • Avocado • Lychee • Longan • Linkage mapping

---

R.S. Arias, Ph.D. (✉)
National Peanut Research Laboratory, USDA-ARS,
1011 Forrester Dr., S.E, 39842 Dawson, GA, USA
e-mail: renee.arias@ars.usda.gov

J.W. Borrone, MS, Ph.D.
Department of Entomology & Plant Pathology, Oklahoma State University,
Stillwater, OK, USA

C.L. Tondo, Ph.D. • D.N. Kuhn, Ph.D.
Subtropical Horticulture Research Station, USDA-ARS,
Miami, FL, USA

B.M. Irish, BS, MS, Ph.D.
Tropical Agriculture Research Station, USDA-ARS, Mayaguez, PR, USA

R.J. Schnell, Ph.D.
National Germplasm Repository for Tropical/Subtropical Fruit Crops,
USDA-ARS, Miami, FL, USA
e-mail: Ray.Schnell@effem.com

R.J. Schnell and P.M. Priyadarshan (eds.), *Genomics of Tree Crops*,
DOI 10.1007/978-1-4614-0920-5_7,

Introduction

Genetic improvement of tropical fruit trees[1] has lagged far behind the progress achieved in temperate fruit trees. The reasons for this are many and include the lack of resources, political stability, and infrastructure in many of the lesser-developed countries where these species are grown. Nevertheless, the value of these tree fruit crops in providing locally available quality nutrition is significant. The commercial potential of these crops is often limited because of logistics involving transportation and storage. The commercial value of these crops could be greatly enhanced if cultivars that were adapted to commercial production, storage, and shipping were developed.

Tropical fruit trees are a major source of carbohydrates and vitamins for much of the developing world population. The world production of the major tropical fruits from trees is expected to reach 65 million tons for the year 2010, with 98% of the global production by developing countries. The production in the year 2010 for the top three tropical fruit trees, mango, papaya, and avocado, is expected to be 30.7, 12.4, and 3.1 million tons, respectively. For the minor tropical fruit trees, such as lychee, durian, rambutan, and guava, the production is smaller, and trading occurs at regional markets; therefore, there are no global statistics for these species (FAO 2003).

Compounds with a vast number of applications are present in tropical fruit trees. To mention some examples, *Garcinia mangostana* Linn. can accumulate up to 56% of oil in seeds (Hawkins and Kridl 1998). Other compounds from *Garcinia* spp. include antioxidants (Obolskiy et al. 2009) used for medicinal purposes, as well as antifungal compounds (Geetha et al. 1997) both extracted from the fruits' pericarp. The fruits of *Crataeva* spp. are high in carotenoids (Englberger et al. 2009), and the fruits of *Mangifera indica* Linn., *Carica papaya* Linn., and *Psidium guajava* Linn. are a source of carotenoids and vitamin C (Oliveira et al. 2010). Antimicrobial compounds are produced by *Diospyros blancoi* A. DC. (Ragasa et al. 2009), *Nephelium longana* Cambess. (Ripa et al. 2010), *Irvingia gabonensis* (Aubrey-Lecomte ex O. Rorke) Baill. (Kuete et al. 2007), and *Inga fendleriana* Benth. (Pistelli et al. 2009). Anticancer activity has been described for several tropical fruit tree species, for example, *Irvingia malayana* Oliv. ex A. W. Benn. has antiangiogenic properties (Ng et al. 2010), compounds from *Annona* spp. arrest cancer cells at G1 (Yuan et al. 2003), and *Morinda citrifolia* Linn. is active against cancer cells (Liu et al. 2001). Other useful applications described for tropical fruit trees are the cardioprotective activity of *Garcinia mangostana* compounds (Devi-Sampath and Vijayaraghavan 2007), the effective gastroenteritis control by *Spondias purpurea* Linn. (Caceres et al. 1993), the boost for the immune system by *Morinda citrifolia* (Palu et al. 2008), the anti-snake venom properties of *Tamarindus indica* Linn. (Ushanandini et al. 2006), and the antifatigue activity of *Dimocarpus longan* Lour. polysaccharides

[1] The tropical fruit tree species discussed in this chapter will generally exclude tree species whose commercial value is derived from non-fruit parts, are predominantly temperate species, are monocots, or are discussed in another chapter in this edition.

(Zheng et al. 2010). The importance and diversity of this group of species warrant a greatly expanded effort in genetics and genomics for compound discovery and improvement of human health.

The use of genomics in tropical fruit crops has largely been confined to the development of isozyme and dominant PCR-based markers and their use for germplasm diversity analysis and clonal fingerprinting. Information from dominant markers is of limited use in genomics applications; nonetheless, it has aided in the creation of a few genetic recombination maps and in our understanding of genetic diversity in germplasm collections. More recently, codominant microsatellite markers, also known as simple sequence repeat (SSR) markers, have been developed for a number of tropical fruit crops and have been used for parentage analysis, clonal fingerprinting, genetic diversity analysis, and development of genetic linkage maps. With the increased sequencing capacity of second- and third-generation pyrosequencing, many transcriptomes (gene space) of these species will be sequenced over the next 510 years. The generation of this new information should lead to increased interest in tropical tree fruit crops and to new opportunities to increase the rate of genetic gain in breeding programs.

Breeding tropical fruit trees is complicated by their reproductive biology. For instance, avocado (*Persea americana* Mill.) has an unusual flower behavior with two complementary types of flowering patterns, called "A" and "B," that promotes outcrossing (Bergh 1969; Davenport 1986). Avocado does not contain any self-incompatibility system, and self-pollination is frequently observed. A mature tree can produce upwards of a million flowers with only 1% of these setting fruit. A single ovary develops into the seed from a single pollination, so the generation of large numbers of seedlings by hand pollination is not practical. In mango (*Mangifera indica*), a similar situation exists; flowering is strongly influenced by weather, and some genotypes flower very irregularly. The flowers of mango are small, and both perfect and staminate (male) flowers occur in the same inflorescence. Hand pollination is possible but difficult to perform on a large scale to generate large numbers of progeny (Sharma et al. 1971 in Singh et al. 1980; Pinto et al. 2004). In addition, polyembryony in mango complicates breeding schemes. In polyembryonic cultivars, seedlings arise from nucellar tissue or from a zygote, but distinguishing between the two can be complicated (Schnell et al. 1994). For many of the tropical fruit tree species in the Sapindaceae, flower polymorphisms (i.e., monoecy, dioecy, androdioecy, gynodioecy, etc.) add complexity to the breeding process. In lychee (*Litchi chinensis* Sonn.), there are three types of flowers appearing in irregular sequence or simultaneously on the same inflorescence. There are male flowers, hermaphrodite flowers that act as females, and hermaphrodite flowers that act as males (Morton 1987). Again, the flowers are small and not amenable to hand pollination (Stern and Gazit 2003). In longan (*Dimocarpus longan*), there are three flower types: staminate (functionally male), pistillate (functionally female), and hermaphroditic (bisexual). Flowering in each panicle occurs in progressive openings of staminate (male) flowers first, then pistillate flowers followed by hermaphroditic flowers functioning as females and then hermaphroditic flowers functioning as males. Pollination is mainly by small insects but also by wind (Blanche et al. 2006).

Lack of genetic diversity in mangosteen (*Garcinia mangostana*) is a consequence of its mode of reproduction as an obligate apomict. Only anecdotal reports of male trees have been made for this dioecious species (Sand et al. 2005). In addition, an extremely long juvenile stage, upwards of 10 years, has been described for this particular fruit tree species (Poerwanto 2002).

In most of these tropical fruit tree species, controlled pollinations are very difficult to achieve, so breeders and horticulturalists have relied on the use of open-pollinated progeny for selection. The use of maternal half-sib families is very inefficient, and for most tropical fruit trees, no recurrent parentage selection has occurred. The ability to generate large numbers of codominant markers quickly using next-generation sequencing has revolutionized what can be accomplished in tropical fruit tree breeding. Identifying full-sib families is relatively simple using SSR markers. By planting trees in a polycross design (Iyer and Schnell 2009) or harvesting from commercial orchards where only a few cultivars are being grown, a breeder can now have large full-sib families to perform selection. As reported later in this chapter, the use of these types of molecular markers has greatly accelerated mango and avocado breeding efforts. Furthermore, recent efforts have focused on the development of new SSR markers for a number of tropical fruit tree species including lychee, longan, rambutan, Spanish lime, sapodilla, mamey, *Annona* spp., mangosteen, *Artocarpus* spp., and star fruit which are available to breeders and geneticists working on these crops.

Applications of Molecular Genetics and Molecular Marker Development

The main genetic tools used for the identification and breeding of cultivars of domesticated species are morphological and molecular markers (Tanksley 1993). Morphological markers in tropical fruit trees take years to be usable, for example, until the plants overcome their juvenility stage besides being subject to epigenetics and environmental factors. Molecular markers, instead, can be used from any tissue at any time during the plant growth, expediting the process of variety identification and breeding, and overcoming the limitations of traditional methods (Azofeifa-Delgado 2006). Furthermore, molecular marker analysis of tropical fruit trees help understand the past domestication of taxa, determine effective population sizes, and assess the value of a decentralized approach for future domestication (Jamnadass et al. 2009). In tropical fruit tree species for which expressed sequence tags (ESTs) are available, these can be used to develop markers such as SSRs and single nucleotide polymorphisms (SNPs) in relation to flavor, color, fragrance, vitamins, fruit softening, and other traits of interest, for example, *Actinidia* spp. (Crowhurst et al. 2008). SNPs are becoming more popular than SSRs as genetic markers in linkage analysis because they are more abundant and suitable for automatic allele calling (Novelli et al. 2004; Selmer et al. 2009). SNPs have been developed only for a few tropical fruit tree species, one in relation to resistance to papaya ringspot virus, PRSV, in *Carica papaya* Linn. (Dillon et al. 2006), another for phylogenetic studies in *Citrus* spp. (Novelli et al. 2004) and for genotyping and linkage mapping in *Theobroma cacao* Linn. (Livingstone et al. 2010). However, at present, the cost of developing and testing SNPs is still higher than for developing SSRs. Nonetheless, SNP discovery is actively being carried out in crops such as avocado and mango.

SSRs have been the most widely employed class of molecular markers used in genetic studies with applications in many fields of genetics including genetic resources conservation, population genetics, molecular breeding, and paternity testing (Ellegren 2004). This range of applications is due to the fact that SSR markers are codominant, multiallelic, and highly reproducible; have high resolution; are amenable to high throughput; and are based on polymerase chain reaction (PCR) (Oliveira et al. 2006). As a convention, SSRs are regions in the genome where a group of bases (1–8 bp long) are repeated in tandem (Richard et al. 2008). These regions can be isolated either by data mining of existing sequences or by generating SSR-enriched libraries (Kijas et al. 1994; Zane et al. 2002). With the exception of *Carica papaya* and *Theobroma cacao* for which the genomes have been sequenced (Ming et al. 2008; Argout et al. 2010), most tropical fruit tree species do not have enough DNA sequence information to use data mining for identifying potential markers. Alternatively, expressed sequence tag (EST) information (i.e., cDNA) may be used to develop markers. We have summarized the number of entries for nucleotides, ESTs, and SSRs listed in the National Center for Biotechnology Information (NCBI) database, GenBank, for some of the most important tropical fruit trees in Table 7.1.

The use of molecular markers in tropical fruit trees can sometimes be hindered by socioeconomic reasons. One limitation is the significant cost associated with the development of markers for each crop. Tropical fruit tree species are distributed in a large number of taxonomic groups (Muchugi et al. 2008), and though transferring markers from other species could be used to reduce costs (Viruel and Hormaza 2004), such transferability is not feasible among distant taxa (Ellis and Burke 2007). In general, funds to study each tropical fruit tree species are scarce resulting in insufficient information on molecular markers. Additionally, the practical implementation of the existing results on molecular markers is often limited by the absence of guidance on how to best apply them (Muchugi et al. 2008). For example, from the top 25 indigenous tropical fruit tree species identified as priority by the International Centre for Research in Agroforestry (ICRAF), only for eight has some work been done using molecular markers (Jamnadass et al. 2009). For most of these species, the amount of genetic information available is negligible or null. A list of the main tropical fruit tree species and the molecular markers that have been developed for each of them is shown in Table 7.2.

SSR Isolation Using Pyrosequencing

The isolation of SSRs from species for which little to no genetic information is available, such as most tropical fruit trees, can be difficult. At the USDA-ARS Mid South Area Genomics Laboratory (MSAGL), an effective pipeline has been created to isolate SSR markers from these species. The process first involved a slight

Table 7.1 Summary of number of entries in NCBI for nucleotides, ESTs, and SSRs for some of the most important tropical fruit trees

Species	Nucleotide	EST	SSR	Plant type
Actinidia deliciosa+A. arguta (kiwi)	230+78	57,751+7,257	30	Vine
Carica papaya	51,217	77,393	45+	
Musa spp. and hybrids	4,210	31,268	550	Monocot
Citrus spp.	2,592	549,188	106	Temperate
Persea americana	493	16,558	0	Tropical tree
Mangifera indica	401	68	462	Tropical tree
Cocos nucifera	382	6	0	monocot
Diospyros kaki	296	9,474	69	Tropical tree
Citrus grandis (*C. maxima*)	202	0	17	Tropical tree
Spondias purpurea	141	0	0	Tropical tree
Dimocarpus longan	137	66	0	Tropical tree
Annona cherimolia	120	0	97	Tropical tree
Inga edulis	95	0	5	Tropical tree
Litchi chinensis	88	0	27	Tropical tree
Psidium sp. (guava)	75	0	24	Tropical tree
Bactris gasipaes	71	0	46	Monocot
Durio spp.	66	0	7	Tropical tree
Garcinia mangostana	64	149	0	Tropical tree
Morinda citrifolia	62	0	0	Tropical tree
Adansonia spp.	55	0	18	Tropical tree
Passiflora edulis (passion fruit)	51	0	10	Vine
Manilkara zapota	34	0	0	Tropical tree
Averrhoa carambola	31	0	0	Tropical tree
Annona muricata	30	0	0	Tropical tree
Anacardium occidentale	30	0	21	Tropical tree
Tamarindus indica	25	0	0	Tropical tree
Irvingia spp.	22	0	0	Tropical tree
Nephelium lappaceum	14	0	0	Tropical tree
Pometia pinnata	14	0	0	Tropical tree
Artocarpus heterophyllus	10	1	0	Tropical tree
Annona squamosa	8	0	4	Tropical tree
Artocarpus altilis	7	1	0	Tropical tree
Nephelium ramboutan-ake	7	0	7	Tropical tree
Annona reticulata	6	0	4	Tropical tree
Melicoccus bijugatus	6	0	0	Tropical tree
Syzygium samarangense	5	0	0	Tropical tree
Premna serratifolia	3	0	0	Tropical tree
Sandoricum koetjape	2	0	0	Tropical tree
Garcinia portoricensis	1	0	0	Tropical tree
Pouteria sapota	1	0	0	Tropical tree
Spondias dulcis	1	0	0	Tropical tree
Dracontomelon vitiense	0	0	0	Tropical tree
Garcinia binucao	0	0	0	Tropical tree
Garcinia cochinchinensis	0	0	0	Tropical tree
Crataeva speciosa	0	0	0	Tropical tree

Species were sorted first by decreasing number of nucleotide entries and then by decreasing number of ESTs. The *orange highlight* corresponds to species considered minor fruit crops, underutilized or rare, depending on the region, by the International Tropical Fruits Network (www.itfnet.org). The *green highlight* corresponds to species economically important that are either called "trees" though they are monocots, that are dicots but not exactly trees, or that are temperate

Table 7.2 Summary table of publications on molecular markers per tropical fruit tree species

Species	Work done on molecular markers
Carica papaya L.	CAPS marker linked to PRSV-P resistance (45); 50 accessions analyzed using nine isozyme systems (Ocampo et al. 2006); high-density genetic map with 712 SSRs and 277 AFLPs (Blas et al. 2009); draft of complete genome (97)
Persea americana Mill.	25 SSRs used on 37 cultivars and wild relatives (11); SNP development based on resequencing of 4 nuclear loci of 21 wild accessions of *P. americana* (Chen et al. 2008); screening of 56 SSR markers on *P. americana* (162)
Mangifera indica L.	15 SSR markers tested on 59 *M. indica* cultivars (137); 28 SSRs tested on 15 *M. indica* cultivars (46); 16 SSRs tested on 28 mango cultivars (157); 19 SSRs used on 307 *M. indica* accessions (Duval et al. 2006); six populations of *M. indica* analyzed by RAPD (Diaz-Matallana et al. 2009); 11 SSRs used in the phylogeographical analysis of *M. indica* (Hirano et al. 2010)
Diospyros kaki L. f.	Six SSRs used to screen 12 genotypes (Guo and Luo et al. 2008)
Citrus maxima L. Osbeck	370 accessions, 76 *C. maxima*, tested with 24 SSR markers (Barkley et al. 2006); genetic map of *Citrus* based on 256 RAPD markers used on 94 hybrids (de-Oliveira et al. 2005); putative SNPs annotated for *Citrus* (Martinez-Godoy et al. 2008)
Spondias purpurea, S. dulcis	Chloroplast spacer *trn*G-*trn*S identified five haplotype varieties under cultivation that were not present and probably were lost to extinction in the wild (Miller and Schaal et al. 2005); 216 individuals of 34 populations with two primer pair RFLPs (Miller and Schaal et al. 2006)
Dimocarpus longan Lour.	28 RAPD markers used to uniquely identify each of 22 accessions (163); partial rbcL 66 AFLP markers used on 41 *D. longan* accessions (82)
Annona cherimolia Mill., A. muricata L., *A. squamosa* L.	Development of monomorphic SSRs as phylogenetic markers of *Annona* species (Chatrou et al. 2009); 13 isozyme loci used to study segregation in 14 self-fertilized cultivars (Perfectti and Pascual et al. 1996); 206 cultivars analyzed with 23 isozyme loci in a germplasm bank (Perfecti and Pascual et al. 2005); 94 new SSRs, 58 of them polymorphic, were tested on 23 cherimoya cultivars (Escribano et al. 2008); Nine accessions discriminated by 14 polymorphic RAPD fragments (Brown et al. 2003); 5 *Anonna* species analyzed with 11 allozymes indicated *A. muricata* more genetically distant from the rest (Samuel et al. 1991)
Inga edulis	ITS sequencing and phylogenetic study of 32 species of the seven sections of Inga (Richardson et al. 2001); five SSR loci to study 189 trees for possible genetic erosion (Hollingsworth et al. 2005); chloroplast trnL-F region and five SSR loci used to study five locations in the Peruvian Amazon (Dawson et al. 2008)
Litchi chinensis Sonn.	12 SSRs tested on 21 *Litchi* cultivars (159); 16 SSR markers tested on 58 *Litchi* cultivars (81)
Psidium spp.	23 SSRs tested on 16 accessions of *Psidium guajava* and three other species (Risterucci et al. 2005); RAPD markers associated to quercetin (Feria-romero et al. 2009); AFLP on 48 guava cultivars (Hernández-Delgado et al. 2007); revision of molecular tools applied to *Psidium* (Rai et al. 2010)
Durio spp. Adans.	Ten *Durio* species were analyzed for phylogenetic relationships using RFLP on two chloroplast genes ndhC-trnV and rbcL (Santoso et al. 2005); phylogenetic relationships inferred from analysis of ndhF and ITS sequences (Nyffeler & Baum 2000)

(continued)

Table 7.2 (continued)

Species	Work done on molecular markers
Garcinia mangostana, G. cochinchinensis, G. portoricensis, G. binucao	Random amplified DNA fingerprinting (RAF) used on 37 accessions of *G. mangostana* (Ramage et al. 2004); 21 trees analyzed with 5 primers for RAPD analysis (Sobir et al. 2007)
Morinda citrifolia	11 individuals of 3 varieties of *M. citrifolia* analyzed to determine geographical origin by using nrETS, nrITS, rps16 and trnT-F sequence data (Razafimandimbison et al. 2010)
Adansonia spp. L.	AFLP analysis of 137 individuals from six populations (Assogbadjo et al. 2006); 18 SSR markers to analyze 214 individuals of *Adansonia digitata* and 30 individuals of other *Adansonia* species (Larsen et al. 2009); 11 populations of 4 countries in West Africa analyzed by AFLP (Kyndt et al. 2009); phylogeography of 344 individuals from 74 populations analyzed by chloroplast DNA (Tsy et al. 2009)
Manilkara zapota (L.) van Royen	Analysis of four populations of *M. zapota* from Mexico using 4 polymorphic RAPD markers (Heaton et al. 1999); 12 polymorphic SSR loci developed for *M. huberi* (Azevedo et al. 2005)
Averrhoa carambola L.	None found
Anacardium occidentale	Fingerprinting of 19 *A. occidentale* accessions using 50 RAPD primers, 12 ISSRs, and 6 AFLPs (Archak et al. 2003); 91 individuals of *A. occidentale* analyzed by AFLP (Archak et al. 2009); 21 polymorphic SSRs from *A. occidentale* tested on other *Anacardium* species (Croxford et al. 2006)
Tamarindus indica L.	RAPDs used to analyze ten populations of *T. indica* (Diallo et al. 2007)
Irvingia spp.	Eight CAPS primers used to analyze *I. gabonenensis* and *I. wombolu* (Lowe et al. 1998); 130 individuals of *I. gabonensis* and *I. wombolu* from West Africa were analyzed by RAPDs (Lowe et al. 2000); AFLP analysis of 15 accessions of *I. gabonensis* (Ude et al. 2006)
Nephelium lappaceum and *N. ramboutan-ake*	Seven SSR markers from *Litchi chinensis* had amplification in *N. ramboutan-ake* (Sim et al. 2005)
Pometia pinnata	None found
Artocarpus altilis (Parkinson), *A. heterophyllus* Lam.	26 *A. heterophyllus* accessions analyzed by AFLP using 12 primer pairs (136); Six populations of *A. altilis* analyzed by AFLP and 15 morphological traits (Sreekumar et al. 2007); Eight primer pairs used for AFLP analysis of 50 *A. heterophyllus* accessions (Shyamalamma et al. 2008); AFLP analysis of 200 breadfruit samples of *A. camansi* and *A. mariannensis* (166)
Melicoccus bijugatus Jacq.	None found
Syzygium samarangense	Isozymes used on a related rainforest species *S. nervosum* (Shapcott et al. 1999); 8 SSRs developed for *S. sayeri* (Hillyer et al. 2007)
Premna serratifolia L.	None found
Sandoricum koetjape	None found
Pouteria sapota	20 cultivars of *M. zapota* analyzed by RAPD markers (Meghala et al. 2005)
Dracontomelon vitiense	None found
Crataeva speciosa Volkens	None found

The order of species is the same as in Table 7.1, in which NCBI entries are summarized

modification to the DNA extraction method, given the presence of copious latex and phenolics in the vegetative tissues. Second, for generating SSR-enriched libraries, the method previously developed by Techen et al. (2010) was modified to adapt to high-throughput pyrosequencing with a Roche 454 GS-FLX (F. Hoffmann-La Roche Ltd., Basel, Switzerland). One modification employs two adapters (Techen et al. 2010) that allow simultaneous loading of pairs of samples in the same region of picotiter plates; the adapters act as bar coding to separate the samples via bioinformatics. Another modification is reducing the number of PCR cycles during library preparation to minimize redundant sequences. SSRs were isolated from the following species/crop groups: *Nephelium lappaceum* Linn. (rambutan), *Manilkara zapota* Linn. (sapodilla), *Pouteria sapota* Jacq. (sapote), *Litchi chinensis* (lychee), *Melicoccus bijugatus* Jacq. (Spanish lime), *Annona squamosa* Linn. (sugar apple), *Dimocarpus longan* (longan), *Averrhoa carambola* Linn. (star fruit), *Artocarpus altilis* (Parkinson) Fosberg (breadfruit), and *Garcinia mangostana* (mangosteen).

For the combined species, a total of 2,510,291 reads were assembled into 224,815 contigs in which a total of 49,898 SSR repeats (not including mononucleotides) were detected. A total of 10,310 primer pairs were designed under stringent conditions (Tm 65°C and 3′ GC clamp). Only 384 primer sets were tested per species. This was done to identify and select the SSR markers that were most readily amplifiable, reproducible, and which detected the highest allelic diversity for variety/cultivar identification and plant breeding. The SSR development process was accomplished in a short amount of time (less than 6 months), and greater than 96% of the markers resulted in SSR amplification. A detailed protocol for the SSR development procedure is available in Methods in Molecular Biology, Humana Press (Arias et al. 2010).

Specific Examples of Genomics in Tree Fruits

Avocado

The avocado (*Persea americana*) is an evergreen subtropical tree that is native from Mexico to northern South America and produces a fruit that is unique and nutritious. This fruit was known by the Aztecs as "ahuacacuauhitl," which was later shortened by the Spaniards to "aguacate." In the United States, avocado was introduced to Florida in 1833, California in 1848, and Hawaii by 1855 (Nakasone and Paull 1998). Major commercial production of avocado in the United States is limited to California and Florida. In 2000, global production exceeded 2.4 MMT, and the major producers were Mexico, Indonesia, South Africa, and the USA (Anonymous 2002).

Persea americana has been subdivided into three horticultural groups: Mexican (*P. americana* var. *drymifolia* (Schect. & Cham.) Blake), Guatemalan (*P. americana* var. *guatemalensis* Wms.), and West Indian (*P. americana* var. *americana* Mill.)

races. The West Indian race is known to be from the lowland areas of the Pacific coast of Central America and not the West Indies, while the Guatemalan and Mexican races are native to specific highland areas within each country (Scora and Bergh 1992). The three racial groups can be distinguished by the percentage of oil content in the fruit, with the West Indian cultivars ranging from 2.5% to 8.0%, Guatemalan accessions from 10% to 13%, and Mexican accessions ranging from 15% to 20% (Knight 2002). The racial classes also vary phenotypically for characters such as fruit size and shape, skin thickness, skin color, seed size, and fruit ripening (Lahav and Lavi 2002). Sterility barriers do not exist between or among the three racial types (Lahav and Lavi 2002). Avocado has a distinct flowering habit known as protogynous, diurnally synchronous dichogamy (Bergh 1969). This type of reproductive behavior promotes outcrossing; however, significant amounts of self-pollination are known to occur in commercial plantings (Davenport et al. 1994; Borrone et al. 2008; Schnell et al. 2009). Named cultivars often originate from open-pollinated seedlings. The unknown pollen parent has often been estimated based on the flower types of available donor trees. Morphological characters have been used to infer parentage, although these are influenced by environmental factors and may not unambiguously distinguish closely related genotypes or interracial hybrids. Many of the cultivars grown in Florida are interracial hybrids between Guatemalan and West Indian types, while those grown in California are hybrids between Mexican and Guatemalan types.

The haploid genome size of avocado has been estimated to be 8.83×10^8 bp (Arumuganathan and Earle 1991a). Avocado contains 24 chromosomes with bivalent pairing at meiosis indicating that $n = 12$ (Darlington and Wylie 1945). Avocado has been proposed to be the result of an ancient polyploid event (Chanderbali et al. 2008). However, evidence from analyses of germplasm collections and seedling populations using SSR markers has demonstrated diploid inheritance for most markers, although some primer pairs have amplified more than one locus (Schnell et al. 2003; Borrone et al. 2007, 2008, 2009).

Gene Discovery

Recently, a large number of sequences have been generated from the transcriptome of *P. americana* from two projects, the Floral Genome Project (Albert et al. 2005) and the Ancestral Angiosperm Genome Project (Wall et al. 2008). The Floral Genome Project, initiated to investigate the evolutionary development of floral patterning (Chanderbali et al. 2009), produced two directionally cloned, non-normalized, Sanger-sequenced cDNA libraries and are deposited in the EST database of GenBank (16,558 sequences). Both libraries, Pam01 (8,735 sequences) and Pam01b (7,823 sequences), were developed from premeiotic flower buds. The Pam01 library was the source of the EST-SSRs used to evaluate outcrossing rates (Borrone et al.

2007, 2008; Schnell et al. 2009) and to develop a moderately dense genetic map of avocado (Borrone et al. 2009).

The Ancestral Angiosperm Genome Project continued where the Floral Genome Project left off using pyrosequencing (Wall et al. 2008). A total of 12 libraries, developed from various tissue types and stages, generated 1,698,670 "passing" sequences covering 490.7 Mb. The unprocessed, unfiltered, raw data is deposited in GenBank as 14 Sequence Read Archives (SRAs). The processed (filtered, trimmed, and quality checked) sequence information is readily obtained from the Ancestral Angiosperm Genome Project website (http://ancangio.uga.edu/content/aagp-home). The website includes a detailed description of each library, a number of tools allowing comparison of the libraries with one another, and retrieval of individual reads and contigs. The entire dataset derived from pyrosequencing has been assembled with the cDNA libraries (Pam01 and Pam01b, developed from the Floral Genome Project) in a "Sanger-454 hybrid" assembly of 115,151 contigs. Polymorphic regions, SNPs and SSRs, have been identified in the assembled sequences as part of the pipeline process. The entire assembly is available for download at the site. Additional cDNA libraries have been developed from *Persea americana* var. *drymifolia* but are not yet publically available (López-Gómez et al. 2007).

An additional resource, not yet deposited in the NCBI database, is sequences of small RNAs (Blake Myers and Pamela Green, Delaware Biotech Institute, University of Delaware; NSF Plant Genome Program Award #0638525). Small RNAs, 21–24 nucleotides long, consist of two major types known as small interfering RNA (siRNA) and microRNA (miRNA) and are implicated in regulation of gene transcription/translation. These sequences were generated by pyrosequencing (Accerbi et al. 2010). A total of 12,689,456 sequences, 5,918,358 identified as being distinct, have been generated from three separate libraries: leaves (3,800,961 total; 1,263,610 distinct), flower (5,325,501 total; 2,870,885 distinct), and fruit (3,562,994 total; 1,783,863 distinct). The sequences are available through the Comparative Sequencing of Plant Small RNAs website (http://smallrna.udel.edu/).

Marker Development

Previously, there was a limited set of SSR markers for *P. americana* (Sharon et al. 1997; Ashworth et al. 2004), some of which do not consistently amplify in all varieties (Ashworth and Clegg 2003; Ashworth et al. 2004). For example, 14 of 39 SSR markers tested were suitable to fingerprint diverse collections of *P. americana* (Schnell et al. 2003). To increase the number of informative SSR markers, publicly available *P. americana* ESTs were screened (Borrone et al. 2007, 2009) bringing the number of markers available to upwards of 300. Data-mining SSRs from expressed sequence tags (ESTs) has proven effective for generating markers for fingerprinting, genetic mapping, and comparative mapping among species (Varshney et al. 2005).

Linkage Mapping

The first linkage map of avocado was produced by Sharon et al. (1997) from 50 progeny of a cross between 'Pinkerton' and 'Ettinger' using 50 SSR markers, 17 random amplified polymorphic DNA (RAPD) markers, and 23 minisatellite DNA fingerprint (DFP) markers. Twelve linkage groups with 34 mapped loci covering 352.6 centimorgans (cM) were identified. Seven linkage groups contained two markers, two linkage groups contained three markers, one linkage group contained four markers, and two linkage groups contained five markers each. A larger population and an increased number of genetic markers were needed to produce a linkage map useful for quantitative trait loci (QTL) discovery. The development of over 300 SSR markers (Sharon et al. 1997; Ashworth et al. 2004; Borrone et al. 2007, 2009) enabled the development of a partially saturated genetic linkage map and the potential identification of QTLs controlling horticultural traits of interest in avocado. Mapping populations have been developed in California (Ashworth et al. 2007; Chen et al. 2007) and in Spain (Viruel et al. 2007) by producing full-sib families that are similar in size to those used by Sharon et al. (1997) and are focused upon Guatemalan-Mexican hybrids.

In Florida, a large population of seedlings from a commercial grove interplanted with two cultivars of opposite flowering types, 'Tonnage' (type B) and 'Simmonds' (type A), in approximately equal numbers was screened to determine the outcrossing rate in avocado under south Florida conditions (Borrone et al. 2008). Eight fully informative SSR markers identified 870 seedlings as progeny of a reciprocal cross between 'Tonnage' and 'Simmonds'. Using these seedlings, the first large mapping population and linkage map for QTL discovery were developed in avocado, focusing upon West Indian-Guatemalan hybrids (Borrone et al. 2009). The final linkage map (Fig. 7.1, Table 7.3) was constructed from 163 markers generated by 135 primer pairs, 112 designed from EST-SSRs and 23 SSR primers previously developed by Sharon et al. (1997). Twenty primer pairs amplified more than one locus, with 15 amplifying two loci, four amplifying three loci, and one amplifying six loci. 'Tonnage' was heterozygous for 92.0% of the total markers used in its sub-composite map, whereas 'Simmonds' was heterozygous for 41.0% of the total markers in its sub-composite map, comparable with the previous estimate for heterozygosity for each cultivar using 12 SSR markers (Borrone et al. 2008). Eighteen markers were fully informative with four polymorphic alleles between the two parents, and the other 145 markers were partially informative. One hundred nineteen markers were scored on both sub-composite maps. An additional 43 markers were scored only on the 'Tonnage' × 'Simmonds' progeny, and one additional marker was scored only on the 'Simmonds' × 'Tonnage' progeny.

Twelve linkage groups (LG) representing the haploid set of 12 chromosomes of *P. americana* were generated from 163 markers, at a minimum logarithm (base 10) of odds (LOD) score of 4.0 (Fig. 7.1, Table 7.3). Linkage groups ranged in size, longest to shortest, from 157.3 (LG2) to 2.4 cM (LG12), and the number of markers mapped per group ranged from 29 (LG1) to 2 (LG12). The total length of the Florida

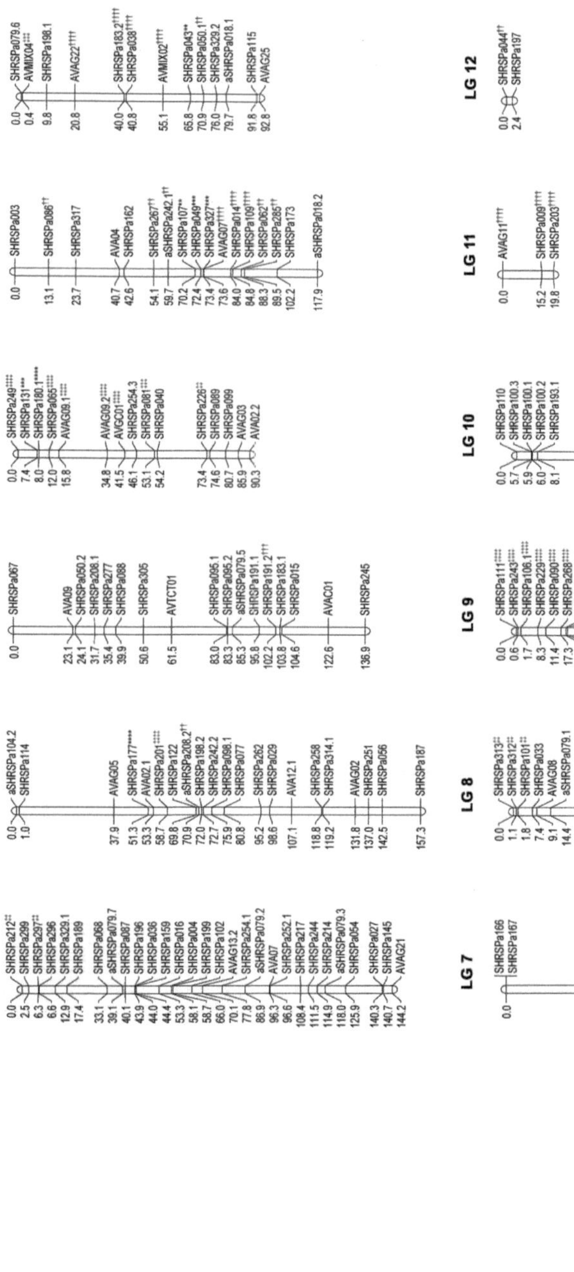

Fig. 7.1 Avocado linkage map constructed from 163 markers generated by 135 primer pairs

Table 7.3 *Persea americana* linkage mapping results for the Florida mapping population. Mapping results of the reciprocal cross (sub-composite maps) and parental maps

Map	LG	1		2	3	4	5			6	7	8	9	10	11	12	Total
'Tonnage' × 'Simmonds'							**5a**[a]	**5b**	**5c**								
456 individuals	No. loci	29		21	16	15	12	3	2	13	15	11	12	8	3	2	162
	Length (cM)	143.9		170.1	138.7	95.7	117.3	17.6	29.3	93.7	119.6	95.3	46.6	71.3	23.7	3.1	1,119.0
	cM/locus	5.0		8.1	8.7	6.4	9.8	5.9	14.7	7.2	8.0	8.7	3.9	8.9	7.9	1.6	7.5
'Tonnage' maternal	No. loci	27		18	16	15	15			12	13	11	11	8	3	2	149
	Length (cM)	146.5		154.7	138.6	96.3	116.5			92.1	121.1	95.4	47.0	70.9	23.8	3.1	1,106
	cM/locus	5.4		8.6	8.7	6.4	7.8			7.7	9.3	8.7	4.3	8.9	7.9	1.6	7.1
'Simmonds' paternal							**5a**	**5b**									
	No. loci	16		11	3	4	3	3		7	5	0	6	2	0	0	60
	Length (cM)	144.8		113.6	38.5	35	19.9	41.9		100.8	68.6	–	48.4	46.9	–	–	658.4
	cM/locus	9.1		10.3	12.8	8.8	6.6	14.0		14.4	13.7	–	8.1	23.5	–	–	12.1
'Simmonds' × 'Tonnage'							**5a**	**5b**									
259 individuals	No. loci	20		16	11	13	10	2		9	12	8	8	6	3	2	120
	Length (cM)	141.1		132.2	115.5	79.6	84.4	14.8		89.2	99.1	74.6	39.8	68.9	13.0	1.2	953.4
	cM/locus	7.1		8.3	10.5	6.1	8.4	7.4		9.9	8.3	9.3	5	11.5	4.3	0.6	7.4
'Simmonds' maternal		**1a**	**1b**														
	No. loci	11	2	8	0	4	6			6	3	0	5	0	0	0	45
	Length (cM)	69.1	16.5	92.6	–	41.3	103.6			92.3	42.6	–	44.3	–	–	–	502.3
	cM/locus	6.3	8.3	11.6	–	10.3	17.3			15.4	14.2	–	8.9	–	–	–	11.5
'Tonnage' paternal	No. loci	18		14	11	13	10			9	11	8	8	6	3	2	113
	Length (cM)	141.1		124.6	115.4	77.5	106.3			85.6	99.5	74.6	37.1	68.8	13	1.2	944.7
	cM/locus	7.8		8.9	10.5	6.0	10.6			9.5	9.0	9.3	4.6	11.5	4.3	0.6	7.7

[a]LG 5a represents the entire length of the LG in the 'Tonnage' × 'Simmonds' map and was used to calculate the overall length of the linkage map. For all others, LG5 are smaller-sized fragments of the entire LG

F_1 map was 1,087.4 cM. This Florida F_1 map is three times the size of the map reported by Sharon et al. (1997) and contains five times as many markers. Comparison of markers shared between the Florida F_1 map and the F_1 map of Sharon et al. (1997) allowed the identification of several analogous linkage groups. Ten linkage groups of the 12 described by Sharon et al. (1997) contained SSR markers. Nine of the ten corresponded to seven linkage groups in the Florida F_1 map. Distances between SSR markers within LGs reported by Sharon et al. (1997) and distances between these same markers placed in the sub-composite maps and the Florida F_1 map corresponded well, which is remarkable given the disparity in the population sizes and the numbers of markers between the two maps. The 'Tonnage' × 'Simmonds' linkage groups were longer than the 'Simmonds' × 'Tonnage' linkage groups (Table 7.3). This can be attributed to the larger number of markers mapped in the 'Tonnage' × 'Simmonds' sub-composite map versus its reciprocal (162 vs. 120) and to the larger number of individuals in this cross (456), slightly less than double that of the 'Simmonds' × 'Tonnage' cross (259). Also, LG5 formed three linkage groups in the 'Tonnage' × 'Simmonds' map and two linkage groups in the 'Simmonds' × 'Tonnage' map. The parts of this linkage group reformed a single linkage group in the composite map. Five linkage groups (LGs 3, 4, 8, 11, and 12) were composed almost entirely from markers polymorphic in the 'Tonnage' parent. The formation of linkage groups for all maps was straightforward except for a small number of markers due to the high homozygosity of 'Simmonds'.

The number of SSR markers now available for avocado allowed the development of this first moderate-density map composed of 163 loci. Linkage maps developed using molecular markers, such as this Florida F1 map, enable the detection and use of QTLs affecting traits of economic importance. The two parental cultivars 'Simmonds' and 'Tonnage' differ for many phenotypic traits useful in humid lowland environments like south Florida. 'Simmonds' is of the West Indian race and believed to be a seedling of 'Pollock'. It was selected in south Florida, propagated commercially in 1921, and is still a major commercial cultivar some 90 years later. 'Simmonds' has a light green, oblong-oval to pear-shaped large fruit with medium-sized seed and low oil content (3–6%), imparts tolerance to *Phytophthora* root rot (PRR) in its progeny, has the "A" flowering type, and is a high-yielding, early-season (June–July) cultivar (Campbell and Malo 1978; Ploetz et al. 2002). West Indian selections were the only important commercial cultivars in Florida until the 1920s when competition from Cuba depressed the market for Florida avocados. A number of Guatemalan-West Indian hybrids had since been selected which ripened in the fall and winter, extending the season and plantings shifted to include these hybrids. One of these was the cultivar 'Tonnage' which is now considered a minor cultivar in south Florida. 'Tonnage', a seedling of 'Taylor', was first propagated commercially in 1930. It has a dark green, pear-shaped fruit with medium seed size and moderate oil content (8–15%), does not impart tolerance to PRR in its progeny, is a "B" flowering type, and is a late-season (August–September) cultivar (Ploetz et al. 2002).

Recently, the USDA-ARS Subtropical Horticulture Research Station (SHRS) in Miami, FL, has produced another large mapping population of 'Hass' × 'Bacon' (~1,000 individual seedlings). 'Hass' is the most important avocado cultivar worldwide (Ashworth et al. 2007), and growers receive a premium price for 'Hass' fruit.

A commercial California 'Hass' avocado orchard was identified with adjacent 'Bacon' pollinizer rows. 'Hass' is an interracial hybrid of mostly Guatemalan ancestry, and 'Bacon' is of the Mexican landrace; they are of opposite flowering types (A and B, respectively) and, thus, expected to outcross. The full-sib family members from this 'Hass' × 'Bacon' cross were identified by genotyping with five SSR markers, and seedling trees, product of self-pollination, were excluded. However, a map has not yet been produced. Presently, the limiting factor is the availability of sufficient molecular markers to make a saturated, high-resolution map for QTL mapping of important agronomic traits such as oil content and composition, flowering type, flowering time, and cold tolerance. Once the field data for agronomic traits is collected on this population, this map will be useful for detection and use of QTLs.

SNP Discovery

SNP discovery from transcriptome and genomic sequence data is needed for obtaining sufficient molecular markers for complete linkage map saturation. The following strategy has been used at SHRS to generate and validate new SNP and SSR markers. The strategy takes advantage of two next-generation sequencing platforms, Illumina GAII (Illumina, Inc., San Diego, CA) and Roche 454, and uses both transcriptome and genomic sequence data to identify and validate SSRs and SNPs. Once sufficient SNPs have been identified (5–10 K), the production of an Illumina Infinium oligonucleotide array for genotyping of mapping populations may be employed. 'Hass', the most important commercially grown avocado cultivar in the world, has been selected as the reference cultivar for the transcriptome sequencing.

Transcriptome Sequencing

RNA was isolated from leaves, unopened flowers, female flowers, and male flowers from 'Hass', 'Bacon', 'Simmonds', and 'Tonnage', from tissue collected from genotyped clones at the SHRS. The RNA from the 'Hass' tissue was pooled for Roche 454 sequencing to generate the reference transcriptome. In addition, a large EST library was produced by Roche 454 sequencing of developing mesocarp tissue of 'Hass' and that data was used to expand the 'Hass' reference transcriptome (John Ohlrogge, Michigan State Univ., personal communication 2010). The RNA from 'Bacon', 'Simmonds', and 'Tonnage' leaves and flowers was pooled for each clone and sequenced by Illumina GAII. Illumina reads were aligned to the 'Hass' reference transcriptome, and a variant report with single-nucleotide polymorphisms and indels generated. The variant report was filtered for heterozygous SNP loci, as a locus that is heterozygous in any parent of the mapping populations can be mapped. SSR motifs are identified from the 'Hass' transcriptome sequences as described earlier in this chapter. The SNP and SSR discovery will yield a sufficient number of markers to completely saturate the genetic maps for both populations.

Reduced Representational Sequencing of the 'Hass' Genome

Reduced representational sequencing of the genome of 'Hass' from hypomethylated DNA is being carried out using Roche 454 pyrosequencing to obtain a ~2–4X coverage of the avocado gene space. Hypomethylated DNA is actively transcribed, and this should enrich for gene-containing regions of the genome. In the cacao SNP and SSR discovery project (http://www.cacaogenomedb.org), the genomic sequence has proved essential for identifying intron-exon junctions in the transcriptome sequences prior to SNP oligonucleotide design or SSR primer design. In addition, the assembled gene space of 'Hass' will be made available to the avocado genome sequencing group, producing a hypomethylated library of the Mexican landrace cultivar by an alternative method. Such a comparison will be of great interest for future genome sequencing projects targeting only the gene space.

Genome Sequencing Project

The avocado genome sequencing project has been under way for over a year in Mexico at the Plant Biotechnology Unit, Centre for Research and Advanced Studies (CINVESTAV; Luis Herrera-Estrella and Gustavo Hernandez), as well as by the biofuels research on non-seed oils group in the Plant Lipid Metabolism lab at Michigan State University (John Ohlrogge). The Mexican avocado genome sequencing project is using a highly homozygous individual from the Mexican landrace (*P. americana* var. *drymifolia*). Using flanking sequences of the 163 SSR markers mapped in the Florida F1 population, CINVESTAV has designed overgo probes to anchor their physical map to the genetic recombination map. The current version of the avocado physical map has 5,008 contigs, and the estimated coverage is 5X. Using the 163 markers will certainly help in the anchoring process, but thousands of markers would allow anchoring and correct ordering of the entire physical map of avocado. The Illumina Infinium chip being designed from the SNP development project will saturate the avocado genetic recombination map and anchor physical map contigs.

Mango

Mango (*Mangifera indica*) is a significant tree fruit crop grown commercially in tropical and subtropical areas of many countries. It has been under cultivation in India for at least 4,000 years, and over 1,000 varieties are known to exist in that country

(Mukherjee 1953). India is the largest producer of mangos with 10.8 million MT accounting for 41% of the world's mango production (Gunjate 2009, Proc VIII int. mango sym.). Most commercial mango cultivation is in tropical and subtropical regions; however, mangos are also cultivated in a wide range of marginal climatic areas throughout the world. Mango genotypes are divided into two distinct categories or types based on their origin: monoembryonic mangos, which are mostly subtropical (Indian types), and polyembryonic mangos, which are mostly tropical (Southeast Asian types). The seeds of Indian types characteristically contain a zygote embryo, and the fruit skin is highly colored (mixes of red, purple, and yellow), while the seeds of Indo-Chinese types contain several nucellar embryos, and the skin is soft or pale in color (green to light green to yellow) (Iyer and Degani 1997; Viruel et al. 2005).

Information on the cytology of mango is quite limited. Only *Mangifera* species *M. indica, M. caloneura, M. sylvatica, M. foetida, M. caesia, M. odorata,* and *M. zeylanica* have been studied, and these were found to have chromosome numbers of $2n=40$ and $n=20$ (Mukherjee 1950, 1957, 1963). Chromosome numbers and ploidy status of other *Mangifera* species have yet to be studied (Bompard and Schnell 1997). Mango has a small haploid genome size (0.91 pg), which is three times larger than *Arabidopsis thaliana* (L.) Heynh. and comparable to that of rice (Arumuganathan and Earle 1991b). Mango has been referred to as an allopolyploid based on the conclusions drawn by Mukherjee (1950). Due to the presence of secondary associations at metaphase of meiosis, he suggested that the basic chromosome number of *Mangifera* is $n=8$. In addition, the high number of somatic chromosomes and the correspondingly high number of nucleolar chromosomes led him to conclude that mango is an allopolyploid. However, the evidence used to arrive at this conclusion is not unequivocal. In fact, the molecular marker evidence is antithetical to this conclusion. Results from Duval et al. (2005), Schnell et al. (2005, 2006), and Viruel et al. (2005) all resulted in normal diploid segregation for SSR markers.

Applications of Molecular Genetics

Isozymes were the first markers to be used for fingerprinting mango cultivars to determine self- vs. cross-pollination and to estimate genetic relationships (Degani et al. 1990; Knight and Schnell 1994). RAPD markers were also used to fingerprint cultivars and estimate genetic relationships in mango (Schnell et al. 1995). In that research, a group of 'Haden' seedlings and a random group of seedlings were evaluated using 11 RAPD primers. This study supported the 'Haden' parentage of 'Eldon', 'Lippens', 'Tommy-Atkins', and 'Zill'; however, the parentage of 'Glenn' and 'Osteen' was questioned. Adato et al. (1995) used DNA fingerprinting (DFP) to evaluate genetic relationships between 26 mango cultivars and 14 rootstocks. They provided a pedigree that further confirmed the relationship between many of the 'Haden' seedlings. Lopez-Valenzuela et al. (1997) used RAPD markers to estimate genetic diversity among 15 rootstock cultivars using 13 markers and identified a specific RAPD band associated only with the polyembryonic types. Eiadthong et al.

(1999) utilized anchored simple sequence repeat markers to analyze 22 mango cultivars; they were able to distinguish genotypes. However, the authors were unable to find markers unique to either monoembryonic or polyembryonic types or for the Thai cultivars selected for green harvest (crispy mango) from the cultivars selected for ripe fruit production. Pandit et al. (2007) also used inter-simple sequence repeat (ISSR) markers to evaluate 60 elite Indian mango cultivars and 10 non-Indian cultivars. They were not able to distinguish Indian cultivars from northern and southern India, and they concluded that ISSR markers could not be considered a comprehensive marker system for mango. Kashkush et al. (2001) utilized amplified fragment length polymorphisms (AFLP) to estimate genetic relationships between 16 cultivars and 7 rootstock cultivars. They also analyzed 29 progeny from a cross of 'Tommy-Atkins' and 'Keitt' and produced a crude linkage map that identified 13 of the 20 linkage groups. The start codon targeted (SCoT) markers have also proved useful in evaluating the relationship among the Xiang Ya Mango type cultivars in China (Luo et al. 2011). SCoT markers are useful for the confirmation of identical genotypes and clones with different names. SCoT and other dominant markers are useful for germplasm collection management, and they are technically simple and inexpensive to implement.

SSR Markers

Viruel et al. (2005) developed the first reported set of 16 SSR markers for mango, of which 14 produced the expected one or two amplification products per genotype. These 14 SSRs were used to evaluate 28 mango genotypes that included 14 Florida cultivars. Discrimination of all 28 genotypes was possible, and the average number of alleles per locus was 5.3. Previously known pedigree information for the 'Haden' family of mangos was confirmed and was in agreement with published RAPD and DFP analyses (Adato et al. 1995; Schnell et al. 1995) with one exception. Viruel's clone of 'Zill' was not resolved as a seedling of 'Haden'. Schnell et al. (2005) developed a second set of 15 SSR markers and analyzed 59 Florida cultivars and four related species. Two of the SSRs were monomorphic among the Florida cultivars; the other 13 had an average number of alleles per locus of 4.2 with polymorphism information content (PIC) values varying from 0.21 to 0.63.

Schnell et al. (2006) used 25 SSR loci to estimate genetic diversity among 203 unique mangos (*M. indica*), two *M. griffithii* Hook. f. & Thomson, and three *M. odorata* Griff. accessions maintained at the National Germplasm Repository (NGR) and at the Fairchild Tropical Garden (FTG) in Miami, Florida. The 25 SSR loci had an average of 6.96 alleles per locus and an average PIC value of 0.552 for the *M. indica* population. The total propagation error in the collection (i.e., plants that had been incorrectly labeled or grafted) was estimated to be 6.13%. When compared by origin, the Florida cultivars were more closely related to Indian than to Southeast Asian cultivars. Unbiased gene diversity (H_{nb}) of 0.600 and 0.582 was found for Indian and Southeast Asian cultivars, respectively, and both were higher than H_{nb} among Florida cultivars (0.538). When compared by horticultural type, H_{nb} was higher among the polyembryonic types (0.596) than in the monoembryonic types (0.571).

Until recently, a total of 62 SSR markers had been developed for mango (Duval et al. 2005; Honsho et al. 2005; Schnell et al. 2005; Viruel et al. 2005). This number is more than adequate for genetic diversity studies and for parentage analysis as has been demonstrated by Schnell et al. (2006); however, these numbers are not sufficient to develop a saturated linkage map for the 20 linkage groups of mango. Over 300 new SSR markers have been developed at the USDA-ARS Mid South Area Genomics Laboratory (MSAGL) using pyrosequencing and verified using a mango diversity panel of 11 cultivars, using the method described earlier in this chapter. Another additional 1,800 SSR markers have been produced in the Australian mango genomics program (Ian Bally, personal communication 2010).

Parentage Analysis

The SSR markers that have been developed for mango are easily used to verify parentage using a software package like CERVUS (Marshall et al. 1998). When caging trees to exclude foreign pollen or using the polycross mating design, it is now simple to identify the male parent from a set of potential male parents. This has been very useful in cacao breeding where mistakes in pollination have led to the estimation of unreliable breeding values for parental clones.

Among the 64 Florida cultivars evaluated in the parentage analysis by Schnell et al. (2006), the genetic background was found to be based on as few as four Indian cultivars and the polyembryonic cultivar 'Turpentine'. Two Indian cultivars, 'Mulgoba' and 'Sandersha', are in the background of most Florida types with 'Amini', 'Bombay', 'Cambodiana', 'Long', 'Julie', 'Turpentine', and 'Nam doc Mai' making lesser contributions. The seedling races of Cuba and Florida were considered the same by Popenoe (1920) who called them the West Indian race commonly known as 'Turpentine' in Florida. In the parentage analysis, 'Turpentine 10' was identified as a most probable paternal parent for 'Haden'. 'Haden' was reported as the maternal parent for 10 cultivars included in the analysis, but based on the parentage analysis, 31 cultivars were found to have 'Haden' as one of the most likely parents. Likewise, the other important early Florida selection 'Brooks' is the parent of seven cultivars. 'Haden', 'Brooks', and seedlings of 'Haden' and 'Brooks' have contributed disproportionately to the genetic background for many of the cultivars in the Florida group.

Linkage Mapping

The first genetic linkage map in mango was reported by Kashkush et al. (2001), utilizing AFLP markers and 29 progeny from a cross of 'Tommy-Atkins' × 'Keitt' in Israel. They were able to map 34 AFLP loci and produced a crude linkage map that identified 13 of the 20 linkage groups covering 160 cM. A second map has been produced using 60 progeny from a cross of 'Keitt' × 'Tommy-Atkins' in China using AFLP markers. Eighty-one markers with the correct segregation ratios were identified, and 39 of these were used to identify 15 linkage groups. The average distance between two adjacent markers was 14.74 cM. Improvement of the mango recombination map requires the development of more codominant molecular markers. Using Roche 454 sequencing and the SSR discovery pipeline discussed earlier in this chapter, new SSR markers have been developed and verified and are now being validated at SHRS on mapping populations. A suitable number of SSR markers are being identified to develop a moderately saturated recombination map for mango.

Florida Mapping Populations

Two experimental populations have been developed and planted in the field at SHRS as mapping populations. The first population is an F_2 population derived from self-pollination of 'Tommy-Atkins' consisting of 168 seedlings that were planted in the field in 1995. The second population is an F_2 population derived from self-pollination of 'Haden'. A total of 224 seedlings from a single isolated 'Haden' tree have been in the field for 3 years. Phenotypic data collection is in progress for both of these populations. The development of a saturated linkage map and the identification of QTL for important traits are objectives for the USDA-ARS program at SHRS in Miami for the next 5 years.

Australian Mapping Populations

In the Australian program, supported by Agri-Science Queensland, three mapping populations have been developed using controlled pollinations (Ian Bally, Queensland DEEDI, personal communication). The largest populations are 'Irwin' × 'Kensington Pride', 'Tommy-Atkins' × 'Kensington Pride', and 'Creeper' × 'Kensington Pride'. 'Kensington Pride' is the primary commercial mango cultivar in Australia and has a fine flavor and good productivity. A consensus linkage map has been produced using 700 AFLP and 10 SSR markers. Phenotypic data is currently being collected on the mapping populations for QTL analysis, and additional SSR markers are being added to the maps. These maps have not been published.

Marker-Assisted Selection

To develop a marker-assisted selection (MAS) program for mango, more extensive linkage maps need to be developed and mapping populations field evaluated. The progeny size of both of the Florida mapping populations is small (168 and 224). In the Australian program, sizable populations have been developed from controlled pollination. These populations are the best candidates for QTL identification in mango. The populations developed from isolated groves by the USDA-ARS SHRS in Florida together with the Australian populations will be used in the next few years to produce a comprehensive linkage map and to identify QTL for disease resistance and for important horticultural traits.

Holton (2010) reported that the Queensland Primary Industries and Fisheries has invested in a gene discovery project for mango with the goal of discovering genes controlling consumer and grower traits, such as fruit quality and tree architecture. Using a multidisciplinary approach, they are sequencing expressed genes via ESTs and serial analysis of gene expression (SAGE), using next-generation sequencing producing low-pass genome converge, identifying candidate genes from fruit quality and tree architecture, and identifying aroma volatiles from fruit.

Lychee

The lychee, *Litchi chinensis*, is the sole member of the genus *Litchi* in the soapberry family Sapindaceae. It is a tropical fruit tree native to southern China, which is also the largest lychee-producing country and the center of diversity for this genus. It is a diploid with $x=15$ and $2n=30$. A number of manuscripts have been published using RAPD and AFLP markers to estimate genetic diversity in lychee including, Tongpamanak et al. (2002) working with Thai cultivars, Kumar et al. (2006) with Indian cultivars, and Jones et al. (2006) working with the USDA-ARS germplasm collection at the Pacific Area Basin Research Center (PBARC) in Hilo, Hawaii. Substantial genetic diversity has been found among cultivars and a high level of heterozygosity detected in these studies. A limited number of SSR markers were developed for lychee (Viruel and Hormaza 2004; Li et al. 2006) and used to evaluate germplasm collections. In contrast to the RAPD and AFLP marker studies, Viruel and Hormaza (2004) analyzed 21 lychee cultivars using SSR markers, and their results suggested a very narrow genetic base. Li et al. (2006) also found limited genetic diversity when evaluating 58 lychee cultivars from Hainan Island using SSR markers. Using the SSR-enrichment method of Techen et al. (2010) in combination with pyrosequencing, 384 markers were developed and are being verified for this species.

Linkage Map

A high-density recombination map was constructed using AFLP and RAPD molecular markers using 73 individuals from a cross of 'Maguili' × 'Jiaohesanyuehong' (Liu et al. 2010). These individuals are highly divergent, as 'Maguili' is a very late ripening cultivar, while the male 'Jiaohesanyuehong' is a very early ripening cultivar.

Using AFLP and RAPD markers that segregated in the double pseudo-testcross configuration, 425 maternal specific markers, 357 paternal specific markers, and 257 that were segregating in both parents were identified. Using JoinMap 3.0, a map was developed with 312 markers, covering a total genetic distance of 1,040 cM and identifying 16 linkage groups. Among the linkage groups, MS1 and MS2 had a large number of markers relative to the other linkage groups, 102 and 76, respectively. The remaining 14 linkage groups contained between 12 and 4 markers (Liu et al. 2010). The progeny of this cross were also evaluated for the length of the juvenile phase and for fruit maturation date and fruit growth duration. Twenty markers closely related to fruit maturation were discovered; however, QTLs were not identified for any of the traits (Fu et al. 2008).

Longan

Longan, *Dimocarpus longan*, is native to southern China in Yunnan province and is an important tree fruit crop in the Sapindaceae family. It is a diploid with $x=15$ and $2n=30$. The fruit has a bright black seed, and the literal meaning of the word longan is "eye of the dragon." Isozymes were first used to identify longan cultivars in the 1980s (Chen and Ke 1989); however, this method was not very useful in distinguishing cultivars. RAPDs were the first PCR-based marker developed for longan, allowing the separation of different groups and the identification of cultivars (Chen and Liu 2001; Lin et al. 2005; Zhong et al. 2007). Yonemoto et al. (2006), using RAPD markers, was able to distinguish three different groups of germplasm from Taiwan, China, and Malaysia. AFLPs were used by Yi et al. (2003), and they were able to distinguish 11 clusters among the 46 cultivars studied. ISSR markers (Hong 2007) have also been developed and used for cultivar identification and diversity analysis. Using the SSR-enrichment method of Techen et al. (2010) in combination with pyrosequencing, 384 markers were developed and are being verified for this species.

Linkage Map

Guo et al. (2010) constructed a linkage map using 94 progeny from a cross of 'Fengliduo' and 'Dawuyuan'. 'Fengliduo' is a high-quality cultivar, while 'Dawuyuan' has a large fruit size. Using RAPD, ISSR, sequence-related amplified polymorphism (SRAP), and AFLP markers in and a pseudo-testcross design, JoinMap 3.0 was used to construct the linkage maps. For the cultivar 'Fengliduo', 183 loci were mapped and 21 linkage groups were identified covering 965 cM. For the cultivar 'Dawuyuan', 251 markers were mapped into 22 linkage groups covering 1,064 cM. This is the first reported map on longan which needs to be enhanced with codominant markers.

Summary

The ease of development of molecular markers, especially those amenable to high throughput in a MAS program will greatly enhance the rate of genetic gain for many of these tropical fruit species. The development of second- and third-generation pyrosequencing combined with the SSR and SNP discovery pipelines has removed many of the constraints on the use of molecular markers in these tropical fruit species. Thousands of SSR markers and hundreds of thousands of SNP markers can now be discovered with just a few Roche 454 or Illumina runs. With the removal of the limitation on numbers of markers, much of the difficulty in applying these techniques has been greatly reduced. The more challenging problem is the development of large mapping populations and the accumulation of phenotypic data to identify QTL regions associated with traits of interest. Once these populations are made and evaluated, the application of MAS to tropical fruit crops is straightforward.

Complete genome sequencing for most of these species is still a distant possibility. *De novo* assembly of second-generation sequence data into pseudo-molecules representing complete chromosomes requires a significant investment of time and funds. Considering the limited commercial value of many of these tropical species, complete genome sequencing and assembly are not likely using current technology. However, the ability to generate thousands of markers and to apply MAS will greatly accelerate tropical fruit breeding programs.

References

Accerbi M, Schmidt SA, De Paoli E, Park S, Jeong D-H, Green PJ (2010) Methods for isolation of total RNA to recover miRNAs and other small RNAs from diverse species. In: Meyers BC, Green PJ (eds) Plant microRNAs. Methods in molecular biology, vol 592. Humana Press. doi: 10.1007/978-1-60327-005-2_3

Adato A, Sharon D, Lavi U (1995) Application of DNA fingerprints for identification and genetic analyses of mango (*Mangifera indica*) genotypes. J Am Soc Hortic Sci 120:259–264

Albert VA, Soltis DE, Carlson JE, Farmerie WG, Wall PK, Ilut DC, Solow TM, Mueller LA, Landherr LL, Hu Y, Buzgo M, Kim S, Yoo M-J, Frohlich MW, Perl-Treves R, Schlarbaum SE, Bliss BJ, Zhang X, Tanksley SD, Oppenheimer DG, Soltis PS, Ma H, dePamphilis CW, Leebens-Mack JH (2005) Floral gene resources from basal angiosperms for comparative genomics research. BMC Plant Biol 5(1):5

Anonymous (2002) FAOSTAT online database at http://www.fao.org/

Archak S, Gaikwad AB, Gautam D, Rao EV, Swamy KR, Karihaloo JL (2003) Comparative assessment of DNA fingerprinting techniques (RAPD, ISSR and AFLP) for genetic analysis of cashew (*Anacardium occidentale* L.) accessions of India. Genome 46:362–369

Archak S, Gaikwad AB, Swamy KR, Karihaloo JL (2009) Genetic analysis and historical perspective of cashew (*Anacardium occidentale* L.) introduction into India. Genome 52:222–230

Argout X, Salse J, Aury J-M, Guiltinan MJ, Droc G, Gouzey J, Allegre M, Chaparro C, Legavre T, Maxomova SN, Abrouk M, Murat F, Fouet O, Poulain J, Ruiz M, Roguet Y, Rodier-Goud M, Fernandes Barbosa-Neto J, Sabot F, Kudrna D, Ammiraju JSS, Schuster SC, Carlson JE, Sallet E, Schiex T, Dievart A, Kramer M, Gelley L, Shi Z, Berard A, Viot C, Boccara M, Risterucci AM, Guignon V, Sabau X, Axtell MJ, Ma Z, Zhang Y, Brown S, Bourge M, Golser W, Song X, Clement D, Rivallan R, Tahi M, Akaza JM, Pitollat B, Gramacho K, D'Hont A, Brunel D,

Infante D, Kebe I, Costet P, Wing R, McCombie WR, Guiderdoni E, Quetier F, Panaud O, Wincker P, Bocs S, Lanaud C (2010) The genome of *Theobroma cacao*. Nat Genet. doi:10.1038/ng 736

Arias RS, Ballard LL, Duke MV, Simpson SA, Liu XF, Irish BM, Scheffler BE (2010) Batch isolation of microsatellites from tropical species using pyrosequencing. In: Methods in molecular biology (in press)

Arumuganathan K, Earle ED (1991a) Estimation of nuclear DNA content of plants by flow cytometry. Plant Mol Biol Rep 9:229–233

Arumuganathan K, Earle ED (1991b) Nuclear DNA content of some important plant species. Plant Mol Biol Rep 9:208–218

Ashworth VETM, Clegg MT (2003) Microsatellite markers in avocado (*Persea americana* Mill.): genealogical relationships among cultivated avocado genotypes. J Hered 94:407–415

Ashworth VETM, Kobayashi MC, De La Cruz M, Clegg MT (2004) Microsatellite markers in avocado (*Persea americana* Mill.): development of dinucleotide and trinucleotide markers. Scientia Hortic 101:255–267

Ashworth VETM, Chen H, Clegg MT (2007) 17 Avocado. In: Kole C (ed) Genome mapping and molecular breeding in plants, fruits and nuts, vol 4. Springer, Berlin, pp 325–329

Assogbadjo AE, Kyndt T, Sinsin B, Gheysen G, van Damme P (2006) Patterns of genetic and morphometric diversity in baobab (*Adansonia digitata*) populations across different climatic zones of Benin (West Africa). Ann Bot 97:819–830

Azevedo VCR, Vinson CC, Ciampi AY (2005) Twelve microsatellite loci in *Manilkara huberi* (Ducke) Standl (Sapotaceae), an Amazonian timber species. Mol Ecol Notes 5:13–15

Azofeifa-Delgado A (2006) Uso de marcadores moleculares en plantas; aplicaciones en frutales del tropico. Agronomia Mesoamericana 17:221–242

Barkley NA, Roose ML, Krueger RR, Federici CT (2006) Assessing genetic diversity and population structure in a citrus germplasm collection utilizing simple sequence repeat markers (SSRs). Theor Appl Genet 112:1519–1531

Bergh BO (1969) Avocado (*Persea americana* Mill.) In: Ferwerda FP, Wit F (eds) Outlines of perennial crop breeding in the tropics. Misc. Paper 4. Landbouwhogeschool, Wageningen, pp 23–51

Blanche KR, Ludwig JA, Cunningham SA (2006) Proximity rainforest enhances pollination and fruit set in orchards. J Appl Ecol 43:1182–1187

Blas AL, Yu Q, Chen C, Veatch O, Moore PH et al (2009) Enrichment of a papaya high-density genetic map with AFLP markers. Genome 52:716–725

Bompard JM, Schnell RJ (1997) Taxomomy and systematics. In: Litz RE (ed) The mango, botany, production and uses. CABI International, New York, pp 21–48

Borrone JW, Schnell RJ, Violi HA, Ploetz RC (2007) Seventy microsatellite markers from *Persea americana* Miller (avocado) expressed sequence tags. Mol Ecol Notes 7(3):439–444

Borrone JW, Olano CT, Kuhn DN, Brown JS, Schnell RJ, Violi HA (2008) Outcrossing in Florida avocados as measured using microsatellite markers. J Am Soc Hortic Sci 133:255–261

Borrone J, Brown JS, Tondo CT, Mauro-Herrera M, Kuhn DN, Violi HA, Sautter RT, Schnell RJ (2009) An EST-SSR-based linkage map for *Persea americana* Mill (avocado). Tree Genet Gen 5:553–560. doi:10.1007/s11295-009-0208-y

Brown J, Laurentín H, Dávila M (2003) Genetic relationships between nine *Annona muricata* L. accessions using RAPD markers. Fruits 58:255–259

Caceres A, Fletes L, Aguilar L, Ramirez O, Figueroa L et al (1993) Plants used in Guatemala for the treatment of gastrointestinal disorders. 3. Confirmation of activity against enterobacteria of 16 plants. J Ethnopharmacol 38:31–38

Campbell CW, Malo SE (1978) Review of methods for measuring avocado maturity in Florida. Proc Am Soc Hortic Sci Trop Reg 22:16–26

Chanderbali AS, Albert VA, Ashworth VETM, Clegg MT, Litz RE, Soltis DE, Soltis PS (2008) *Persea americana* (avocado): bringing ancient flowers to fruit in the genomics era. Bioessays 30(4):386–396

Chanderbali AS, Albert VA, Leebens-Mack J, Altman NS, Soltis DE, Soltis PS (2009) Transcriptional signatures of ancient floral developmental genetics in avocado (*Persea americana*; Lauraceae). Proc Natl Acad Sci USA 106(22):8929–8934

Chatrou LW, Escribano MP, Viruel MA, Maas JW, Richardson JE, Hormaza JI (2009) Flanking regions of monomorphic microsatellite loci provide a new source of data for plant species-level phylogenetics. Mol Phylogenet Evol 53:726–733

Chen X, Ke GW (1989) A preliminary study on group classification of Fujian longan varieties examined by the analytical technique of peroxidase isozymes. J China Fruit 1:18–21

Chen YZ, Liu CM (2001) Analysis and identification of cultivated varieties of longan by RAPD molecular markers. J China Fruit 4:28029

Chen H, Ashworth VETM, Xu S, Clegg MT (2007) Quantitative analysis of growth rate in avocado. J Am Soc Hortic Sci 132(5):691–696

Chen H, Morrell PL, de la Cruz M, Clegg MT (2008) Nucleotide diversity and linkage disequilibrium in wild avocado (*Persea americana* Mill.). J Hered 99:382–389

Crowhurst RN, Gleave AP, MacRae EA, Ampomah-Dwamena C, Atkinson RG et al (2008) Analysis of expressed sequence tags from *Actinidia*: applications of a cross species EST database for gene discovery in the areas of flavor, health, color and ripening. BMC Genomics 9:351

Croxford AE, Robson M, Wilkinson MJ (2006) Characterization and PCR multiplexing of polymorphic microsatellite loci in cashew (*Anacardium occidentale* L.) and their cross-species utilization. Mol Ecol Notes 6:249–251

Darlington CD, Wylie AP (1945) Chromosome atlas of flowering plants. George Allen & Unwin Ltd., London, p 16

Davenport TL (1986) Avocado flowering. Hort. Rev. 8:257–289

Davenport TL, Parnitzki P, Fricke S, Hughes MS (1994) Evidence and significance of self-pollination of avocados in Florida. J Am Soc Hortic Sci 119:1200–1207

Dawson IK, Hollingsworth PM, Doyle JJ, Kresovich S, Weber JC et al (2008) Origins and genetic conservation of tropical trees in agroforestry systems: a case study from the Peruvian Amazon. Conserv Genet 9:361–372

de Oliveira RP, Cristofani M, Machado MA (2005) Integrated genetic map of citrus based on RAPD markers. Fruits 60:187–193

Degani C, El-Batsri R, Gazit S (1990) Enzyme polymorphism in mango. J Am Soc Hortic Sci 115:844–847

Devi-Sampath P, Vijayaraghavan K (2007) Cardioprotective effect of alpha-mangostin, a xanthone derivative from mangosteen on tissue defense system against isoproterenol-induced myocardial infarction in rats. J Biochem Mol Toxicol 21:336–339

Diallo BO, Joly HI, McKey D, Hossaert-McKey M, Chevallier MH (2007) Genetic diversity of *Tamarindus indica* populations: any clues on the origin from its current distribution? Afr J Biotechnol 6:853–860

Diaz-Matallana M, Schuler-Garcia I, Ruiz-Garcia M, de Jaramillo EH (2009) Analysis of diversity among six populations of Colombian mango (*Mangifera indica* L. cvar. Hilacha) using RAPDs markers. Elect J Biotechnol 12:1–8

Dillon S, Ramage C, Ashmore S, Drew RA (2006) Development of a codominant CAPS marker linked to PRSV-P resistance in highland papaya. Theor Appl Genet 113:1159–1169

Duval MF, Bunel J, Sitbon C, Risterucci AM (2005) Development of microsatellite markers for mango (*Mangifera indica* L.). Mol Ecol Notes 5:824–826

Duval MF, Bunel J, Sitbon C, Risterucci AM, Calabre C, Le Bellec F (2006) Genetic diversity of Caribbean mangoes (*Mangifera indica L.*) using microsatellite markers. Presented at proceedings of the eighth international mango symposium, Sun City, South Africa

Eiadthong W, Yonemori K, Sugiura A, Utsunomiya N, Subhadrabandhu S (1999) Identification of mango cultivars of Thailand and evaluation of their genetic variation using the amplified fragments by simple sequence repeat-(SSR-) anchored primers. Scientia Hortic 82:57–66

Ellegren H (2004) Microsatellites: simple sequences with complex evolution. Nat Rev Genet 5:435–445

Ellis JR, Burke JM (2007) EST-SSRs as a resource for population genetic analyses. Heredity 99:125–132

Englberger L, Schierle J, Hofmann P, Lorens A, Albert K et al (2009) Carotenoid and vitamin content of Micronesian atoll foods: Pandanus (*Pandanus tectorius*) and garlic pear (*Crataeva speciosa*) fruit. J Food Compos Anal 22:1–8

Escribano P, Viruel MA, Hormaza JI (2008) Development of 52 new polymorphic SSR markers from cherimoya (*Annona cherimola* Mill.): transferability to related taxa and selection of a reduced set for DNA fingerprinting and diversity studies. Mol Ecol Resour 8:317–321

FAO (2003) Medium-term prospects for agricultural commodities. Projections to the year 2010. Rome, Italy, p 193

Feria-Romero I, Astudillo-de la Vega H, Chavez-Soto MA, Rivera-Arce E, López M et al (2009) RAPD markers associated with quercetin accumulation in *Psidium guajava*. Biol Plant 53:125–128

Fu JX, Guo YS, Zhato HY, Xie JH, Liu CM, Hu YL, Ouyang R, Zhao YH, Zhou J, Huang SS (2008) A study on the genetic effect and MAS exploration of juvenile phase and fruit maturation period in a mapping population of litchi. In: Qui Dongliang (ed) 3rd IS on longan, lychee and other fruit trees in Spindaceae family. Acta Hortic 863:195–198

Geetha G, Banumathi B, Suresh G (1997) Evaluation of the antifungal activity of natural xanthones from *Garcinia mangostana* and their synthetic derivatives. J Nat Prod 60(5):519–524

Gunjate RT, Kumbhar AR, Thimaiah IM, Amin SM (2009). Acta Horticulturae No. 820 pp. 463–467

Guo DL, Luo ZR (2008) Microsatellite isolation and characterization in Japanese persimmon (*Diospyros kaki*). Biochem Genet 46:323–328

Guo YS, Liu CJ, Zhao YH, Xie JH, Ren PR, Huang TL, Fu JX, Lu BB, Liu CM (2010) Construction of a molecular genetic linkage map for longan based on RAPD, ISSR, SRAP and AFLP markers. Acta Hortic 863:141–147

Hawkins DJ, Kridl JC (1998) Characterization of acyl-ACP thioesterases of mangosteen (*Garcinia mangostana*) seed and high levels of stearate production in transgenic canola. Plant J 13:743–752

Heaton HJ, Whitkus R, Gomez-Pompa A (1999) Extreme ecological and phenotypic differences in the tropical tree chicozapote (*Manilkara zapota* (L.) P. Royen) are not matched by genetic divergence: a random amplified polymorphic DNA (RAPD) analysis. Mol Ecol 8:627–632

Hernández-Delgado S, Padilla-Ramírez JS, Nava-Cedillo A, Mayek-Pérez N (2007) Morphological and genetic diversity of Mexican guava germplasm. Plant Genet Res 5:131–141

Hillyer MJ, Boulter SL, Kitching RL, Hughes JM (2007) Isolation and characterization of eight polymorphic microsatellite loci in the rainforest canopy tree, *Syzygium sayeri* (Myrtaceae). Mol Ecol Notes 7:1199–1201

Hirano R, Oo TH, Watanabe KN (2010) Myanmar mango landraces reveal genetic uniqueness over common cultivars from Florida, India and Southeast Asia. Genome 53:321–330

Hollingsworth PM, Dawson IK, Goodall-Copestake WP, Richardson JE, Weber JC et al (2005) Do farmers reduce genetic diversity when they domesticate tropical trees? A case study from Amazonia. Mol Ecol 14:497–501

Holton T (2010) Mango genomics. Abs PAG XVIII, Jan 2010, San Diego

Hong ZT (2007) Inter-simple sequence repeat (SSR) analysis of different cultivars and seed-abortive character in longan (*Dimocarpus longan* Lour) D. Fuzhou: Fujian Agr For Univ 56

Honsho C, Nishiyama K et al (2005) Isolation and characterization of new microsatellite markers in mango (*Mangifera indica*). Mol Ecol Notes 5:152–154

Iyer CPA, Degani C (1997) Classical breeding and genetics. In: Litz RE (ed) The mango, botany, production and uses. CAB International, New York, pp 49–68

Iyer CPA, Schnell RJ (2009) Classical breeding and genetics. In: Litz RE (ed) The mango, botany, production and uses, vol 2. CAB International, New York, pp 67–96, Chapter 4

Jamnadass R, Lowe A, Dawson IK (2009) Molecular markers and the management of tropical trees: the case of indigenous fruits. Trop Plant Biol 2:1–12

Jones M, Zee FT, Moore PH, Kim MS, Ming R (2006) Genetic diversity of litchi germplasm assessed by AFLP marker. Plant Animal and Microbe Genomes Conference XIV P489, p. 223

Kashkush K, Jinggui F, Tomer E, Hillel J, Lavi U (2001) Cultivar identification and genetic map of mango (*Mangifera indica*). Euphytica 122:129–136

Kijas JM, Fowler JC, Garbett CA, Thomas MR (1994) Enrichment of microsatellites from the citrus genome using biotinylated oligonucleotide sequences bound to streptavidin-coated magnetic particles. Biotechniques 16:656–660

Knight RJ (2002) History distribution and uses. In: Whiley AW, Schaffer B, Wolstenholme BN (eds) The avocado: botany production and uses. CAB International, Wallingford

Knight RJ, Schnell RJ (1994) Mango introduction in Florida and the Haden cultivar's significance to the modern industry. Econ Bot 48(2):139–145

Kuete V, Wabo GF, Ngameni B, Mbaveng AT, Metuno R et al (2007) Antimicrobial activity of the methanolic extract, fractions and compounds from the stem bark of *Irvingia gabonensis* (Ixonanthaceae). J Ethnopharmacol 114:54–60

Kumar M, Gupta M, Shrivastava D, Prasad M, Passar US, Sarin NB (2006) Genetic relatedness among Indian litchi accessions (*Litchi chinensis* Sonn.) by RAPD markers. Int J Agri Res 1:390–400

Kyndt T, Assogbadjo AE, Hardy OJ, Glele Kakai R, Sinsin B et al (2009) Spatial genetic structuring of baobab (*Adansonia digitata*, Malvaceae) in the traditional agroforestry systems of West Africa. Am J Bot 96:950–957

Lahav E, Lavi U (2002) Genetics and classical breeding. In: Whiley A, Schaffer B, Wolstenholome B (eds) The avocado, botany, production, and uses. CAB International, New York, pp 39–69

Larsen A, Vaillant A, Verhaegen D, Kjær E (2009) Eighteen SSR-primers for tetraploid *Adansonia digitata* and its relatives. Conserv Genet Resour 1:325–328

Li MF, Zheng XQ, Zhu YQ, Wang XS, Liang SY et al (2006) Development and characterization of SSR markers in lychee (*Litchi chinensis*). Mol Ecol Notes 6:1205–1207

Lin T, Lin Y, Ishiki K (2005) Genetic diversity of *Dimocarpus longan* in China revealed by AFLP markers and partial rbcL gene sequences. Sci Hortic 103:489–498

Liu G, Bode A, Ma WY, Sang S, Ho CT, Dong Z (2001) Two novel glycosides from the fruits of *Morinda citrifolia* (noni) inhibit AP-1 transactivation and cell transformation in the mouse epidermal JB6 cell line. Cancer Res 61:5749–5756

Liu CM, Guo YS, Liu R, Zhang B, Xie JH, Hu YL, Ouyang R, Zhao YH, Fu JC (2010) Construction of a high density molecular linkage map for lychee based on AFLP and RAPD markers. Acta Hortic 863:87–94

Livingstone DS, Motamayor JC, Schnell RJ, Cariaga K, Freeman B, Meerow AW, Brown JS, Kuhn DN (2010) Mol Breed. doi:10.1007/s11032-010-9416-2

López-Gómez R, Torres-Cárdenas Y, Chávez-Moctezuma M, Salgado-Garciglia R, Jiménez-Moraila B, Corona-Armenta G, Herrera-Estrella L (2007) Genomics of Avocado Criollo Fruit (*Persea americana* var. *drymifolia*). In: Proceedings VI World Avocado Congress (Actas VI Congreso Mundial del Aguacate) 2007. Viña Del Mar, Chile, 12–16 Nov 2007

Lopez-Valenzuela JA, Martinez O, Paredes-Lopez O (1997) Geographic differentiation and embryo type identification in *Mangifera indica* L. cultivars using RAPD markers. Hortscience 32:1105–1108

Lowe AJ, Russell JR, Powell W, Dawson IK (1998) Identification and characterization of nuclear, cleaved amplified polymorphic sequence (CAPS) loci in *Irvingia gabonensis* and *I. wombolu*, indigenous fruit trees of west and central Africa. Mol Ecol 7:1786–1788

Lowe AJ, Gillies AC, Wilson J, Dawson IK (2000) Conservation genetics of bush mango from central/west Africa: implications from random amplified polymorphic DNA analysis. Mol Ecol 9:831–841

Luo C, He XH, Chen H, Ou SJ, Gao MP, Brown JS, Tondo CT, Schnell RJ (2011) Genetic diversity of mango cultivars estimated using SCot and ISSR markers. Current Science (in press)

Marshall TC, Slate J, Kruuk L, Pemberton JM (1998) Statistical confidence for likelihood-based paternity inference in natural populations. Mol Ecol 7:639–655

Martinez-Godoy MA, Mauri N, Juarez J, Marques MC, Santiago J et al (2008) A genome-wide 20 K citrus microarray for gene expression analysis. BMC Genomics 9:318

Meghala R, Ravishankar KV, Anand L, Rekha A (2005) Genetic diversity of Indian sapota (*Manilkara zapota*) cultivars characterized by RAPD markers. PGR Newsletter 142:43–46: FAO-Bioversity

Miller A, Schaal B (2005) Domestication of a Mesoamerican cultivated fruit tree, *Spondias purpurea*. Proc Natl Acad Sci USA 102:12801–12806

Miller AJ, Schaal BA (2006) Domestication and the distribution of genetic variation in wild and cultivated populations of the Mesoamerican fruit tree *Spondias purpurea* L. (Anacardiaceae). Mol Ecol 15:1467–1480

Ming R, Hou SB, Feng Y, Yu QY, Dionne-Laporte A et al (2008) The draft genome of the transgenic tropical fruit tree papaya (*Carica papaya* Linnaeus). Nature 452:991–U997

Morton J (1987) Lychee. In: Fruits of warm climates. Miami, pp 249–259. http://www.hort.purdue.edu/newcrop/morton/lychee.html

Muchugi A, Kadu C, Kindt R, Kipruto H, Lemurt S et al (2008) Molecular markers for tropical trees. A practical guide to principles and procedures. World Agroforestry Centre, Nairobi

Mukherjee SK (1950) Mango: its allopolyploid nature. Nature 166:196–197

Mukherjee SK (1953) The mango – its botany, cultivation, uses and future improvement, especially as observed in India. Econ. Bot. 7:130–62

Mukherjee SK (1957) Cytology of some Malayan species of *Mangifera*. Cytologia 22:239–241

Mukherjee SK (1963) Cytology and breeding of mango. Punjab Hortic J 3:107–115 (observed in India. *Economic Botany* 7)

Nakasone HY, Paull RE (1998) Avocado. In: Tropical fruits. CAB International, New York, pp 76–102

Ng KW, Salhimi SM, Majid AM, Chan KL (2010) Anti-angiogenic and cytotoxicity studies of some medicinal plants. Planta Med 76:935–940

Novelli VM, Takita MA, Machado MA (2004) Identification and analysis of single nucleotide polymorphisms (SNPs) in citrus. Euphytica 138:227–237

Nyffeler R, Baum DA (2000) Phylogenetic relationships of the durians (Bombacaceae-Durioneae or/Malvaceae/Helicteroideae/Durioneae) based on chloroplast and nuclear ribosomal DNA sequences. Plant Systemat Evol 224:55–82

Obolskiy D, Pischel I, Siriwantanametanon N, Heinrich M (2009) *Garcinia mangostana* L. a phytochemical and pharmacological review. Phytother Res 23:1047–1065

Ocampo J, dEeckenbrugge GC, Bruyère S, de Lapeyre de Bellaire L, Ollitrault P (2006) Organization of morphological and genetic diversity of Caribbean and Venezuelan papaya germplasm. Fruits 61:25–37

Oliveira EJ, Pádua JG, Zucchi MI, Vencovsky R, Vieira MLC (2006) Origin, evolution and genome distribution of microsatellites. Genet Mol Biol 29:294–307

Oliveira DD, Lobato AL, Ribeiro SM, Santana AM, Chaves JB, Pinheiro-Sant'ana HM (2010) Carotenoids and vitamin C during handling and distribution of guava (*Psidium guajava* L.), mango (*Mangifera indica* L.), and papaya (*Carica papaya* L.) at commercial restaurants. J Agric Food Chem 2010:4

Palu AK, Kim AH, West BJ, Deng S, Jensen J, White L (2008) The effects of *Morinda citrifolia* L. (noni) on the immune system: its molecular mechanisms of action. J Ethnopharmacol 115:502–506

Pandit SS, Mitra S, Giri AP, Pujari KH, Bhimarao PP, Jambhale ND, Gupta VS (2007) Genetic diversity analysis of mango cultivars using inter simple sequence repeat markers. Curr Sci 93(8):1135–1141

Perfectti F, Pascual L (1996) Segregation distortion of isozyme loci in cherimoya (*Annona cherimola* Mill). Theor Appl Genet 93:440–446

Perfectti F, Pascual L (2005) Genetic diversity in a worldwide collection of cherimoya cultivars. Genet Resour Crop Evol 52:959–966

Pinto ACQ, Andrade SRM, Ramos VHV, Cordeiro MCR (2004) Intervarietal hybridization in mango (*Mangifera indica* L.): techniques, main results and their limitations. Acta Hortic (ISHS) 645:327

Pistelli L, Bertoli A, Noccioli C, Mendez J, Musmanno RA et al (2009) Antimicrobial activity of *Inga fendleriana* extracts and isolated flavonoids. Nat Prod Commun 4:1679–1683

Ploetz R, Schnell RJ, Haynes J (2002) Variable response of open-pollinated seedlings progeny of avocado to Phytophthora root rot. Phytoparasitica 30(3):262–268

Poerwanto R (2002) Nurse stock plant – a new technique to enhance mangosteen (*Garcinia mangostana*) growth. Acta Hortic 575:751–756

Popenoe W (1920) Manual of tropical and subtropical fruits. Macmillan, New York

Ragasa CY, Puno MR, Sengson JM, Shen CC, Rideout JA, Raga DD (2009) Bioactive triterpenes from *Diospyros blancoi*. Nat Prod Res 23:1252–1258

Rai M, Asthana P, Jaiswal V, Jaiswal U (2010) Biotechnological advances in guava (*Psidium guajava* L.): recent developments and prospects for further research. Trees – Struct Func 24:1–12

Ramage CM, Sando L, Peace CP, Carroll BJ, Drew RA (2004) Genetic diversity revealed in the apomictic fruit species *Garcinia mangostana* L. (mangosteen). Euphytica 136:1–10

Razafimandimbison SG, McDowell TD, Halford DA, Bremer B (2010) Origin of the pantropical and nutriceutical *Morinda citrifolia* L. (Rubiaceae): comments on its distribution range and circumscription. J Biogeogr 37:520–529

Richard GF, Kerrest A, Dujon B (2008) Comparative genomics and molecular dynamics of DNA repeats in eukaryotes. Microbiol Mol Biol Rev 72:686–727

Richardson JE, Pennington RT, Pennington TD, Hollingsworth PM (2001) Rapid diversification of a species-rich genus of neotropical rain forest trees. Science 293:2242–2245

Ripa FA, Haque M, Bulbul IJ (2010) In vitro antibacterial, cytotoxic and antioxidant activities of plant *Nephelium longan*. Pak J Biol Sci 13:22–27

Risterucci AM, Duval MF, Rohde W, Billotte N (2005) Isolation and characterization of microsatellite loci from *Psidium guajava* L. Mol Ecol Notes 5:745–748

Samuel R, Pinsker W, Balasubramaniam S, Morawetz W (1991) Allozyme diversity and systematics in Annonaceae. Plant Systemat Evol 178:125–134

Sand L, Peach C, Ramage C, Carroll BJ, Drew R (2005) Assessment of genetic diversity in Australian-grown mangosteen (*Garcinia mangostana* L.) and its wild relatives. Acta Hortic 692:143–151

Santoso PJ, Saleh GB, Saleh NM, Napis S (2005) Phylogenetic relationships amongst 10 Durio species based on PCR-RFLP analysis of two chloroplast genes. Indones J Agric Sci 6:20–27

Schnell RJ, Knight RJ, Harkins DM, Zill G (1994) Eliminating zygotic seedlings in 'Turpentine' mango rootstock populations by visual roguing. Hortscience 29(4):319–320

Schnell RJ, Ronning CM, Knight RJ (1995) Identification of cultivars and validation of genetic relationships in *Mangifera indica* L. using RAPD markers. Theor Appl Genet 90:269–271

Schnell RJ, Olano CT, Campbell RJ, Brown JS (2001) AFLP analysis of genetic diversity within a jackfruit germplasm collection. Sci Hortic 91:261–274

Schnell RJ, Brown JS, Olano CT, Power EJ, Krol CA, Kuhn DN, Motamayor JC (2003) Evaluation of avocado germplasm using microsatellite markers. J Am Soc Hortic Sci 128(6):881–889

Schnell RJ, Olano CT, Quintanilla WE, Meerow AW (2005) Isolation and characterization of 15 microsatellite loci from mango (*Mangifera indica* L.) and cross-species amplification in closely related taxa. Mol Ecol Notes 5:625–627

Schnell RJ, Brown JS, Olano CT, Meerow AW, Campbell RJ, Kuhn DN (2006) Mango genetic diversity analysis and pedigree inferences for Florida cultivars using microsatellite markers. J Am Soc Hortic Sci 131(2):214–224

Schnell RJ, Tondo CL, Brown JS, Kuhn DN, Ayala-Silva T, Borrone JW, Davenport TL (2009) Outcrossing rates between 'Bacon' pollinators and adjacent 'Hass' trees in a commercial California avocado grove estimated using microsatellite markers and the identification of two new lethal mutants. Hortscience 44(6):1522–1526

Scora RW, Bergh BO (1992) Origin of and taxonomic relationship within the genus *Persea*. In: Proceedings of the 2nd World Avocado Congress, vol 2, Riverside, pp 505–514

Selmer KK, Brandal K, Olstad OK, Birkenes B, Undlien DE, Egeland T (2009) Genome-wide linkage analysis with clustered SNP markers. J Biomol Screen 14:92–96

Shapcott A (1999) Vagility and monsoon rainforest archipelago of Northern Australia: patterns of genetic diversity in *Syzygium nervosum* (Myrtaceae). Biotropica 31:579–590

Sharon D, Cregan PB, Mhameed S, Kusharska K, Hillel J, Lahav E, Lavi U (1997) An integrated genetic linkage map of avocado. Theor Appl Genet 95:911–921

Shyamalamma S, Chandra SBC, Hedge M, Naryanswamy P (2008) Evaluation of genetic diversity in Jackfruit (*Artocarpus heterophyllus* Lam.) based on amplified fragment length polymorphism markers. Genet Mol Res 7:645–656

Sim CH, Mahani MC, Choong CY, Salma I (2005) Transferability of SSR markers from lychee (*Litchi chinensis* Sonn.) to pulasan (*Nephelium ramboutan-ake* L.). Fruits 60:379–385

Singh RN, Sharma DK, Majumder PK (1980) An efficient technique of mango hybridization. Sci Hortic 12:299–301

Sobir PR (2007) Mangosteen genetics and improvement. Int J Plant Breed 1:105–111

Sreekumar V, Binoy A, George S (2007) Genetic and morphological variation in breadfruit (*Artocarpus altilis* (Park.) Fosberg) in the Western Ghats of India using AFLP markers. Genet Resour Crop Evol 54:1659–1665

Stern RA, Gazit S (2003) The reproductive biology of the lychee. Hortic Rev 28:393–453

Tanksley SD (1993) Mapping polygenes. Annu Rev Genet 27:205–233

Techen N, Arias RS, Glynn NC, Pan Z, Khan I, Scheffler BE (2010) Optimized construction of microsatellite-enriched libraries. Mol Ecol Resour 10:508–515

Tongpamanak P, Asalaha P, Peerasak S, Chalongchai B (2002) Determination of genetic diversity and relationships among Thai litchi accessions by RAPD and AFLP markers. Kasetsart J (Nat Sci) 36:370–380

Tsy JMLP, Lumaret R, Mayne D, Vall AOM, Abutaba YIM et al (2009) Chloroplast DNA phylogeography suggests a West African centre of origin for the baobab, *Adansonia digitata* L. (Bombacoideae, Malvaceae). Mol Ecol 18:1707–1715

Ude GN, Dimkpa CO, Anegbeh PO, Shaibu AA, Tenkouano A et al (2006) Analysis of genetic diversity in accessions of *Irvingia gabonensis* (Aubry-Lecomte ex O'Rorke) Baill. Afr J Biotechnol 5:219–223

Ushanandini S, Nagaraju S, Kumar KH, Vedavathi M, Machiah DK et al (2006) The anti-snake venom properties of *Tamarindus indica* (leguminosae) seed extract. Phytother Res 20:851–858

Varshney RK, Graner A, Sorrells ME (2005) Genic microsatellite markers in plants: features and applications. Trends Biotechnol 23(1):48–55

Viruel MA, Hormaza JI (2004) Development, characterization and variability analysis of microsatellites in lychee (*Litchi chinensis* Sonn., Sapindaceae). Theor Appl Genet 108:896–902

Viruel MA, Escribano P, Barbieri M, Ferri M, Hormaza JI (2005) Fingerprinting, embryo type and geographic differentiation in mango (*Mangifera indica* L., Anacardiaceae) with microsatellites. Mol Breed 15:383–393

Viruel MA, Gross-Germann E, Barceló A (2007) Desarrollo de un mapa genético con marcadores SSRs y AFLPs en aguacate. VI World Avocado Congress, Viña del Mar, Chile, 12–16 Nov 2007

Wall PK, Leebens-Mack J, Muller K, Field D, Altman N, dePamphilis CW (2008) PlantTribes: a gene and gene family resource for comparative genomics in plants. Nucleic Acids Res 36:D970–D976, http://ancangio.uga.edu/content/persea-americana

Yi GJ, Tan WP, Hur HQ (2003) Studies on the genetic diversity and relationship of longan cultivars by AFLP analysis. Acta Hortic Sin 30:272–276

Ying ZT, Davenport TL, Zhang TF, Schnell RJ, Tondo CL (2009) Selection of highly informative microsatellite markers to identify pollen donors in 'Hass' avocado orchards. Plant Mol Biol Rep 27:374–380

Yonemoto Y, Chowdhury AK, Kato H, Macha MM (2006) Cultivars identification and their genetic relationships in *Dimocarpus longan* subspecies based on RAPD markers. Sci Hortic 109:147–152

Yuan SS, Chang HL, Chen HW, Yeh YT, Kao YH et al (2003) Annonacin, a mono-tetrahydrofuran acetogenin, arrests cancer cells at the G1 phase and causes cytotoxicity in a Bax- and caspase-3-related pathway. Life Sci 72:2853–2861

Zane L, Bargelloni L, Patarnello T (2002) Strategies for microsatellite isolation: a review. Mol Ecol 11:1–16

Zerega NJC, Ragone D, Motley TJ (2004) Complex origins of breadfruit (*Artocarpus altilis*, Moraceae): implications for human migrations in Oceania. Am J Bot 91:760–766

Zheng SQ, Jiang F, Gao HY, Zheng JG (2010) Preliminary observations on the antifatigue effects of longan (*Dimocarpus longan* Lour.) seed polysaccharides. Phytother Res 24:622–624

Zhong FL, Pan DM, Guo Z (2007) RAPD analysis of longan germplasm resources. J Chin Agric Sci Bull 23:558–563

Chapter 8
Papaya Genome and Genomics

Ray Ming, Qingyi Yu, and Paul H. Moore

Abstract Papaya is a major fruit crop of the tropics with minor production in the subtropics. The papaya genome is small (372 Mbp) and has evolutionarily primitive sex chromosomes. These characters justify papaya genomics programs. Recently, a draft of the papaya genome has been sequenced, and the male-specific region of the Y chromosome (MSY) and its corresponding region of the X chromosome have been fully sequenced. Sequencing the papaya genome and the MSY will enhance our capacity to explore the origin and evolution of dioecy in the family of Caricaceae, expand our knowledge on genome evolution by serving as an outgroup for the intensively studied family Brassicaceae, identify candidate genes for target traits, and provide genome-wide DNA markers for papaya improvement.

Keywords *Carica papaya* • Expressed sequence tags • Genetic mapping • Genome sequencing • Physical mapping • Sex chromosomes • Whole genome shotgun sequencing

---

R. Ming, Ph.D. (✉)
Department of Plant Biology, University of Illinois at Urbana-Champaign,
1201 W. Gregory Drive, Urbana, IL 61801, USA
e-mail: rming@life.illinois.edu

Q. Yu, Ph.D.
Texas Agrilife Research Center, Department of Plant Pathology and Microbiology,
Texas A&M University, Weslaco, TX 78596, USA

P.H. Moore, Ph.D.
Hawaii Agriculture Research Center,
P.O. Box 100, Kunia, HI 96759, USA

R.J. Schnell and P.M. Priyadarshan (eds.), *Genomics of Tree Crops*,
DOI 10.1007/978-1-4614-0920-5_8,

Introduction

Papaya (*Carica papaya* L.) is a fast-growing, soft-wooded, perennial herb-like tree in the family Caricaceae. Although papaya trees can live as long as 25 years, those in home gardens and commercial fields are generally grown for less than 3 years to avoid the high cost of harvesting and reduced productivity of older trees. Papaya trees are generally single-stemmed, unless the top has been damaged. The main stem or trunk is straight and tapers from a 20–30 cm base to a 5–7 cm width at the crown. The trunk is hollow between nodes and is covered with a thin bark that is smooth with prominent leaf scars. Under optimal conditions, papaya trees are fast growing, producing 2–4 large palmate leaves per week at the apex of the trunk and losing a similar number toward the base of the tree. Mature papaya trees may be crowned with 15–30 mature leaves alternating spirally on the upper 2 m of the stem. Papaya leaves have stout petioles attached directly to the main stem.

Papaya flowers occur on cymose inflorescences that arise in the axils of mature leaves. The type of inflorescence produced depends on the sex of the tree. Varieties typically are either dioecious (with unisexual flowers and exclusively male and female plants) or gynodioecious (with bisexual flowers on hermaphrodite plants and unisexual flowers on female plants). Male trees produce 60–90-cm long, pendulous inflorescences bearing numerous slender flowers lacking a pistil. Female trees produce short inflorescences with a single flower or a cluster of 2–3 flowers bearing large functional pistils without stamens. Hermaphroditic trees have short inflorescences like those of the female, but they bear bisexual flowers that can be sexually variable.

Fruits are ready to harvest 5–6 months after flowering, which occurs 3–8 months after seed germination. Papaya fruit are fleshy berries from 200 g to 12 kg in weight, hanging from peduncles attached to the upper trunk. The youngest fruit, one to three per node, occur nearer the top and self-pruning results in fewer mature fruit lower down where the leaves have abscised. Fruit shape is a sex-linked character and varies from spherical to ovoid from female flowers to pyriform (pear-shaped) from hermaphrodite flowers. Fruit is normally composed of five carpels fused to form a central ovarian cavity that is lined with the placenta containing numerous gray-black, spherical seeds about 5 mm in diameter. The ovarian cavity is larger in female than in hermaphrodite fruit. The shape of the cavity seen in a transverse cut ranges from a star shape with 5–7 furrows, to smooth and circular (Chan and Paull 2007). Ripe papaya fruit have a smooth, thin, yellow-orange colored skin. Depending on the variety, the flesh color ranges from pale yellow to red, and flesh thickness from 1.5 to 4 cm.

Papaya is a major fruit crop in the tropics. The fruit is primarily consumed fresh, but it is also used in drinks, jams, and as a dried and crystallized fruit candy. Green fruit, leaves, and flowers can be cooked and consumed as a vegetable (Watson 1997). Papaya fruit is rich in vitamins A and C and is a good source for the minerals K, Mg, and B. One medium papaya fruit contains the adult minimum daily requirements of vitamins A and C established by the U.S. Food and Nutrition Board (USDA 2001).

Since papaya grows relatively easily and quickly from seed, it is commonly grown in small gardens for home consumption or local trade in many tropical countries. The commercially reported production of papaya in 61 countries reached 9.1 million metric tons (mmt) harvested from 386,376 ha in 2008 (FAOSTAT 2008). The largest producer was India with 2.69 mmt, followed by Brazil (1.81), Mexico (0.92), Nigeria (0.77), and Indonesia (0.62).

Papaya has several industrial uses based primarily on its proteolytic enzyme, papain (EC: 3.4.22.2), a major component of the mixture of enzymes extracted from the latex of green, unripe fruit (note that ripe papaya fruit contains no latex or papain). Evolutionarily, papain may be associated with protection from frugivorous predators and herbivores (El Moussaui et al. 2001). Commercially, papain is used directly for protein digestion such as a red meat tenderizer, chill proofing of beer, and the external treatment of human hard tissues such as warts and scars (Irwig et al. 2002; Kennedy and Pike 1981; Mekkes et al. 1997). Indirect pharmaceutical applications of papain include the development of selective inhibitors to the animal cysteine proteases that exhibit abnormal activity in a variety of diseases including muscular dystrophy, osteoporosis, pulmonary emphysema, and tumor growth (Czaplewski et al. 1999).

Several characteristics of papaya contribute to its being used as an experimental model for tree crops. Papaya trees are small, generally requiring less than 5 m^2 per plant for a field density of 1,200–2,000 trees per hectare, depending on the variety and where grown. As previously mentioned, papaya trees develop rapidly resulting in a short 3–8-month juvenile phase (from germination to flowering) and only a 9–14-month generation time (from seed of one generation to seed of the next generation). Flowering and fruiting are more abundant during warm weather than during the cold, but they are mostly continuous throughout the year with the production of one to three ripe fruit per week and hundreds of fruit over the life of the tree. Fruit production may occur following cross-pollination, self-pollination, or parthenocarpy (without fertilization), depending on the cultivars and whether the lines are dioecious or gynodioecious. For example, when outcrossing of the Hawaii solo papaya cultivars, Sunrise and Kapho, was limited experimentally by bagging flowers to prevent pollen flow, more than 90% of the flowers of both the cultivars set fruit (Rodriquez-Pastor et al. 1990). However, data showed 65% of "Sunrise" fruit was from selfing and 35% was parthenocarpic. All fruits produced by "Kapoho" were from selfing. Thus, hermaphrodite trees are mostly selfing. However, papaya flower anthers and stigma are large by the time of anthesis, so that it is easy to conduct controlled crossing. Each fruit matures in about 4–5 months, and the hermaphrodite fruit will contain about 800 seeds and the well-pollinated female fruit will contain about 1,000 seeds to provide an abundance of progeny for genetic studies. Genetic analyses are also facilitated by the fact that papaya is easily cloned from cuttings to allow growing the same individual across multiple environments. Various aspects of genomics are relatively easy with papaya, because it is diploid with nine pairs of chromosomes and has a comparatively small haploid genome of 372 Mb (86% as large as the rice genome, Arumuganathan and Earle 1991) and an established transformation system (Fitch et al. 1992).

Papaya is among the few plant species recognized as polygamous with three sexes – female, male, and hermaphrodite. Yampolsky and Yampolsky (1922) reported in an extensive catalog of sexuality among plant species that only 7% of the 120,000 species examined were gynodioecious, which is the condition of the more tropical papaya varieties, while 4% were strictly dioecious, which is the condition of the less tropical papaya varieties (primarily grown in Queensland, Australia). A more recent survey of 250,000 plant species reported that 6% are dioecious (Renner and Ricklefs 1995). Hypotheses to explain the polygamous nature of papaya sex have evolved over the past seven decades to include: a single gene with three alleles, a group of closely linked genes, a genic balance of sex chromosomes over autosome, classical XY chromosomes, and regulatory elements of the flower development pathway (Hofmeyr 1938; Storey 1938a, b). Most recently, we have shown that sex determination in papaya is controlled by a pair of primitive sex chromosomes showing typical characteristics of chromosome rearrangements and suppression of recombination around the sex determination gene (Liu et al. 2004). Papaya, by having multiple sex types in a single species, offers a rare opportunity to use genomics for developing an understanding of the evolutionary history of plant sexuality.

Genetic Diversity

Cultivated papaya, *Carica papaya* L., belongs to Caricaceae, a small family of six genera and 35 species distributed throughout tropical and subtropical regions primarily of the Americas (Ming et al. 2005). Studies over the past two decades have resulted in dividing the former genus *Carica* into two genera, *Vasconcellea* with 21 species and *Carica* with only one species, *C. papaya* (Badillo 2000). While few studies have evaluated genetic diversity within cultivated papaya, it appears that cultural preference and geographic isolation have forced selection of cultivated papaya from a relatively narrow genetic base to result in extremely low genetic diversity (Aradhya et al. 1999; Jobin-Décor et al. 1997; Kim et al. 2002; Van Droogenbroeck et al. 2002). Hawaiian "solo" papayas, for instance, were developed from a single introduction from Barbados in 1910 (Storey 1969). Although a wide range of morphological characters is visible in the field, only ca. 12% amplified fragment length polymorphisms (AFLP) among 63 accessions (Kim et al. 2002). Of the 63 accessions analyzed, 82% of the pair-wise comparisons exhibited genetic similarity greater than 0.85 and fewer than 4% showed less than 0.80. The genetic variation that exists in cultivated papayas is attributed to natural outcrossing events, and the genetic similarity reported is attributed to each accession's origin according to a specific breeding or selection program (Kim et al. 2002). Van Droogenbroeck et al. (2002) included six accessions of cultivated papaya in a study of genetic relationships among 95 accessions representing three genera and 11 species of the Caricaceae from Ecuador. AFLP analysis of these accessions showed low genetic variation (0.99 average similarity) and also showed cultivated papaya to be very distinct from the other genera tested, with an average genetic similarity of only 0.23, supporting the idea that *C. papaya* diverged early from its wild relatives and proceeded to evolve in isolation (Aradhya et al. 1999; Van Droogenbroeck et al. 2002).

The low genetic variation within cultivated papaya indicates a need to introgress desirable traits from wild relatives. However, the genetic dissimilarity between papaya and its wild relatives prevents natural hybridization. While many of papaya's wild relatives are intercompatible, spontaneously producing natural hybrids in areas of overlapping distributions, cultivated papaya requires embryo rescue to produce intergeneric hybrids from its wild relatives (Manshardt and Wenslaff 1989; Manshardt and Drew 1998). Isozyme and randomly amplified polymorphic DNA (RAPD) analysis revealed 73% and 69% dissimilarity, respectively, between cultivated papaya and highland papayas of the genus *Vasconcellea*, formerly regarded as a section of *Carica* (Jobin-Décor et al. 1997). AFLP analysis of a chloroplast DNA (cpDNA) intergenic spacer region also grouped cultivated papaya in a separate clade away from *Vasconcellea* with a bootstrap analysis confidence level of 64% (Aradhya et al. 1999). Use of an intergeneric hybrid between *C. papaya* and *V. quercifolia* backcrossed to *C. papaya*, in a joint Australia-Philippines project, has thus far produced three lines with resistance to Australian isolates of Papaya ringspot virus (PRSV) and six lines with resistance to Philippine PRSV isolates. Infertility and incompatibility problems are no longer a concern by the second backcross generation (Drew et al. 2006). Refinement of these techniques for intergeneric hybridization and embryo-rescue will further facilitate future introgression of desirable traits and genetic diversity into *C. papaya*.

Cytogenetics

The nine pairs of papaya chromosomes were defined by conventional chromosome karyotyping (Heilborn 1921). Its close relative *Vasconcellea* species also have the same number of chromosomes (Heilborn 1921; Storey 1976). Papaya chromosomes are small and were thought in early cytogenetic studies to be similar in size (Kumar et al. 1945; Storey 1953). Recent analyses of papaya metaphase chromosomes demonstrated variation in chromosome sizes (Wai et al. 2010). The physical length of each chromosome was obtained from measuring pachytene chromosomes. The papaya sex chromosomes are the second largest chromosome pair, but within the range of standard deviation comparing with the largest chromosomes. The sex chromosome has been designated as chromosome 1 for its distinctive features as a recently evolved sex chromosome and its role on sex determination. The remaining eight chromosomes were numbered from chromosomes 2–9 based on their physical size from the largest to the smallest (Zhang et al. 2010).

Numerous attempts in the early days of papaya genetic research were made to identify sex chromosomes, but no heteromorphic sex chromosomes were found (Meurman 1925; Suguira 1927; Lindsay 1930; Hofmeyr 1938; Storey 1941). However, a pair of precociously separated chromosomes at meiosis anaphase I of pollen mother cells was observed in males and hermaphrodites (Kumar et al. 1945;

Storey 1953). It was suggested that this pair of chromosomes could be the sex chromosomes (Kumar at al. 1945). Following the discovery of the nascent sex chromosomes in papaya (Liu et al. 2004), florescent in situ hybridization (FISH) of bacterial artificial chromosomes (BACs) to the sex chromosomes revealed distinctive features including five heterochromatic Knobs, one shared between the X and Y chromosomes and four that are Y-specific (Zhang et al. 2008). Notably, the centromere of the Y chromosome appeared to be between Knobs 4 and 5, perhaps recently evolved since no typical centromere repeats were found in this region. The 10–15% of the male-specific region not pairing between the X and Y chromosomes could partly explain the earlier observed precocious separation of one pair of chromosomes during meiosis because of their weaker binding compared to the other eight pairs of chromosomes.

Genetic Mapping

A high-density genetic map is essential for the integration of genetic and physical maps and for assigning sequence scaffolds to papaya chromosomes. It is the first step toward isolating and cloning genes of interest via a map-based cloning approach. High-density genetic maps are also important for genomic dissection of complex traits, comparative analysis of plant genomes, and marker-assisted selection.

The first genetic map of papaya, constructed 70 years ago, consisted of only three morphological markers: sex form, flower color, and stem color (Hofmeyr 1939). The ensuing 50 years failed to produce a more detailed genetic map due to the lack of morphological markers. When DNA markers were first used for linkage mapping in the 1980s, restriction fragment length polymorphism (RFLP) markers were explored in an attempt to construct a genetic map. That attempt was without success for two reasons: (1) The Southern filters of papaya could be used only twice, not 20 times like for other plant species, which made the RFLP marker system inefficient and costly for papaya (Ming unpublished). The reason for this unusual phenomenon is unknown. (2) The polymorphism rate among parental lines of papaya mapping population is very low due to the inbreeding nature of hermaphrodite papaya and the narrow gene pool used in papaya breeding (Kim et al. 2002).

When randomly amplified polymorphic DNA (RAPD) markers emerged in the early 1990s, they were quickly adopted for papaya mapping and the second genetic map was constructed using 62 RAPD markers (Sondur et al. 1996). The sex determination gene *Sex1* was mapped on linkage group 1 where it was flanked by OPT12 and OPT1C approximately 7 cM away on each side. These two markers were further analyzed, and sequence-characterized amplified region (SCAR) markers were developed to show that T1 is present in all papaya samples, so it could be used as a positive control for PCR analysis. T12 along with another SCAR marker W11 showed co-segregation with sex (Deputy et al. 2002; Liu et al. 2004).

Automation with a Li-Cor sequencer of amplified fragment length polymorphisms (AFLP, Vos et al. 1995) made it possible to construct a high-density genetic map of papaya. A total of 1,778 AFLP markers was generated leading to the construction of the third genetic map of papaya consisting of 1,498 AFLP markers, the papaya ringspot virus coat protein marker, morphological sex type, and fruit flesh color (Ma et al. 2004). These markers were mapped onto 12 linkage groups covering a total length of 3,294 cM, with an average distance of 2.2 cM between adjacent markers. The sex determination gene was mapped on linkage group 1 (LG1) with 225 co-segregating AFLP markers that accounted for 67% of the 334 markers on LG1 and 16% of the markers of the entire genome. The large group of co-segregating markers provided strong evidence in support of the hypothesis that recombination is suppressed around the sex determination locus in papaya (Storey 1953).

Simple sequence repeats (SSR) were mined from the whole genome shotgun (WGS) sequence and the BAC end sequences of papaya. Over 11,000 SSR markers were surveyed across parental lines SunUp and AU9 used for developing an F2 mapping population. Seven hundred and thirteen (713) markers were mapped, including 712 SSR markers and one morphological marker (Chen et al. 2007). This fourth genetic map consists of nine major linkage groups corresponding to the nine chromosomes plus three minor linkage groups that initially failed to be integrated.

This map was then enriched with AFLP markers in an attempt to bridge the gaps of three minor linkage groups with better map saturation. The more comprehensive genetic map spanned 945.2 cM, covering nine major and five minor linkage groups containing 712 SSR, 277 AFLP, and one morphological marker. AFLP markers were distributed throughout the 14 linkage groups and resulted in several large locus order rearrangements within the nine major linkage groups. However, the 277 additional AFLP markers failed to merge any major and minor linkage groups (Blas et al. 2009). Integration of the AFLP markers did provide tighter linkage association between loci, leading to a reduction in map distance on LGs 1, 2, and 4 that had been inflated in the previous map. The AFLPs also corrected marker order on LG8. Suppression of recombination in the male-specific Y (MSY) region of LG1 was further validated by the addition of 27 sex co-segregating AFLP markers.

Two of the three minor groups were merged to two major groups based on assembled genome sequence and molecular cytogenetic evidence (Ming et al. 2008). The third minor linkage group was recently merged to a major linkage group based on FISH mapping data, resulting in nine linkage groups representing the nine chromosomes (Wai et al. 2010).

Physical Mapping

A BAC-based physical map of papaya was constructed using a high information-content fingerprinting approach (Yu et al. 2009). The BAC library used for the fingerprinted physical map was constructed from a hermaphrodite plant of the transgenic cultivar, “SunUp.” This BAC library consists of 39,168 BAC clones with an average insert size of 132 kb, providing 13.7× genome equivalents (Ming et al. 2001). The entire BAC library of 39,168 BAC clones was fingerprinted and a total of 38,522 successful fingerprints were obtained. However, 7,698 (ca. 20%) of the successful fingerprints could not be used for the map due to their being empty or having small inserts, or containing cross-contaminated fragments. The remaining 30,824 fingerprints were used to construct the fingerprinted contig map (FPC). A total of 26,466 BAC clones were assembled into 963 contigs; 4,358 clones remained as singletons. The total length of the papaya FPC map is 224,354 consensus band (CB) units; the average number of CB units per contig is 233.0. The average band size of the papaya FPC map was estimated at 1.6 kb based on 22 nonoverlapping FPC contigs of 7,503 CB units covering 12,256,987 bp. Several contigs containing papaya organelle genome sequences were identified and these were excluded for estimating the genome coverage of this FPC map. The total length of the remaining contigs of 222,808 CB units was estimated at 356.5 Mb, covering 95.8% of the papaya genome.

The papaya physical map was integrated with the genetic map and genome sequence using BAC end sequences and a sequence-tagged high-density genetic map. For the integrated map, 535 (55.6%) FPC contigs containing 21,371 (81.2%) BAC clones and 168,217 CBs were anchored on the genetic linkage map. A total of 255 shotgun scaffolds covering 233 Mb were anchored to the genetic map. Overall, 63% of the papaya genome sequences were placed on the genetic map. The integrated genetic and physical map allowed estimates of physical distances between genetic markers and provided the framework for assembling the whole genome shotgun sequences. Meanwhile, the assembled genome sequence provided precise physical distance between genes and DNA markers in gapless regions. These three genomic resources complement one another and correct errors from each individual source. The combined information enhances the capacity for map-based cloning and identification of underlying genes controlling quantitative traits in papaya (Yu et al. 2009).

A total of 1,181 overgos representing conserved sequences of *Arabidopsis* and genetically mapped *Brassica* loci were anchored on the papaya integrated genetic and physical map and the draft genome sequence of papaya. Among these overgos, 756 (64.0%) hit single contigs; the average number of clones per overgo is 6.0. These overgos are direct DNA links between papaya, *Arabidopsis,* and *Brassica* genomes for comparative genomic research among species within the order Brassicales. The overgos were designed from single-copy genes and sequences of *Arabidopsis* and *Brassica.* These anchored overgos further improve the quality of papaya's physical map. Along with the FPC contigs and WGS sequences, overgo markers could help identify synteny and rearrangements in target regions of these genomes, particularly in the recently duplicated genomes of *Arabidopsis* and *Brassica* (Yu et al. 2009).

EST Sequencing

Five papaya flower cDNA libraries were constructed, three from pre-meiosis (<4 mm) flower buds (male, hermaphrodite, and female) and two from mature flower buds (hermaphrodite and female). ESTs from these five libraries were sequenced from the 5′-end to produce 31,652 clean sequences with a minimum length of 200 nucleotides. The average read length of a clean sequence was 486 nucleotides with a minimum quality score of 20. The final clean sequences were used in clustering and assembly using a paracel transcript assembler. Contaminant sequences from *E. coli*, mitochondria, chloroplast, cloning vector, and RNA were filtered during the cleanup stage. Repeat sequences were masked and annotated. EST sequences were then clustered based on local similarity scores of pairwise comparison using 88% similarity over 100 nucleotides (nt). Clusters containing only one sequence were grouped as singletons. The EST clusters were assembled into contigs (contiguous sequence) by multiple-sequence alignment that generates a consensus sequence for each of the clusters, with the criterion of 95% identity over 30 nt overlap. A unigene set of 8,571 EST contigs and singletons was assembled. Blast analysis indicated that about 82% of the unigenes from these papaya libraries have homologous sequences in the protein database of *Arabidopsis* (Yu et al. unpublished). In addition, a normalized and subtractive cDNA library was constructed using pooled RNA samples isolated from roots, leaves, seeds, Cali, three sex types of flowers, and three ripening stages of fruit. Over 50,000 EST sequences were generated from this library, yielding additional unigenes with a total of 16,432 unigenes for genome annotation (Ming et al. 2008).

Papaya Genome Sequencing

For reasons discussed earlier, papaya has been recognized as an excellent model system for studying sex chromosome evolution and for tropical fruit tree genomics. It is a member of the order Brassicales sharing a common ancestor with *Arabidopsis* about 72 million years ago (mya) (Wikström et al. 2001), an excellent out-group to study genome evolution in the family Brassicaceae.

The agricultural importance and the unique biological feature of the nascent sex chromosome justified the sequencing of the papaya genome. The transgenic variety SunUp female genomic DNA was chosen for genome sequencing because of its impact on the papaya industry and to avoid complications of genome assembly in the heterozygous male-specific region of the Y chromosome. In addition, SunUp's non-transgenic progenitor is Sunset, a Solo variety that has undergone more than 25 generations of inbreeding, an ideal homozygous genotype for a genome sequencing project.

The genome of a SunUp female was sequenced using whole genome shotgun (WGS) approach with Sanger sequencers (Ming et al. 2008). It was assembled into contigs containing 278 Mb and scaffolds spanning 372 Mb including embedded gaps. The estimated residual heterozygosity of SunUp is 0.06%, confirming the highly inbred nature of this Solo variety. Of the 16,362 unigenes derived from ESTs, 15,219 (92.5%) matched this assembly. Among 706 BAC end and WGS sequence-derived SSR markers on the genetic map, 652 (92.4%) could be used to anchor 167 Mb of contigs or 235 Mb of scaffolds to papaya linkage groups in the current genetic map. Papaya chromosomes contain heterochromatin knobs, concentrated in the centromeric and pericentromeric regions. The heterochromatic regions account for approximately 17% of the genome, representing about 30–35% of the genomic DNA due to their highly condensed nature. A large portion of the heterochromatic DNA was likely not covered by WGS sequence, as evident by the absence of centromere-specific repeats from the shotgun sequences. The 278 Mb of contig sequence was estimated to represent about 75% of the papaya genome and more than 90% of the euchromatic regions, which is in line with the 92.5% of the EST and 92.4% of genetic markers covered by the assembled genome.

The assembled genome was masked using a *de novo* papaya repeat database for genome annotation. Gene predictions were combined with spliced alignments of proteins and transcripts to produce a reference gene set of 27,950 gene models (revised from the 28,038 when the genome sequence was published). A total of 20,067 (71.8%) of the predicted papaya genes with an average length of 1,102 bp shared similarity to proteins in the nonredundant (NR) database from the National Center for Biotechnology Information (NCBI), and 9,642 (48.0%) of them supported by papaya unigenes. Among 7,971 genes with an average length of 307 bp that had no hits to the nonredundant protein database in the GenBank, only 647 (8.1%) were supported by papaya unigenes, implying that the number of predicted papaya-specific genes was inflated. If the 647 genes with unigene support represent 48.0% of the total, then 1,348 predicted papaya-specific genes may be real, and the number of predicted genes in the assembled papaya genome would be 21,415. Considering that the assembled genome covers 92.5% of the unigenes and 92.4% of the mapped genetic markers, the number of predicted genes in the papaya genome could be 7.5% higher, or 23,151, about 25% less than *Arabidopsis* (Arabidopsis Genome Initiative 2000; Hanada et al. 2007), 38% less than rice (International Rice Genome Sequencing Project 2005), 49% less than poplar (Tuskan et al. 2006), and 24% less than grape (Jaillon et al. 2007). This number is likely the upper limit for papaya genes, because EST-based unigenes and predicted genes from WGS sequence may each be fragmented and counted multiple times, as demonstrated by initially inflated gene numbers estimated from rice WGS draft sequences.

The papaya genome consists of about 52% repetitive sequences, including 43.4% of the papaya genome that is homologous to identifiable transposable elements (TEs) and additional 8.5% repetitive sequences that are currently unannotated, but are likely to be novel TEs. Most of the >600 types of repeats in Repbase[12] are represented in papaya, with the dominant class being retrotransposons (40% of the genome) and some of the abundant types being *Ty3-gypsy* (27.8%) and *Ty1-copia* (5.5%) retrotransposons. An interesting feature of papaya is the relatively low abundance of known DNA transposons (0.20%) compared to other plant genomes. The papaya genome is dominated by papaya-specific TE families, accounting for 38% of the genome sequences.

Papaya is the fifth angiosperm genome to be sequenced and the first transgenic crop to be characterized at the whole genome level. Major findings from the papaya genome sequence include:

- Papaya has fewer genes than *Arabidopsis*, with reductions in most gene families and biosynthetic pathways, making it an excellent system in which to study the function of complex biosynthetic pathways and networks.
- The lower gene number is largely because, unlike *Arabidopsis*, the papaya genome contains no recent genome-wide duplication, with fewer opportunities for subfunctionalization, implying that papaya genes may be more representative of ancestral angiosperms than *Arabidopsis* genes. This lack of a genome-wide duplication event makes papaya a valuable outgroup for comparative genomics of the Brassicaceae.
- Under the assumption that a generalized angiosperm plant could potentially require only the types and minimal numbers of genes that are shared among divergent plant species, we estimate that a minimal angiosperm genome would contain about 13,311 genes.
- Papaya contains significantly fewer (only 25%) disease resistance gene analogs than *Arabidopsis*, suggesting that papaya may have evolved alternative defense mechanisms.
- Papaya also contains significantly fewer *P450* genes than *Arabidopsis*, with some subfamilies expanded, some completely absent, and others novel to the papaya genome.
- Despite reduced gene numbers in most biosynthetic pathways, the number of predicted MADS-box family members is strikingly higher (171 vs. 78 in rice and 141 in *Arabidopsis*) in papaya than in other sequenced plant genomes.
- Papaya has fewer members of gene families involved in fruit ripening, with the exception of starch synthase, possibly reflecting a need for starch storage in the stem and during early fruit development.
- Tremendous amplification in papaya of genes related to volatile development implies strong natural selection for enhanced attractants that may be key to fruit (seed) dispersal by animals and aboriginal peoples.
- Papaya contains fewer circadian clock and light-signaling genes than either poplar or *Arabidopsis*, suggesting that papaya does not require the same level of control for daily and seasonal timing.
- Genome-wide searches for transgenic sequences revealed only three insertions, including a functional cassette with the intact PRSV coat protein gene, a fragment of the *nptII* gene, and a fragment of the *tetA* gene. None of the insertions disrupted functional genes.

Sex Chromosomes

Papaya is trioecious with three sex forms, male, female, and hermaphrodite. Sex determination in papaya is an intriguing system that is primarily under genetic control, although epigenetic influences are also evident as sex reversals in hermaphrodite and male are observed under favorable (for male) or stressed (for hermaphrodite) conditions. Hermaphrodites and males are heterogametic, whereas females are homogametic. Seeds from self-pollinated hermaphrodite trees and the occasional male flowers with recovered carpels always segregate into hermaphrodite to female, or male to female, at the ratio 2:1. This segregation ratio is not the typical Mendelian segregation ratio of 3:1, because the combination of homozygous male or hermaphrodite sex determination factors is lethal as postulated previously (Storey 1938; Hofmeyr 1938). Seeds from female trees segregate hermaphrodite to female or male to female at the ratio of 1:1, depending on the pollen source.

Identification of Nascent Sex Chromosomes in Papaya

Segregation of sex types in any mapping population make it the focal point to map the sex determination genes in genetic mapping projects of papaya, because of the biological significance of sex determination and the potential implication on papaya production. The first and second linkage maps of papaya placed sex determination genes in a linkage group (Hofmeyr 1939; Sondur et al. 1996). Mapping the sex determination gene in a linkage group defied the notion of sex chromosomes in papaya, because classical heteromorphic sex chromosomes do not recombine across majority of the chromosomes. A high-density linkage map of the papaya genome was constructed to further characterize the papaya sex determination locus using 1,498 AFLP markers, the PRSV coat protein marker, morphological sex type, and fruit flesh color (Ma et al. 2004). The sex determination locus was mapped to the middle of a large linkage group (LG1) having a large cluster of 225 sex co-segregating markers. This nonrecombinant block accounted for 67% of the 342 markers on LG1 and 16% of all markers mapped on the genome. This map clearly demonstrated severe suppression of recombination at or around the sex determination locus as proposed previously (Storey 1953).

Fine mapping of the sex determination locus was carried out using six DNA markers with known physical distance on 4,380 informative chromosomes, two each from 2,190 female and hermaphrodite plants of three F_2 and one F_3 populations. Despite the large populations mapped, not a single recombination event was detected, showing complete suppression of recombination in this region (Liu et al. 2004).

The non-recombining (NR) region was physically mapped using a 13× BAC library (Ming et al. 2001) to produce a 2.5-Mb physical map containing 57% of the random sex co-segregating markers developed from co-segregating AFLP markers. Random subclones from nonredundant BACs on the physical map were sequenced to assess the genomic features of the NR region. Sequencing results revealed that the NR region has lower gene density and higher percentage of repetitive sequences than the recombining autosomal parts of the genome. This data coupled with the NR region suppression of recombination in a 4–5-Mb segment and the high degree of sequence divergence between homologous chromosomes in this region is compatible with features of primitive sex chromosomes as envisioned by evolutionary biologists (Charlesworth and Charlesworth 1978; Charlesworth 1991). It was concluded that sex determination in papaya was controlled by a pair of recently evolved sex chromosomes and that the NR region is the male-specific region of the Y chromosome (MSY) (Liu et al. 2004).

An added twist to this story is that two slightly different Y chromosomes exist in papaya; the one controlling males is designated as Y and the other controlling hermaphrodites is designated as Y^h (Ming et al. 2007). The lethal effect of any combination of the Y and Y^h chromosomes is likely due to the loss of function of essential regulatory genes on the Y and Y^h chromosomes.

Molecular Cytogenetics of Sex Chromosomes

High-density genetic mapping placed the MSY of the Y^h chromosome near the middle of LG1 (Ma et al. 2004). Because most papaya chromosomes are metacentric, the MSY might be in the vicinity of the centromere. To physically locate the MSY on the Y^h chromosome, two MSY BACs, 54H01 and 76M08, were hybridized on interphase, prometaphase, metaphase, and anaphase chromosomes using florescent in situ hybridization (FISH) (Yu et al. 2007). Both BACs located on or near the centromere. BAC 54H01 hybridized strongly on the Y^h chromosome and weakly on the X chromosome. BAC 76M08 hybridized only on the Y^h chromosomes but not on the X chromosome, suggesting more extensive sequence divergence between the X and Y^h chromosomes in this region. More detailed FISH analysis revealed five Knobs in the MSY, including four MSY-specific Knobs, and the centromere of the Y chromosome appeared to be embedded in the MSY between Knobs 4 and 5 (Zhang et al. 2008). The MSY was highly methylated.

Physical Mapping of the MSY and Its X Counterpart

Physical mapping of the MSY was initiated from the male-specific marker W11 (Deputy et al. 2002; Liu et al. 2004). Screening the 13× hermaphrodite BAC library with the sex-linked SCAR marker W11 produced four positive BACs. A contig map was constructed by cloning the BAC ends and hybridizing the ends to the four positive BACs. The two outermost ends were used to screen the BAC library to identify and confirm two groups of positive BACs. One large BAC contig spanning 990 kb was constructed by this stepwise chromosome walking process. In addition, 42 AFLP-derived SCAR markers were hybridized to the BAC library and generated four more contigs. After exhausting the genomic resources available at the time, the first MSY physical map was constructed spanning 2.5 Mb and consisting of two major and three smaller contigs.

The second phase of the MSY physical mapping began with fingerprinting all 39,168 clones of the papaya hermaphrodite BAC library (see section "Physical

Mapping" above). Previously identified MSY BACs were confirmed by FISH mapping. The positive BAC clones were used to detect contigs from the genome-wide physical map. Chromosome walking extended the contigs. The relative positions of a set of MSY BACs were verified by fiber FISH and pachytene FISH mapping.

Sex co-segregating SSR markers from genetic mapping were used for physical mapping. These SSR markers frequently fell within already-established contigs, but occasionally an SSR marker would provide a new starting point on the MSY or on the corresponding region of the X chromosome. To date, about 8.4 Mb of the MSY region and 5.4 Mb of its X counterpart have been mapped with one gap each remaining in the MSY and X physical map. The gap on the MSY was the Knob 1, which is also on the X chromosome where it was mapped on the X physical map spanning about 1 Mb. The gap on the X corresponds to the MSY between Knobs 4 and 5, which raises the question whether it is the centromere of the X chromosome. Remarkably, this gap was filled in the MSY, suggesting that the physical maps of MSY and X counterpart are near completion. The 4 Mb of extra sequence of the MSY (excluding the 1-Mb Knob 1 shared between X and Y) is largely accounted for by the four MSY-specific knobs due to accumulation of transposable elements and possible local duplication (Yu et al. unpublished data).

Sequencing of X- and Y^h-BACs

We sequenced seven MSY and two X BACs to examine the genomic features of the MSY region (Yu et al. 2007, 2008a). None of these BACs contained known centromere-specific sequences, but they each contained abundant gypsy retroelements and several copia elements, which are features typical of the pericentromeric regions of plant chromosomes. Expression analysis failed to reveal any genes in five of the seven BACs, thus demonstrating the extreme gene paucity in the MSY. Without a papaya-specific repeat database, only 20% of the sequences of the MSY were classified as repetitive. When a papaya-specific repeat database became available from the papaya genome sequencing project, the repetitive sequence content jumped to 85% of the MSY and 58% of the X sequences compared to a 52% average of the entire papaya genome. This difference between the MSY and the rest of the papaya genome indicates a rapid accumulation of repetitive sequences in the MSY (Ming et al. 2008).

Direct comparison of homologous X and Y^h BAC sequences provided quantitative data for documenting the process of the Y chromosome degeneration and for estimating the time of divergence between the X and Y chromosomes. Two pairs of X and Y^h BACs were sequenced and direct alignment of their sequences revealed three inversion events on the MSY (Yu et al. 2008a). Further analysis of the aligned sequences of the two X and Y^h BAC pairs showed 9.6–35.2% DNA sequence expansion on the MSY. Gene expression analyses indicated seven genes on the two X-BACs and four genes on the two Y-BACs. All four genes on the Y BACs had X counterparts. One of the three unmatched genes appeared to have been either deleted or translocated to another part of the MSY or to autosomes since this gene located within the matched regions of the X and Y BACs. The other two unmatched genes located in the unaligned region of the X BACs. The time of divergence between the X–Y^h gene pairs was estimated to be between 0.5 and 2.2 million years ago (mya), supporting the concept of recent origin of the sex chromosomes in papaya (Yu et al. 2008a).

Sequencing of Y^h and Y BACs

A pair of dioecious X- and Y-specific BACs were sequenced and their sequences were compared to corresponding gynodioecious X- and Y^h-specific BACs (Yu et al. 2008b). Numerous chromosomal rearrangements were detected between the X- and Y-specific BACs, including inversions, deletions, insertions, and duplications. DNA sequence expansion was documented on the Y BAC as happened on the homologous Y^h BAC. Dioecious X- and gynodioecious X^h-specific BACs were virtually identical sharing 99.97% sequence identity with only seven single nucleotide polymorphism and five single nucleotide indels. The Y- and Y^h-specific BACs shared high degree (98.6%) of DNA sequence identity, while the X and Y BACs shared about 84.4% sequence identity. Local chromosomal rearrangements between Y and Y^h BACs were detected, as the consequence of suppression of recombination in the male-specific region and the isolation of Y and Y^h chromosomes enforced by the male lethal effect. Analysis of sequence divergence between three dioecious X and Y gene pairs resulted in an estimated age of divergence from 0.6 to 2.5 million years, to support the hypothesis of a recent origin of the papaya sex chromosomes. The estimated age of divergence between Y and Y^h chromosomes was approximately 73,000 years ago, a time prior to the origin of agriculture about 1,000 years ago (Gupta 2004). Therefore, the hermaphrodite Y^h chromosome likely evolved in nature from an ancestral Y chromosome, and not from human selection as was once suggested (Storey 1976).

Conclusions and Perspective

Considerable genetic diversity and evolutionary innovation have taken place in the tropics, yet tropical plants remain underexplored in their basic biology even though they should have enormous potential for novel discovery. The rapid progress in papaya genomic research clearly demonstrates the tremendous impact that genomic and DNA sequencing technologies can have on minor crops where the financial resources for research are quite limited. Within a decade, papaya genomic resources expanded from a small number of DNA markers to a high-density sequence tagged linkage map, a physical map, a large collection of ESTs, a draft genome, and the complete sequence of the MSY and its X corresponding region. These genomic resources will have profound impact on papaya improvement through providing a better understanding of relevant biology for direct application of genomic tools in breeding programs. The genomic resources of papaya are valuable for studying unique biological features of this trioecious species and for improving the quality and productivity of this nutritious tropical fruit. The lack of recent genome duplication makes papaya a valuable resource for studying Brassicales genome evolution, perhaps clarifying what appears to have been a tumultuous evolutionary history in *Arabidopsis*.

The draft genome sequence of papaya will help explore its medicinal and nutritional applications and enhance the value of this tropical fruit tree species to benefit both farmers and consumers. Papaya leaves, flowers, fruits, and seed extracts have been used in folk medicine and modern medicine. Papain is used to develop selective inhibitors to the animal cysteine proteases that exhibit abnormal activity in a variety of diseases, including muscular dystrophy, osteoporosis, pulmonary emphysema, and tumor growth. Nutritionally, papain is used in beer brewing for chill proofing and in meat tenderization by its action on connective tissue and muscle protein.

Sequencing of the male-specific region of the Y chromosome (MSY) and its X chromosome counterparts will have direct applications in papaya production. Hermaphrodites are preferred in most production regions of the world for their higher productivity since every hermaphrodite tree will produce fruit, whereas female tree fruit production systems lose 6–10% of land area that has to be devoted to growing male trees to pollinate the females. In hermaphrodite fruit production systems, the lack of true breeding hermaphrodite varieties results in reduced productivity due to sex segregation among the seedlings, resulting in unwanted female trees. Identification of sex determination genes in papaya will ultimately lead to the engineering of hermaphrodite plants lacking a Y chromosome and thus be true breeding varieties to improve papaya fruit productivity.

Acknowledgments The authors thank the following agencies and programs for funding relevant parts of the research: NSF Plant Genome Research Program, USDA-ARS Cooperative Agreements with the Hawaii Agriculture Research Center, USDA T-STAR program through the University of Hawaii at Manoa, and the University of Illinois at Urbana-Champaign.

References

Arabidopsis Genome Initiative (2000) Analysis of the genome sequence of the flowering plant *Arabidopsis thaliana*. Nature 408:796–815

Aradhya MK, Manshardt RM, Zee F, Morden CW (1999) A phylogenetic analysis of the genus *Carica* L. (Caricaceae) based on restriction fragment length variation in a cpDNA intergenic spacer region. Genet Res Crop Evol 46:579–586

Arumuganathan K, Earle ED (1991) Nuclear DNA content of some important plant species. Plant Mol Biol Rep 9:208–218

Badillo VM (2000) *Carica* L. vs. *Vasconcella* St. Hil. (Caricaceae): con la rehabilitación de este último. Ernstia 10:74–79

Blas AL, Yu Q, Chen C, Veatch O, Moore PH, Paull RE, Ming R (2009) Enrichment of a papaya high-density genetic map with AFLP markers. Genome 52:716–725

Chan YK, Paull RE (2007) Papaya Carica papaya L., Caricaceae. In: Janick J, Paull RE (eds) Encyclopedia of Fruit and Nuts. CABI, Wallingford, United Kingdom, pp 237–247
Charlesworth B (1991) The evolution of sex chromosomes. Science 251:1030–1033
Charlesworth B, Charlesworth D (1978) A model for the evolution of dioecy and gynodioecy. Am Nat 112:975–997
Chen C, Yu Q, Hou S, Li Y, Eustice M, Skelton RL, Veatch O, Herdes R, Diebold L, Saw J, Feng Y, Bynum L, Wang L, Moore PH, Paull RE, Alam M, Ming R (2007) Construction of a sequence-tagged high density genetic map of papaya for comparative structural and evolutionary genomics in Brassicales. Genetics 177:2481–2491
Czaplewski C, Grzonka Z, Jaskolski M, Kasprzykowski F, Kozk M (1999) Binding modes of a new epoxysuccinyl-peptide inhibitory of cysteine proteases. Where and how do cysteine proteases express their selectivity? Biochem Biophys Acta 1431:290–305
Deputy JC, Ming R, Ma H, Liu Z, Fitch MMM, Manshardt R, Stiles JI (2002) Molecular markers for sex determination in papaya (*Carica papaya* L.). Theor Appl Genet 106:107–111
Drew RA, Siar SV, O'Brien CM, Sajise AGC (2006) Progress in backcrossing between *Carica papaya* × *Vasconcellea quercifolia* intergeneric hybrids and *C. papaya*. Aust J Exp Agric 46:419–424
El Moussaoui A, Nijs M, Paul C, Wintjens R, Vincentelli J, Azarkan M, Looze Y (2001) Revisiting the enzymes stored in the laticifers of Carica papaya in the context of their possible participation in the plant defence mechanism. Cell Mol Life Sci 58:556–570
FAOSTAT (2008) http://faostat.fao.org
Fitch MMM, Manshardt RM, Gonsalves D, Slightom JL, Sanford JC (1992) Virus resistant papaya derived from tissues bombarded with the coat protein gene of papaya ringspot virus. Biotechnology 10:1466–1472
Gupta AK (2004) Origin of agriculture and domestication of plants and animals linked to early Holocene climate amelioration. Curr Sci 87:54–59
Hanada K, Zhang X, Borevitz JO, Li W-H, Shiu S-HA (2007) Large number of novel coding small open reading frames in the intergenic regions of the *Arabidopsis thaliana* genome are transcribed and/or under purifying selection. Genome Res 17:632–640
Heilborn O (1921) Taxonomical and cytological studies on cultivated Ecuodorian species of Carica. Ark Bot 17:1–16
Hofmeyr JDJ (1938) Genetical studies of *Carica papaya* L. I. The inheritance and relation of sex and certain plant characteristics. II. Sex reversal and sex forms. South Africa, Department of Agriculture, Science Bulletin No. 187, 64 pp
Hofmeyr JDJ (1939) Sex-linked inheritance in *Carica papaya* L. S Afr J Sci 36:283–285
International Rice Genome Sequencing Project (2005) The map-based sequence of the rice genome. Nature 436:793–800
Irwig MS, El-Sohemy A, Baylin A, Rifai N, Campos H (2002) Frequent intake of tropical fruits that are rich in beta-cryptoxanthin in associated with higher plasma beta-cryptoxanthin concentrations in Costa Rican adolescents. J Nutr 132:3161–3167
Jaillon CO et al (2007) The grapevine genome sequence suggests ancestral hexaploidization in major angiosperm phyla. Nature 449:463–467
Jobin-Décor MP, Graham GC, Henry RJ, Drew RA (1997) RAPD and isozyme analysis of genetic relationships between *Carica papaya* and wild relatives. Genet Res Crop Evol 44:471–477
Kennedy JF, Pike VW (1981) Papain, chymotrypsin and related proteins – a comparative study of their beer chill-proofing abilities and characteristics. Enzyme Microb Technol 3:59–63
Kim MS, Moore PH, Zee F, Fitch MMM, Steige DL, Manshardt RM, Paull RE, Drew RA, Sekioka T, Ming R (2002) Genetic diversity of *Carica papaya* as revealed by AFLP markers. Genome 45:503–512
Kumar LSS, Abraham A, Srinivasan VK (1945) The cytology of *Carica papaya* Linn. Indian J Agric Sci 15:242–253
Lindsay RH (1930) The chromosomes of some dioecious angiosperms. Am J Bot 17:152–174

Liu A, Moore PH, Ma H, Ackerman CM, Makandar R, Yu Q, Pearl HM, Kim MS, Charlton JW, Stiles JI, Zee FT, Paterson AH, Ming R (2004) A primitive Y chromosome in papaya marks incipient sex chromosome evolution. Nature 427:348–352

Ma H, Moore PH, Liu Z, Kim MS, Yu Q, Fitch MMM, Sekioka T, Paterson AH, Ming R (2004) High-density linkage mapping revealed suppression of recombination at the sex determination locus in papaya. Genetics 166:419–436

Manshardt RM, Drew RA (1998) Biotechnology of papaya. Acta Hortic 461:65–73

Manshardt RM, Wenslaff TF (1989) Inter-specific hybridization of papaya with other species. J Am Soc Hortic Sci 114:689–694

Mekkes JR, le Poole IC, Das PK, Kammeyer A, Westerhof W (1997) In vitro tissue-digesting properties of krill enzymes compared with fibrinolysin/DNAse, papain and placebo. Int J Biochem Cell Biol 29:703–706

Meurman O (1925) The chromosome behavior of some dioecious plants and their relatives with special reference to the sex chromosomes. Soc Sci Fennica Comm Biol 2:105p

Ming R, Moore PH, Zee F, Abbey CA, Ma H, Paterson AH (2001) Construction and characterization of a papaya BAC library as a foundation for molecular dissection of a tree-fruit genome. Theor Appl Genet 102:892–899

Ming R, Van Droogenbroeck B, Moore PH, Zee FT, Kynd T, Scheldeman X, Sekioka T, Gheysen G (2005) Molecular diversity of *Carica papaya* and related species. In: Sharma AK, Sharma A (eds) Plant genome: biodiversity and evolution, vol 1B: Phanerograms. Science Publishers, Enfield, pp 229–254

Ming R, Yu Q, Moore PH (2007) Sex determination in papaya. Semin Cell Dev Biol 18:401–408

Ming R, Hou S, Feng Y, Yu Q, Dionne-Laporte A, Saw JH, Senin P, Wang W, Ly BV, Lewis KLT, Salzberg SL, Feng L, Jones MR, Skelton RL, Murray JE, Chen C, Qian W, Shen J, Du P, Eustice M, Tong E, Tang H, Lyons E, Paull RE, Michael TP, Wall K, Rice D, Albert H, Wang M-L, Zhu YJ, Schatz M, Nagarajan N, Agbayani R, Guan P, Blas A, Wai CM, Ackerman CM, Ren Y, Liu C, Wang J, Wang J, Na J-K, Shakirov EV, Haas B, Thimmapuram J, Nelson D, Wang X, Bowers JE, Gschwend AR, Delcher AL, Singh R, Suzuki JY, Tripathi S, Neupane K, Wei H, Irikura B, Paidi M, Jiang N, Zhang W, Presting G, Windsor A, Navajas-Pérez R, Torres MJ, Feltus FA, Porter B, Li Y, Burroughs AM, Luo M-C, Liu L, Christopher DA, Mount SM, Moore PH, Sugimura T, Jiang J, Schuler MA, Friedman V, Mitchell-Olds T, Shippen DE, dePamphilis CW, Palmer JD, Freeling M, Paterson AH, Gonsalves D, Wang L, Alam M (2008) The draft genome of the transgenic tropical fruit tree papaya (*Carica papaya* Linnaeus). Nature 452:991–996

Renner SS, Ricklefs RE (1995) Dioecy and its correlates in the flowering plants. Am J Bot 82:596–606

Rodriquez-Pastor MC, Galan-Sauco V, Herrero-Romero M (1990) Evaluation of papaya autogamy. Fruits 45:387–391

Sondur SN, Manshardt RM, Stiles JI (1996) A genetic linkage map of papaya based on randomly amplified polymorphic DNA markers. Theor Appl Genet 93:547–553

Storey WB (1938a) The primary flower types of papaya and the fruit types that developed from them. Proc Am Soc Hort Sci 35:80–82

Storey WB (1938b) Segregations of sex types in Solo papaya and their application to the selections of seed. Proc Am Soc Hortic Sci 35:83–85

Storey WB (1941) The botany and sex relations of the papaya. Hawaii Agric Exp Stn Bull 87:5–22

Storey WB (1953) Genetics of papaya. J Hered 44:70–78

Storey WB (1969) Papaya. In: Ferwerda FP, Wit F (eds) Outlines of perennial crop breeding in the tropics. H Veenman & Zonen N.V, Wageningen, pp 21–24

Storey WB (1976) Papaya. In: Simmonds NW (ed) The evolution of crop plants. Longman, London, pp 21–24

Suguira T (1927) Some observations on the meiosis of the pollen mother cells of *Carica papaya*, *Myrica rubra*, *Acuba japonica*, and *Beta vulgaris*. Bot Mag 41:219–224

Tuskan GA et al (2006) The genome of black cottonwood, *Populus trichocarpa* (Torr. & Gray). Science 313:1596–1604

USDA (2001) USDA national nutrient database for standard reference, release 17. Papayas, raw: measure 3 (whole papaya, edible portion). http://www.nal.usda.gov/fnic/foodcomp/Data/SR17/reports/sr17fg09.pdf

Van Droogenbroeck B, Breyne P, Goetghebeur P, Romeijn-Peeters E, Kyndt T, Gheysen G (2002) AFLP analysis of genetic relationships among papaya and its wild relatives (Caricaceae) from Ecuador. Theor Appl Genet 105:289–297

Vos P, Hogers R, Bleeker M, Reijans M, van de Lee T, Hornes M, Friters A, Pot J, Paleman J, Kuiper M, Zabeau M (1995) AFLP: a new technique for DNA fingerprinting. Nucleic Acids Res 23:4407–4414

Wai CM, Ming R, Moore PH, Paull RE, Yu Q (2010) Development of chromosome-specific cytogenetic markers and merging of linkage fragments in papaya. Tropical plant biology 3:171–181

Watson B (1997) Agronomy/Agroclimatology notes for the production of papaya. Ministry of Agriculture Forests Fisheries and Meterology, Australia

Wikström N, Savolainen V, Chase MW (2001) Evolution of the angiosperm: calibrating the family tree. Proc Biol Sci 268:2211–2220

Yampolsky C, Yampolsky H (1922) Distribution of sex forms in the phanerogamic flora. Bibl Genet 3:1–62

Yu Q, Hou S, Hobza R, Feltus FA, Wang X, Jin W, Skelton RL, Blas A, Lemke C, Saw JH, Moore PH, Alam M, Jiang J, Paterson AH, Vyskot B, Ming R (2007) Chromosomal location and gene paucity of the male specific region on papaya Y chromosome. Mol Genet Genomics 278:177–185

Yu Q, Hou S, Feltus FA, Jones MR, Murray J, Veatch O, Lemke C, Saw JH, Moore RC, Thimmapuram J, Liu L, Moore PH, Alam M, Jiang J, Paterson AH, Ming R (2008a) Low X/Y divergence in four pairs of papaya sex-liked genes. Plant J 53:124–132

Yu Q, Navajas-Pérez R, Tong E, Robertson J, Moore PH, Paterson AH, Ming R (2008b) Recent origin of dioecious and gynodioecious Y chromosomes in papaya. Top Plant Biol 1:49–57

Yu Q, Tong E, Skelton RL, Bowers JE, Jones MR, Murray JE, Hou S, Guan P, Acob RA, Luo M-C, Moore PH, Alam M, Paterson AH, Ming R (2009) A physical map of the papaya genome with integrated genetic map and genome sequence. BMC Genomics 10:371

Zhang W, Wang X, Yu Q, Ming R, Jiang J (2008) DNA methylation and heterochromatinization in the male specific region of the primitive Y chromosome of papaya. Genome Res 18:1938–1943

Zhang W, Wai CM, Ming R, Yu Q, Jiang J (2010) Integration of genetic and cytological maps and development of a pachytene chromosome-based karyotype in papaya. Tropical plant biology 3:166–170

Chapter 9
Genomics of *Hevea* Rubber

Thakurdas Saha and P.M. Priyadarshan

Abstract *Hevea brasiliensis* is the most recent domesticated tree species from Amazonian rain forest producing latex of commercial utility. Major hurdles for genetic improvement of rubber tree were attributed to its perennial nature, long juvenile period, and a narrow genetic base. Further, the limited availability of *Hevea* genomic resources/information is another impediment to genomics-assisted crop improvement. Improvement of rubber tree in terms of latex production through breeding was the major focus of the scientific community dealing with the crop. Due to unidirectional selection for yield, other secondary attributes of rubber plants were lost during the process of developing high-yielding clones. Work on plant genomics gained momentum only after whole genome sequencing of *Arabidopsis thaliana* in 2000 (Arabidopsis Genome Initiative, 2000) followed by rice (International Rice Genome Sequencing Project, 2002) and poplar, the first tree genome (International Populus Genome Consortium, 2004). However, rubber genomics is still in its infancy. Initial molecular work started in the 1990s with cloning and characterization of latex biosynthesis genes followed by the studies on gene expression influenced by various biotic and abiotic stresses, tapping panel dryness (TPD), and ethylene stimulation of latex production. Simultaneously, different genetic markers were established in rubber for understanding the inheritance and diversity of natural variation existing among the Wickham and wild populations. Genetic markers were used successfully to generate linkage map for QTLs involving disease tolerance. During the last decade, transgenic research also progressed significantly with the development of transgenic *Hevea* clones with overexpressed

T. Saha, Ph.D. (✉)
Genome Analysis Laboratory, Rubber Research Institute of India,
Kottayam 686009, Kerala, India
e-mail: saha@rubberboard.org.in

P.M. Priyadarshan, Ph.D.
Rubber Research Institute of India, Central Experiment Station,
Chethackal, Thompikandom, Ranni, Kerala 689676, India
e-mail: rriipriya@gmail.com

R.J. Schnell and P.M. Priyadarshan (eds.), *Genomics of Tree Crops*,
DOI 10.1007/978-1-4614-0920-5_9,

MnSOD gene effective against TPD and drought stress. In recent years, with the advent of new-generation sequencing techniques, large-scale EST generation in rubber had been possible, which provided insights into genomic architecture and helped to elucidate genes involved in biological processes like latex production. In the absence of whole genome sequence information, the available transcriptome sequences form a potential resource to be utilized in genetic enhancement of rubber tree. The future challenge is to translate and integrate available genomic knowledge into appropriate methodologies, which we believe will revolutionize future *Hevea* breeding program.

Keywords Genomics • *Hevea brasiliensis* • Latex transcriptome • Linkage map • Para rubber tree

Introduction

Hevea (the Para rubber tree), which produces natural rubber, is a deciduous tree of 30–40 m high in the Amazonian forest (its natural habitat), belonging to the family Euphorbiaceae. One of the species, *Hevea brasiliensis* (Willd. ex Adr. de Juss.) Muell. Arg., is exclusively cultivated over 11.33 million hectares in the world for providing the industry with natural rubber (10.4 million tons in 2010)*. Natural rubber 1,4 *cis*-polyisoprene is a renewable ("green") elastomer being used mainly in tire sector (70%), in latex products (12%), and in many other industrial applications.

Natural rubber is produced in Southeast Asia (92%), Africa (6%), and Latin America (2%). The main natural rubber producing countries are Thailand (3.25 million tons in 2010), Indonesia, Malaysia, India, Vietnam and Côte d'Ivoire, China, Sri Lanka, Brazil, Philippines, Liberia, Cambodia, Nigeria, Cameroon, Guatemala, Myanmar, Ghana, Democratic Republic of Congo, Gabon, and Papua New Guinea. Natural rubber, collected by tapping the bark of the tree, was used by the native people of West Indies. It was brought to Europe by Columbus in 1493, Cortes in 1528, la Neuville in 1723, la Condamine in 1736, Fresneau in 1751, and Fusée Aublet in 1775. Its industrial use was first developed by Charles Macintosh in 1823 and reached its full potency with the invention of vulcanization by Charles Goodyear in 1839. Natural rubber was increasingly associated with transport with the invention and development of tires by Dunlop in 1888, Michelin in 1895, and Ford in 1910.

As per directions of Sir Clements Markham, Sir Henry Wickham collected 70,000 seeds from Rio Tapajoz region of Upper Amazon (Boim district) and transported it to Kew Botanic Gardens during June 1876 (Wycherley 1968; Schultes 1977; Baulkwill 1989). Of the 2,700 seeds germinated, 1,911 were sent to Botanical Gardens, Ceylon, during 1876, and 90% of them survived. During September 1877, 100 *Hevea* plants specified as "Cross material" were sent to Ceylon. Earlier, in June 1877, 22 seedlings, not specified either as "Wickham" or "Cross", were sent from Kew to Singapore, which were distributed in Malaya and formed the prime source of 1,000 tappable trees found by Ridley during 1888. An admixture of "Cross" and

Source: International Rubber Study Group (IRSG), Singapore.

"Wickham" materials might have occurred, as the 22 seedlings were unspecified (Baulkwill 1989). One such parent tree planted during 1877 was available in Malaysia even after 100 years (Schultes 1987). Seedlings from Wickham collection of Ceylon were also distributed worldwide. As a matter of fact, rubber trees covering millions of hectares in Southeast Asia are derived from a very few plants of Wickham's original stock from the banks of the Tapajoz (Imle 1978).

The first commercial planting with bud-grafted plants was undertaken during 1918 in Sumatra's east coast. Ct3, Ct9, and Ct38 were the first clones identified by Cramer (Dijkman 1951; Tan et al. 1996). Commercial ventures gradually spread to China, Thailand, and Vietnam, and rubber became an integral part of the economy of Southeast Asia towards latter half of the twentieth century.

Breeding was initiated with a very strict mass selection among the trees at the beginning of the twentieth century. With the introduction of bud grafting, "generative" and "vegetative" selection methodologies were simultaneously used that resulted in seedlings and grafted clones (Dijkman 1951). Around 1950, the advantages of grafted clones proved to be overwhelming for yield potential compared to genetically improved seedlings, and the focus shifted to derivation of clones for latex productivity. Progress in yield improvement in *Hevea* resulted in a gradual increment, from 650 kg/ha in unselected seedlings during the 1920s to 1,600 kg/ha in best clones during the 1950s. The yielding potential was further enhanced to 2,500 kg/ha in PB, RRIM, RRII, RRIC, IRCA, BPM, and RRIV clones during the 1990s. During these 70 years of rigorous breeding and selection, notable clones like RRIM 501, RRIM 600, RRIM 712, PB 217, PB 235, PB 260, RRII 105, RRIC 100, IRCA 18, IRCA 230, IRCA 331, and BPM 24 were derived (Tan 1987; Simmonds 1989; Clément-Demange et al. 2001; Priyadarshan 2003a, b). Some of the primary clones like PB 56, Tjir 1, Pil B84, Pil D65, Gl 1, PB 6/9, and PB 86 selected during the aforesaid period became parents of improved clones. It must also be highlighted that primary clones like GT 1 and PR 107 are still widely used, although their identification traces back to the 1920s.

In Latin America, every breeding effort is focused on derivation of clones having acceptable yield together with durable resistance to South American leaf blight (SALB) (Dean 1987). It must be emphasized that this disease represents a permanent threat for the whole rubber industry (Davies 1997). In the more humid areas of Asia, susceptibility to Corynespora leaf fall disease has become important for the breeders and rubber industry.

Rubber is currently planted in the form of grafted trees, at a density of about 450 trees per hectare (Fig. 9.1). The buds are collected from budwood grown in the budwood gardens, which are developed for the recommended clones. The plants produced in the nurseries can be budded stumps grown in the soil or budded plants grown in plastic bags. Rootstocks can be also grown directly in the plantation field at standard density, with budding carried out at field level. Rubber tree experiences an immature phase that may vary from 5 to 9 years, depending on climate, soil conditions, and management. When the trunk girth of the trees reaches 50 cm, tapping is initiated that may last between 15 and 30 years. The tapping, a periodically renewed cut incised in the bark of the trunk, generates latex (cell cytoplasm containing rubber

Fig. 9.1 Rubber plantation at the Rubber Research Institute of India, Kerala

particles) throughout the year (Jacob et al. 1995). The tapping intensity is a result of the combination of tapping frequency (tapping every 2, 3, 4, or 5 days) and of chemical stimulation intensity by the application of ethephon, an ethylene releaser (Abraham et al. 1968). In Asia, rubber wood has become an increasingly important economic product, and it represents a new challenge for breeders, which was first addressed by RRIM (Othman et al. 1995). Many of the clones issued from the RRIM 2000 series claim to be latex-timber clones. The trunk and the branches are used for varied transformation (furniture, plywood, particle wood, fuel wood). Latex and wood are two complementary ways for atmospheric carbon sequestration (d'Auzac 1998).

The genus *Hevea* is basically composed of ten species: *H. brasiliensis, H. guianensis, H. benthamiana, H. pauciflora, H. spruceana, H. microphylla, H. rigidifolia, H. nitida, H. camporum,* and *H. camargoana* (Webster and Paardekooper 1989; Wycherley 1992; Schultes 1990). Seven species are found in the upper Rio Negro region, considered to be the center of origin of the genus. *Hevea brasiliensis* is found in southern areas outside this center, in the upper Rio Madeira, where five other species are represented. It has generally been assumed that the species are freely intercompatible (Baldwin 1947). Pires (1981) observed natural hybrids of *H. camargoana* × *H. brasiliensis*, and Gonçalves et al. (1982) analyzed progenies issued from hand pollination from this type of crossing. Consequently, *Hevea* species might be considered as a species complex due to the absence of a strict barrier to recombination between species. Many efforts led to the identification of certain types, which were formerly presented as other possible species. *H. paludosa* was identified in Brazil by Ule in 1905 and is often considered as an 11th species (Gonçalves et al. 1990; Priyadarshan and Gonçalves 2003).

All *Hevea* species have $2n=36$ chromosomes, with the exception of one triploid clone of *H. guianensis* ($2n=54$) and the existence of one genotype of *H. pauciflora* with $2n=18$ (Baldwin 1947; Majumder 1964). Although *Hevea* behaves as a diploid, it is believed to be an amphidiploid ($2n=36$; $x=9$) that stabilized during the course of evolution. This contention is supported by the observance of tetravalents during meiosis (Raemer 1935; Wycherley 1976). *In situ* hybridization studies revealed two distinct 18S-25S rDNA loci and one 5S rDNA locus, suggesting a possible allotetraploid origin with the loss of 5S rDNA during the course of evolution (Leitch et al. 1998). But locus duplications are infrequent in *Hevea* genome, and they could have occurred due to chromosomal modifications posterior to the polyploidization event (Seguin et al. 2003); consequently, the two unknown ancestral genomes of *Hevea* would have strongly diverged.

Low and Bonner (1985) characterized *Hevea* nuclear genome as containing 48% of slowly annealing DNA (putative single copy) and 32% middle repetitive sequences with remaining highly repetitive or palindromic DNA. Estimated haploid genome size of *H. brasiliensis* is 4×10^9 base pairs considering its disomic nature (Roy et al. 2004). Mean molecular size of chloroplast DNA (cpDNA) is predicted to be 152 kb (Fong et al. 1994). Differentiation of the genus into species appears to be linked with the evolution of the Amazonian forest over the last one hundred thousand years. Alternations of humid and semiarid periods responsible for the forest extension or fragmentation resulted in the formation of forest islets. These are assumed to have become zones of protection and differentiation under local selection pressures.

Genetic as well as genomic resources play equally important role in crop improvement. In conventional breeding, genetic resources contribute significantly to develop improved varieties. In the last century, *Hevea* breeding made significant strides in enhancing its productivity. However, the progress in manipulating quantitative traits is still far behind with limited success. The major constraint is the lack of knowledge about the genetic interactions and complex biochemical pathways that are involved in plant responses to varying environmental stresses in traditional and nontraditional rubber-growing regions. During the past two decades, there has been an exponential increase in genomic data acquisition such as gene sequences, DNA markers, linkage maps, ESTs, and knowledge about gene expression profiles. Researchers are now better equipped to apply genomics to understand complex biological processes by combining high-throughput genomic techniques with innovative bioinformatic tools, which facilitate development of superior clones suited to different agroclimatic conditions. This chapter gives an overview of the genomic work carried out in rubber during the last two decades, which have far-reaching impact on *Hevea* improvement towards plant health, productivity, and enhanced stress tolerance.

Genetic Resources

Allied species of *Hevea* make up a gene pool for breeding purposes, especially for the identification and introduction of genes of resistance to leaf diseases (Priyadarshan and Gonçalves 2003). Within *Hevea brasiliensis*, the basic species for natural rubber production, a very clear distinction need to be made between "Wickham" population and the series of wild accessions from the Amazonian forest, usually called "Amazonian" population. "Wickham" population has been the basis for rubber domestication and has evolved through a breeding history of one century, with a current high level of adaptation to modern rubber cropping, except in SALB-affected areas (many areas of Latin America). Conversely, the Amazonian populations, still under evaluation, have not been much modified by human selection. They display an average latex yield of around 12% of the level of currently developed Wickham clones (Clément-Demange et al. 2001) and a fairly high resistance to leaf diseases such as *Microcyclus* or *Corynespora cassiicola* (Berk. & Curt.) Wei.

Different expeditions for the collection and transfer of allied species and Amazonian accessions have been organized since 1890. During 1951–1952, 1,614 seedlings of five *Hevea* species (*H. brasiliensis, H. guianensis, H. benthamiana, H. spruceana, and H. pauciflora*) were introduced to Malaysia (Tan 1987). In Sri Lanka, 11 clones of *H. brasiliensis* and *H. benthamiana* and 105 hybrid materials were imported during 1957–1959 through triangular collaboration of USDA, Instituto Agronomico do Norte (IAN) (Brazil), and Liberia. Many of these clones were later given to Malaysia (Tan 1987). Introductions to the germplasm collection of CNRA in Côte d'Ivoire, with CIRAD cooperation, are made of 40 accessions from the French-Brazilian collection of 1974 from Acre and Rondonia, 19 accessions from a Firestone collection in the Madre de Dios basin in Peru (MDF accessions), 24 accessions given by the Brazilian Research Centre, Embrapa, in Manaus (CNSAM accessions), and 10 accessions from allied *Hevea* species. Part of the collections made by R.E. Schultes has also been rescued from two conservation sites in Columbia, thanks to a France-Columbia agreement, with 302 accessions from Calima site and 41 accessions from Palmira site, which were transferred to Côte d'Ivoire in 1987 after a quarantine period in Martinique island. Between 1945 and 1982, collections from Brazil (mostly Rondonia) have been undertaken at least ten times (Goncalves et al. 1983).

During 1981, due to initiative taken by IRRDB, 63,768 seeds, 1,413 m of budwood from 194 high-yielding trees, and 1,160 seedlings were collected from Brazilian Amazonia (Tan 1987; Simmonds 1989). This collection was performed over three states, namely, Acre, Rondonia, and Mato Grosso, in 16 different districts and in 60 different locations overall. Of this, 37.5% of the seeds were sent to Malaysia and 12.5% to Côte d'Ivoire. Half of the collections were maintained in Brazil. The accessions from budwood collection were brought to Malaysia and Côte d'Ivoire after quarantine against SALB. After the establishment of two IRRDB germplasm centers in Malaysia and in Côte d'Ivoire, other IRRDB member countries were supplied with material according to their request.

Crosses between Wickham and Amazonian accessions are relevant due to possible introgression of more variation. A specific program was undertaken by Côte d'Ivoire and France (CIRAD and CNRA) for the characterization and utilization of Amazonian accessions from 1985 to 1997. Evaluation of the wild Amazonian germplasm for latex yield showed that the wild origins annually produce around 10% of the currently used Wickham clones, which means about 200–300 kg/ha. It was found that the average latex yield in Wickham × Amazonian crosses was rather low, ranging between 30% and 50% of the level of GT1, probably due to the important gap lying between the two populations. Conversely, a wide variability was found within these crosses for growth, with probable heterotic effects enabling the selection of very vigorous Wickham × Amazonian clones. However, the breeders of Rubber Research Institute of India could show significant yield increase (14–82%) in Wickham × Amazonian hybrids (Sankariammal and Mydin 2011). In 1997, a hybridization program was conducted by the breeders of Rubber Research Institute of India to broaden the narrow genetic base of cultivated rubber incorporating two popular Wickham clones, namely, RRII 105 and RRIM 600, as females and seven wild accessions as males. Out of 27 hybrid clones generated, 5 hybrids showed higher yield than the RRII 105 and other superior secondary attributes. These promising hybrids were selected for the next phase of evaluation in participatory trials.

Genomic Resources

Molecular Markers

Application of molecular tools in rubber tree improvement was lagging behind because of limited knowledge of the genome. The genetic base of the cultivated rubber tree, *Hevea brasiliensis,* is assumed to be narrow. It is from the "Wickham gene pool" that a spectacular yield improvement of about ten times has been achieved. The genetic variability of *H. brasiliensis* is high at the center of origin, and knowledge on such variability is fundamental for conservation, breeding, and commercial production of this species. Most often, specific phenotypes of discrete variation have been used as morphological markers. In addition, molecular markers could be highly beneficial as a tool in assisting genetic characterization and breeding (Nodari et al. 1997; Brondani et al. 1998). The perennial nature, the long breeding and selection cycle, and the difficulties in raising F_2 progeny make conventional genetic analysis in *Hevea* difficult (Varghese et al. 1998). Thus, the genetics of rubber tree has been poorly investigated.

Since the development of DNA marker technology in the 1980s, tremendous advancement has taken place in terms of marker development, genetic map construction, functional and comparative genomic linkages, genome sequencing, and development of low-cost technologies. Consequently, a variety of molecular techniques were introduced to study the extent of genetic relationship between wild and cultivated *Hevea* clones.

Restriction Fragment Length Polymorphisms (RFLPs)

Initially, hybridization-based RFLP markers providing codominant information were used to characterize *Hevea* germplasm. RFLP technique was proved to be useful for genetic diversity study in wild and cultivated *Hevea* accessions using low copy number nuclear probes (Besse et al. 1994). Wild populations appeared more polymorphic than the cultivated clones. Rondonian and Mato Grosso populations were found more variable than Acre. RFLP analysis of organelle genomes of *Hevea* was also performed for establishing evolutionary relationships as these two genomes could reflect true evolution because of their uniparental inheritance (Luo et al. 1995). RFLP analysis with mitochondrial DNA probes revealed considerable variations among the accessions from Brazil, Colombia, Peru, and cultivated clones from Wickham collections. Chloroplast DNA-RFLP analysis revealed low level of genetic variation, indicating conserved nature of chloroplast genome than the mitochondrial genome.

Random Amplified Polymorphic DNAs (RAPDs)

The random amplified polymorphic DNA (RAPD) technique, described by Williams et al. (1990), has provided a useful approach for evaluating population genetic differentiation. Using RAPD analysis, Varghese et al. (1998) analyzed 24 cultivated *Hevea brasiliensis* clones to estimate genetic distance. Subsequently, Venkatachalam et al. (2002) described the genetic relationships for 37 *Hevea clones* using RAPD markers, and the clones were classified into seven major groups. Venkatachalam et al. (2004) identified a dwarf-specific RAPD marker and studied inheritance pattern among F1 hybrid progenies. Mathew et al. (2005) studied the phylogenetic relationship among three species of rubber, *Hevea brasiliensis, H. benthamiana,* and *H. spruceana,* employing different molecular marker techniques, namely, RAPD, chloroplast DNA PCR-RFLP, and heterologous chloroplast microsatellites. RAPD analysis clearly indicated a high degree of polymorphism among the three species. Analysis of interrelationships among the species clearly revealed that the clones of *H. brasiliensis* (>50% genetic dissimilarity) are closer to *H. benthamiana* than to *H. spruceana* (>70% genetic dissimilarity). Species-specific RAPD markers were identified for each species, and their locus specificity was proved through hybridization. Venkatachalam et al. (2006) identified two DNA markers in *Hevea* sequencing, and one of them (1.4-kb RAPD marker) revealed homology with *Saccharomyces cerevisiae* proline-specific permease gene. RAPD analysis was used to examine the genetic diversity and structure of the IRRDB'81 germplasm (Lam et al. 2009). A total of 59 accessions from 13 different districts of the Brazilian states, namely, Acre, Rondonia, and Mato Grosso, were studied using few primers. However, low interdistrict differentiation was noticed.

Microsatellite Markers

Microsatellites are known also as simple sequences or simple sequence repeats and are of 1–6 nucleotides. These repeats are subject to a high rate of single-motif insertion and deletion mutations through the process of replication slippage (Levinson and Gutman 1987). They appear to be ubiquitous in higher organisms, although the frequency of microsatellites varies between species. These are abundant, dispersed throughout the genome, and show higher levels of polymorphism than other genetic markers (Schlotterer and Tautz 1992). These features coupled with their ease of detection through PCR using flanking primers have made them useful molecular markers. Their potential for automation and their inheritance in a codominant manner are additional advantages (Morgante and Olivieri 1993; Thomas and Scott 1993). Microsatellite markers are found throughout both the transcribed and nontranscribed regions of a genome (Varshney et al. 2005). Their role in gene regulation and genome evolution has also been discussed widely (Aishwarya and Sharma 2007). The genotyping results can be used for the registration of clones and the protection of breeders' rights (Bocharova et al. 2009). Microsatellite analysis helps in better germplasm management and for devising strategies for identifying core selection (Upadhyay et al. 2010).

There are two approaches for the identification of SSR-containing sequences: (1) molecular and (2) computational. The molecular approach for the development of SSRs is to construct genomic libraries (with or without enrichment for SSRs), screen the libraries, sequence candidate clones, and identify SSR motifs either manually or using computer programs. The computational or bioinformatics approaches take advantage of the available sequences such as those in the public databases by scanning through them and then identifying the ones that contain SSRs. They supplement the molecular approaches by identifying SSR repeats in candidate sequences derived from the libraries.

DNA fingerprints in *H. brasiliensis* using heterologous minisatellite probes from humans were reported by Besse et al. (1993). Low et al. (1996), for the first time, detected microsatellites in the *Hevea* genome through the database search of some *Hevea* gene sequences. The construction of a microsatellite-enriched library in *Hevea brasiliensis* was reported by Atan et al. (1996). Studies to identify SSRs in *Hevea* by Roy et al. (2004) revealed the presence of 67 microsatellites having characteristic simple and compound repeats. They showed the prevalence of $(AG)_n$ and $(AC)_n$ repeats in *Hevea* genome. Besides dinucleotide repeat motifs such as TG/AC, AG/TC, and TA/AT, trinucleotides (AAG, AGG, ATT), tetranucleotides (GAAA, AAGG, ATCC, TAAA, AAAT), and pentanucleotide (GAAAT) repeats were also found. AG repeat motifs occurred at higher frequency as a component repeat. Microsatellite markers developed from the above study were successfully used to identify 27 *Hevea brasiliensis* clones (Saha et al. 2005). The polymorphic microsatellite loci isolated and characterized from an enriched genomic library of *H. brasiliensis* were highly useful in understanding genetic diversity and gene flow among *Hevea* species (Souza et al. 2009). Using 15 highly polymorphic microsatellite loci, Le Guen et al. (2009) assessed genetic diversity of 307 clonally propagated individuals of the wild *Hevea brasiliensis*. Moderate differentiation among wild population was explained based on a subsample of 220 individuals from 14 populations. Among the wild *Hevea* population, Mato Grosso populations were genetically more distant from all other populations. Three population clusters that match the boundaries of hydrographical basins of the main Amazon River tributaries were identified. For genomic studies in *H. brasiliensis*, Le Guen et al. (2010) introduced 296 new polymorphic microsatellite markers through screening of an enriched genomic library. More than 100 *Hevea* microsatellite sequences were registered with the NCBI GenBank by Genome Analysis Laboratory of the Rubber Research Institute of India, and markers were also generated for linkage mapping in rubber (unpublished results).

EST-SSR Markers

Data mining of microsatellites from ESTs has proven effective for generating markers for fingerprinting, genetic mapping, and comparative mapping among species (Varshney et al. 2005). Large-scale SSR-mining projects in plants have aimed at developing microsatellite markers, especially in economically important crop plants. Sequences from many genomes are continuously made freely available in the public databases, and mining of these sources using computational approaches permits rapid and economical marker development. Expressed sequence tags (ESTs) are ideal candidates for mining SSRs not only because of their availability in large numbers but also due to the fact that they represent expressed genes.

By analysis of 10,018 ESTs out of 10,829 for *Hevea brasiliensis*, available in public domain DNA databases, 799 SSR loci were found in the 643 nonredundant SSR-containing ESTs (Feng et al. 2009). Out of 799 SSRs in these ESTs, 84.2% contained simple repeat motifs while 15.8% represented compound motif types, and among the total EST-SSRs, 42.2% were dinucleotide repeats. Genetic variability among 60 *Hevea* genotypes consisting of Asiatic, Amazonian, African, and IAC clones was estimated with 68 selected polymorphic SSRs generated through data mining of 470 reads from GenBank by Gouveia et al. (2010). Recently available next-generation transcriptome sequencing data set (NCBI database – accession number GSE26514), submitted by Xia et al. (2011) was analyzed in the author's laboratory for large-scale SSR mining. The repeat number threshold was designated as more than five for dinucleotide, four for trinucleotide, and three for tetranucleotide repeat motifs. Consequently, 698 dinucleotide, 867 trinucleotide, and 72 tetranucleotide repeat sequences were identified from 48,768 unigenes (unpublished).

Genic Microsatellites

In rubber, SSRs were also identified at the 5′ and 3′ UTR of mRNA sequences. Gene sequences like HMG-CoA reductase (*HMGR*), MnSOD, and β-1,3-glucanase also contain repeat sequences at the untranslated region of mRNA or in the introns in the genomic sequences. Dinucleotide $(CT)_n$ repeats detected in MnSOD had been used as SSR markers for genetic relationship studies by Lespinasse et al. (2000a) and Lekawipat et al. (2003). HMG-CoA reductase encoded by the gene *HMGR* is a key

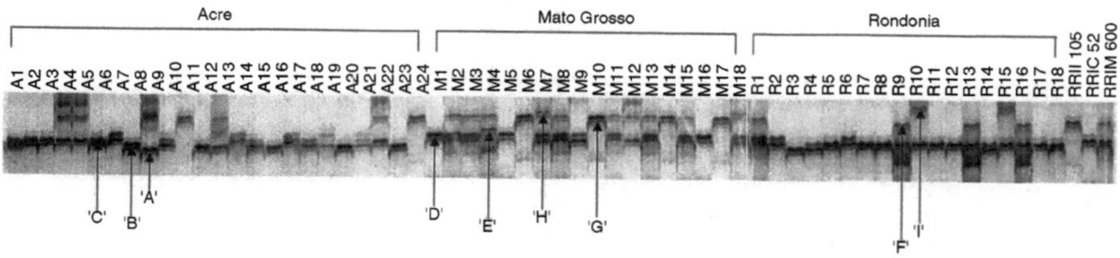

Fig. 9.2 Autoradiogram showing the allelic variation at the locus *HMGR* among wild *Hevea* germplasm accessions from three different provinces of Brazil: Acre, Rondonia, and Mato Grosso. Three cultivated popular clones, RRII 105, RRIC 52, and RRIM 600, were also genotyped. Nine microsatellite alleles ("*A*" to "*I*") were identified at this locus (Saha et al. 2007)

enzyme involved in latex biosynthesis in rubber. It was detected with dinucleotide repeats $(AG)_n$ at the 3′ UTR of mRNA (Saha et al. 2005). SSR polymorphism at this locus was successfully used for studying the allelic diversity in wild accessions of rubber by Saha et al. 2007 (Fig. 9.2). Cross-species amplification of the markers developed for *H. brasiliensis* was also found successful in the wild *Hevea* species *H. guianensis, H. rigidifolia, H. nitida, H. pauciflora, H. benthamiana, and H. camargoana* (Saha et al. 2005; Souza et al. 2009), revealing a high degree of sequence homology at the microsatellite flanking regions of these species. The SSR loci developed are considered as potential tool for studies of population genetics, genetic diversity, and gene flow among *Hevea* species.

Single Nucleotide Polymorphisms

Single nucleotide polymorphisms (SNPs) are the most abundant form of DNA polymorphism in a genome and a resource for mapping complex traits (Rafalski 2002). Bini et al. (2010) identified SNPs in popular *Hevea brasiliensis* clones at the 3′ untranslated regions (3′ UTRs) of 12 genes responsible for complex biochemical traits including latex biosynthesis (Fig. 9.3). Out of 12 loci, 5 loci, (1) geranylgeranyl diphosphate synthase, (2) farnesyl diphosphate synthase, (3) mevalonate kinase, (4) ubiquitin precursor, and (5) latex patatin homolog, were detected with 40 nucleotide substitutions and four indels. Average frequency of SNPs was found to be one in every 90 bases. Heterozygosity for SNPs could also be detected in some of the genotypes/clones.

Retroelements

Retroelements are dispersed as interspersed repetitive sequences throughout the host genome and exploited as genetic tools for plant genome analysis. A reverse transcriptase (*RT*) gene fragment of *Hevea* was cloned indicating the presence of

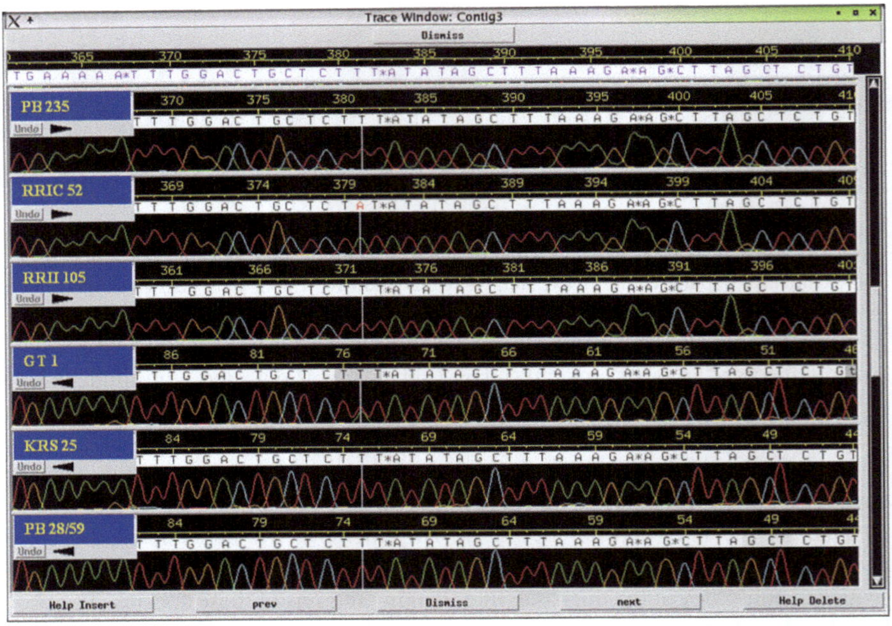

Fig. 9.3 Screen-shot illustration of Consed window showing alignment of partial chromatograms of the locus geranylgeranyl diphosphate synthase from *Hevea* genotypes for identifying SNP at the nucleotide position 381. Heterozygosity could clearly be detected in GT1

retrotransposons – a class of mobile genetic elements in the *Hevea* genome for the first time (Saha et al. 2006) (Fig. 9.4). A *Hevea* genomic library was screened for retroelements using reverse transcriptase (*RT*) gene fragment as the probe, and consequently 23 positive clones were identified. Sequence analysis of positive clones, screened for retroelements, showed homology of eight clones with nucleotide sequences of putative non-LTR retrotransposon *RT* in *Arabidopsis thaliana*, *RT* in *Medicago truncatula*, Ty3-Gypsy type of retrotransposons in *Oryza sativa*, viral gag/pol polyprotein from *Pisum sativum,* and polyprotein of *Ananas comosus* suggesting abundance of retroelements in rubber genome.

Genetic Linkage Mapping

Genetic linkage map presents the linear order of markers (genes and other identifiable DNA sequences) in their respective linkage groups depicting the relative chromosomal locations of DNA markers by their patterns of inheritance. The linkage map allows revelation of more and more restricted segments of the genome and undoubtedly enhances our understanding in many areas of plant systematics. A genetic map for *Hevea* spp. was constructed using a population derived from an interspecific

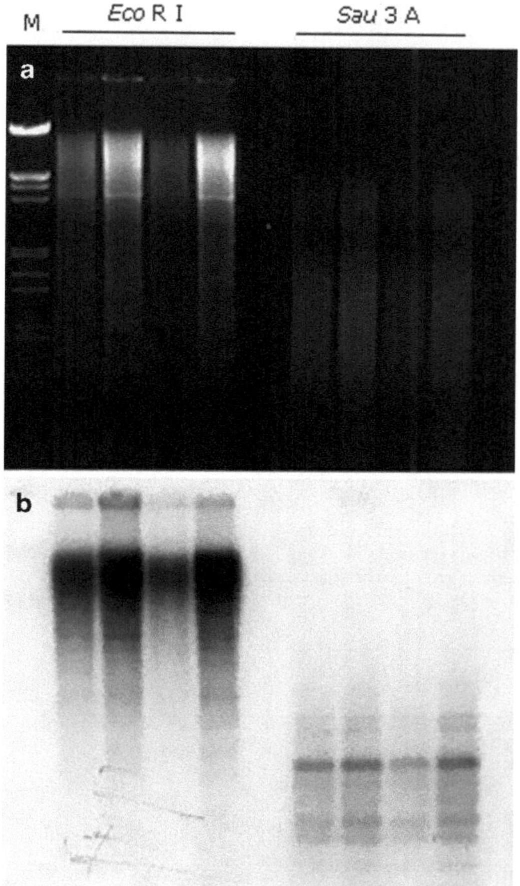

Fig. 9.4 Abundance of retrotransposons in *Hevea* genome was confirmed through Southern hybridization of genomic DNA digested with *Eco*RI and *Sau*3AI against partial reverse transcriptase gene (*RT*) probe generated from rubber. (**a**) Digested DNA on agarose gel; (**b**) autoradiogram of the hybridized blot; *M* molecular weight marker (Lambda DNA/*Eco*RI+*Hin*dIII)

cross between PB 260 (*H. brasiliensis*) and RO 38, an interspecific hybrid clone (*H. brasiliensis*×*H. benthamiana*), following the pseudotestcross strategy (Lespinasse et al. 2000a). The markers were assembled into 18 linkage groups (Fig. 9.5), thus reflecting the basic chromosome number, and covered a total distance of 2,144 cm. A total of 717 loci constituted the synthetic map, including 301 restriction fragment length polymorphisms, 388 amplified fragment length polymorphisms, 18 microsatellites, and 10 isoenzymes. Homologous linkage groups between the two parental maps were merged using bridge loci. Average marker density was 1 per 3 cm. Lespinasse et al. (2000b) mapped quantitative trait loci (QTL) for resistance to South American leaf blight (SALB), a disease of the rubber tree caused by the fungus *Microcyclus ulei* using the same cross combination (PB 260, a susceptible clone, and

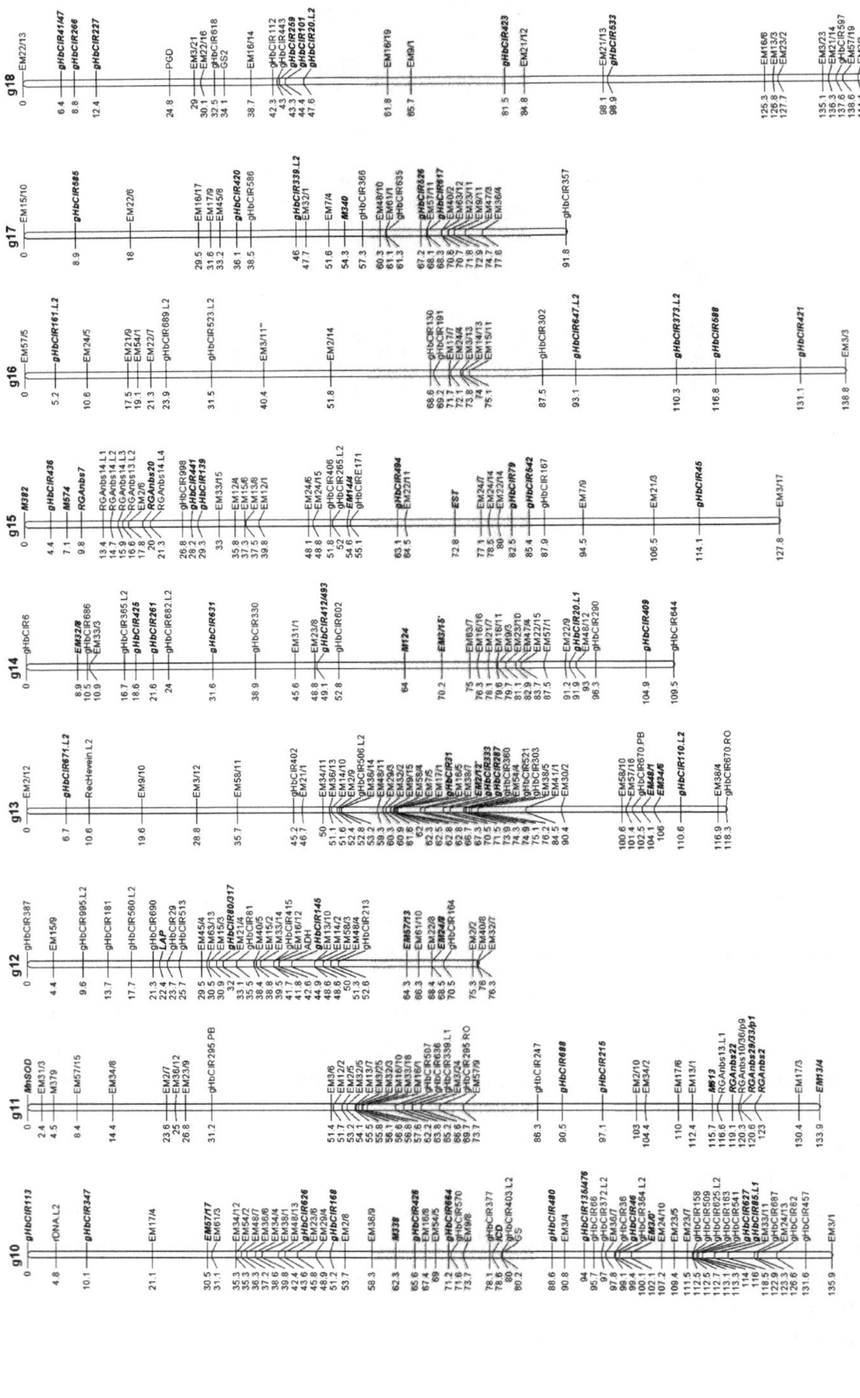

Fig. 9.5 Genetic linkage map *Hevea brasiliensis*. Markers are distributed in 18 linkage groups

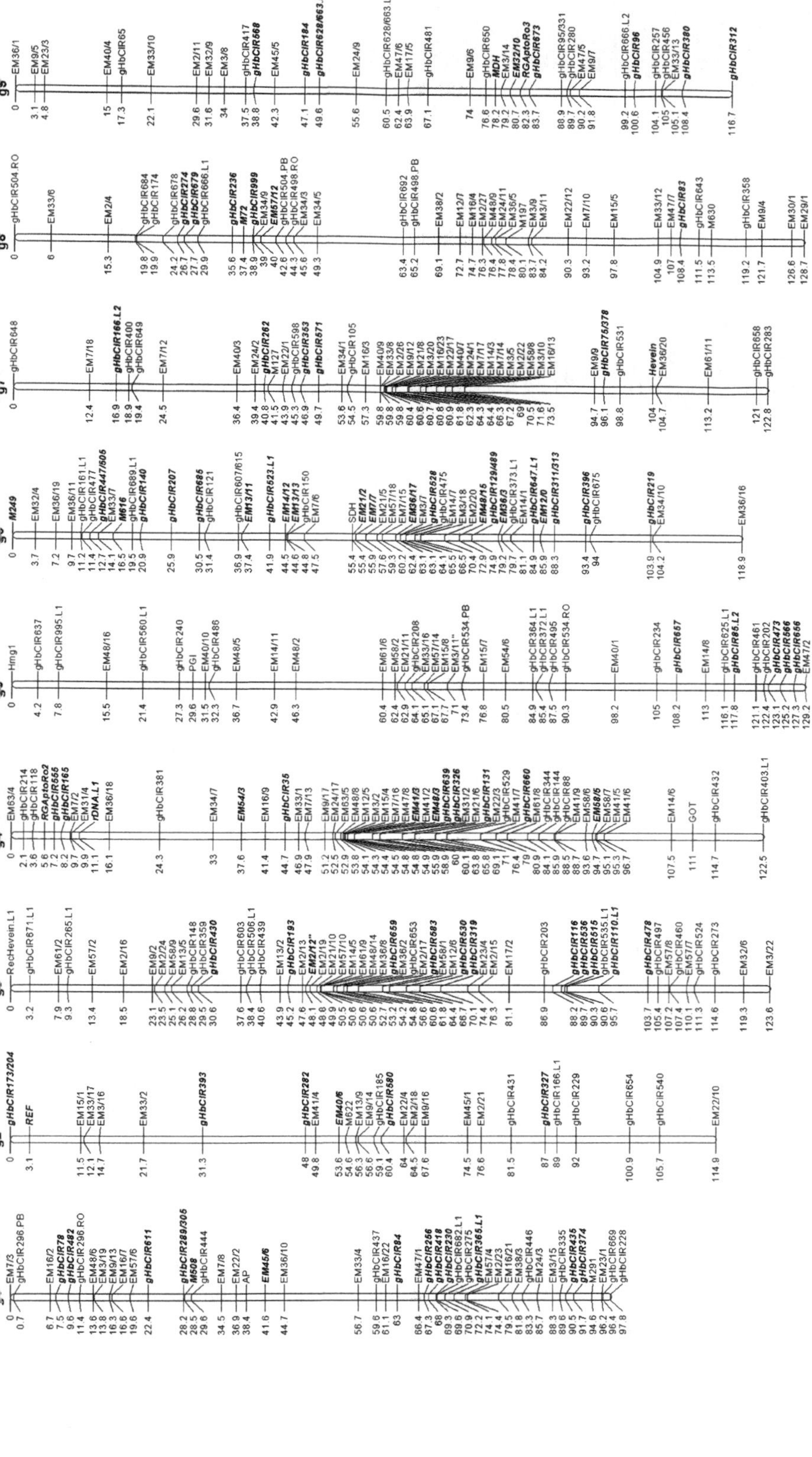

Fig. 9.5 (Continued)

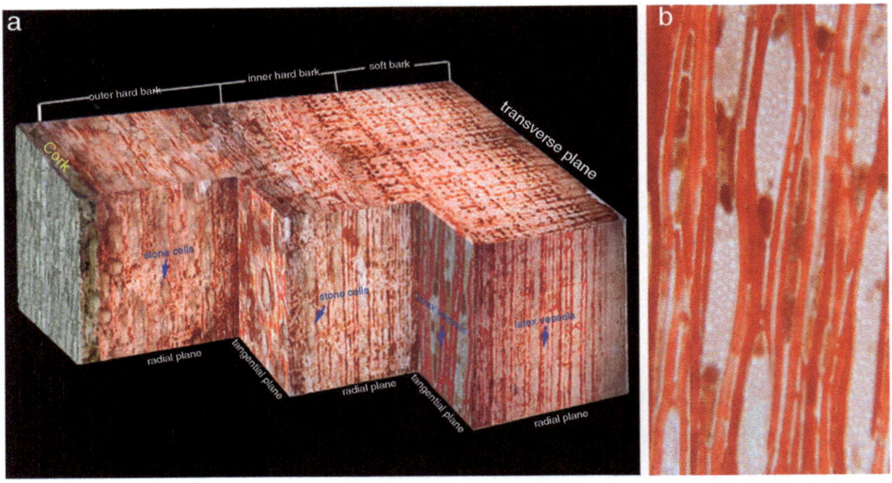

Fig. 9.6 (**a**) Three-dimensional picture of *H. brasiliensis* bark anatomy; (**b**) Enlarged view of the laticiferous system in tangential section of soft bark tissue stained with Oil Red O (Omman and Reghu 2003)

RO 38, a SALB-resistant clone). Eight QTLs for resistance were identified on the RO 38 map, whereas only one QTL was detected on the PB 260 map.

In the author's laboratory, also an effort was made to construct a linkage map of rubber using a segregating progeny population obtained from a cross between two popular cultivated *Hevea* clones: RRII 105 and RRII 118. A total of 227 markers comprising of 96 RAPD, 79 AFLP, 47 SSRs, and five SNP-based markers were utilized for the construction of a genetic linkage map (unpublished).

Gene Discovery

Latex Biosynthesis Genes

Latex is produced in specialized cells known as laticifers or latex vessels, located adjacent to the phloem of the rubber tree. These laticifers form a very complex laticiferous system by anastomosis between tubular cells in the tree (Fig. 9.6). Isoprenoid biosynthesis is brought about through the mevalonate-dependent metabolic pathway (Hepper and Audley 1969; Gronover et al. 2011) (Fig. 9.7). Although it is known that biosynthesis of natural rubber takes place by a mevalonate pathway, molecular biological characterization of related genes has not been adequate. Initial understanding on the regulation of gene expression in the laticifers of *H. brasiliensis* came from the study of Kush et al. (1990), who demonstrated for the first time that transcript levels of genes involved in rubber biosynthesis and genes induced by wounding and ethylene treatment were higher in laticifers than in leaves. Rubber particle

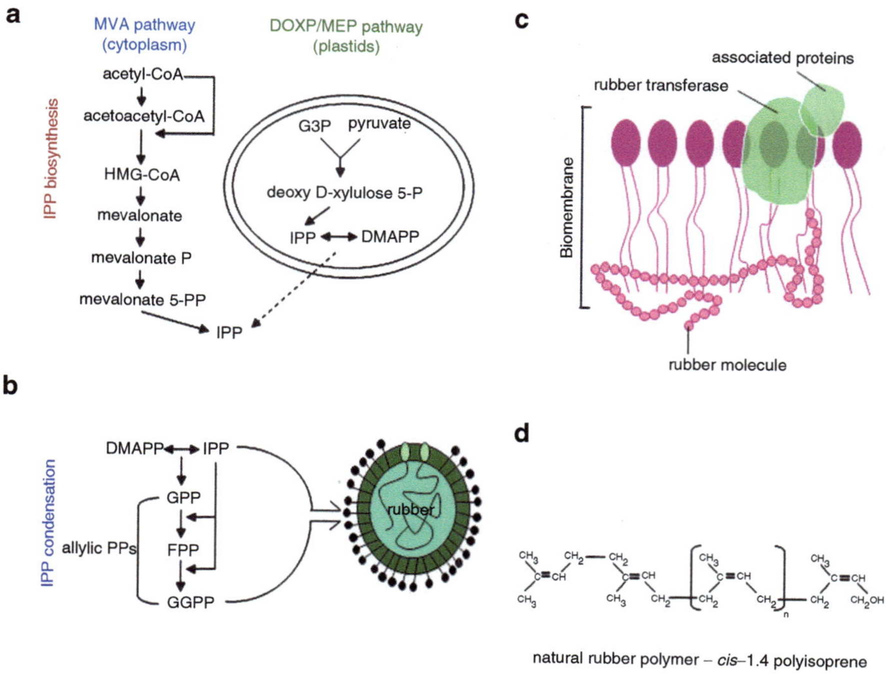

Fig. 9.7 Schematic representation of *cis*-1,4-polyisoprene biosynthesis in plants. (**a**) The monomeric subunit of natural rubber IPP is synthesized by MVA pathway from acetyl CoA in cytosol and chloroplastic DOXP/MEP pathway from G3P and pyruvate. (**b**) IPP condensation to allylic diphosphates for natural rubber synthesis. Each new molecule of dis-1,4-polyisoprene requires an allylic diphosphate initiator before the isoprene units from IPP are polymerized in rubber particles. (**c**) Natural rubber is synthesized by the activity of rubber transferase and other associated proteins at the monolayer biomembrane surface of rubber particles. (**d**) *cis*-1,4 polyisoprene. IPP isopentenyl diphosphate, MVA mevalonate, DMAPP dimethyallyl diphosphate, DOXP/MEP 2-1-deoxy-D-xylulose-5-phosphate/2-C-methyl-D-erythritol-4-phosphate, G3P glyceraldehyde-3-phosphate, GPP geranly diphosphate, FPP farnesyl diphosphate, GGPP geranylgeranyl diphosphate (modified after Gronover et al. 2011)

in the laticifers is the site of rubber (*cis*-1,4-polyisoprene) biosynthesis. A 14-kilodalton protein, rubber elongation factor (REF), is associated with the rubber particle. To obtain more information concerning the function of REF and its synthesis and assembly in the rubber particle, Goyvaerts et al. (1991) isolated cDNA clones encoding REF and characterized the same. Biosynthesis of natural rubber is known to take place biochemically by a mevalonate pathway including six steps catalyzed by corresponding enzymes (Sando et al. 2008). A key enzyme involved in rubber biosynthesis is HMG-CoA synthase, which catalyzes the condensation of acetyl-CoA with acetoacetyl-CoA to form HMG-CoA (Suwanmanee et al. 2002, 2004; Sirinupong et al. 2005). Reduction of HMG-CoA to mevalonic acid, catalyzed by HMG-CoA reductase, is considered as a rate-limiting factor in rubber biosynthesis and thereby regulating the biosynthesis of natural rubber. These two enzymes possibly function in concert in response to the supply of substrates for rubber biosynthesis (Suwanmanee et al. 2002). HMGS mRNA transcript accumulation was found to be more in laticifers than in leaves. A positive correlation was also observed between the activity of *HMGS* and dry rubber content of the latex. Two members of *HMGS* from *Hevea brasiliensis hmgs-1* and *hmgs-2* were cloned and characterized. *hmgs-1* was found to be higher in laticiferous cells than in leaves, whereas the abundance of *hmgs-2* was more in laticifer and petiole than in leaves. In the case of *HMGR*, three genes, *hmg-1*, *hmg-2,* and *hmg-3*, were identified of which *hmg-1* was reported to be involved in rubber biosynthesis (Chye et al. 1991, 1992).

The main precursor for IPP in the pathway is phosphorylated mevalonate and is synthesized by mevalonate kinase (Archer and Audley 1987). Oh et al. (2000) isolated and characterized a cDNA clone encoding IPP isomerase from *H. brasiliensis* and showed involvement of IPP isomerase in rubber biosynthesis through *in vitro* assay. The major enzymes involved in the condensation of IPP are geranylgeranyl diphosphate synthase (GGPP synthase) and farnesyl diphosphate synthase (FDP synthase). Takaya et al. (2003) identified GGPP synthase catalyzing the condensation of IPP with allylic diphosphates to produce (all-E)-GGPP. Adiwilaga and Kush (1996) isolated a full-length cDNA encoding a 47-kDa FDP synthase from *Hevea* and suggested a dual role for FDP synthase in the biosyntheses of rubber and other isoprenoids. Its expression level increased with the regeneration of latex due to tapping.

The gene responsible for the *cis*-1,4 polymerization of isoprene units was isolated and characterized by Asawatreratanakul et al. (2003). Predominance of the transcripts of *Hevea cis*-prenyltransferase was detected in latex as compared with other *Hevea* tissues examined.

Sando et al. (2008) isolated full-length cDNA of genes encoding enzymes catalyzing the six steps of MVA pathway. They characterized three acetyl-CoA acetyltransferase genes, two HMG-CoA synthase genes, and four HMG-CoA reductase genes and one each of mevalonate kinase (*MVK*), phosphomevalonate kinase (*PMK*), and mevalonate diphosphate decarboxylase (*MVD*), which were highly expressed in latex. According to Sando et al. (2008), *MVK* and *PMK* were found to be involved in other isoprenoid biosynthesis in addition to MVA pathway, as their expression level was found to be the same in both laticifers and xylem (which contains no rubber). Venkatachalam et al. (2009) also cloned and characterized a full-length cDNA as well as genomic fragment for *hmgr1* gene from an elite rubber clone RRII 105. The nucleotide sequence of a genomic clone comprised of four exons and three introns, giving a total length of 2,440 bp. The sequence of 42 bp 5′ UTR and 69 bp of 3′ UTR was also determined.

Rubber biosynthesis could also follow a mevalonate-independent pathway as evidenced by transcriptome studies (Ko et al. 2003). The alternative metabolic pathway for IPP synthesis is 1-deoxy-D-xylulose-5-phosphate/2-C-methyl-D-erythritol-4-phosphate (DOXP/MEP) pathway, which is located in the plastid (Rohmer et al. 1996). Both the pathways coexist in laticifers and need sucrose as a precursor for rubber synthesis (d'Auzac 1964; Chow et al. 2007). Therefore, sucrose should cross the plasma membrane through specific sucrose transporters before being metabolized in the laticifers. Two isoforms of the sucrose transporter SUT1, *HbSUT1A,* and

HbSUT2A cloned from latex-specific cDNA library were found to play an essential role in sucrose import into laticifers of virgin *Hevea* trees (Dusotoit-Coucaud et al. 2009), although the relative importance of these sucrose transporters in determining latex yield is unknown. These genes were upregulated by ethylene application (essential for ethylene-stimulated latex production), and their localization in the latex cell was confirmed by *in situ* hybridization. Tang et al. (2010) functionally characterized another *Hevea* SUT member, *HbSUT3*. This isoform was found to be the predominant member expressed in the rubber-containing cytoplasm (latex) of laticifers compared to other *Hevea* SUT genes.

Ethylene-Regulated Genes for Latex Production

Biosynthesis of natural rubber, like other secondary metabolites, is affected by various plant hormones. The latex flow rate and duration are the first intrinsic factors known to limit rubber yield – the faster and the longer the latex flow, the higher the yield (d'Auzac et al. 1989). In the extensively studied plant hormones, only ethylene was identified to stimulate the latex production, which is applied as ethephon (an ethylene releaser). Bark treatment with ethephon is known to increase the latex yield by 1.5–2-fold in rubber tree (Coupé and Chrestin 1989; Pujade-Renaud et al. 1994). Even though the exact mechanism of ethylene action is poorly understood on the rubber tree, progress has been made in physiological and biochemical aspects. Compared with the physiology and biochemistry of ethylene stimulation on latex production, the progress in understanding the molecular mechanism is at a slow pace. Till date, only a few genes responding to ethylene have been characterized in *H. brasiliensis*.

Kush et al. (1990) reported laticifer-specific genes induced by ethylene in *H. brasiliensis.* Hevein, a lectin-like protein involved in the coagulation of latex, was mediated by ethylene (Broekaert et al. 1990; Sivasubramaniam et al. 1995; Gidrol et al. 1994). Ethylene could upregulate activity of glutamine synthetase (GS), a key enzyme of nitrogen metabolism and its transcript levels in *H. brasiliensis* latex cells, suggesting involvement of GS in stimulation of rubber production with ethylene (Pujade-Renaud et al. 1994). Higher expression of MnSOD, regulated by ethephon application, prevented lutoid disruption by superoxide radicals (Miao and Gaynor 1993), leading to increased rate of latex flow.

Both the HMGS and HMGR were known to be involved in early steps of rubber biosynthesis. Of the three HMGR genes, *hmg1, hmg2,* and *hmg3*, only *hmg1*, responsible for rubber biosynthesis, was induced by ethylene (Chye et al. 1991, 1992); *hmg2* was involved in defense reactions against wounding and pathogens, and *hmg3* was possibly involved in other isoprenoid biosynthesis for housekeeping purposes (Chye et al. 1992; Wititsuwannakul 1986). Ethephon also influenced the expression of the HMG-CoA synthase gene activity (Suwanmanee et al. 2004; Sirinupong et al. 2005). However, the expression level of FDP synthase, catalyzing the synthesis of the last common substrate isopentenyl pyrophosphate (IPP) in the isoprenoid biosynthesis, was not affected by ethylene treatment (Adiwilaga and

Kush 1996). Luo et al. (2009) and Zhu and Zhang (2009) documented that ethephon had no effect on the gene expression and the activity of RuT (a cis-prenyltransferase) needed for rubber biosynthesis.

A specific and significant activation of the cytosolic glutamine synthetase (GS) in the laticiferous cells after ethylene treatment parallels the increase of latex yield. A marked accumulation of the corresponding mRNA was found, but in contrast, a slight and variable increase of the polypeptide level was noticed (Pujade-Renaud et al. 1994). The GS response to ethylene might be mediated by ammonia that increases in latex cytosol following ethylene treatment.

Zhu and Zhang (2009) reported prolonged latex flow and acceleration of sucrose metabolism due to ethylene stimulation and considered the main reasons for enhanced latex yield. The rapid water exchanges with surrounding liber cells probably occur via the aquaporin pathway as the mature laticifers are devoid of plasmodesmata. Two full-length aquaporin cDNAs (*HbPIP2;1* and *HbTIP1;1*, for plasma membrane intrinsic protein and tonoplast intrinsic protein, respectively) known to facilitate water and/or small neutral solute fluxes across cell membranes were cloned and characterized from rubber (Tungngoen et al. 2009). Through their study, it was evident that aquaporin *HbPIP2;1* was effective in increasing plasmalemma water conductance than HbTIP1;1, and their expression was noticed in all liber tissues in the young stem, including the laticifers. *HbPIP2;1* was upregulated in both liber tissues and laticifers, whereas *HbTIP1;1* was downregulated in liber tissues but upregulated in laticifers in response to bark Ethrel treatment. Increase in latex yield in response to ethylene was related with water circulation between the laticifers and their surrounding tissues as well as with the probable maintenance of liber tissue turgor, which together favor prolongation of latex flow. Ethylene stimulation of latex production results in high sugar flow from the surrounding cells of inner bark towards the latex cells. Dusotoit-Coucaud et al. (2010) studied the expression pattern of sugar transporters (*HbSUTs*) and hexose transporter (*HbHXT1*) in two *Hevea* clones, PB 217 and PB 260, under different physiological conditions. The *HbSUT1*, one of the most abundant isoforms, displayed the greatest response to ethylene treatment. Ethylene treatment led to a higher accumulation of *HbSUT1B* in latex cells than in the inner bark tissues of the high-yielding PB 217 clone. Tang et al. (2010) reported induced expression of *HbSUT3* by the latex stimulator Ethrel and positive correlation with latex yield. Application of ethylene also enhanced transcription of cysteine protease *HbCP1* isolated from rubber (Peng et al. 2008).

Defense/Stress-Related Genes

Defense/stress-related genes, namely, MnSOD, *HEVER*, hevein, chitinase, and β-1,3-glucanase, are expressed in laticifers of *Hevea*. Miao and Gaynor (1993) isolated MnSOD gene, which was found to express in all tissues, *i.e.*, leaf, petiole, root, latex, and callus, and highest level expression was noticed in young leaves through northern analysis. A novel stress-induced gene, *HEVER* (*Hevea* ethylene-responsive), from the rubber tree was isolated and characterized (Sivasubramaniam et al. 1995). A multigene family encodes HEVER. *HEVER* transcript and protein were induced by stress treatment with salicylic acid and ethephon. β-1,3-glucanase gene was identified from a cDNA library derived from latex by Chye and Cheung (1995), and its higher expression level was noticed in latex compared to leaf. Thanseem et al. (2003, 2005) also cloned and characterized the same gene from Indian *Hevea* clones and demonstrated prolonged accumulation of β-1,3-glucanase transcripts in abnormal leaf fall-tolerant RRII 105. Hevein, a lectin-like protein, belonging to a multigene family was found to play a crucial role in the protection of wound sites from fungal attack through latex coagulation (Broekaert et al. 1990; Pujade-Renaud et al. 2005). Overexpression of chitinase was noticed during fungal infection and by ethylene stimulation. In addition to the role in defense responses, expression of hevein and chitinase is linked to the characteristics of the latex flow. The products of hevein ("procoagulant") and chitinase ("anticoagulant") genes, which compete for the same site (the N-acetyl-glucosamine moiety) of the hevein receptor to induce or inhibit the process of coagulation, could be used as molecular markers for assessing yield potential of rubber clones. Such markers are of help in early selection of high-yielding and stimulation-responsive rubber clones (Chrestin et al. 1997). The full-length cDNA encoding a cysteine protease, designated *HbCP1*, was isolated for the first time from *Hevea brasiliensis* (Peng et al. 2008). The predicted HbCP1 protein possessed a putative repeat in toxin (RTX) domain at the N-terminal and a granulin (GRAN) domain at the C-terminal. In plants, cysteine proteases are involved in diverse physiological and developmental processes including biotic and abiotic stresses. Transcription pattern analysis revealed that *HbCP1* had high transcription in laticifer and low transcription in bark and leaf.

Resistance Gene Analogues (RGAs)

Isolation and characterization of the NBS-LRR-encoding genes are of significance in understanding plant-pathogen interactions for effective management of diseases. Sequences analogous to plant resistance genes of NBS-LRR class (Fig. 9.8) were cloned from the genomic DNA of two species of rubber: *Hevea brasiliensis* (clones RRII 105 and RRIM 600) commercially cultivated for latex production and *H. benthamiana*, a noncultivated species known for its tolerance to fungal diseases (Saha et al. 2010a). From the genomic RGA library of rubber, different RGAs were identified, structurally each having an open reading frame and characteristic motifs. Sequence analysis revealed that RGAs are highly diverged in rubber. Functional RGAs from *Corynespora*-challenged leaf samples of RRII 105 were identified through RT-PCR using degenerated primers. A comparison of these two types of RGAs revealed that a large group of closely related genomic RGAs, except a few, had no function against *Corynespora* leaf disease as they did not show perfect homology with any of the RT-RGAs on the basis of deduced amino acid sequences. Characterization of these RT-RGAs is a significant step towards understanding plant response to *Corynespora* infection in rubber.

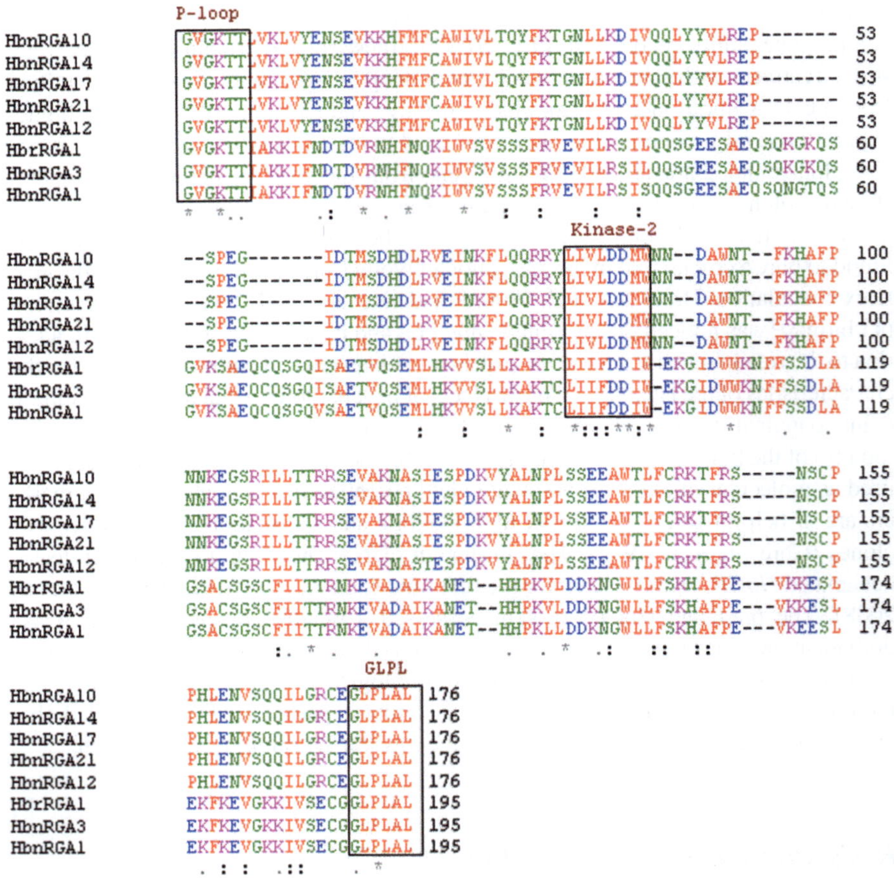

Fig. 9.8 Multiple alignment of representative amino acid sequence (conceptual translation) of resistance gene analogues (RGA) isolated from rubber showing three conserved motifs: P-loop, Kinase-2, and GLPL in NBS region. "HbnRGA" and "HbrRGA" refer to RGAs derived from *Hevea benthamiana* and *H. brasiliensis*

Genes for Flowering

To understand the genetic and molecular mechanisms underlying the reproductive process in rubber trees, Dornelas and Rodriguez (2005) characterized FLORICAULA/LEAFY (FLO/LFY) orthologue *HbLFY* from *H. brasiliensis* (RRIM 600), regulating flower and inflorescence development. Expression patterns of *HbLFY* were analyzed during vegetative and reproductive development. *HbLFY* is expressed in lateral meristems that give rise to inflorescence and in all flower meristems, consistent with a role in reproductive development.

Genes Involved in Signaling Pathways

Coronatine-insensitive 1 protein (COI1) is essentially involved in jasmonic acid signaling pathway regulating defense responses against stress in plants. A cDNA encoding coronatine-insensitive 1 protein (HbCOI1) from rubber was cloned by Peng et al. (2009) to study jasmonic acid signaling for the genes involved in latex biosynthesis. Transcription of *HbCOI1* in latex was induced by jasmonate and tapping. Three MADS-box genes from *Hevea brasiliensis*, *HbMADS1*, *HbMADS2*, and *HbMADS3*, encoding polypeptides consisting of 245, 217, and 239 amino acids, respectively, were cloned and characterized by Li et al. (2011). All of them contained conserved MADS-box motifs at N-terminus. Transcript abundance of all these three genes was noticed in the laticifer cells. The transcriptions of *HbMADS1* and *HbMADS3* were induced by jasmonic acid. Ethephon was not effective in inducing their expression. It was observed that these three genes were differentially expressed during somatic embryogenesis of rubber tree.

Promoter Research

Interests in promoter research for tissue-specific expression of desired gene are increasing in rubber as rubber tree has great potential to produce foreign proteins in the latex, which can easily be purified from the serum of the latex. Therefore, main interest was to identify the promoter region at the upstream of the genes involved in latex biosynthesis. Pujade-Renaud et al. (2005) isolated promoter regions of two *hevein* genes and analyzed in rice through transgenic approach. They showed that the longest promoter sequence (*PHev2.1*) regulated high level of expression of the transgene. Priya et al. (2006) characterized promoter sequence of *REF* gene. A gene construct containing *REF* promoter sequence and the GUS coding *uidA* as reporter gene were transformed to tobacco and *Arabidopsis* to understand regulatory role of *REF* promoter. Results suggested that isolated promoter sequence was capable of regulating gene expression. The promoter sequences of two aquaporin genes were cloned and found to harbor ethylene-responsive and other chemical-responsive (auxin, copper, and sulfur) elements known to increase latex yield (Tungngoen et al. 2009).

EST Sequencing

Expressed sequence tags (ESTs), short partial cDNA sequences, are currently the most widely sequenced nucleotide element from the plant genome with respect to the number of sequences and the total number of nucleotide available to researchers. EST provides a robust sequence resource that can be exploited for gene discovery, genome annotation, and comprehensive genomics. In rubber, EST-sequencing approach has enabled laticifer gene expression analysis on a large scale. Chow et al. (2007) reported a collection of 10,040 ESTs from latex to analyze genes involved in latex biosynthesis. Among these ESTs, they identified 1,380 consensus sequence and 2,061 singletons through progressive assembly. Functional analysis revealed that 26.2% of the 3,441 unique transcripts could be assigned known gene identities. Among these sequences, there was a dominance of ESTs involved in rubber biosynthesis, *i.e., REF* and *SRPP*, followed by latex abundant protein, and ring-zinc finger protein. Very recently, Xia et al. (2011) adopted the next-generation massively parallel sequencing technologies to gain a comprehensive overview of the *H. brasiliensis* transcriptome. They reported 48,768 unigenes through *de novo* transcriptome assembly – the most comprehensive sequence resource available for the study of rubber trees. In total, 37,432 unigenes were successfully annotated, of which 24,545 (65.5%) aligned to *Ricinus communis* proteins.

Stress-Related ESTs

Understanding of stress adaptation process at molecular level is essential for improvement of abiotic stress tolerance in rubber tree for extending its cultivation to nontraditional cold-prone areas. Genes that are differentially expressed during cold acclimation in rubber were identified (Saha et al. 2010b, c). Transcript profiling in two relatively stress-tolerant *Hevea* clones PR 261 and RRII 208 in relation to cold stress was performed (Fig. 9.9). Sequencing of 131 cDNA sequences (59 downregulated and 72 upregulated cDNAs) revealed 110 unique sequences comprising of 13 clusters/contigs and 97 singletons. However, several differentially expressed genes, *i.e.*, catalase, phosphatidylinositol/phosphatidylcholine transfer protein, NADH dehydrogenase, MYB transcription factor, downward leaf-curling protein, epimerase/dehydratase, Na^+/H^+ antiporter, chloroplast *ycf2*, and chloroplast *ftsH* protease, involved in cold adaptation process were also identified along with the unique transcripts.

A subtracted cDNA library was constructed from the cold-stressed leaf sample (Saha et al. 2010b). High-quality sequences of 156 subtracted cDNA clones (ScDNA) were subjected to "contig analysis" to assemble similar sequences in groups. Thirty-one contigs containing 90 clones (2–8 clones per contig) and 66 singletons (single sequences) were identified. All sequences were subjected to BLASTX search to know about the homology with the gene sequences existing in GenBank from rubber or other plant species. Transcripts/clones were assigned to the category based on the shared structural elements and (or) inferred functions. All these ESTs, except a few with unknown functions, relevant to cold responsiveness were grouped into the following categories for which interesting functions in relevance to stress response could be inferred. These groups are (1) osmoprotection/detoxification, (2) oxidoreductases, (3) cell wall and polysaccharide metabolism, (4) protein/amino acid metabolism, (5) transport and secretion, and (6) transcription

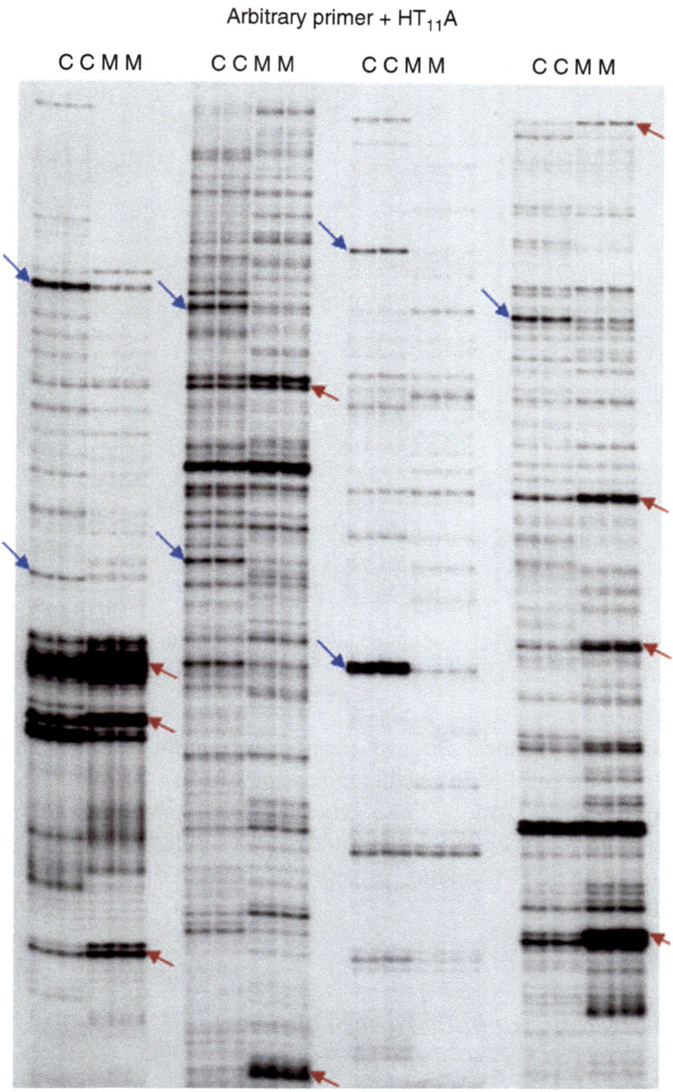

Fig. 9.9 Autoradiogram of transcript profiling through DDRT-PCR of stress-tolerant clone RRII 208 of *Hevea brasiliensis* grown at Kottayam (C: control) and at Munnar (M), a cold-prone high-altitude area to identify differentially expressed stress-responsive genes. Samples were loaded twice to avoid loading error. Blue and brown arrows indicate down-regulation and up-regulation of transcripts under low temperature, respectively. Four arbitrary primers were used in combination with $HT_{11}A$

factors. Ninety-six stress-responsive cDNA clones (31 contigs + 65 singletons) were subjected to reverse northern dot-blot analysis to screen for truly differentially expressed cDNA fragments. Duplicate blots of the 96 stress-responsive cDNA clones (subtracted) were hybridized with labeled cDNA probes from cold-treated and control RNA samples to screen for truly differentially expressed cDNA

Fig. 9.10 (**a**) Healthy tapping panel of rubber tree showing latex flow; (**b**) cessation of latex flow due to tapping panel dryness (TPD)

fragments. Out of 96 clones, 56 gave quantitatively more signals with the cDNA probe from cold-treated PR 261 plants indicating overexpression of the respective genes under cold stress. Among these genes, carbonic anhydrase, glutathione peroxidase, metallothionein, chloroplastic Cu/Zn SOD, serine/threonine protein kinase, transcription factor, DNA-binding protein, etc., showed significant increase in expression levels.

Genes Involved in Tapping Panel Dryness (TPD)

TPD is the appearance of partial dry zones without latex flow along the tapping panel (Fig. 9.10). In later stage, the tapping panel becomes completely dry, and other symptoms such as browning, thickening, and flaking of bark can occur (Sookmark et al. 2002). A great deal of work was performed to reveal the nature and molecular mechanisms leading to TPD. It was initially presumed that TPD might be caused by pathogens (Zheng and Chen 1982; Soyza 1983), which could not be confirmed (Nandris et al. 1991a, b). Later involvement of viroid in etiology of TPD was reported (Ramachandran et al. 2000).

Research on the physiological aspects suggested that the TPD syndrome was a complex physiological disorder resulting from excessive tapping and overexploitation/overstimulation with ethylene (Chrestin 1989; de Faÿ and Jacob 1989; Faridah et al. 1996). In TPD tree, the contents of protein, nucleic acid, thiols, and ascorbic acid decreased (Fan and Yang 1995), whereas the activities of RNase and proteinase increased in general (Tupy 1969; Fan and Yang 1995; Zeng 1997). In addition, the levels of variable peroxidase and superoxide dismutase (SOD) also decreased (Xi and Xiao 1988). Krishnakumar et al. (1997) reported reduced level of cytokinin in bark tissues of TPD-affected trees. Increased bark respiration due to TPD was demonstrated by Krishnakumar et al. (2001). Proteins related to TPD were identified by comparing the expression patterns between healthy and TPD-affected trees (Dian et al. 1995; Lacrotte et al. 1995; Sookmark et al. 2002), but their functional relations with TPD still remain unknown. Therefore, it is necessary to identify the genes associated with TPD to unravel the molecular mechanisms involved.

Chen et al. (2003) reported lower expression of *HbMyb1*, a key transcription factor in bark and latex of TPD trees compared to healthy trees. Functional analyses further indicated that *HbMyb1* negatively regulated programmed cell death (PCD) in transgenic tobacco plants. In another study, Venkatachalam et al. (2007) identified 134 genes associated with TPD in rubber tree by SSH method. They analyzed expression patterns of partial genes and discussed the relationship between differentially expressed genes and TPD. Later, the same group through mRNA differential display technique claimed involvement of a gene *HbTOM20*, which played an important role in the alteration of mitochondrial metabolism, resulting in impaired latex biosynthesis (Venkatachalam et al. 2009).

Li et al. (2010) also identified the genes associated with TPD using suppression subtractive hybridization (SSH) method. Among 237 unique genes, 205 were reported to be related to TPD in rubber tree. Of different functional categories, the large numbers of genes related to TPD were associated with transcription and post-transcription, metabolism and energy, protein metabolism, or stress/defense response. Systematic analyses of the genes related to TPD suggested that the production and scavenging of reactive oxygen species (ROS), ubiquitin proteasome pathway, programmed cell death, and rubber biosynthesis play important roles in TPD. However, much more information is needed for understanding TPD in rubber tree at the molecular level.

Genome Sequencing

The past several years have witnessed major advances in our understanding of plant genomes and genomic information through whole genome sequencing. The increasing availability of data from several plant genome-sequencing projects provides a promising direction for investigating genes and their functional and sequence homologs involved in plant development (Avraham et al. 2008). Although genome-sequencing projects lead to the identification of the complete catalogue of genes of an organism, they do not consider the gene expression patterns. Large-scale end sequencing of cDNA library generates ESTs, representing genes expressed in particular tissues or under particular developmental or environmental conditions. They have also been the target of sequencing in many of the projects and found invaluable for genome assembly and annotation. Whole genome sequence information helps in many aspects of plant-trait improvement through gene discovery to transgenesis and use of molecular markers in breeding. *Hevea* genome-sequencing project had already been launched jointly by Tun Abdul Razak Research Centre (TARRC) of the Malaysian Rubber Board in the UK and newly established. The Genome Analysis Centre (TGAC) at Norwich, UK. Although completion of whole genome sequencing of a Malaysian latex/timber clone RRIM 928 performed jointly by TARRC and TGAC was announced in 2010 by the Malaysian Rubber Development Board, sequence information has not been released in the public domain till date. International Rubber Research and Development Board (IRRDB) also proposed to do whole genome sequencing of rubber in biotechnology meeting at CIRAD in France in 2009. However, quantum of *Hevea* genome-sequencing work is a monumental task as the haploid genome size is enormous ($\sim 4 \times 10^3$ Mbp as per our calculation based on the DNA content measured by Leitch et al. 1998), and also rubber possesses a high-complexity genome with >60% repetitive sequences making the sequencing effort more challenging.

Methylation Dynamics in *Hevea* Genome

Cytosine methylation is a fundamental epigenetic mechanism for gene expression regulation and development in plants. Identification of DNA methylation patterns and their putative relationship with abiotic stress in *H. brasiliensis* was reported by Uthup et al. (2011). Significant variations in the methylation pattern were observed at core DNA-binding motifs within the regulatory sequences of four major genes involved in the mevalonate pathway and one general defense-related gene of three high-yielding popular rubber clones, RRII 105, RRIM 600 and PB 260, grown at two different agroclimatic conditions. Several consistent clone-specific and location-specific methylation patterns were identified (Fig. 9.11).

Transgenesis

Transgenesis is referred to as the introduction of heterologous or homologous DNA into plant genome resulting in its stable integration and expression. The technology has played a critical role in defining the *in vivo* functions of plant genes. In recent years, with the rapid increase in gene sequence information, systematic transgenic approaches have been adopted to characterize large number of genes in both reverse and forward genetic studies. As one of the experimental methods in functional genomics, transgenesis has the advantage of revealing the direct link between gene sequence and function; such results not only provide further understanding of basic biological question but also facilitate exploitation of genomic information for crop improvement (Dixon et al. 2007). There are many variations of gene transfer methods

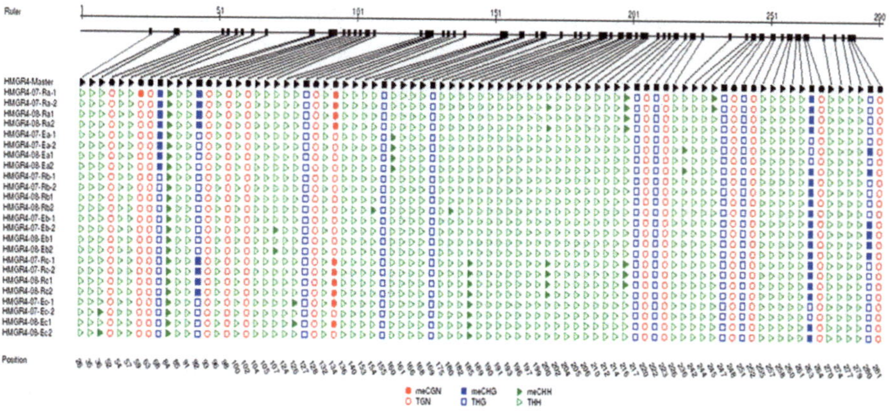

Fig. 9.11 DNA methylation pattern of a 288-bp partial promoter region of HMGR gene from control and test plants belonging to the three different clones having differential responses toward cold stress. The data was generated through bisulfite sequencing followed by CyMATE analysis of the region of interest. Clone and location wise groupings of each class of methylation are identified at the positions: 68, 92, 134, 160, 185, and 200 bp. The figure shows all the possible sites of methylation in the given sequence under the three categories. The blocked symbols represent the sites where actual methylation has taken place. R: control plants grown at RRII campus, E: test plants grown at Elappara (cold-prone area), year of sampling is suffixed to the gene (*HMGR*4-07 and *HMGR*4-08). Clone designation – a: RRII 105, b: RRIM 600, c: PB 260

to introduce transgenes into the plant genome. Genetic transformation offers a potential tool to breeders for introducing valuable traits to crop plants, leading to the development of elite clones in a relatively short period of time. The most widely used methods are *Agrobacterium*-mediated gene transfer and biolistic transformation.

Conventional rubber breeding takes more than 25 years to develop a new clone. The first transformation report in *Hevea brasiliensis* was published in 1991 (Arokiaraj and Wan 1991) through *Agrobacterium*-mediated transformation. The first transgenic *Hevea* plants, using anther-derived calli as the explant of the clone Gl 1, were successfully developed by Arokiaraj et al. (1994) following biolistic transformation method. Subsequently, transgenic plant was developed using *Agrobacterium*-mediated gene transfer of anther-derived calli (Arokiaraj et al. 1996, 1998). Transformation efficiency could significantly be enhanced when the friable callus was treated with calcium chloride and cultured on calcium-free medium prior to transformation (Montoro et al. 2000). Inner integument tissue of the immature fruit of the clone PB 260 was used as the explant for genetic transformation (Montoro et al. 2003). Transgenic plants of *H. brasiliensis* PB 260 were developed through *Agrobacterium*-mediated transformation by Blanc et al. (2005). It was further reported that anther-derived embryogenic callus was the most suitable explant for genetic transformation (Rekha et al. 2006). Earlier transformation events were only with various marker genes. Later, the experiments were focused on transferring various agronomically important genes into *Hevea* with enhanced tolerance to abiotic stresses, production of recombinant proteins, etc. Subsequently, attempts were made to increase the SOD enzyme activity by overexpression of the same genes in

Hevea. Transgenic plants were developed with SOD gene under the control of CaMV 35S and FMV 34S promoters (Jayashree et al. 2003; Sobha et al. 2003a). Biochemical analysis of the transgenic embryogenic calli of *Hevea* with SOD indicated significant increase in the activity of superoxide dismutase, catalase, and peroxidase as compared to the control (Sobha et al. 2003a, b). Jayashree et al. (2003) reported successful development and establishment of transgenic rubber plant with SOD gene for their further evaluation. Genetic transformation experiment to overexpress *hmgr1* gene, involved in latex biosynthesis, in *Hevea* was performed by Arokiaraj et al. (1995). They could generate transgenic embryos, which failed to produce any transgenic plant. However, they showed enhanced *hmgr* activity in the transformed calli.

Experiments were also undertaken for the production of foreign proteins in the latex of *Hevea*. The Para rubber tree, which produces enormous volume of latex upon tapping, could easily be exploited without any destruction for the production of foreign proteins in the latex throughout the year. Recombinant protein may be expressed in the specific parts of the plants or in specific organelle within the plant cell using tissue-specific promoters. Human serum albumin protein was expressed in transgenic *Hevea* plants by Arokiaraj et al. (2002). To characterize tissue-specific promoters derived from latex biosynthesis genes, transgenic approaches were adopted by Priya et al. (2006). They cloned and characterized promoter sequence of the rubber elongation factor gene. A significant achievement towards antibiotic marker-free *Hevea* transgenic development avoiding the constraints of GMO regulations was made by Leclercq et al. (2010). They developed an efficient genetic transformation procedure in the clone PB 260 using a recombinant green fluorescent protein (GFP). They showed GFP selection is less time-consuming in terms of callus subculturing and offered the possibility of producing antibiotic-resistant marker-free transgenic plant.

Conclusion

The past decade has witnessed tremendous advancement in our understanding of plant genomics. Genomic information available is being utilized for accelerating crop improvement program through gene discovery, transgenesis, and use of molecular markers. The accelerating pace of technology development promises much more than can easily be applied into traditional plant improvement programs. Therefore, the genomic information is not exploited fully as anticipated earlier. The major challenge of the postgenomic biology is to translate and integrate the knowledge into application tools and methodologies for the full understanding of gene functions at the organismal level.

Although significant achievements in the area of genomics were made in other crops including some tree crops, very little work on the genomics of *Hevea* was initiated till the end of the twentieth century. This is evident with the availability of limited number of GenBank submissions (12,385 ESTs, 1,499 nucleotide sequences, and 246 GSS as on August 15, 2011) compared to other tree crops. However, a large quantum of genomic information is expected to be available in the public domain soon due to the initiatives taken by various laboratories with regard to transcriptome sequencing and whole genome sequencing using next-generation sequencing platform. In fact very recently, next-generation transcriptome sequencing data generated using Illumina RNA-Seq technology by Xia et al. (2011) was made available in the public domain (NCBI database – GSE26514; public on Sept. 01, 2011). The enormous information thus generated recently will hopefully be translated by the concerted efforts from all the rubber-growing countries for the improvement of rubber tree in terms of crop health and productivity.

Acknowledgments We thank Dr. C. Bindu Roy, Senior Scientist, Molecular Plant Pathology, RRII for assistance with artwork and manuscript preparation and Dr. C. P. Reghu, Deputy Director, Germplasm Division, RRII for providing bark anatomy figures. We also thank Dr. James Jacob, Director of Research, RRII for his support and inspiration.

References

Abraham PD, Wycherley PR, Pakianathan SW (1968) Stimulation of latex flow in *Hevea brasiliensis* by 4-amino-3,5,6-trichloropicolinic acid and 2-chloroethane-phosphonic acid. J Rubber Res Inst Malays 20:291–305

Adiwilaga K, Kush A (1996) Cloning and characterization of cDNA encoding farnesyl diphosphate synthase from rubber tree (*Hevea brasiliensis*). Plant Mol Biol 30:935–946

Aishwarya V, Sharma PC (2007) UgMicroSat*db:* database for mining microsatellites from unigenes. Nucleic Acids Res 1:1–4

Archer BL, Audley BG (1987) New aspects of rubber biosynthesis. Bot J Linn Soc 94:181–196

Arokiaraj P, Wan AR (1991) *Agrobacterium*-mediated transformation of *Hevea* cells derived from *in vitro* and *in vivo* seedling cultures. J Nat Rubber Res 6:55–61

Arokiaraj P, Jones H, Cheong KF, Coomber S, Charlwood BV (1994) Gene insertion into *Hevea brasiliensis*. Plant Cell Rep 13:425–430

Arokiaraj P, Jaafar H, Hamzah S, Yeang HY, Wan AR (1995) Enhancement of *Hevea* crop potential by genetic transformation: HMGR activity in transformed tissue. In: Proceedings of IRRDB symposium on physiological and molecular aspects of the breeding of *Hevea brasiliensis*, Penang, Malaysia, pp 74–82

Arokiaraj P, Jones H, Jaafar H, Coomber S, Charlwood BV (1996) *Agrobacterium*-mediated transformation of *Hevea* anther calli and their regeneration into plantlets. J Nat Rubber Res 11:77–87

Arokiaraj P, Yeang HF, Cheong KF, Hamzah S, Jones H, Coomber S, Charlwood BV (1998) CaMV 35S promoter directs β-glucuronidase expression in the laticiferous system of transgenic *Hevea brasiliensis*. Plant Cell Rep 17:621–625

Arokiaraj P, Ruker F, Obermayr E, Shamsul Bahri AR, Hafsah J, Carter DC, Yeang HY (2002) Expression of human serum albumin in transgenic *Hevea brasiliensis*. J Nat Rubber Res 5:157–166

Asawatreratanakul K, Zhang YW, Wititsuwannakul D, Wititsuwannakul R, Takahashi S, Rattanapittayaporn A, Koyama T (2003) Molecular cloning, expression and characterization of cDNA encoding cis-prenyltransferases from *Hevea brasiliensis*. A key factor participating in natural rubber biosynthesis. Eur J Biochem 270:4671–4680

Atan S, Low FC, Saleh NM (1996) Construction of a microsatellite enriched library from *Hevea brasiliensis*. J Nat Rubber Res 11:247–255

Avraham S, Tung CW, Ilic K, Jaiswal P, Kellogg EA, McCouch S, Pujar A, Reiser L, Rhee SY, Sachs MM et al (2008) The plant ontology database: a community resource for plant structure and developmental stages controlled vocabulary and annotations. Nucleic Acids Res 36:D449–D454

Baldwin JJT (1947) *Hevea*: a first interpretation. A cytogenetic survey of a controversial genus, with a discussion of its implications to taxonomy and to rubber production. J Hered 38:54–64

Baulkwill WJ (1989) The history of natural rubber production. In: Webster CC, Baulkwill WJ (eds) Rubber. Longman Scientific and Technical, Essex, pp 1–56

Besse P, Lebrun P, Seguin M, Lanaud C (1993) DNA fingerprints in *Hevea brasiliensis* (rubber tree) using human minisatellite probes. Heredity 70:237–244

Besse P, Seguin M, Lebrun P, Chevallier MH, Nicolas D, Lanaud C (1994) Genetic diversity among wild and cultivated populations of *Hevea brasiliensis* assessed by nuclear RFLP analysis. Theor Appl Genet 88:199–207

Bini K, Madhavan M, Ravindran M, Thomas KU, Roy Bindu C, Saha T (2010) Single nucleotide polymorphisms in rubber (*Hevea brasiliensis*) and their inheritance in RRII 400 series hybrid clones. In: Proceedings of plantation crops symposium – PLACROSYM XIX, RRII, Kottayam, India

Blanc G, Baptiste C, Oliver G, Martin F, Montoro P (2005) Efficient *Agrobacterium tumefaciens* mediated transformation of embryogenic calli and regeneration of *Hevea brasiliensis* Mull. Arg. plants. Plant Cell Rep 24:724–733

Bocharova R, Kovaliova IA, Mazurenko LS (2009) Identification of grapevine clone genotypes by use of microsatellite markers. Cytol Genet 43:371–378

Broekaert I, Lee HI, Kush A, Chua NH, Raikhel N (1990) Wound-induced accumulation of mRNA containing a hevein sequence in laticifers of rubber tree (*Hevea brasiliensis*). Proc Natl Acad Sci USA 87:7633–7637

Brondani RPV, Brondani C, Tarchini R, Grattapaglia D (1998) Development, characterization and mapping of microsatellite markers in *Eucalyptus grandis* and *E. urophylla*. Theor Appl Genet 97:816–827

Chen S, Peng S, Huang G, Wu K, Fu X, Chen Z (2003) Association of decreased expression of a Myb transcription factor with the TPD (tapping panel dryness) syndrome in *Hevea brasiliensis*. Plant Mol Biol 51:51–58

Chow KS, Wan KL, Mat IMN, Bahari A, Tan SH, Harikrishna K, Yeang HY (2007) Insights into rubber biosynthesis from transcriptome analysis of *Hevea brasiliensis* latex. J Exp Bot 58:2429–2440

Chrestin H (1989) Biochemical aspects of bark dryness induced by overstimulation of rubber tree with Ethrel. In: d'Auzac J, Jacob JL, Chrestin H (eds) Physiology of rubber tree latex. CRC Press, Boca Raton, pp 341–441

Chrestin H, Gidrol X, Kush A (1997) Towards a latex molecular diagnostic of yield potential and the genetic engineering of the rubber tree. Euphytica 96:77–82

Chye ML, Cheung KY (1995) beta-1,3-Glucanase is highly expressed in the laticifers of *Hevea brasiliensis*. Plant Mol Biol 29:397–402

Chye ML, Kush A, Tan CT, Chua NH (1991) Characterization of cDNA and genomic clones encoding 3-hydroxy-3-methylglutaryl CoA reductase from *Hevea brasiliensis*. Plant Mol Biol 16:567–577

Chye ML, Tan CT, Chua NH (1992) Three genes encode 3-hydroxy-3-methylglutaryl-coenzyme A reductase in *Hevea brasiliensis*: hmg1 and hmg3 are differentially expressed. Plant Mol Biol 3:473–484

Clément-Demange A, Legnate H, Seguin M, Carron MP, Le Guen V, Chapuset T, Nicolas D (2001) Rubber tree. In: Charrier A, Jacquot M, Hamon S, Nicolas D (eds) Tropical plant breeding. CIRAD-ORSTOM, Montpellier, pp 455–480 (Collection Repères)

Coupé M, Chrestin H (1989) The hormonal stimulation of latex yield: physico-chemical and biochemical mechanisms of hormonal (ethylene) stimulation. In: d'Auzac J, Jacob J-L, Chrestin H (eds) Physiology of rubber tree latex. CRC Press, Boca Raton, pp 295–319

d'Auzac J (1964) Mise en evidence de la glycolyse et de ses relations avec la biosynthe'se du caoutchouc au sein du latex d' *Hevea brasiliensis*. Revue Ge´ne´rale des Caoutchoucs et Plastiques 41:1831–1834

d'Auzac J (1998) From sucrose to rubber: Hevea as a "green factory". In: Proceedings of IRRDB symposium on natural rubber (*Hevea brasiliensis*), vol. I. General, soils and fertilization and breeding and selection, Ho Chi Minh City, 14–15 Oct 1997, pp 10–27

d'Auzac J, Jacob JL, Chrestin H (1989) In: d'Auzac J, Jacob JL, Chrestin H (eds) Physiology of the rubber tree latex. CRC Press, Boca Raton

Davies W (1997) The rubber industry's biological nightmare. Fortune 136:86

de Faÿ E, Jacob JL (1989) Symptomatology, histological and cytological aspects of the Bark dryness disease. In: d'Auzac J, Jacob JL, Chrestin H (eds) Physiology of rubber tree latex. CRC Press, Boca Raton, pp 407–430

de Gonçalves PS, Fernando DM, Rossetti AG (1982) Interspecific crosses in the genus *Hevea*. A preliminary progeny test of SALB resistant dwarf hybrids. Pesq Agropec Brasileira 17(5):775–781

de Goncalves PS, Paiva JR, Sewza RA (1983) Retrospectiva e Atualidade do Melhoramento Genetico da Seringueira (*Hevea* sp.) no Brasil e em paises Asiatics. EMBRAPA-CNPSD. Documentos 2, Manaus

de Gonçalves PS, Cardoso M, Ortolani AA (1990) Origin, variability and domestication of *Hevea*. Pesq Agropec Brasileira 25:135–156

Dean W (1987) Brazil and the struggle for rubber. Cambridge University Press, New York

Dian K, Sangare A, Diopoh JK (1995) Evidence for specific variation of protein pattern during tapping panel dryness condition development in *Hevea brasiliensis*. Plant Sci 105:207–216

Dijkman MJ (1951) *Hevea*: thirty years of research in the Far East. University Miami Press, Coral Gables

Dixon RA, Bouton JH, Narasimhamoorthy B, Saha M, Wang Z-Y, May GD (2007) Beyond structural genomics for plant science. Adv Agron 95:77–161

Dornelas MC, Rodriguez APM (2005) The rubber tree (*Hevea brasiliensis* Muell. Arg.) homologue of the Leafy/Floricaula gene is preferentially expressed in both male and female floral meristems. J Exp Bot 56:1965–1974

Dusotoit-Coucaud A, Brunel N, Kongsawadworakul P, Viboonjun U, Lacointe A, Julien JL, Chrestin H, Sakr S (2009) Sucrose importation into laticifers of *Hevea brasiliensis*, in relation to ethylene stimulation of latex production. Ann Bot 104:635–647

Dusotoit-Coucaud A, Kongsawadworakul P, Maurousset L, Viboonjun U, Brunel N, Pujade-Renaud V et al (2010) Ethylene stimulation of latex yield depends on the expression of a sucrose transporter (HbSUT1B) in rubber tree (*Hevea brasiliensis*). Tree Physiol 30:1586–1598

Fan SW, Yang SQ (1995) Tapping panel dryness induced by excessive tapping is a local senescence phenomenon. Chin J Trop Crops Res 19:15–22

Faridah Y, Siti Arija MA, Ghandimathi H (1996) Changes in some physiological latex parameters in relation to over exploitation and onset of induced tapping panel dryness. J Nat Rubber Res 10:182–186

Feng SP, Li WG, Huang HS, Wang JY, Wu YT (2009) Development, characterization and cross-species/genera transferability of EST-SSR markers for rubber tree (*Hevea brasiliensis*). Mol Breed 23:85–97

Fong CK, Lek KC, Ping CN (1994) Isolation and restriction analysis of chloroplast DNA from *Hevea*. J Nat Rubber Res 9:278–288

Gidrol X, Chestin H, Tan HL, Kush A (1994) Hevein, a lectin-like protein from *Hevea brasiliensis* (rubber tree) is involved in the coagulation of latex. J Biol Chem 269:9278–9283

Gouvea LRL, Rubiano LB, Chioratto AF, Zucchi MI, de Gonçalves PS (2010) Genetic divergence of rubber tree estimated by multivariate techniques and microsatellite markers. Genet Mol Biol 33(2):308–318

Goyvaerts E, Dennis M, Light D, Chua NH (1991) Cloning and sequencing of the cDNA encoding the rubber elongation factor of *Hevea brasiliensis*. Plant Physiol 97:317–321

Gronover CS, Wahler D, Prufer D (2011) Natural rubber biosynthesis and physico-chemical studies on plant derived latex. In: Magdy Elnashar (ed) Biotechnology of biopolymers. Tech Open Access, Rijeka, pp 75–88

Hepper CM, Audley BG (1969) The biosynthesis of rubber from ß-hydroxy-ß-methylglutaryl-coenzyme A in *Hevea brasiliensis* latex. Biochem J 114:379–386

Imle EP (1978) *Hevea* rubber: past and future. Econ Bot 32:264–277

Jacob JL, Prévot JC, Lacrotte R, Clément A, Serres E, Gohet E (1995) Clonal typology of laticifer functioning in *Hevea brasiliensis*. Plantations Recherche Développement 2:48–49

Jayashree R, Rekha K, Venkatachalam P, Uratsu SL, Kumari Jayasree P, Kala RG, Priya P, Sushma Kumari S, Sobha S, Ashokan MP, Sethuraj MR, Thulaseedharan A, Dandekar AM (2003) Genetic transformation and regeneration of rubber tree (*Hevea brasiliensis* Muell. Arg.) transgenic plants with a constitutive version of an anti-oxidase stress super oxide dismutase gene. Plant Cell Rep 22:201–209

Ko JH, Chow KS, Han KH (2003) Transcriptome analysis reveals novel features of the molecular events occurring in the laticifers of *Hevea brasiliensis* (para rubber tree). Plant Mol Biol 53:479–492

Krishnakumar R, Sasidhar VR, Sethuraj MR (1997) Influence of TPD on cytokinin level in *Hevea* bark. Indian J Nat Rubber Res 10:107–109

Krishnakumar R, Annamalainathan K, Jacob J (2001) Tapping panel dryness syndrome in *Hevea* increases dark respiration but not ATP status. Indian J Nat Rubber Res 14:14–19

Kush A, Goyvaerts E, Chye ML, Chua NH (1990) Laticifer-specific gene expression in *Hevea brasiliensis* (rubber tree). Proc Natl Acad Sci USA 87:1787–1790

Lacrotte R, Gidrol X, Vichitcholchai N et al (1995) Protein markers of tapping panel dryness. Plantations Recherche Développement 2:40–45

Lam LV, Thanh T, Chi VTQ, Tuy LM (2009) Genetic diversity of *Hevea* IRRDB'81 collection assessed by RAPD markers. Mol Biotechnol 42:292–298

Le Guen V, Doaré F, Weber C, Seguin M (2009) Genetic structure of Amazonian populations of *Hevea brasiliensis* is shaped by hydrographical network and isolation by distance. Tree Genet Genomes 5:673–683

Le Guen V, Gay C, Xiong TC, Souza LM, Rodier-Goud M, Seguin M (2010) Development and characterization of 296 new polymorphic microsatellite markers for rubber tree (*Hevea brasiliensis*). Plant Breed 130:294–296

Leclercq J, Lardet L, Martin F, Chapuset T, Oliver G, Montoro P (2010) The green fluorescent protein as an efficient selection marker for *Agrobacterium tumefaciens*-mediated transformation in *Hevea brasiliensis* (Müll. Arg.). Plant Cell Rep 29:513–522

Leitch AR, Lim KY, Leitch IJ, O'Neill M, Chye ML, Low FC (1998) Molecular cytogenetic studies in rubber *Hevea brasiliensis*. Muell. Arg. (Euphorbiaceae). Genome 41:64–467

Lekawipat N, Teerawatannasuk K, Rodier-Goud M, Seguin M, Vanavichit A, Toojinda T, Tragoonrung S (2003) Genetic diversity analysis of wild germplasm and cultivated clones of *Hevea brasiliensis* Muell. Arg. by using microsatellite markers. J Rubber Res 6:36–47

Lespinasse D, Rodier-Goud M, Grivet L, Leconte A, Legnaté H, Seguin M (2000a) A saturated genetic linkage map of rubber tree (*Hevea* spp.) based on RFLP, AFLP, microsatellite and isozyme markers. Theor Appl Genet 100:127–138

Lespinasse D, Grivet L, Troispoux V, Rodier-Goud M, Pinard F, Seguin M (2000b) Identification of QTLs involved in the resistance to South American leaf blight (*Microcyclus ulei*) in the rubber tree. Theor Appl Genet 100:975–984

Levinson G, Gutman GA (1987) High frequencies of short frameshifts in poly-CA/TG tandem repeats borne by bacteriophage M13 in *Escherichia coli* K-12. Nucleic Acids Res 15:5323–5338

Li D, Deng Z, Chen C, Xia Z, Wu M, He P, Chen S (2010) Identification and characterization of genes associated with tapping panel dryness from *Hevea brasiliensis* latex using suppression subtractive hybridization. BMC Plant Biol 10:140. doi:10.1186/1471-2229-10-140

Li HL, Wang Y, Guo D, Tian WM, Peng SQ (2011) Three MADS-box genes of *Hevea brasiliensis* expressed during somatic embryogenesis and in the laticifer cells. Mol Biol Rep 38:4045–4052

Low FC, Bonner J (1985) Characterization of the nuclear genome of *Hevea brasiliensis*. In: Proceedings of international rubber conference, Kuala Lumpur, Malaysia, pp 1–9

Low FC, Atan S, Jaafar H, Tan H (1996) Recent advances in the development of molecular markers for *Hevea* studies. J Nat Rubber Res 11:32–44

Luo H, van Coppenolle B, Seguin M, Boutry M (1995) Mitochondrial DNA polymorphism and phylogenetic relationships in *Hevea brasiliensis*. Mol Breed 1:51–63

Luo MW, Deng LH, Yi XP, Zeng HC, Xiao SH (2009) Cloning and sequence analysis of a novel cis-prenyltransferases gene from *Hevea brasiliensis*. J Trop Subtrop Bot 17:223–228

Majumder SK (1964) Chromosome studies of some species of *Hevea*. J Rubber Res Inst Malays 18:269–273

Mathew R, Roy BC, Ravindran M, Nazeer MA, Saha T (2005) Phylogenetic relationships of *Hevea* species based on molecular markers. Indian J Nat Rubber Res 18:14–25

Miao Z, Gaynor JJ (1993) Molecular cloning, characterization and expression of Mn-superoxide dismutase from the rubber tree (*Hevea brasiliensis*). Plant Mol Biol 23:267–277

Montoro P, Tcinscree N, Rattana W, Kongsawadworakul P, Michaux-Ferriere N (2000) Effect of exogenous calcium on *Agrobacterium tumefaciens*-mediated gene transfer in *Hevea brasiliensis* (rubber tree) friable calli. Plant Cell Rep 19:851–855

Montoro P, Rattana W, Pugade-Renaud V, Michaux-Ferriere N, Monkolsook Y, Kanthapura R, Adunsadthapong S (2003) Production of *Hevea brasiliensis* transgenic embryogenic callus lines by *Agrobacterium tumefaciens*: roles of calcium. Plant Cell Rep 21:1095–1102

Morgante M, Olivieri AM (1993) PCR-amplified microsatellites as markers in plant genetics. Plant J 3:175–182

Nandris D, Chrestin H, Noirot M, Nicole M, Thouvenel JC, Geiger JP (1991a) The phloem necrosis of the trunk of rubber tree in Ivory Coast. (1) Symptomatology and biochemical characteristics. Eur J For Pathol 21:325–339

Nandris D, Thouvenel JC, Nicole M, Chrestin H, Rio B, Noirot M (1991b) The phloem necrosis of the trunk of the rubber tree in Ivory Coast. (2) Etiology of the disease. Eur J For Pathol 21:340–353

Nodari RO, Ducroquet JP, Guerra MP, Meler K (1997) Genetic variability of *Feijoa sellowiana* germplasm. Acta Hortic 452:41–46

Oh SK, Kang H, Shin DH, Yang J, Han KH (2000) Molecular cloning and characterization of a functional cDNA clone encoding isopentenyl diphosphate isomerase from *Hevea brasiliensis*. J Plant Physiol 157:549–557

Omman P, Reghu CP (2003) Staining procedure for laticiferous system of *Hevea brasiliensis* using Oil Red O. Indian J Nat Rubber Res 16:41–44

Othman R, Arshgad NL, Huat OS, Hashim O, Benong M, Wanchik MG, Aziz MZA, Ghani ZA, Ghani MNA (1995) Potential *Hevea* genotypes for timber production. In: Proceedings of rubber growers' conference sustainability & competitiveness, pp 340–360

Peng SQ, Zhu JH, Li HL, Tian WM (2008) Cloning and characterization of a novel cysteine protease gene (*HbCP1*) from *Hevea brasiliensis*. J Biosci 33:681–690

Peng SQ, Xu J, Li HL, Tian WM (2009) Cloning and molecular characterization of HbCOI1 from *Hevea brasiliensis*. Biosci Biotechnol Biochem 73:665–670

Pires JM (1981) Euphorbiaceae: *Hevea camargoana* sp. In: Notas de herbario I. Belem, Museu Emilio Goeldi, pp 4–8

Priya P, Venkatachalam P, Thulaseedharan A (2006) Molecular cloning and characterization of the rubber elongation factor gene and its promoter sequence from rubber tree (*Hevea brasiliensis*): a gene involved in rubber biosynthesis. Plant Sci 171:470–480

Priyadarshan PM (2003a) Breeding *Hevea brasiliensis* for environmental constraints. Adv Agron 79:351–400

Priyadarshan PM (2003b) Contributions of weather variables for specific adaptation of rubber tree (*Hevea brasiliensis* Muell. Arg.) clones. Genet Mol Biol 26:435–440

Priyadarshan PM, de Gonçalves PS (2003) *Hevea* gene pool for breeding. Genet Res Crop Evol 50:101–114

Pujade-Renaud V, Clement A, Perrot-Recbenmann C, Prevot JC, Chrestin H, Jacob JL et al (1994) Ethylene induced increase in glutamine synthetase activity and mRNA levels in *Hevea brasiliensis* latex cells. Plant Physiol 105:127–132

Pujade-Renaud V, Sanier C, Cambillau L, Arokiaraj P, Jones H, Ruengsri N, Tharreau D, Chrestin H, Montoro P, Narangajavana J (2005) Molecular characterization of new members of the

Hevea brasiliensis hevein multigene family and analysis of their promoter region in rice. Biochemica Biophysica Acta 1727:151–161
Raemer H (1935) Cytology of Hevea. Genetics 17:193
Rafalski A (2002) Applications of single nucleotide polymorphisms in crop genetics. Curr Opin Plant Biol 5:94–100
Ramachandran P, Mathur S, Francis L, Varma A, Mathew NM, Sethuraj MR (2000) Evidence for association of a viroid with tapping panel dryness syndrome of rubber (*Hevea brasiliensis*). Plant Dis 84:1155
Rekha K, Jayashree R, Kumary Jayasree P, Venkatachalam P, Thulaseedharan A (2006) An efficient protocol for *Agrobacterium*-mediated genetic transformation in rubber tree (*Hevea brasiliensis*). Plant Cell Biotechnol Mol Biol 7:155–158
Rohmer M, Seemann M, Horbach S, Bringer-Meyer S, Sahms H (1996) Glyceraldehyde 3-phosphate and pyruvate as precursors of isoprenic units in an alternative non-mevalonate pathway for terpenoid biosynthesis. J Am Chem Soc 118:2564–2566
Roy CB, Nazeer MA, Saha T (2004) Identification of simple sequence repeats in rubber (*Hevea brasiliensis*). Curr Sci 87:807–811
Saha T, Roy CB, Nazeer MA (2005) Microsatellite variability and its use in the characterization of cultivated clones of *Hevea brasiliensis*. Plant Breed 124:86–92
Saha T, Roy CB, Ravindran M, Bini K, Nazeer MA (2006) Existence of retroelements in rubber (*Hevea brasiliensis*) genome. J Plant Crops 34:546–551
Saha T, Roy CB, Ravindran M, Bini K, Nazeer MA (2007) Allelic diversity revealed through SSR polymorphisms at the locus encoding HMG-CoA reductase in rubber (*Hevea brasiliensis*). Silvae Genetica 56(2):86–92
Saha T, Roy CB, Ravindran M (2010a) Characterization of a family of disease resistant gene analogues (RGAs) in rubber (*Hevea brasiliensis*) and their relationship with functional RGAs in response to *Corynespora* infection. In: Proceedings of the national symposium on molecular approaches for management of fungal diseases of crop plants, IIHR, Bangalore, India
Saha T, Alam B, Ravindran M, Thomas KU (2010b) Cold-induced gene expression profiling in rubber (*Hevea brasiliensis*). In: Proceedings of the international workshop on climate change and rubber cultivation. RRII, Kottayam, India
Saha T, Thomas M, Sathik MBM, Thulaseedharan A (2010c) Molecular characterization on abiotic stress tolerance and development of transgenic plants of *Hevea brasiliensis*. In: Proceedings of the International Rubber Research and Development Board (IRRDB) conference on climate change, China
Sando T, Takaoka C, Mukai Y, Yamashita A, Hattori M, Ogasawara N, Fukusaki E, Kobayashi A (2008) Cloning and characterization of mevalonate pathway genes in a natural rubber producing plant, *Hevea brasiliensis*. Biosci Biotechnol Biochem 72:2049–2060
Sankariammal L, Mydin KK (2011) Heterosis for growth and test tap yield in Wickham × Amazonian hybrids of *Hevea brasiliensis*. In: International Rubber Research and Development Board (IRRDB) international workshop on tree breeding, Michelin Plantations, Bahia, Brazil
Schlötterer C, Tautz D (1992) Slippage synthesis of simple sequence DNA. Nucleic Acids Res 20:211–215
Schultes RE (1977) A new infrageneric classification of *Hevea*. Bot Mus Leafl Harv Univ 25:243–257
Schultes RE (1987) Studies in the genus *Hevea* VIII. Notes on intrageneric variants of *Hevea brasiliensis* (Euphorbiaceae). Econ Bot 41:125–147
Schultes RE (1990) A brief taxonomic view of the genus *Hevea*, MRRDB monograph 14. MRRDB, Kuala Lumpur
Seguin M, Flori A, Legnaté H, Clément-Demange A (2003) Rubber tree. In: Hamon P, Seguin M, Perrier X, Glaszmann JC (eds) Genetic diversity of tropical crops. CIRAD, France and Science Publ., Enfield, pp 277–306
Simmonds NW (1989) Rubber breeding. In: Webster CC, Baulkwill WJ (eds) Rubber. Longman Scientific and Technical, Essex, pp 85–124

Sirinupong N, Suwanmanee P, Doolittle RF, Suvachitanont W (2005) Molecular cloning of a new cDNA and expression of 3-hydroxy-3-methylglutaryl- CoA synthase gene from *Hevea brasiliensis*. Planta 221:502–512

Sivasubramaniam S, Vanniasingham VM, Tan CT, Chua NH (1995) Characterisation of *HEVER*, a novel stress-induced gene from *Hevea brasiliensis*. Plant Mol Biol 29:173–178

Sobha S, Sushamakumari S, Thanseem I, Kumari Jayasree P, Rekha K, Jayashree R, Kala RG, Asokan MP, Sethuraj MR, Dandekar AM, Thulaseedharan A (2003a) Genetic transformation of *Hevea brasiliensis* with the gene coding for superoxide dismutase with FMV 34S promoter. Curr Sci 85:1767–1773

Sobha S, Sushamakumari S, Thanseem I, Rekha K, Jayashree R, Kala RG, Kumari Jayasree P, Asokan MP, Sethuraj MR, Dandekar AM, Thulaseedharan A (2003b) Abiotic stress induced over-expression of superoxide dismutase enzyme in transgenic *Hevea brasiliensis*. Indian J Nat Rubber Res 16:45–52

Sookmark U, Pujade-Renaud V, Chrestin H, Lacote R, Naiyanetr C, Seguin M, Romruensukharom P, Narangajavana J (2002) Characterization of polypeptides accumulated in the latex cytosol of rubber trees affected by the tapping panel dryness syndrome. Plant Cell Physiol 43:1323–1333

Souza LM, Mantello CC, Santos MO, de Souza Gonçalves P, Souza AP (2009) Microsatellites from rubber tree (*Hevea brasiliensis*) for genetic diversity analysis and cross-amplification in six *Hevea* wild species. Conserv Genet Resour 1:75–79

Soyza AG (1983) The investigation of the occurring rule and distributing pattern of brown blast disease of rubber tree in Sri Lanka. J Rubber Res Inst Sri Lanka 61:1–6

Suwanmanee P, Suvachittanont W, Fincher GB (2002) Molecular cloning and sequencing of a cDNA encoding 3-hydroxy-3-methylglutaryl coenzyme A synthase from *Hevea brasiliensis* (HBK) Mull. Arg. Sci Asia 28:29–36

Suwanmanee P, Sirinupong N, Suvachittanont W (2004) Regulation of the expression of 3-hydroxy-3-methylglutaryl-CoA synthase gene in *Hevea brasiliensis* (B.H.K.) Mull. Arg. Plant Sci 166:531–537

Takaya A, Zhanga Y, Asawatreratanakulb K, Wititsuwannakulc D, Wititsuwannakuld R, Takahashia S, Koyamaa T (2003) Cloning, expression and characterization of a functional cDNA clone encoding geranylgeranyl diphosphate synthase of *Hevea brasiliensis*. Biochim Biophys Acta 1625:214–220

Tan H (1987) Strategies in rubber tree breeding. In: Abbott AJ, Atkin RK (eds) Improving vegetatively propagated crops. Academic, London, pp 28–54

Tan H, Khoo SK, Ong SH (1996) Selection of advanced polycross progenies in *Hevea* improvement. J Nat Rubber Res 11:215–225

Tang C, Huang D, Yang J, Liu S, Sakr S, Li H et al (2010) The sucrose transporter HbSUT3 plays an active role in sucrose loading to laticifer and rubber productivity in exploited trees of *Hevea brasiliensis* (para rubber tree). Plant Cell Environ 33:708–1720

Thanseem I, Venkatachalam P, Thulaseedharan A (2003) Sequence characterization of β-1,3-glucanase gene from *Hevea brasiliensis* through genomic and cDNA cloning. Indian J Nat Rubber Res 16:106–114

Thanseem I, Joseph A, Thulaseedharan A (2005) Induction and differential expression of β-1,3-glucanase mRNAs in tolerant and susceptible *Hevea* clones in response to infection by *Phytophthora meadii*. Tree Physiol 25:1361–1368

Thomas MR, Scott NS (1993) Microsatellite repeats in grapevine reveal DNA polymorphisms when analyzed as sequenced-tagged sites (STSs). Theor Appl Genet 86:985–990

Tungngoen K, Kongsawadworakul P, Viboonjun U, Katsuhara M, Brunel N, Sakr S, Narangajavana J, Chrestin H (2009) Involvement of HbPIP2;1 and HbTIP1;1 aquaporins in ethylene stimulation of latex yield through regulation of water exchanges between inner liber and latex cells in *Hevea brasiliensis*. Plant Physiol 151:843–856

Tupy J (1969) Nucleic acid in latex and production of rubber in *Hevea brasiliensis*. J Rubber Res Inst Malays 21:468–476

Upadhyay A, Kadam US, Chacko P, Karibasappa GS (2010) Microsatellite and RAPD analysis of grape (*Vitis* spp.) accessions and identification of duplicates/misnomers in germplasm collection. Indian J Hortic 67:8–15

Uthup TK, Ravindran M, Bini K, Saha T (2011) Divergent DNA methylation patterns associated with abiotic stress in *Hevea brasiliensis*. Mol Plant 4:996–1013

Varghese YA, Knaak C, Sethuraj MR, Ecke W (1998) Evaluation of random amplified polymorphic DNA (RAPD) markers in *Hevea brasiliensis*. Plant Breed 116:47–52

Varshney RK, Graner A, Sorrells ME (2005) Genic microsatellite markers in plants: features and applications. Trends Biotechnol 23:48–55

Venkatachalam P, Thomas S, Priya P, Thanseem J, Gireesh T, Saraswathyamma CK, Thulaseedharan A (2002) Identification of DNA polymorphism among clones of *Hevea brasiliensis* Muell. Arg. using RAPD analysis. Indian J Nat Rubber Res 15:172–181

Venkatachalam P, Priya P, Saraswathyamma CK, Thulaseedharan A (2004) Identification, cloning and sequence analysis of a dwarf genome specific RAPD marker in rubber tree (*Hevea brasiliensis*). Plant Cell Rep 23:237–332

Venkatachalam P, Priya P, Gireesh T, Saraswathyamma CK, Thulaseedharan A (2006) Molecular cloning and sequencing of a polymorphic band from rubber tree (*Hevea brasiliensis* Muell. Arg.): the nucleotide sequence revealed partial homology with proline-specific permease gene sequence. Curr Sci 90:1510–1515

Venkatachalam P, Thulaseedharan A, Raghothama KG (2007) Identification of expression profiles of tapping panel dryness (TPD) associated genes from the latex of rubber tree (*Hevea brasiliensis* Muell. Arg.). Planta 226:499–515

Venkatachalam P, Thulaseedharan A, Raghothama KG (2009) Molecular identification and characterization of a gene associated with the onset of tapping panel dryness (TPD) syndrome in rubber tree (*Hevea brasiliensis* Muell. Arg.) by mRNA differential display. Mol Biotechnol 41:42–52

Webster CC, Paardekooper EC (1989) Botany of the rubber tree. In: Webster CC, Baulkwill WJ (eds) Rubber. Longman Scientific and Technical, Essex, pp 57–84

Williams JGK, Kumbelik AR, Livakk J, Rafalsk JA, Tingey SV (1990) DNA polymorphisms amplified by arbitrary primers are useful as genetic markers. Nucleic Acids Res 18: 6531–6535

Wititsuwannakul R (1986) Diurnal variation of HMGR activity in latex of *Hevea brasiliensis* and its relation to rubber content. Experientia 42:44–50

Wycherley PR (1968) Introduction of *Hevea* to the orient. The Planter 4:1–11

Wycherley PR (1976) Rubber. In: Simmonds NW (ed) Evolution of crop plants. Longman, London, pp 77–80

Wycherley PR (1992) The genus *Hevea:* botanical aspects. In: Sethuraj MR, Mathew NM (eds) Natural rubber: biology, cultivation and technology. Elsevier, Amsterdam, pp 50–66

Xi WX, Xiao XZ (1988) Study on peroxidase isozyme and syperoxyde dismutase isozyme of TPD hevea trees. Chin J Trop Crops 9:31–36

Xia Z, Xu H, Zhai J, Li D, Luo H, He C, Huang X (2011) RNA-Seq analysis and de novo transcriptome assembly of *Hevea brasiliensis*. Plant Mol Biol 77:299–308

Zeng RZ (1997) The relation between contents of nucleic acid and tapping panel dryness in latex from Hevea brasiliensis. Chin J Crops 18:10–15

Zheng GB, Chen MR (1982) Study of the cause for brown blast disease. Chin J Trop Crops 3:57–61

Zhu J, Zhang Z (2009) Ethylene stimulation of latex production in *Hevea brasiliensis*. Plant Signal Behav 4:1072–1074

Chapter 10
Coconut, Date, and Oil Palm Genomics*

Alan W. Meerow, Robert R. Krueger, Rajinder Singh, Eng-Ti L. Low, Maizura Ithnin, and Leslie C.-L. Ooi

Abstract The palm family, consisting of over 2,500 species arrayed among ca. 200 genera, is the third most economically important family of plants after the grasses and legumes. Three palm species account for the large majority of the family's economic importance: coconut (*Cocos nucifera*), African oil palm (*Elaeis guineensis*), and date palm (*Phoenix dactylifera*). Of the three, genomics has been least developed in the coconut, where molecular tools have largely been used to characterize germplasm, and, to a lesser extent, develop quantitative trait loci (QTL). Both date palm and oil palm have recently had their genomes sequenced. The application of genomic tools to these palm species will result in enormous advances in the genetic improvement of all three crops.

Keywords Palm • Arecaceae • Microsatellite DNA • Genome sequencing • QTL • Marker-assisted selection • Germplasm • Crop improvement

---

*Alan W. Meerow authored the introduction and the section on coconut; Robert R. Krueger authored the section on date palm; and Rajinder Singh, Eng-Ti L. Low, Maizura Ithnin, and Leslie C.-L. Ooi wrote the section on African oil palm.

A.W. Meerow, Ph.D. (✉)
USDA-ARS-SHRS, National Germplasm Repository,
13601 Old Cutler Road, Miami, FL 33158, USA
e-mail: alan.meerow@ars.usda.gov

R.R. Krueger, Ph.D.
USDA-ARS National Clonal Germplasm Repository for Citrus & Dates,
1060 Martin Luther King Blvd, Riverside, CA 92507-5437, USA

R. Singh, Ph.D. • E.-T.L. Low, Ph.D. • M. Ithnin, Ph.D. • L.C.-L. Ooi, M.Sc.
Advanced Biotechnology and Breeding Centre, Malaysian Palm Oil Board,
No. 6, Persiaran Institusi, Bandar Baru Bangi, 43000 Kajang, Selangor, Malaysia

R.J. Schnell and P.M. Priyadarshan (eds.), *Genomics of Tree Crops*,
DOI 10.1007/978-1-4614-0920-5_10, © Springer Science+Business Media, LLC 2012

The palm family, consisting of over 2,500 species arrayed among ca. 200 genera (Dransfield et al. 2008), is known to botanists as the Arecaceae, though the old name Palmae is still sometimes used in horticulture. Over 90% of the diversity within the family is contained within the world's tropics, and the utility of many palms in human industry at both the subsistence and world market levels makes the Arecaceae the third most economically important family of plants after the grasses and legumes.

Three palm species account for the large majority of the family's economic importance. *Cocos nucifera* L. (Harries and Paull 2008) and *Elaeis guineensis* L. are major export crops throughout the world's tropics, as is the date palm (*Phoenix dactylifera* L.) in subtropical arid zones. Equally significant are the myriad uses to which local palms are put by indigenous human cultures wherever palms are found naturally (Balick and Beck 1990; Johnson 1998; Schultes and Raffauf 1990). These include exploitation for food, oil, fiber, and construction, as well as medicinal and ceremonial use.

The degree to which genomic approaches have been applied toward the improvement of palm tree crops has trailed that of cereals, vegetables, and many temperate fruit crops, probably for no other more important reason than the two most economically important palm species, African oil palm and coconut, are crops of the developing world, with limited resources available for the often expensive technology required to acquire these data. This has begun to change as certain tropical nations enjoy expansive economic growth and the cost of genomic research plummets.

Coconut

Cocos nucifera L., the coconut, is not only the universal symbol of the tropics but is one of the most important economic plants in the low-latitude developing world. *C. nucifera* is pantropically distributed, a present-day range significantly influenced both by a seed well-adapted to oceanic dispersal and the species' importance to humans (Gruezo and Harries 1984; Harries 1978, 1995). Copra, the dried endosperm of the seed, which is the source of coconut oil, and coir, the fibers of the mesocarp, are both significant export products for Third World economies (Harries and Paull 2008), and there are many subsistence uses for almost all parts of the coconut (Balick and Beck 1990). For millennia, humankind has found myriad uses for almost all parts of the coconut palm, and this history has played an important role in shaping the phenotypic diversity of this cultigenic species (Harries 2001). Coconut cultivars are generally classified into the Tall and Dwarf types. The tall type is primarily out-crossing, while the dwarf type is mainly selfing (with a few exceptions). The coconut is diploid, with $2n = 16$ chromosomes (Read 1966). Genome size is estimated at $2C = 6.96$ Gbp (Beaulieu et al. 2007).

Despite the economic importance of the coconut, there has not been a large amount of genomic data generated, and at this writing, no next-generation sequencing of either transcriptomes or the entire genome are visibly underway anywhere in the world. Marker-assisted selection programs in coconut breeding have also not yet been documented, despite as many as six mapping populations worldwide and a number of QTL (Ashburner 1999; Baudouin et al. 2005). Nonetheless, a large body of molecular marker studies of genetic variation, some sequence-based phylogenetic studies, and a few candidate gene and QTL discoveries have been accomplished. These will be reviewed in turn.

Biochemical and Molecular Marker Studies of Genetic Variation

Biochemical Markers

Before the development and widespread application of DNA-based markers, isozyme analysis – genetic analysis of the protein products of genes, visualized with gel electrophoresis (Tanksley and Orton 1983) – was used to investigate genetic variation among coconut varieties and has continued even into the genomic era (Carpio 1982; Benoit 1979; Benoit and Ghesquière 1984; Canto-Canché et al. 1992; Asmono et al. 1993; Hartana et al. 1993; Ashburner 1995; Fernando 1995; Fernando and Gajanayake 1997; Cardeña et al. 1998; Geethalakshiui et al. 2004, 2005; Upadhyay et al. 2004; Niral et al. 2007). Ultimately, the low polymorphisms of isozyme phenotypes in coconut limited their usefulness (Ashburner 1999; Lebrun et al. 2005). Leaf polyphenols assayed via HPLC (high performance liquid chromatography) were used as well for coconut cultivar characterization (Jay et al. 1988), but this latter approach was later determined to have low repeatability (Lebrun et al. 2005).

Dominant Molecular Markers

Successive dominant molecular marker applications in coconut have included randomly amplified polymorphic DNA (RAPD; Ashburner et al. 1997; Wadt et al. 1999; Cardeña et al. 2003; Upadhyay et al. 2004; Manimekalai and Nagarajan 2006a, 2010), organeller restriction fragment length polymorphisms (RFLP; LeBrun et al. 1998a, b), amplified fragment length polymorphism (AFLP; Perera et al. 1998; Teulat et al. 2000), inter simple sequence repeats (ISSR; Manimekalai and Nagarajan 2006b), and inverse sequence-tagged repeats (ISTRs; Rohde et al. 1995; Duran et al. 1997).

RAPDs have not been used extensively in coconut due to their non-specific nature and low repeatability (Lebrun et al. 2005). ISSR, like RAPD, is a multilocus technique, thus there is the possibility of non-homology of similar-sized fragments. It can also have similar reproducibility problems. Little mitochondrial and no chloroplast polymorphism was found in coconut RFLPs derived from these organelles (Lebrun et al. 1998a, b, 1999; Perera 2002). AFLP have been used primarily to more fully saturate genetic maps (Lebrun et al. 2001; Riedel et al. 2009). ISTRs are of greatest use at the population level and have not been informative for diversity studies across numerous cultivars (Lebrun et al. 2005).

Co-dominant Molecular Markers

Such markers include RFLPs from the nuclear genome (LeBrun et al. 1998a, b, 1999), microsatellites (simple sequence repeats, SSRs), and single nucleotide polymorphisms (SNPs). The only SNPs developed for coconut were those of Mauro-Herrera et al. (2006, 2007), visualized with single-strand conformational polymorphisms (SSCP). One would assume that SNPs will become much more of a focus in coconut genetic studies with the advent of next-generation sequencing platforms.

RFLPs are robust markers, though time-consuming to develop. Those derived from the nuclear genome proved variable and fairly informative (LeBrun et al. 1998a, b, 1999).

Three groups have developed SSR primers from *C. nucifera* (Karp 1999; Perera et al. 1999, 2000; Rivera et al. 1999; Teulat et al. 2000). Teulat et al. (2000) used 37 SSR loci and AFLP to measure genetic diversity in 31 individual plants representing 14 varieties of diverse geographic origin. Perera et al. (2000) used 8 loci across 130 individuals representing 94 different varieties. In both cases, SSRs were very successful in distinguishing among coconut genotypes. Ultimately, a 14-locus microsatellite kit was developed and promoted for the reliable identification of coconut cultivars and to promote consistency in diversity studies (Baudouin and Lebrun 2002). Microsatellites have been the most widely utilized molecular markers for diversity analysis in coconut (Karp 1999; Perera et al. 1999, 2000, 2003; Rivera et al. 1999; Teulat et al. 2000; Meerow et al. 2003; Devakumar et al. 2006; Zizumbo-Villarreal et al. 2006; Mauro-Herrera et al. 2007, 2010; Dasanayaka et al. 2009; Martinez et al. 2010).

Based on the results of these studies, a classification was proposed by Lebrun et al. (2005). The classification proposed two major cultivar groups: the Pacific group with five sub-groups (Southeast Asia, Melanesia, Micronesia, Polynesia, and the Pacific coast of Central and South America) and the Indo-Atlantic group, originating in India and subsequently dispersed to West Africa, the Atlantic coast of Latin America, and East Africa (Lebrun et al. 2005).

Diversity studies have strongly supported a common ancestry/domestication region for dwarfs from India, Malaysia, and the Philippines (Lebrun et al. 1998a; Teulat et al. 2000; Meerow et al. 2003; Perera et al. 2003; Harries et al. 2004; Upadhyay et al. 2004; Lebrun et al. 2005; Devakumar et al. 2006; Mauro-Herrera et al. 2010), all of which appear to be heavily self-pollinating and thus inbred (Teulat et al. 2000; Meerow et al. 2003). The South-Pacific 'Niu Leka' dwarf is the exceptional case, as it is much more heterozygous and outbred (Meerow et al. 2003; Lebrun et al. 2005), and is thought to represent a separate domestication event (Harries 1978).

However, Mauro-Herrera et al. (2010) examined the consequences of small sample size and method of analysis in marker-based coconut diversity analyses. Many coconut studies have evaluated just between one and three (maximum seven) individuals for most of their cultivars (Manimekalai and Nagarajan 2006a, b; Perera et al. 1998, 1999, 2000; Rivera et al. 1999; Teulat et al. 2000), which may be a critical factor when evaluating relationships between tall (out-crossing) genotypes. Furthermore, only recently, Baudouin and Lebrun (2009), Baudouin et al. (2008), and Lebrun et al. (2008) assessed more than 20 and up to 104 individuals for specific cultivars ('Malayan Yellow Dwarf' and 'Panama Tall') and showed the prevalence of cross-contamination/admixture in those cultivars.

Method of analysis of marker data revealed ambiguity in the relationships of the Tall cultivars (Mauro-Herrera et al. 2010). With Neighbor-joining (NJ) clustering, 'Atlantic Tall' and 'Pacific Tall' coconuts clustered together with high bootstrap support. With unweighted pair group method (UPGMA), the distinction between the Indo-Atlantic and Pacific groups championed by Lebrun et al. (2005) was more or less resolved, albeit with much lower bootstrap support. NJ differs from the UPGMA by not assuming equal rates of evolution across all lineages, and is considered a more accurate clustering method for genetic distances (Saitou and Nei 1987). Except for studies by Perera et al. (1998, 2000, 2003), most of the other coconut studies on genetic diversity have used the UPGMA method (e.g., Ashburner et al. 1997; Rivera et al. 1999; Teulat et al. 2000; Manimekalai and Nagarajan 2006a, b) to present the relationships between cultivars (Lebrun et al. (1998b) used factorial analysis of correspondences (FAC) instead of cluster analysis). An earlier study of the 'Atlantic Tall'('Jamaica Tall') and 'Panama Tall' cultivars along with African and Caribbean coconuts reported dissimilar results depending on the marker system used (Manimekalai and Nagarajan 2006a, b). The RAPD analysis (Manimekalai and Nagarajan 2006a) paralleled the accepted origin theory by grouping the 'Atlantic Tall' with an African (Nigerian) and Indian cultivars in one main cluster and the 'Panama Tall' with cultivars from the South Pacific and South-East Asia in another (though 'Panama Tall' was paired with a cultivar from North East India). The Nigerian cultivar was genetically most similar to a Caribbean cultivar. Alternatively, the ISSR study (Manimekalai and Nagarajan 2006b) found highest similarity between 'Panama Tall' and the Nigerian cultivar, placing them both in a cluster that included the 'Atlantic Tall' and Caribbean cultivars as well as other Indian and South-East Asian genotypes. Both studies used the UPGMA clustering method and only the ISSR results were reported with bootstrap values. Zizumbo-Villarreal et al. (2006) studied Atlantic and Pacific genotypes from Mexico and analyzed them with both clustering methods (UPGMA and NJ); with NJ, tall and dwarf cultivars were separated into two main groups, though their 'Malayan Dwarf' were grouped with some of their 'Panama Tall' and, with some exceptions, their talls were divided between those of Atlantic origin and those from the Pacific. With UPGMA, their 'Malayan Dwarf' were also grouped with a subgroup of 'Pacific Tall,' and the Atlantic and Pacific Talls were grouped in two separate clusters. The fact that they found a consistent grouping of their 'Malayan Dwarf' with a subgroup of 'Pacific Tall' may be indicative of admixture between those Pacific genotypes, which can play a significant role in the clustering patterns of the different genetic groups in the dendrograms. Such discrepancies with the same genotypes make it difficult to conclude that any one scenario of tall coconut varietal relationships is accurate.

Non-linkage Association of Markers with Phenotypic Traits

Only two studies have associated molecular markers with phenotypic traits of interest, outside of formal genetic linkage mapping studies and QTL identification (see below). Cardeña et al. (2003) associated a subset RAPD markers to lethal yellowing (LY), a highly destructive phytoplasma disease to which coconut shows variable susceptibility. Three coconut populations were genotyped, one consisting of resistant 'Malayan Yellow Dwarf' (MYD), survivors from a diseased population of the generally considered susceptible 'Atlantic Tall' (AT), and the susceptible 'West African Tall' (WAT). A particular RAPD band was considered associated with LY resistance if its frequency were high in MYD and surviving AT, and low in WAT. Twelve markers were identified that fit these criteria, and their use in marker-assisted breeding (MAS) was suggested. However, without any quantitative analysis and the precision of linkage mapping, their conclusions are best considered anecdotal. A more sophisticated study was conducted by Shalini et al. (2007) on coconut mite (*Aceria guerreronis* Keifer) resistance. An assortment of both mite-susceptible and mite-resistant accessions in India were genotyped with 32 SSRs and seven RAPD markers. Nine SSR and four RAPD markers were identified that could be associated with mite resistance. In stepwise multiple regression analysis of each marker, a combination of six SSRs had 100% association with mite infestation, while three RAPD markers accounted for 83.86% of mite-resistant genotypes. A combination of five markers (three SSRs and two RAPD) explained 100% of the association with mite resistance.

Linkage Mapping and QTL Identification

The first coconut linkage map was constructed in the Philippines using hybrids of a 'Malayan Yellow Dwarf' (MYD) × 'Laguna Tall' (LAGT) cross (Rohde et al. 1999). Three hundred eighty-two (382) markers were placed on 16 linkage groups, and six QTL for early flowering were identified (Herran et al. 2000). The total map length was 2,226 cM. Ritter et al. (2000), using the same mapping population, mapped QTL for leaf production (3) and girth height (7).

The second linkage map was based on a MYD × 'Rennell Island Tall' in Côte d'Ivoire, with 227 markers in 16 linkage groups. Nine QTL associated with fruit yield were found (Lebrun et al. 2001). A total length of about 2,000 cM was assigned to the coconut genome (Herran et al. 2000; Lebrun et al. 2001). This second mapping population, with the addition of 53 new SSR markers, was later used to identify another 52 QTL for 11 fruit traits including fruit component weight, endosperm moisture content, and fruit production (Baudouin et al. 2006).

Baudouin et al. (2005) reported that two additional mapping populations were created: 'East African Tall' (EAT) × 'Pemba Red Dwarf' (PRD) and EAT × RIT in Tanzania as part of a European joint research project on oil palm and coconut (INCO LINK2PALM project). The latter was used to map 46 QTL for the cuticular wax biosynthesis (Riedel et al. 2009). This linkage map contained a total of 1,000 markers, a combination of SSR and AFLP.

Candidate Genes and Other Genomic Tools

WRKY Loci

WRKY transcription factors are predominantly plant-specific proteins of diverse functions including stress response (Dong et al. 2003; Ülker and Somssich 2004; Garcia et al. 2005; Luo et al. 2005; Wu et al. 2005; Eulgem and Somssich 2007), broadly distributed across the genome (Zhang and Wang 2005), but absent from the genomes of fungi and animals (Zhang and Wang 2005). There is a well-developed classification for the family (Eulgem et al. 2000; Wu et al. 2005; Xie et al. 2005; Zhang and Wang 2005). Intron variability distinguishes among diverse WRKY loci isolated from a single species and, in many cases, allows for the design of primers specific to each (Borrone et al. 2004). Moreover, the coding regions tend to be SNP rich, and, thus, a source of allelic variation (Borrone et al. 2004; Mauro-Herrera et al. 2006). Mauro-Herrera et al. (2006) isolated 21 WRKY loci from coconut and developed SNP markers from 13 of them (Mauro-Herrera et al. 2006, 2007). It was hoped that they might provide reliable markers for lethal yellowing (LY) resistance, but that possibility remains undeveloped at present.

Cyclin-Dependent Kinase Gene

The CDKA gene has been implicated in the control of cell proliferation (Martinez et al. 1992; Hemerly et al. 1993). Montero-Cortés et al. (2010) isolated this gene from coconut, and performed a detailed expression analysis during the process of somatic embryogenesis in tissue culture. They found that the highest expression profile coincided with callus formation and progressively decreased as somatic embryos developed.

Expressed Sequence Tags

To date, only a single cDNA library has been constructed from coconut, that of Li and Fan (2008) from endosperm ESTs. Genes believed to be associated with cell division, protein synthesis, and growth regulation were identified, and sequence identity for some was identical to putative homologs from African oil palm (*Elaeis guineensis*).

MicroRNAs

MicroRNAs *(miRNAs)* are a group of small (20–24 nt), endogenously expressed, non-coding RNAs that play important regulatory roles in plants and animals. So far, more than 800 miRNAs have been identified from angiosperms (Hewezi et al. 2008). miRNAs function by targeting mRNA for cleavage or translational repression and thus affect many functions in plants, including leaf (Ori et al. 2007), shoot, root (Guo et al. 2005), and floral development (Allen et al. 2005), as well as stress response (Sunkar et al. 2006). Li et al. (2009) compared the expression profiles of miRNAs between two different stages in the development of endosperm in coconut. They found 32 miRNAs that showed differential expression, thus implicating miRNAs in coconut endosperm development.

Lysophosphatidyl Acyltransferase

The coding region of lysophosphatidyl acyltransferase (LPAAT) has been cloned and sequenced from coconut (Knutzon et al. 1995). Xu et al. (2010) amplified the upstream region of this endosperm-specific expressed gene from the coconut genome, and characterized a number of putative promoter elements. Transgenic rice plants were then created with inserts of varying sizes from this upstream region and used to test the expression of gene function believed to be driven by the promoters. Expression was limited to the endosperm of rice seeds, and was highest with the full-length upstream insert versus the smaller fragments. The authors suggest that the promoter region from coconut could be used to improve cereal crops.

Phylogenetic Analysis of DNA Sequences

Gunn (2004) produced the first molecular phylogenetic study with an explicit goal of elucidating the relationships of *C. nucifera* using the nuclear gene *prk* across a broad sample of the tribe Cocoseae, but several studies with chloroplast genes, while establishing the broader relationships of coconut, were ambiguous in terms of its sister relationships (Asmussen et al. 2000, 2006; Hahn 2002a, b). Gunn was only able to show weak sister relationship of *Cocos* with the Andean genus *Parajubaea* using maximum likelihood analysis; parsimony and Bayesian approaches failed to resolve the exact phylogenetic position of coconut within the tribe. Meerow et al. (2009) successfully applied a super-matrix of seven WRKY loci across the palm tribe Cocoseae subtr. Attaleinae within which *Cocos* is classified (Dransfield et al. 2008), and robustly identified the genus *Syagrus* as sister to *Cocos*, thereby indicating that the deep ancestry of the coconut lay in South America. They also resolved the South-Pacific coconut variety 'Niu Leka' as sister to all other cultivars included ('Atlantic Tall,' 'Pacific Tall,' 'Red Spicata,' 'Green Malayan Dwarf,' and 'Red Malayan Dwarf'), which is congruent with a Pacific Ocean dispersal scenario for coconut (Harries 1978). Unfortunately, beyond the sister relationship of 'Niu Leka' to all other varieties, the well-resolved clade of coconut cultivars, while demonstrating the utility of WRKY loci at infra-specific levels was not able to be interpreted strictly because SSR studies indicated that three of the six individuals ('Atlantic Tall,' 'Pacific Tall,' and 'Red Spicata'), used were introgressed with other cultivars (Mauro-Herrera et al. 2010). However, the degree of resolution suggested that WRKY loci could be successfully applied to a phylogeographic study of *C. nucifera*.

The Future of Coconut Genomics

With the advent of new DNA sequencing platforms that achieve an ever-increasing degree of speed, coverage, and sharply decreasing costs for obtaining the data, we can expect to see many coconut transcriptome libraries, SNP chips, and, ultimately, a complete genome sequencing effort. A coconut bacterial artificial chromosome (BAC) library was reported to exist at CIRAD, with a total length of five times that of the coconut genome (Baudouin et al. 2005), but no publications have as yet appeared utilizing this important genomic tool.

With a complete genome available for the related *E. guineensis* (see below), the potential for comparative genomic studies across two of the most important crops in the developing world is at hand. In fact, at this writing, an international consortium for the sequencing of the *C. nucifera* genome is being organized (M. Edwards, 2011). With the development of well-saturated linkage maps and an exponential increase in the number of QTL for important traits such as disease resistance and productivity, a true marker-assisted breeding program for the coconut may finally see fruition.

Date Palm

Date palm (*Phoenix dactylifera* L.) is one of the oldest fruit crops grown in the arid regions of the Arabian Peninsula, North Africa, and the Middle East (Chao and Krueger 2007, 2008). The most probable area of origin of the date palm was in or near what is now the country of Iraq, but date cultivation spread to many countries beginning in ancient times. Dates are a major food source and income source for local populations in the Middle East and North Africa, and play significant roles in the economy, society, and environment in these areas (Chao and Krueger 2007, 2008). In addition to serving directly as a food source, dates are packed and processed in a number of ways, and other parts of the tree are used for various purposes.

The date palm is a diploid, perennial, dioecious, and monocotyledonous plant adapted to arid environments. It has unique biological and developmental characteristics that necessitate special propagation, culture, and management techniques. Thousands of date palm cultivars and selections exist in different date-growing countries. Different genetic marker systems have been used to study genetic relationships among date palm cultivars. The long life cycle, long period of juvenility, and dioecism of date palms make breeding challenging. Worldwide date production has grown from 1,809,091 t in 1962 to 6,924,975 t in 2005; it will continue to grow, especially in the Middle East, despite current and future challenges (Chao and Krueger 2007).

The date palm is the type species for the genus *Phoenix* L., which in turn is the type genus for the tribe Phoeniceae of subfamily Coryphoideae in the Arecaceae (Dransfield et al. 2008). The tribe Phoeniceae is widely distributed throughout the Old World but appears in the New World only due to transport by human beings. Estimates of the number of species of *Phoenix* have varied, with most researchers recognizing ca. 12–17 spp. (Moore 1963, 1971; Barrow 1998; Dransfield et al. 2008). The species problem in *Phoenix* is compounded by the genus's dioecy and free interspecific hybridization (DeMason and Sekhar 1988).

Basic Genetics of Date Palms

Phoenix spp. are dioecious and diploid with $2n = 36$ (Darlington and Wylie 1955). There have been some anomalous reports of different chromosome numbers. This is discussed in a recent review (Al-Ani et al. 2010). Date palm cultivars as well as other *Phoenix* spp. have chromosomes that are similar in morphology, which may account for the cross-compatibility and hybridization between different cultivars and spp. (Beal 1937). Sarkar (1957) reported that in *P. sylvestris* and *P. rupicola*, male trees had one heteromorphic pair of chromosomes, which allowed differentiation between male and female trees. In *P. dactylifera* itself, Siljac-Yakovlev et al. (1996) reported that female trees had a pair of GC-rich homomorphic chromosomes, whereas the analogous pair in male trees was heteromorphic.

As recently as 1996, Siljac-Yakovlev et al. (1996) noted that the genetics of date palms were largely unknown. This is due to the factors noted above as well as the long generative time of the date palm, typically 5–7 years between pollination and fruiting. This situation has recently improved somewhat due to the development of molecular markers (see next section). However, most knowledge in the area of the genetics of date palms and inheritance of characteristics has been based upon observation rather than experimentation. Although date growers in the Middle East and North Africa have made anecdotal observations about date palms for millennia, scientists have only recently been able to document these observations. Most of the documentation of genetic characteristics of date palms was developed by the USDA Date and Citrus Station in Indio, California, during a breeding project conducted between 1948 and 1978 (Nixon 1959; Nixon and Furr 1965; Barrett 1973; Ream and Carpenter 1975; Carpenter and Ream 1976).

A few of the documented observations concerned seedling variability due to hybridization. This was thought to have been the origin of nearly all date palm varieties, though a few may have been derived from "sectorial mutations" that could be clonally propagated from offshoots. Although date palm seedlings are variable, seedlings tend to resemble the mother tree in both vegetative and fruit characteristics. Some phenotypic characteristics were believed to be genetically homozygous based upon observation of seedling character segregation. Some varieties showed marked genotypic-environmental interactions. Increasing homozygosity due to back-crossing resulted in a tendency toward sterility in male progeny, in addition to various other reproductively deleterious consequences. Increasing homozygosity also resulted in decreased vigor in the progeny (Nixon 1959; Nixon and Furr 1965; Barrett 1973; Ream and Carpenter 1975; Carpenter and Ream 1976).

Genetic Analysis of Date Palms

While observations on date palm phenotypes were being made by USDA personnel, the technology to begin exploring the underlying genetics was not available. Beginning in the late 1970s, workers at various institutions began to develop biochemical and later DNA-based marker systems to study various aspects of plant genetics. These techniques were applied to the date palm in due course. They have since been utilized in attempts to understand the genetic diversity present in the date palm and in individual oases systems to identify or fingerprint cultivars, and to identify variation in palms produced by tissue culture. The following section reviews some of the publications in these areas but is by no means exhaustive.

Isozymes

The first marker system developed for use in date palm genetic analysis was based on isozymes. Tisserat and Torres (1979) and Torres and Tisserat (1980) used isozymes to study inheritance in seedling populations with known parents from the USDA breeding program. Five isozymes were coded by seven polymorphic genes with 14 alleles. Additional polymorphism was detected in other *Phoenix* spp. As expected, inheritance followed classical Mendelian genetics and the genetic uniformity of offshoots compared to mother palms was confirmed.

Since the first reports of Tisserat and Torres (1979) and Torres and Tisserat (1980), various researchers have utilized isozyme systems to study different aspects of date palm genetics. An early study by Bennaceur et al. (1991) surveyed genetic diversity of date palms in different areas of Tunisia using 20 enzyme systems.

Of these, seven showed polymorphism as represented by 16 alleles. This allowed the identification of 65% of the cultivars studied. The genetic uniformity of "Deglet Nour" was confirmed and two other cultivars were found to be identical to each other. Genetic diversity was found to be greater in the Western region than in the Eastern region of Algeria. In Morocco, Baaziz and Saaidi (1988) initially utilized only one enzyme system to attempt to identify date palm cultivars. Later refinements and the use of an additional two enzyme systems allowed the identification of 28 reputed cultivars of date palms (Bendiab et al. 1998). This system also demonstrated the lack of identity of approximately two-thirds of the palms in different oases, thus confirming the predominance of seedling populations in these oases. Cultivar composition of different oases was also revealed. Similarly, in Tunisia, Ould Mohamed Salem et al. (2001a, b) used four enzyme systems providing five polymorphic genes and 12 alleles to assess genetic diversity of date palms grown in different geographic areas. They found low genetic differentiation between cultivars from different geographic areas. Identification of most cultivars was possible using this system. Isozymes were also used to differentiate between male "cultivars," with results confirming the identity of clonally derived plants with the mother palm (Al-Jibouri and Adham 1990).

RAPD and RFLP

Similar studies have been performed with DNA-based marker systems. Corniquel and Mercier (1994) reported the development of polymorphic RAPD and RFLP markers for date palm cultivar identification, which allowed the identification of five different cultivars. The markers were also utilized to reveal genetic variation of tissue culture derived calli from the mother offshoots from which they were produced. RFLP analysis was further developed by Corniquel and Mercier (1997), who also reported that comparison with genomic databases indicated similarities with published sequences from other organisms.

Due to the difficulties in extraction of high-quality DNA from date palm tissue, the length of time required to obtain strong hybridization signals, and the use of radiation, RFLPs did not prove suitable for large-scale use in field analysis, and RAPDs became more prevalent in DNA-based genetic analysis of date palms (Bouchireb and Clark 1997). Bouchireb and Clark (1997) reported that out of 50 RAPD primers screened, eight were found to be useful. Cultivars discrimination was possible using the 31 polymorphic bands produced. The technology was also utilized to study genetic diversity in Algerian date palms. Specific cultivars were found to be very similar, whereas other specific cultivars showed more intra-cultivar variation. The latter was taken as evidence for the fact that in many cases, the actual parentage of a reputed cultivar is not definitively known. In this study, regional genetic diversity was often as great as in the country as a whole.

Various other reports of cultivar identification using RAPD have been published. Sedra et al. (1998) screened 37 Moroccan date palm cultivars as well as six from other countries using 19 primers derived from 123 screened pairs. All genotypes analyzed were distinguishable using the markers developed. Saker and Moursy (1999) discovered that four random primers (out of 200 screened) allowed the discrimination of five cultivars from one another. Ben Abdallah et al. (2000) screened 122 RAPD primers and found 11 that produced 53 polymorphic bands. These resulted in the identification of six genotypes. RAPD have also been used to assess genetic diversity and phylogenetic relationships. Sedra et al. (1998) reported that RAPD indicated a low level of polymorphism in Moroccan date palms from different geographic areas. Trifi et al. (2000) assessed genetic diversity of date palms in Tunisia using RAPD. Both Tunisian and introduced varieties clustered into four groups. The introduced varieties clustered along with the Tunisian varieties. This was taken as showing a relatively low level of variation and suggested a common origin for the Tunisian and introduced varieties. The same group reported similar results using RFLP (Sakka et al. 2004). Similarly, in Saudi Arabia, RAPD markers showed most varieties evaluated fell into two main clusters, demonstrating a relatively low genetic base. One variety fell outside the clusters and was clearly different from the bulk of varieties assayed (Al-Khalifa and Askari 2003). The results with RAPD markers were mirrored with ISSR markers (Abdulla and Gamal 2010).

AFLP

The first report of the use of AFLP markers in date palm genetic analysis was actually made by Lacze et al. (2000), but the first widely disseminated report on their use in date palm genetic analysis was by Cao and Chao (2002). Four primer sets were used to detect polymorphisms in 23 date cultivars maintained by the USDA-ARS National Clonal Germplasm Repository for Citrus & Dates (NCGRCD). Based upon the UPGMA-cluster analysis of 328 polymorphic bands, the majority of the cultivars fell into two major groups. A minor group consisting of "Barhee" and "Deglet Noor" was also detected, and "Hayany" and the related species, *P. sylvestris,* fell outside the other accessions. The same group (Devanand and Chao 2003a, b) used AFLP to determine genetic consistency between commercial plantings of "Deglet Noor" and "Medjool." They reported little variation within cultivated "Deglet Noor" but considerable variation within cultivated "Medjool." This led the authors to postulate that "Medjool" represents a landrace in its native Morocco. The results are curious as there were large differences between mother trees and offshoots. Further analysis of Moroccan-derived "Medjool" showed only 79% genetic similarity within 66 collections of "Medjool" from three locations in Morocco, thus supporting the contention that "Medjool" is a landrace (Elhoumaizi et al. 2006). However, the factors could be in play here as with other intra-cultivar variation: lack of knowledge of the parentage or provenance of the accessions collected. The same group used AFLP to analyze Egyptian date palm varieties (El-Assar et al. 2005). The majority of the accessions fell into one major cluster with only a few varieties showing large divergence. It was postulated that this main cluster represented the major group of date palm germplasm in North Africa, which could have been derived from introductions from the center of diversity in the Middle East that had only low levels of genetic diversity.

Diaz et al. (2003) used fiver primer sets generating 310 AFLP fragments to analyze date palm varieties maintained at the Estación Phoenix in Elche, Spain. The technique was useful in analyzing varietal identity and the integrity of tissue culture derived plantlets. Jubrael et al. (2005) used AFLP to analyze date palm germplasm from Iraq. A total of 122 polymorphic AFLP loci were observed, with an average of 17.4 polymorphic loci per primer combination. The use of any of the four combinations was sufficient to identify all 18 of the varieties studied. The varieties were separated into two groups based upon their genetic relationships, which ranged from moderate to diverse.

Microsatellites

As with other crops, microsatellite markers have become a preferred technique for genetic analysis of date palms. Most of the efforts with microsatellites have involved varietal identification, with some work in the area of genetic analysis and phylogeny. Billotte et al. (2004) constructed a $(GA)_n$ microsatellite-enriched library and characterized 16 nuclear SSR loci in *P. dactylifera*. Most of these SSR markers amplified across 11 other *Phoenix* spp., as well as *E. guineensis*, and 17 other palm taxa. These were the first SSR markers to be published for date palm. Similarly, Akkak et al. (2009) isolated 41 SSR from two microsatellite-enriched libraries of *P. dactylifera*. Seventeen selected microsatellite loci were characterized and evaluated on a set of 31 genotypes from NCGRCD and Algeria. All produced amplification products of the expected size, and also across 14 additional *Phoenix* spp. Johnson et al. (2009) analyzed 18 varieties maintained at the NCGRCD using SSR markers originally developed for oil palm (*E. guineensis* Jacq). Three primer pairs were identified that unambiguously identified each of the date palm varieties. The polymorphic bands were used to design new internal primers. However, all amplifications with the new primers yielded only polymorphic bands, indicating that the variation among these date palm varieties was at or near the original primer sites, with internal sequences preserved. The fact that the primers successfully used by Johnson et al. (2009) were derived from another genus of palm was further proof that SSR markers can be used across palm genera.

In Tunisia, Zehdi et al. (2002) used ISSR markers to examine phylogenetic relationships in Tunisian date palm accessions. Overall, considerable genetic diversity was observed, with results comparable to those of Trifi et al. (2000) using RAPD. However, the ISSR markers indicated that introductions from outside Tunisia, as well as male accessions, clustered within the main Tunisian cultivars. The same group also examined a larger group of Tunisian date palms using SSR makers (Zehdi et al. 2004). They observed a high level of polymorphism among the 49 accessions studied from three different geographic areas, with little relationship to the area of collection. This was taken to indicate the existence of one ancestral date palm population, which was consistent with the work of El-Assar et al. (2005) as well as non-molecular accounts (Wrigley 1995). Due to the specificity of the SSR markers developed, Zehdi et al. (2004) speculated that it would be possible to develop a fingerprinting system for date palms. This system was published by Zehdi et al. (2006). Also, working with Tunisian date palm cultivars, Hamza et al. (2011) analyzed fruit maturity (early-late) and fruit consistency (soft, semi-dry, dry) subpopulations with SSR markers. They observed the various groups of the subpopulations to segregate based upon the markers as well as the morphological or fruit quality characteristics. In contrast to Zehdi et al. (2004), Hamza et al. (2011) speculated that these results indicated that the Tunisian date palm population consisted of sub-populations derived from different origins. The various investigations into the diversity of Tunisian date palm germplasm have been recently summarized by Rhouma et al. (2010). They concluded that SSR markers offer the most specific information on genotype identification and that the fingerprinting possible with SSR markers will be useful in the conservation of genetic diversity through tissue culture propagation of endangered or "fruity" genotypes (those with little or no offshoot production).

SSR markers have been utilized in other instances to characterize date palm genotypes in various other countries, including Oman (Al-Ruqaishi et al. 2008), Qatar (Ahmed and Al-Qaradawi 2009, 2010), and Saudi Arabia (Abdulla and Gamal 2010). Ahmed and Al-Qaradawi (2010) also utilized ISSR markers in their analyses. The work of Sakina Elshibli with date palm germplasm in Sudan is particularly interesting (Elshibli 2009). Date cultivars from northern Sudan were found to be diverse in regard to fruit morphology and chemical composition as well as SSR polymorphisms (Elshibli and Korpelainen 2009). SSR markers were utilized to assess genetic diversity in 37 female and 23 male date palm accessions from Sudan as well as 8 female accessions from Morocco. High levels of expected heterozygosity were observed among Sudanese females, Sudanese males, and Moroccan females. The genetic groups of the Sudanese groups did not show clear geographic patterns but were distinct from the Moroccan group. The weak clustering association observed was attributed to the effects of pollination methods and selection (Elshibli and Korpelainen 2008a). A more extensive study analyzed 200 individuals from 19 locations throughout Sudan (Elshibli and Korpelainen 2008b). High levels of heterozygosity were again observed. Within-population variation accounted for over 95% of the total genetic variation, and significant differences between the type groups was observed. Three factors were identified as being significant in determining the genetic differentiation of the trees: geographical isolation, the biological nature of the tree, and environmental conditions. Significant isolation by distance was detected, but the spatial effects were stated to have been complicated by the exchange and introduction of different types of plant materials as well as seed dispersal. The authors concluded that the humans probably have had the highest impact on structuring the date palm genome worldwide.

Most analyses using SSR markers have utilized the 16 developed by Billotte et al. (2004). The availability of a larger number of SSR markers would allow a more robust analysis of genetic relationships in date palm. Utilizing sequence information generated by the Weill Cornell Medical College – Qatar Date Palm Genome Project, Hamwieh et al. (2010) designed 1,090 new microsatellite markers. A preliminary test of 50 primer pairs revealed that 56% were functional and 38% yielded polymorphic PCR products.

The works of Elshibli cited above are one of the few that address human influences on genetic diversity of date palms. The molecular studies cited as well as others are useful in studying oases ecosystem and social systems. Krueger (2011) studied oases in three date-growing areas (Egypt, Morocco, and Tunisia) in light of archival materials on varietal composition as supplemented with modern observations and molecular studies. The overall conclusion supported the concept that the genetic diversity of the oases was due to seedlings originating locally and the importation of desirable elite varieties from other oases. Genetic diversity was preserved due to the presence of the seedlings and the isolation of different oases. Importation of elite varieties was traditionally very difficult due to a lack of infrastructure and technology in those areas.

Molecular Markers and Phylogenetics of Date Palms

Molecular markers have been utilized to investigate phylogenetic relationships between different *Phoenix* spp. The taxonomic treatment of *Phoenix* by Barrow (1998) was based upon morphological and molecular analysis, the latter involving the sequencing of the intergeneric spacer of 5S ribosomal DNA. In the Canary Islands, González-Pérez et al. (2004) and González-Pérez and Sosa (2009) used RAPD and isozymes to demonstrate that palms morphologically intermediate between the endemic *P. canariensis* and the exotic *P. dactylifera* were indeed hybrids. Henderson et al. (2006) used SSR markers across *Phoenix* spp. to investigate the relationships between *P. dactylifera*, *P. canariensis*, *P. sylvestris*, and *P. atlantica*. The various spp. proved to be distinct and geographically separated in the Cape Verde Islands. Overall, the evidence was that *P. atlantica* is a distinct species and isolated from other *Phoenix* spp. on the Cape Verde Islands. Gaut et al. (1992, 1996) observed that the substitution rate at two chloroplast loci was lower in *P. reclinata* than in most other plant taxa, indicating that the rate of evolution of *P. reclinata* would be slower than for other taxa.

Detection of Off-Types in Tissue Culture Derived Date Palms Using Molecular Markers

Off-type plants can be a problem in tissue culture propagated plants due to somaclonal variation, which has both genetic and epigenetic aspects (Kaeppler et al. 2000). Date palms produced by tissue culture often exhibit abnormal growth and development patterns, this being more common in plants produced by asexual somatic embryogenesis than those produced by organogenesis. Some abnormalities, which may affect yield, may not manifest for years; thus, there has been great interest in developing marker assays that could identify off-types at an early stage (Kunert et al. 2003).

The use of isozymes for detection of off-type date palms has been reported by Salman et al. (1988), Saker et al. (2000), and Azeqour et al. (2002). Both these reports found relatively low levels of variation in tissue-cultured date palms. Saker et al. (2000) also utilized RAPD to detect variation in tissue culture derived date palms and found them to be more sensitive than isozymes. Gurevich et al. (2005) reported that AFLPs were reported to be useful for the detection of variant tissue culture derived date palms. They found that AFLP showed more genetic variation in tissue culture produced "Medjool" than in tissue culture produced "Barhee," and that the differences correlated with morphological variation. Saker et al. (2006) also used AFLP to detect variation and found that it was more sensitive in this regard than RAPD. Conversely, Al-Khalifa and Askari (2007) found RAPD to be highly efficient in this regard. Kumar et al. (2010) used both RAPD and ISSR markers to validate their technique as producing true-to-type date palms from tissue culture. They considered the use of two marker systems to be superior in this regard, since they amplify different regions of the genome and thus should be better able to detect genetic variation. Vorster et al. (2002) used Representational Difference Analysis (RDA) to detect variation in "Medjool" and "Barhee." This was proposed by Kunert et al. (2000, 2003) as a potential means of detecting genomic variation in tissue-cultured date palms due to its ability to screen a greater fraction of the total genome in a single experiment than other techniques.

Molecular Markers for Specific Characteristics

There has been little work in the area of developing molecular markers associated with specific characteristics for date palm. Most efforts have involved either phyto-pathology (markers associated with resistance, etc.) or sex determination. The latter would be useful in breeding programs since, due to juvenility, it is approximately 5–7 years before the sex of the palm can be determined. RAPDs have been reported to be useful in sex discrimination by Younis et al. (2008) and Mohsenzadeh and Pasalari (2010). Younis et al. (2008) also found ISSR useful for determination of the sex of date palm.

One of the most devastating of all diseases of date palm is Bayoud, caused by the fungal pathogen *Fusarium oxysporum* f. sp. *albedensis* (Killian & Maire) Gordon (Carpenter and Elmer 1978; Djerbi 1983; Michielse and Rep 2009; El Modafar 2010). Bendiab et al. (1993) reported that Bayoud-resistant cultivars or selections exhibited a high level of isozyme polymorphism. Ouenzar et al. (2001) reported a mitochondrial marker for Bayoud resistance based on the plasmid-like circular DNAs isolated from date palm mitochondria by Benslimane et al. (1994, 1996) (see next section). The simultaneous presence of the *R* plasmid and absence of the *S* plasmid was a reliable marker of Bayoud resistance, whereas the absence of the *R* plasmid and the presence of the *S* plasmid correlated with susceptibility to Bayoud (Ouenzar et al. 2001). This system was used to infer susceptibility of Mauritanian date palm cultivars to Bayoud by Ould Mohamed Salem et al. (2007).

A recently reported and emerging disease of date palms in North Africa is "Maladie des Feuilles Cassantes" or Brittle Leaf Disease (BLD) (Triki et al. 2003). BLD is associated with low concentrations of Mn in the leaves and a small double-stranded chloroplastic RNA of host origin (Triki et al. 2003; Namsi et al. 2007). No biotic factor has been identified, although the epidemiology in some ways resembles a disease with a biotic cause (Triki et al. 2003). The chloroplastic RNA was later shown to be present both as ss- and ds-RNA and is diagnostic for BLD (Marqués et al. 2008). These chloroplastic RNAs were used to develop a DIG-labeled probe by Namsi et al. (2006). Saidi et al. (2009, 2010) constructed subtractive cDNA libraries from BLD-affected and non-affected trees to identify genes that were upregulated in BLD-affected trees. BLD-affected trees exhibited ESTs related to genes associated with stress response, metabolism, protein synthesis, and signal transduction, whereas non-BLD-affected trees exhibited ESTs related to photosynthesis, protein synthesis, and ion transport.

Date Palm Genomics

The date palm was thought to have a relatively small genome of approximately 250 Mb with genes distributed throughout approximately 41% of the total genome and the remainder being non-coding regions; this is a wider distribution than found in other monocots (Barakat et al. 1999). However, recent work suggests that it is 550–650 Mbp in size (Malek 2010). Sakka et al. (2000) provided an early report of the construction of a DNA library from date palm. Little work has apparently been done on the chloroplastic or mitochondrial genomes of date palm. Benslimane et al. (1994) identified and characterized two minicircular plasmid-like DNAs from the date palm mitochondria: plasmid *U* (1,160 bp) and plasmid *R* (1,346 bp). These showed some homology to each other but not to nuclear, chloroplastic, or mitochondrial date palm genomes, nor other higher plant plasmid-like DNAs. An additional plasmid-like mitochondrial DNA was reported by Benslimane et al. (1996). This plasmid-like DNA (the plasmid *S*) was 98.8% homologous with plasmid *R*. The *R* and *S* plasmids were never present in the same mitochondria and the pattern of their presence was related to Bayoud resistance (Ouenzar et al. 2001) (see previous section). Analysis of progenies from controlled crosses suggested that the date palm mitochondrial genome was transmitted strictly maternally (Ould Mohamed Salem et al. 2007).

There have been various proposals or efforts to sequence the date palm genome (Donnelly et al. 2007). However, the maximum progress till date has been achieved by two groups. The Sino-Saudi Date Palm Genome Project (http://english.big.cas.cn/ic/ip/200908/t20090824_33815.html) is a cooperative project between various institutions in the Kingdom of Saudi Arabia, including the King Abdulaziz City for

Science and Technology and King Faisal University, and in China, including the Beijing Institute of Genomics of the Chinese Academy of Sciences. The overall objective is to sequence the genome of the date palm and its pest, the Red Palm Weevil (*Rhynchophorus ferrugineus* Olivier). This group recently published the complete chloroplast genome of the date palm (Yang et al. 2010). This was accomplished by extracting 369,022 chloroplast sequencing reads from whole-genome-shotgun data, and validating with PCR-based verification and sequencing. The date palm chloroplast genome is 158,462 bp in length, with large single copy (86,198 bp) and small single copy (17,712 bp) regions separated by two inverted repeats (27,276 bp). These inverted repeat regions contain 112 unique genes and 19 duplicated fragments. In addition, 78 SNPs were identified, along with 18 polycistronic transcription units and three highly expression-biased genes.

The other major effort in date palm genomics has been taken by the Weil Cornell Medical College – Qatar Date Palm Genome Project (http://qatar-weill.cornell.edu/research/datepalmGenome/index.html). This group recently submitted a complete genome generated by shotgun sequencing using massively parallel sequencing of a "Khalas" female date palm (Al-Dous et al. 2011; Malek 2010). The date palm genome was estimated to be 658 Mbp in size. In addition, eight additional genomes were sequenced, including "Deglet Noor" and "Medjool" females and backcrossed males maintained at NCGRCD. A total of 28,890 gene models were predicted, of which over 85% of the protein-encoding genes showed homology with sequences in the NR database at NCBI. Approximately 1.7 M SNPs were called in 381 MB of sequence for a heterozygosity rate of 0.46%. More detailed information and annotations are available at the website.

Changes in DNA methylation on deoxycytidine residues have been shown to be involved in the regulation of plant development by way of gene expression at the transcriptional level and in response to environmental stresses (Finnegan et al. 2000). In date palms, variation in the methylation status of mother plants and offshoots has been documented (Fang and Chao 2007). Variation during offshoot development was primarily due to demethylation as compared to the mother palm, which might be related to gene expression during offshoot development. Differences in methylation status have also been suggested as being involved in somaclonal variation in tissue culture derived date palms (Cohen et al. 2007).

Date Palm Proteomics

There have recently been a few reports on proteomics of the date palm. Sghaier-Hammami et al. (2009a, b) analyzed zygotic and somatic embryos of date palm using electrophoretic techniques. Zygotic embryos from seeds showed increasing protein levels at different stages of development, with different qualitative expression at the different stages. MS analysis showed that the proteins were in the functional categories: enzymes of glycolysis, the TCA cycle, carbohydrate biosynthesis, protein translation, storage, and stress-related proteins. Somatic embryos generated from tissue culture showed quantitative and qualitative differences in protein expression. Proteins associated with the glycolysis pathway were more abundant in somatic embryos, whereas zygotic embryos were characterized by the presence of a family of proteins associated with carbohydrate biosynthesis, storage, and stress-related proteins. Gómez-Vidal et al. (2009) studied the proteomics of date palms inoculated and not inoculated with three species of entomophagous fungi. There were quantitative and qualitative differences in protein expression between the inoculated and uninoculated trees. Inoculated trees expressed proteins associated with plant defenses or stress responses. Expression of proteins associated with photosynthesis and energy metabolism were also affected by the inoculation.

Conclusion and Prospects

Despite its importance as a crop in the Middle East and North Africa, the genetics of the date palm has not received much attention in comparison with many other crops. The genetic basis of inheritance of important traits is known mostly from observation. More recently, marker systems of various types have been developed and these have increased our knowledge of the population structure of various date palm populations. The more recent sequencing of the nuclear and chloroplastic genomes of the date palm will allow the development of additional tools for studying not only population structure but also the genetic basis of various traits and their inheritance. The most important and productive era of date palm genetics is just emerging.

Oil Palm

The oil palm, *Elaeis guineensis* Jacq., is an economically important tree crop in many Third World countries of the lowland, humid tropics (Soh et al. 2008). High yielding and highly profitable as well, it is produced both by large plantations and small farmers. The mesocarp and seed of the fruit are each a source of vegetable oil varying in physical and chemical properties, such that a multiplicity of food, cosmetic, and industrial products are derived from them (Soh et al. 2008).

E. guineensis is believed to have originated in West Africa (Zeven 1965). Only one other species is recognized in the genus, *E. oleifera*, endemic to Central and South America (de Blank 1952; Schultes 1990). A South American origin was also advocated for *E. guineensis* by some researchers (Corner 1966), on the basis that most of the related palm taxa are indigenous to that region. However, it is now generally accepted that *E. guineensis* is of African origin, with Hartley (1977) proposing that *E. guineensis* was in fact introduced from Africa to South America through the movement of slaves. The two species are closely related and they can be easily intercrossed to produce an interspecific hybrid (Hardon and Tan 1969).

In West Africa, *E. guineensis* occurs as wild and semi-wild groves along the coast in a continuous belt from Senegal to Angola and toward the central regions in Congo, Uganda, and Tanzania (Zeven 1967; Rajanaidu 1990). Early observations indicate that in its natural habitat, *E. guineensis* was often found in association with *Raphia*, or if alone, in fresh water swamps (Waterson 1953). It is usually not found in tropical rain forest or even secondary forest, due to competition from other forest species and also due to limited availability of light caused by the forest canopy (Chevalier 1934; Latiff 2000). *E. oleifera*, on the other hand, is naturally confined to Honduras, Nicaragua, Costa Rica, Panama, Colombia, Suriname, Brazil, and Peru in Central and South America (Rajanaidu 1985).

Oil palm is one of the most important commercial crops, forming the backbone of some countries in South-East Asia, particularly Malaysia and Indonesia. Currently, oil palm is the major source of vegetable oil consumed in the world. The emerging techniques of biotechnology, especially in the area of genomics, are necessary to fortify and maintain the competitiveness of oil palm in the face of major challenges such as declining availability of arable land and production losses due to biotic and abiotic stresses. The full realization of the genomic revolution in agricultural crops like oil palm depends on continual successful innovation in research and development.

Although oil palm is one of the most important commodity crops worldwide, research in understanding and unraveling its genome has lagged behind other vegetable oil crops such as soybean and rapeseed. The perennial nature of the palm has most likely contributed to the lack of in-depth understanding of the molecular genetics of the crop. With its long lifecycle and large land requirement for field planting, the oil palm is in fact an ideal candidate for the application of genomic tools for value enhancement. Genomics research in oil palm has largely been geared toward the generation of diagnostic tools for application in breeding and tissue culture. The application of molecular markers in diversity analysis and genetic mapping, generation of expressed sequence tags (ESTs), and, more recently, whole genome sequencing, are some of the methods employed to enhance the understanding of the genetics of oil palm. These modern genomics tools are generating substantial new knowledge on the biological basis of phenotypic traits that can be exploited in improving productivity and obtaining better product quality.

The Genome of Oil Palm

Cytological analysis of oil palm has shown that *E. guineensis* contains $2n=32$ chromosomes (Madon et al. 1998) with an estimated genome size of $2C=3.86$ pg (Madon et al. 2008). The other species, *E. oleifera,* also contains $2n=32$ chromosomes and they are divided into three groups: Group I consisting of pair no. 1, Group II consisting of pair nos. 2–9, and Group III consisting of pair nos. 10–16. Chromosome pair no. 1 is the longest pair, pair nos. 2–9 are of medium length, and pair nos. 10–16 are medium-short chromosomes. Paired *t*-tests performed for the two species' homologous chromosome pairs showed no significant difference in chromosome lengths between them (Madon et al. 1998). This is expected as the two species can be crossed to generate hybrids.

Although most palm species including oil palm are reported to be diploids, Castilho et al. (2000) in their cytological analysis of oil palm chromosomes observed several pairs of chromosomes with similar in situ patterns, suggesting the possibility that the species is a paleotetraploid. However, as the authors pointed out, further analysis using extensive marker studies and/or identification of trisomic plants using in situ chromosome markers is required before a more definitive conclusion can be made.

Protein Marker Studies of Genetic Variation

The application of markers for genetic diversity studies in oil palm has been reported since the 1980s. Initially, the populations were studied using the simple and inexpensive isozyme assay. Analysis using nine isozyme systems (Ghesquiere 1984, 1985) revealed genetic polymorphism among the oil palm populations prospected from Angola, Zaire, Ivory Coast, Benin, Nigeria and Cameroons, as well as Deli oil palms from Indonesia and Malaysia. Subsequently, isozyme markers were applied in screening oil palm populations from Nigeria (Sapurah 1990; Rajanaidu et al. 1993) and Zaire (Tay 1989; Mohd Noor 1992; Choong et al. 1996). The isozyme markers revealed a level of polymorphism between 16% and 81% among the populations screened. The allelic frequencies were significantly different among the populations analyzed, and the degree of variation increased from the family to the ecotype levels. The variation between the populations seemed to coincide with different zones of climate and altitude, suggesting adaptation of the genotype across diverse environments.

A more comprehensive isozyme analysis of oil palm populations was carried out by Hayati et al. (2004). Six isozyme systems employed on 26 populations originating from Angola, Cameroon, Ghana, Guinea, Madagascar, Nigeria, Senegal, Sierra Leone, Tanzania, and Congo DR indicated high genetic diversity. Hayati et al. (2004) recommended that populations with high variation and unique alleles such as those originating from Nigeria, Cameroon, Sierra Leone, Guinea, Madagascar, and Senegal be given priority for oil palm improvement programs and establishment of core collections.

DNA Markers

RFLPs

The potential application for DNA markers in oil palm was demonstrated by Mayes et al. (1996) and Cheah et al. (1996) with RFLPs used to fingerprint tissue culture materials effectively. Each mother palm (known as ortet) can produce dozens if not hundreds of lines. Since each laboratory tends to tissue culture a few ortets at the same time, culture mix-up due to operator's error is unavoidable, given the large number of lines that are handled daily in the laboratory. RFLP probes developed separately at the two laboratories provided the ability to rectify any culture mix-up that was suspected to have occurred. This "quality control" tool appealed to the tissue culturists, who could now be more confident of the "products" coming out of their laboratories. This simple but effective use of RFLPs demonstrated to the oil palm community the potential practical application of marker systems in this plantation crop, and was subsequently extended to the assessment of genetic variability of oil palm populations.

A diverse set of genotypes of *E.guineensis, E. oleifera,* and their hybrids were evaluated using chloroplast, mitochondrial, and ribosomal probes (Jack et al. 1995). Probes developed from mitochondria and ribosomes were able to identify intra-specific variation and, thus, had potential application in hybrid breeding programs. However, these probes were unable to detect polymorphism within *E. guineensis.* Similarly, the work carried out earlier by Cheah et al. (1993) also indicated that the oil palm cDNA clones used as RFLP probes could distinguish *E. guineensis* and *E. oleifera* but were less capable of detecting variation within *E. guineensis.* However, Mayes et al. (1996) reported that four genomic RFLP probes could distinguish *E. guineensis* genotypes originating from Cameroon and Zaire, thus showing the potential application of RFLP probes in characterization of palms based on origin. The genomic RFLP probes appear to show more promise in their ability in distinguishing oil palm genotypes compared to those developed from cDNA and organelle genome.

The RFLP technique was also applied to study population structure of oil palm. *Tenera, dura,* and *pisifera* types from diverse origins and specific breeding programs were screened for comparative analysis (Mayes et al. 2000). Using 40 genomic RFLP probes, the *dura* palms derived from South-East Asia formed a cluster that was distinctly separated from the AVROS *pisifera* populations. The *tenera* hybrids were grouped at the intermediate position between the other two clusters. The genetic relationship revealed by the RFLP probes was consistent with the pedigree information. Furthermore, the ability to score RFLP bands as co-dominant markers also allowed estimation of heterozygosity levels in the populations. This is important in assisting oil palm breeders to select more uniform palms and draw up hybridization programs to minimize inbreeding depression.

The extent of genetic variability between and within the natural oil palm populations from Africa have also been estimated using RFLP markers (Maizura 1999; Maizura et al. 2006). Fifty-three oil palm wild populations collected from ten African countries, namely, Nigeria, Cameroon, Zaire, Tanzania, Madagascar, Angola, Senegal, Sierra Leone, Gambia, and Guinea, were analyzed with four oil palm cDNA probes. The Nigerian populations exhibited the highest levels of genetic diversity and the most number of rare alleles, suggesting that Nigeria may likely be the center of diversity for wild oil palm. High genetic differentiation within populations was observed, which is common in the case of out-crossing species like oil palm. Oil palm improvement programs as such should emphasize the use of highly variable populations. Such populations can be selected based on measures of polymorphic loci and mean number of alleles per loci obtained from the studies. In the RFLP screening, high variable populations identified include those sampled from Nigeria, Cameroon, Zaire (Congo DR), Tanzania, Angola, Senegal, Sierra Leone, and Guinea. Some of these same populations were also recommended by Hayati et al. (2004) for oil palm improvement and setting up of core collection on the basis of isozyme analysis.

RAPD

The introduction of polymerase chain reaction (PCR) techniques saw the arrival of several new methods for detecting polymorphism in plants. Initially, one of the most popular was RAPD. RAPD markers were also shown to detect genetic variation among African germplasm collection in Malaysia (Shah et al. 1994). Ninety-seven accessions representing four countries, namely, Nigeria, Cameroon, Zaire, and Tanzania, were analyzed with nine RAPD primers. The samples from the Zaire germplasm collection were divided into five sub-populations. The RAPD loci scores were transformed into Shahnon's diversity index. Low index was detected among palms from Cameroon and Population five of Zaire. Due to the costly field trials, Shah et al. (1994) suggested that these populations be given least priority with regard to field data collection and evaluation activities.

Higher numbers of RAPD primers have also been applied on a broader range of germplasm collections (Rajanaidu et al. 2000a), covering *E. guineensis* from 11 countries and *E. oleifera* from six countries. Twenty RAPD primers revealed high level of polymorphism (>80%) among the *E. guineensis* populations (except Gambia) and *E. oleifera* populations from Colombia, Costa Rica, and Panama. Low polymorphism was detected among populations from Gambia, Brazil, Honduras, and Surinam, which could be explained by the limited number of palms included in the study. Higher polymorphism was detected in the study by Rajanaidu et al. (2000a), compared to that reported by Shah et al. (1994). The increased number of RAPD primers probably allowed a wider coverage of the genome, thus increasing the possibility of detecting polymorphism. With regard to the partitioning of the total genetic diversity, RAPD markers demonstrated that the variability within was higher than between populations for both the species. This result is consistent with an earlier analysis using RFLPs (Maizura et al. 2006). The dendrogram constructed clearly separated the populations of *E. guineensis* from those of *E. oleifera.*

An analysis of 175 *E. oleifera* accessions distributed in the Brazilian Amazon Forest and 17 *E. guineensis* palms using RAPD markers was also reported by Moretzsohn et al. (2002). RAPD analysis could again largely differentiate *E. guineensis* and *E. oleifera* collections, which was also reported by Rajanaidu et al. (2000a). Moretzsohn et al. (2002) also revealed that the Brazilian *E. oleifera* collection had lower level of genetic diversity compared to the African oil palm. The study further revealed that among the *E. oleifera* collection analyzed, the total genetic variation was higher within population compared to between populations (4.8%), again similar to the results reported by other groups using RFLP and RAPD markers. UPGMA analysis also revealed that in the Amazon River basin, accessions near the same river were more related compared to accessions from other regions.

AFLPs

The oil palm populations examined via RFLP by Maizura (1999) and Maizura et al. (2006) were also analyzed using AFLPs (Kularatne et al. 2000). Eight AFLP primer combinations generated 377 bands, and about 93% of them were polymorphic. AFLP also showed high diversity among the Nigerian populations, which further supported Nigeria as the center of diversity for natural oil palm. As in the case of the RFLP analysis, AFLP screening also uncovered high genetic differentiation within populations. In terms of clustering, results from both RFLP (Maizura 1999; Maizura et al. 2006) and AFLP (Kularatne et al. 2000) analyses showed separation of populations from Central and East Africa, indicating positive relationship between genetic distance value and the geographical distance among the populations.

Using a combination of DNA and protein markers, namely isozyme and AFLPs, Purba et al. (2000) carried out diversity study on selected oil palm parental lines and their offsprings derived from oil palm populations collected from Cameroon, Ivory Coast, and Zaire. Twenty-eight Deli-based populations were also included. Using only five AFLP primer pairs, Purba et al. (2000) attained fewer scorable bands (158) and lower number of polymorphic loci (61%) than those reported by Kularatne et al. (2000). The isozyme analysis showed high levels of heterozygosity among the Deli populations, despite the exhaustive selection imposed. Similar results were reported when the Malaysian advanced breeding populations were analyzed using SSR markers (Norziha et al. 2008). Among the palms screened, Purba et al. (2000) also detected some illegitimate parental lines, which proved the usefulness of the AFLPs in selecting true-to-type oil palms for subsequent inclusion in breeding programs. The group further recommended that crossing among the populations originating from Africa should be pursued in order to exploit the heterosis expected between distant groups.

Microsatellites (SSRs)

SSRs are generally known to be highly variable, co-dominant, chromosomal-specific, and are easy to use (Parida et al. 2009). As early as 2001, this marker technology has been used successfully for measuring genetic variations among the improved germplasm of oil palm (Billotte et al. 2001; Cochard et al. 2009). SSRs used in these studies were developed from *E. guineensis* enriched SSR libraries (Billotte et al. 2001). An alternative approach is the use of bioinformatics tools to mine SSRs from the oil palm sequence collections, either deposited in the public databases or those available in-house.

At the Malaysian Palm Oil Board (MPOB), oil palm sequences generated from sequencing of expressed sequence tags (ESTs) and genomic clones served as a valuable source for mining of SSRs. SSRs were discovered from ESTs and genomic sequences of both the species, *E. guineensis* and *E. oleifera*. The application of the EST-SSRs developed from *E. guineensis* sequences in assessing genetic diversity among germplasm materials from Africa and South America was reported by Singh et al. (2008a) and Ting et al. (2010). The study revealed that similar to results obtained using RAPD analysis (Moretzsohn et al. 2002), *E. oleifera* collections showed lower level of diversity compared to *E. guineensis*. The results also demonstrated that the EST-SSRs are useful for measuring the genetic variability among the germplasm, thus enabling precise clustering of the populations evaluated. The accuracy of the clustering is comparable to other marker systems (isozyme, RFLP, RAPD, and AFLP) applied previously. A summary of diversity parameters revealed from screening of oil palm using various DNA marker types is presented in Table 10.1. In general, the genomic-SSRs have been found to be more superior in detecting higher number of alleles (A_0), when compared to isozyme, RFLP, AFLP, and EST-SSR. As reported, A_0 of more than 2.85 was detected when the genomic-SSR markers were used to genotype germplasm collected from Zaire, Cameroon, Nigeria, Angola, Benin, Ivory Coast, and Brazil (Billotte et al. 2001; Cochard et al. 2009). Comparatively, isozyme (Purba et al. 2000; Hayati et al. 2004) and RFLP (Maizura et al. 2006) markers revealed A_0 of less than 2.0. Similarly, genomic-SSRs also exhibited higher genetic diversity (Billotte et al. 2001; Cochard et al. 2009), measured in terms of H_e (expected heterozygosity) when compared to EST-SSRs. This could be due to the fact that most of the EST-SSRs are located in the coding regions that are highly conserved, thus explaining the lower level of polymorphism detected. Nevertheless, high polymorphic information content (PIC) was reported (mean = 0.80) for the dinucleotide EST-SSR repeats (Ting et al. 2010), suggesting their potential use for screening genetically narrow breeding materials, such as the commercial breeding material, Deli *dura*. For this purpose, pre-screening probably is required for selection of informative and highly polymorphic dinucleotide EST-SSRs.

Noorhariza et al. (2010) also published a preliminary study on the utilization of *E. oleifera* genomic-SSR for assessing genetic diversity among *E. oleifera* populations collected from Colombia, Costa Rica, Panama, and Honduras. This was one of the first reports on the development of genomic-SSR markers from *E. oleifera*. The results showed that the *E. oleifera* genomic-SSRs are very useful tools in the genetic and evolutionary studies of oil palm. This marker set was particularly useful for evaluating the *E. oleifera* materials where an average of 5.1 alleles per marker was reported. The marker set was more effective in revealing higher number of alleles compared to the *E. guineensis* EST-SSR that detected fewer alleles (<2.2) when used to screen *E. oleifera* germplasm from Colombia, Costa Rica, and Panama. This implies that the allelic diversity among the *E. oleifera* germplasm is more effectively revealed through the use of SSRs developed from *E. oleifera* (as the effect from ascertainment bias). Therefore, the effectiveness in assessing genetic diversity depends on the types of materials and the choice of SSRs (either EST or genomic) and the species from which the SSRs were derived.

Table 10.1 Summary of genetic diversity parameters revealed by different molecular marker types on oil palm populations

Type of marker	No. of primers/cDNA probes/isozyme systems applied	Type of sample screened	A	P	H_o	H_e	F_{is}	References
Isozyme	9	*E. guineensis* germplasm	2.21–2.55	34.3–43.7	0.228–0.465	–	–	Ghesquiere (1985)
		Breeding population	2.31	40.6	0.227	–	–	
	13	*E. oleifera* germplasm	2.21	–	0.18	0.31	0.42	Ghesquire et al. (1987)
	6	*E. guineensis* germplasm	1.43–2.0	42.9–57.1	0.054–0.255	0.109–0.258	−0.328–0.201	Hayati et al. (2004)
		Breeding population	1.43	28.6	0.123	0.186	0.163	
	4	Breeding population	1.25–2.0	–	0.125–0.563	0.102–0.496	−0.232–0.167	Purba et al. (2000)
RFLP	4 cDNA probes, 5 restriction enzymes	*E. guineensis* germplasm	1.5–1.9	41.4–67.2	–	0.168–0.232	–	Maizura et al. (2006)
		Breeding population	1.3	17.2	–	0.085	–	
RAPD	20	*E. guineensis* germplasm	–	66.7–98.7	–	–	–	Rajanaidu et al. (2000a)
		E. oleifera germplasm		8.0–94.0	–	–	–	
AFLP	8	*E. guineensis* germplasm	–	48.4–79.6	–	–	–	Kularatne et al. (2000)
		Breeding population	–	63.9	–	–	–	
	5	Breeding population	–	20.1	–	–	–	Purba et al. (2000)
Genomic SSR	21	*E. guineensis* germplasm	5.25	–	–	0.68	–	Billotte et al. (2001)
		E. oleifera germplasm	5.35	–	–	0.69	–	
	14	*E. guineensis* germplasm	2.85–10.42	–	0.632–0.728	0.580–0.797	0.044–0.140	Cochard et al. (2009)
		Breeding population	3.78	–	0.362	0.510	0.295	
EST-SSR	10	*E. guineensis* germplasm	2.20–3.20	70.0–100.0	0.213–0.460	0.328–0.442	−0.2622–0.3735	Singh et al. (2008a)
		Breeding population	2.20	70.0	0.240	0.243	0.0123	
	9	Breeding population	2.3	94.5	0.621	0.455	–	Norziha et al. (2008)
	15	*E. guineensis* germplasm	–	92.5	0.400	0.437	0.095	Ting et al. (2010)
		E. oleifera germplasm	–	60.8	0.230	0.286	0.192	

A mean number of alleles per locus, P percentage of polymorphic loci, H_o observed heterozygosity, H_e expected heterozygosity, F_{is} inbreeding coefficient

Some selected EST- and genomic-SSRs were tested for cross-amplification in the taxa Arecaceae. With respect to EST-SSRs, all the tested primers could amplify DNA extracted from coconut (*C. nucifera*) and at least one of the six exotic palms (*Euterpe oleracea, Jessenia bataua, Oenocarpus multicaulis, Ptychosperma macarthurii, Cyrtostachys renda,* and *Dictyosperma album*) (Ting et al. 2010). With respect to genomic-SSRs, the primer pairs were amplifiable in *C. nucifera* and *J. bataua*. At the sequence level, the differences in the EST-SSR amplicons were found to be mostly caused by single nucleotide polymorphisms, which had been detected at the SSR sites and the flanking regions. The transferability across oil palm species and taxa reveals the potential of both the EST-SSR and genomic-SSR markers for genetic studies in coconut and exotic palms. Furthermore, they are also suitable candidate markers for comparative studies across members in the Arecaceae.

SNPs

The development and emergence of high-throughput methods for the detection of single nucleotide polymorphisms (SNPs) has led to a revolution in their use as molecular markers. In plant systems, SNPs appear to be very frequent, although their frequency seems to vary from one species to another. In maize, a SNP frequency of one polymorphism in every 60 bp was revealed by Ching et al. (2002). Zhu et al. (2003) reported a frequency of nucleotide change of one SNP for every 270 base pair (bp) in soybean. *In silico* analysis of oil palm ESTs has revealed that this genome has a SNP frequency of 1.36/100 bp (Riju et al. 2007). Riju and Arunachalam (2009) reported that the SNP frequency is higher in *E. oleifera* (17.5/kbp) than *E. guineensis* species (16.8/kbp). The high abundance of SNPs in oil palm strongly indicates the potential of SNP markers to structure genetic diversity. In addition, even though SNPs are mostly bi-allelic, they can be easily used for genetic and association mapping, for genome-wide selection, marker-assisted breeding programs, population studies, and to obtain high-density maps.

Over 380,000 oil palm genomic sequences have been generated at MPOB from several varieties of oil palm, tagging approximately about 90% of the oil palm gene space (Budiman et al. 2005). These oil palm genomic sequences have been exploited to mine for SNPs and then to employ SNP markers to further characterize oil palm genetic resources and genetic mapping populations in oil palm. In order to evaluate the potential of using these SNPs as markers for genetic diversity studies in oil palm, SNP loci were selected for SNP-based CAPs marker development.

Ooi et al. (2010) assayed 24 SNP-based CAPS markers on 317 samples of oil palm originating from 11 African (Angola, Cameroon, Ghana, Gambia, Guinea, Madagascar, Nigeria, Senegal, Sierra Leone, Tanzania, and Zaire) and three South and Central America countries (Colombia, Panama, and Costa Rica). Six markers were informative as they revealed variations in banding patterns in the samples assayed. UPGMA cluster analysis revealed two main clusters: the *E. guineensis* populations (from the 11 African countries) were in one cluster, while the *E. oleifera* populations (South and Central America populations) were in the other.

The study by Ooi et al. (2010) also showed that the lowest genetic distance observed was between Senegal and Gambia, and the highest between Costa Rica and Angola, based on the estimates of mean genetic distances (Nei 1978) among the 14 populations. Genetic divergence among the populations tested had shown that the overall degree of differentiation (F_{ST}) among all the oil palm populations tested was 0.595. The high F_{ST} values observed in this study for these SNP-based CAPS loci indicated the ability of these markers to distinguish the different populations in the oil palm germplasm collections.

SNPs are not only more efficient and cost-effective, but are also amenable to automation and high-throughput approaches to handle large segregating populations. In view of the significance that is being attached to this marker system and in order to evaluate the potential of these technologies in oil palm, MPOB has also designed an Illumina GoldenGate assay (Illumina Inc.) to genotype oil palm germplasm collections and genetic mapping populations with a 96-plex SNP set. Results have also shown that the Illumina GoldenGate assay can be used successfully for high-throughput SNP genotyping of diverse oil palm germplasm collections and breeding populations.

DNA Markers for Breeding

Traditionally, oil palm breeders have made considerable genetic gains by carefully designed breeding and selection strategies. The most common breeding design applied is the North Carolina Mating 1 (NCM1) (Comstock and Robinson 1952), where pollen from the *pisifera* palms are crossed to a number of *dura* mother palms. This design allows the evaluation of the parental palms as well as their progeny performance (Rajanaidu et al. 2000b). One breeding cycle of the oil palm can take 10 years (Oboh and Fakorede 1989), including 3 years of immature phase and another 7 years of yield recording and collecting bunch quality component data. This allows the breeders to select palms for seed production or subsequent rounds of crossing for further improvement. Apart from the long breeding cycle, oil palm is a cross-pollinated crop requiring numerous breeding and selection to fix desirable genes in a population (Rajanaidu et al. 1999).

The long breeding and selection cycle is further confounded by the fact that oil palm requires large tracts of land for field planting and evaluation. In a conventional breeding scheme, usually about 140 palms are planted per hectare (Mayes et al. 1997), seriously limiting the number of breeding trials that can be carried out. Conducting breeding trials is also labor-intensive. These factors make conducting breeding trials for oil palm an extremely slow, laborious, and expensive exercise. Despite these constraints, oil palm breeders have achieved considerable genetic gains over the last 40 years, resulting in yield increases of up to twofold (Corley and

Lee 1992). This is even more impressive taking into account the narrow genetic base of the oil palm breeding materials. However, there remains considerable scope for improvement, especially in achieving the potential theoretical yield of about 18 tons of oil /ha/year (Corley 1983, 1998).

In the oil palm industry, it is recognized that the planting material is the most critical factor in realizing the maximum yield potential (Rajanaidu et al. 1999). As such, achieving further genetic gains through traditional breeding alone to produce the ideal planting material will probably take decades to achieve. This makes oil palm an ideal candidate for marker-assisted selection (MAS), in order to attain maximum yield potential in the shortest period of time and remain competitive in the world's oils and fats market. The use of markers (especially DNA) for selection in a crop like oil palm could greatly reduce the number of breeding cycles. As such, planting materials could be produced faster and with greater precision. The implementation of MAS requires the tagging of DNA markers to traits of interest.

The majority of the agriculturally important traits in most crop species appear to be complex quantitative traits influenced by more than one single gene. The oil palm is no exception. Identifying the individual genetic factors underlying quantitative traits (QTL) will provide the potential for improved oil palm breeding programs (Rance et al. 2001; Billotte et al. 2005).

Naturally, the most important complex trait that the breeders are eager to tag is yield. An important factor to consider here is the heritability of a trait. The concept of heritability determines the relative influence of environmental and genetic factors on a particular trait (Hartley 1988). High heritability translates to a strong genetic influence that could be tagged, and selection progress through traditional or molecular breeding is easy to obtain. In conventional breeding, yield is determined by several components such as bunch number and bunch weight and also in terms of bunch and fruit characteristics such as oil to bunch (O/B), shell to fruit (S/F), mesocarp to fruit (M/F), and kernel to fruit (K/F) ratios. The bunch weight character, at least in the selected populations analyzed, has very low heritability (Hardon et al. 1985), thus complicating the efforts to tag it accurately. Interestingly, other components, such as bunch number, M/F, S/F, and K/F ratios, were found to show relatively high heritability (Jacquemard et al. 1982; Hardon et al. 1985) and, as such, are more amenable to QTL analysis. This fact was exploited by Rance et al. (2001), who detected significant and suggestive QTL for M/F, K/F, and S/F in an F_1 population of oil palm.

Also of interest to oil palm breeders is mapping QTL associated with oil quality. Oil quality in a way is determined by the fatty acid composition of the oil. As the main economic product of oil palm is the oil, it is not surprising that improving oil quality is considered an important objective in an oil palm breeding program. In this respect, there is interest in lowering the saturated fatty acid content, e.g., palmitic acid, and increasing the monounsaturated fatty acid, oleic acid, which, if achieved, has the potential to open up markets for palm oil in the liquid oil and industrial feedstock sectors. Using an interspecific hybrid mapping population, QTL were detected for fatty acid composition (Singh et al. 2009). For oil palm, there are already ongoing efforts to alter fatty acid composition (FAC) through traditional breeding. As such, identification of the QTL associated with FAC could provide useful markers for selection for the above-mentioned program.

The limitation to the studies described above is that the QTL were detected in a single cross with a limited number of palms. Several related crosses linked by common parental palms can be used to overcome this limitation. The multi-parental design was shown to unravel QTL associated with yield and vegetative parameters more efficiently, compared to when only a single cross was used (Billotte et al. 2010). The number of QTL detected (two to six QTL/trait) was higher than that reported when a single cross was used (generally only one or two significant QTL/trait). Highly significant QTL were also detected for the reportedly low heritability trait of bunch weight in the multi-parental design.

The range of QTL reported for oil palm to date is summarized in Table 10.2. The application of these markers, especially in breeding (MAS), is ideally suited to perennial crop like oil palm, in which the economic products are not produced until several years after planting. However, almost all of the markers associated with the QTL have to be validated across materials derived from a wide range of genetic backgrounds planted in different environments. Ultimately, only validated markers would be used for selection of palms carrying favorable alleles, which allow for specific and targeted breeding around the important QTL through MAS.

Sequencing of the Oil Palm Genome: From ESTs to Whole Genome Sequencing

Expressed Sequence Tags (ESTs)

Limited genomic knowledge of oil palm had generally restricted research to phenotype-driven investigations. Prior to reduction in sequencing costs and development of next-generation sequencing (NGS) technologies, the expressed sequence tag (EST) approach was used to identify genes involved in the different regulatory processes of the oil palm. In line with the emergence of first-generation cDNA sequencing technology, cDNA libraries from various tissues of oil palm was constructed and ESTs were generated by partial sequencing of cDNA from the 5′-end. Since then, the number of oil palm ESTs in the EST database (dbEST) at the National Center for Biotechnology Information (NCBI 2010) has increased tremendously. To date, a total of 39,208 ESTs from African oil palm have been deposited in the GenBank database (NCBI 2010). NCBI also has 273 African oil palm complete coding sequences (CDS) (Table 10.3).

The majority of the published oil palm ESTs were groups of transcripts derived from tissue culture materials such as embryogenic callus (EC), non-embryogenic callus (NEC), somatic embryos (EMB), and suspension cell cultures (Ho et al. 2007; Low et al. 2008; Lin et al. 2009; Chan et al. 2010; Roowi et al. 2010). Transcript profiling across these tissues was initiated to unravel the complex molecular mecha-

Table 10.2 QTL identified for important traits at both genome and chromosome-wide significant levels in oil palm using various mapping populations

Mapping family			Marker		Trait	No. of detected QTLs (code)	Reference
cross	Size	Mating design	Type	No.			
dura × *tenera*	61-116	Multiple cross	*E.guineensis* genomic-SSR, *Cocos nucifera* genomic-SSR, morphological marker	93-253	Average bunch number/palm/year at 3-5 years	4 (Bn3_5)	Billotte et al. (2010)
					Average bunch weight at 3-5 years (kg)	5 (Bwt3_5)	
					Fresh fruit bunch yield/palm/year at 3-5 years (kg/palm/year)	3 (FFB3_5)	
					Palm oil yield/palm/year at 3-5 years (ton/ha/year)	2 (PO3_5)	
					Average bunch number/palm/year at 6-9 years	2 (Bn6_9)	
					Average bunch weight at 6-9 years (kg)	5 (Bwt6_9)	
					Fresh fruit bunch yield/palm/year at 6-9 years (kg/palm/year)	3 (FFB6_9)	
					Palm oil yield/palm/year at 6-9 years (ton/ha/year)	2 (PO6_9)	
					Average weight of the analyzed bunch (kg)	5 (aBwt)	
					Average number of fruits per bunch	1 (Fn)	
					Average weight of the fruit (g)	6 (Fwt)	
					Pulp to fruit ratio (%)	5 (%PF)	
					Palm oil to pulp ratio (%)	1 (%POP)	
					Palm oil industrial extraction rate (%)	2 (IER)	
					Iodine value	4 (I)	
					Kernel to fruit ratio (%)	2 (%KF)	
					Stem height (m)	4 (Ht)	
					Average number of leaves per crown	2 (Leaf_n)	
					Average length of the leaf L17 (cm)	6 (L17_L)	
					Petiole average width of the leaf L17 (cm)	4 (P_W)	
					Petiole average thickness of the leaf L17 (cm)	2 (P_T)	
					Average number of leaflets per leaf L17	6 (Lt_n)	
					Leaflet average length of the leaf L17 (cm)	1 (Lt_L)	
					Leaflet average width of the leaf L17 (cm)	2 (Lt_W)	

tenera × *E.oleifera*	110	Single cross	*E.guineensis* cDNA-RFLP, AFLP, genomic-SSR, EST-SSR	252	Iodine value Myristic acid content Palmitic acid content Palmitoleic acid content Stearic acid content Oleic acid content Linoleic acid content	1 (IV) 2 (C14:0) 1 (C16:0) 1 (C16:1) 3 (C18:0) 2 (C18:1) 1 (C18:2)	Singh et al. (2009)
dura × *pisifera*	87	Single cross	*E.guineensis* cDNA-RFLP, AFLP	214-395	Time to first callusing	5 (TFC)	Ting et al. (2006)
tenera × *tenera*	105	Single cross	*E.guineensis* cDNA-RFLP	116	Oil to wet mesocarp (%)	1(O/WM)	Singh et al. (2008b)
tenera × *tenera*	84	Single cross	*E.guineensis* genomic-RFLP, morphological marker	132	Fruit weight (g) Kernel: fruit (%) Shell: fruit (%) Mesocarp: fruit (%) Rachis length (m) Petiole cross section (cm^2) Bunch number/palm/year Bunch weight (kg) Fresh fruit bunch yield (kg) Oil: bunch (%) Leaf Area (m^2)	1 (FWt) 1 (KF) 1 (SF) 1 (MF) 1 (Rach) 1 (PCS) 1 (BN) 1 (BWt) 1 (FFB) 1 (OB) 1 (LeafA)	Rance et al. (2001)

Table 10.3 List of oil palm complete coding sequences (*CDS*) in GenBank

No.	Tissue	No. of complete CDS	Accession no.	Topic of research	References
1	15 weeks after anthesis (WAA) kernel	2	AF261691 AF193433	Kernel-specific genes	Cha and Shah (2001)
2	Zygotic embryo	1	AF250228	Gene expression in zygotic and somatic embryos	Morcillo et al. (2001)
3	Leafy shoots regenerated from tissue culture	1	AF322914	Epigenetic somaclonal variation	Tregear et al. (2002)
4	Young leaf	1	AY583783	Characterization of 1-deoxy-D-xylulose-5-phosphate synthase (DXS) gene	Khemvong and Suvachittanont (2005)
5	Inflorescence	13	AY739698-AY739700 AF411840-AF411848 AF227195	Characterization of MADS box genes	Adam et al. (2006)
6	Leafy shoots regenerated from tissue culture	2	AY556420 AY556421	Somaclonal variation	Morcillo et al. (2006)
7	15 WAA mesocarp	1	EU499363	Regulatory sequences with specific expression in mesocarp and/or senescent leaves	Abdullah and Ramli (2007)
8	Unspecified	1	AY550990	Internal control for RT-PCR	Adam et al. (2007)
9	Zygotic embryos	1	AY691196	Expression studies of AP2 during zygotic and somatic embryogenesis	Morcillo et al. (2007)
10	Zygotic embryos	3	EF622019-EF622021	Developmental differences between zygotic and somatic embryos	Aberlenc-Bertossi et al. (2008)
11	Non-embryogenic callus, embryogenic callus and somatic embryos	196	EU284816-EU284944 EU284946-EU284977 EU284989-EU285023	Gene expression during tissue culture	Low et al. (2008)
12	Inflorescences and callus cultures	3	EU117216-EU117218	Mantled somaclonal variation	Rival et al. (2008)
13	Suspension cell culture	1	EU795363	Characterization of an unknown protein from cell suspension culture	Thuc et al. (2008)

14	Male inflorescence	2	GQ914933 GQ914934	Mantled homeotic flowering abnormality	Beulé et al. (2010)
15	Various tissues	45	AY291337-AY291341 DQ267441-DQ267443 AF147879 DQ333323 AF207699 DQ333324 AF220453 DQ399790 AF236067 DQ400915 AF236068 DQ422858 AF251795 DQ459441 AF295636 DQ497735 AF295637 DQ497736 AY182165 DQ531848 AY254310 EF034155 AY291345 EF466008 AY323927 EU043487 AY550991 EU043488 AY739701 EU561011 AY739702 EU805510 DQ004687 EU805511 DQ090962 FJ196316 DQ267436 FJ196317 DQ267440	–	Unpublished
Total		*273*			

nisms underlying the somatic embryogenesis process in oil palm. Comprehensive analysis of these ESTs has given a global overview of expressed genes at different developmental stages of tissue culture. Several genes with differential gene expression patterns in embryogenic callus and somatic embryos were identified as potential embryogenesis markers. These promising ESTs code for lipid transfer proteins (Low et al. 2008), granule-bound starch synthase, and putative transcription factor Myb1 (Chan et al. 2010).

In addition to tissue culture related ESTs, sequencing was also performed on the inflorescence, normal apex, abnormal apex, zygotic embryo, flower, root, and mesocarp tissues of oil palm (Jouannic et al. 2005; Ho et al. 2007; Nurniwalis et al. 2008; Beulé et al. 2010). ESTs obtained from inflorescence, flower, and shoot apex were used to capture information on flower development and abnormality (Ho et al. 2007; Beulé et al. 2010). As for the ESTs from the 17 weeks after anthesis (WAA) mesocarp cDNA library, the collection of annotated genes was exploited for screening of differentially expressed transcripts in the mesocarp. This study resulted in the identification of an ethylene receptor, which is a mesocarp-specific gene with expression patterns that gradually increases with the development of oil palm fruits (Nurniwalis et al. 2008).

Nevertheless, one of the issues pertaining to random sequencing of cDNA from standard libraries is the lower chances of obtaining rarely expressed genes. Most of the time, the intermediate and highly abundant genes will be selected and sequenced. Therefore, in an attempt to increase the chances of identifying all classes of genes, a normalized cDNA library was constructed for oil palm somatic embryos (Chan et al. 2010). The library was prepared based on the reassociation kinetics reaction (Bonaldo et al. 1996), and 237 new and unique ESTs were identified due to the normalization process. This group of genes were not present in any of the previously sequenced standard cDNA libraries. Figure 10.1 shows the distribution of the oil palm ESTs that were published and submitted to dbEST.

With the burgeoning number of ESTs generated from various tissues of oil palm, there is a need to develop a database to store and manage the sequences. For this purpose, *PalmGenes* was established (Cheah et al. 2003). At this stage, 6240 ESTs have been deposited in the database, with some sequences containing additional information, such as position on genetic linkage maps and results from Northern analysis. This database is accessible at http://palmoilis.mpob.gov.my/palmgenes.html. PalmDNABase, an integrated DNA data management and analysis system, was also developed in MPOB to automatically store and curate oil palm EST sequences (Rozana et al. 2005). The database system analyzes sequencing data using tools, such as PHRED, VecMask, LUCY, CAP3, and BLAST, to validate the quality of sequences, carry out vector masking and trimming, and perform contig assembly and homology searches, and stores the sequence data and analysis results in a systematic manner.

The EST data showed the potential for gene discovery and generated an interest to sequence the oil palm genome, which was estimated to be about 1,700 Mb in size (Rival et al. 1997). A small-scale sequencing project of the oil palm genome using the *GeneThresher* (methylation filtering) technology was thus initiated. The

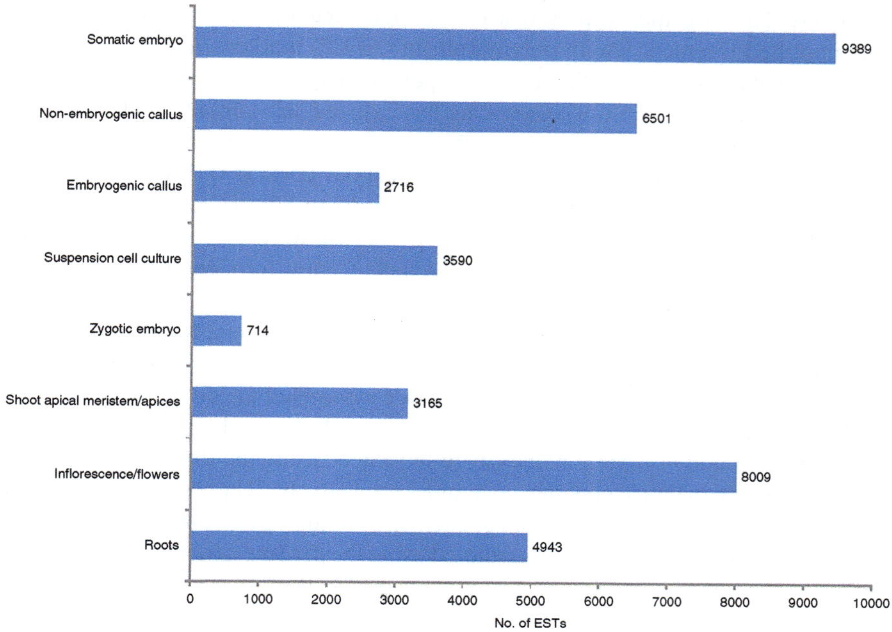

Fig. 10.1 Compilation of ESTs from six oil palm publications. The graph shows the number of ESTs obtained from somatic embryo (Low et al. 2008; Chan et al. 2010), non-embryogenic callus, embryogenic callus (Low et al. 2008), suspension cell culture (Ho et al. 2007; Lin et al. 2009), zygotic embryo, shoot apical meristem/apices, inflorescence/flower (Jouannic et al. 2005; Ho et al. 2007; Beulé et al. 2010), and root (Ho et al. 2007) tissues that were published and is available in dbEST

GeneThresher technology, which was developed by scientists at Cold Spring Harbor Laboratory, allowed the preferential selection of genomic regions that are hypomethylated. These regions encode for the genes and their regulatory regions (Rabinowicz et al. 1999). A total of 383,366 *E. guineensis* (*dura, tenera,* and *pisifera*), *E. oleifera,* and OxG hybrid *GeneThresher* library sequences were generated. A small portion of whole-genome shotgun (WGS) library sequences (15,374) were also generated to calculate the percentage of gene enrichment in the *GeneThresher* library. The sequences were assembled using PHRAP into 139,120 contigs and 99,340 singletons. The unique sequences were compared to the oil palm EST unigenes identified by Low et al. (2008) to determine the percentage of genes tagged by the *GeneThresher* sequences. The results showed that even at a low genome coverage of about 0.15 times, the sequences were able to tag 76% of the EST unigenes at a significance threshold of $1e^{-10}$ using BLASTN. This indicates that the *GeneThresher* sequences were enriched with "genespace" sequences.

The assembled 238,460 unique sequences were mined for genes of interest, SSRs, and SNPs. Data mining of the unique sequences revealed 138,811 SSRs in 86,292 sequences. The sequences also contained 48,381 high-quality SNPs. Selected

SSRs and SNPs, together with the restriction fragment length polymorphism (RFLP) and amplified fragment length polymorphism (AFLP) markers, are being actively used in genetic diversity studies as well as genetic mapping programs.

In recent years, the cost of sequencing has reduced substantially with the introduction of next-generation sequencing (NGS). The cost of sequencing a megabase of sequences using the Sanger methodology is about US$1,000 (Men et al. 2008), while the cost of NGS technologies is from US $1 to US $60, depending on the NGS platform used (Shendure and Ji 2008). This led to the sequencing of the oil palm genome by two private companies (Asiatic Centre for Genome Technology, ACGT, and Sime Darby) and a public research center (MPOB). The private entities sequenced the *dura* and *tenera* fruit types of *E. guineensis,* while MPOB sequenced three oil palm genomes from two species (*pisifera* and *dura* fruit types of *E. guineensis* and *E. oleifera*). All three entities have since assembled the oil palm genome. In MPOB, the three genomes were sequenced at a total of 68-fold genome coverage. The *E. oleifera* and *pisifera* genomes were sequenced using the Roche 454 platform at 17.2- and 16.5-times genome coverage, respectively. The *dura* genome was sequenced using Illumina short read sequencing technology to 34.7-times genome coverage. The *dura* genome was assembled using the *pisifera* genome as a reference genome. The genome has since been annotated with Refseq, rice, and *A. thaliana* genes, and the results can be viewed in MyPalm Genome Viewer, which is a custom oil palm genome browser. New SNP markers, oil palm genes and repeat sequences, as well as markers available in the oil palm genetic maps have also been mapped into the MyPalm Genome Viewer. The genome sequence will allow natural variants in oil palm selected by breeders to be mapped to individual genes. This will provide the materials for biotechnological and breeding improvement of oil palm.

Transcriptome sequencing has also been carried out to identify the gene populations in different oil palm tissues. To date, the transcriptome of 35 oil palm tissues has been completed at MPOB. The transcriptome and genome data has since been used to develop oil palm expression and SNP genotyping microarrays. The expression arrays are being used to identify genes that are associated with different processes in oil palm. These genes would eventually be selected as markers for various important biologically processes and used as selection tools to accelerate breeding programs. Genetic mapping, genome-wide association (GWA), and methylation studies can definitely be accelerated with the availability of the genomic and SNP arrays. In other plants, such as *Arabidopsis thaliana*, rice, and barley, the availability of such high-throughput platforms is already beginning to reveal its rewards through the identification of markers/genes that have strong association with agronomic traits and phenotypes (Aranzana et al. 2005; Atwell et al. 2010; Cockram et al. 2010; Huang et al. 2010). As the cost of NGS library construction and sequencing continues its downward trend and bioinformatics tools for sequence analysis become routinely available to all researchers, whole genome sequencing itself may very well be the platform that will be routinely used to identify DNA polymorphism for application in crop improvement. With the integration of bioinformatics, advanced genomic tools, and a genomics-guided breeding program, improvement of the oil palm will be further expedited.

References

Abdulla M, Gamal O (2010) Investigation on molecular phylogeny of some date palm (*Phoenix dactylifera* L.) cultivars by protein, RAPD and ISSR markers in Saudi Arabia. Aust J Crop Sci 4:23–28

Abdullah SNA, Ramli Z (2007) Regulatory sequences for regulation of gene expression in plants and other organisms, and composition, products and methods related thereto. United States Patent US 7,173,120-B2, 6 Feb 2007

Aberlenc-Bertossi F, Chabrillange N, Duval Y, Tregear J (2008) Contrasting globulin and cysteine proteinase gene expression patterns reveal fundamental developmental differences between zygotic and somatic embryos of oil palm. Tree Physiol 28:1157–1167

Adam H, Jouannic S, Morcillo F, Richaud F, Duval Y, Tregear JW (2006) MADS box genes in oil palm (*Elaeis guineensis*): patterns in the evolution of the SQUAMOSA, DEFICIENS, GLOBOSA, AGAMOUS, and SEPALLATA subfamilies. J Mol Evol 62:15–31

Adam H, Jouannic S, Orieux Y, Morcillo F, Richaud F, Duval Y, Tregear JW (2007) Functional characterization of MADS box genes involved in the determination of oil palm flower structure. J Exp Bot 58:1245–1259

Ahmed TA, Al-Qaradawi AY (2009) Molecular phylogeny of Qatari date palm genotypes using simple sequence repeats markers. Biotechnology 8:126–131

Ahmed TA, Al-Qaradawi AY (2010) Genetic diversity of date palm genotypes in Qatar as determined by SSR and ISSR markers. Acta Hortic 882:279–286

Akkak A, Scariot V, Torello Marinoni D, Boccacci P, Beltramo C, Botta R (2009) Development and evaluation of microsatellite markers in *Phoenix dactylifera* L. and their transferability to other Phoenix species. Biol Plant 53:164–166

Al-Ani B, Zaid A, Shabana H (2010) On the status of chromosomes of the date palm (*Phoenix dactylifera* L.). Acta Hortic 882:253–268

Al-Dous EK, George B, Al-Mahmoud ME, Al-Jaber MY, Wang H, Salameh YM, Al-Azwami EK, Chaludavi S, Pontaroli AC, DeBarry J, Ohlrogge J, Arondel V, Saie IJ, Suliman-Elmcer KM, Bennetzen JL, Krueger RR, Malek JA (2011) De novo genome sequencing and comparative genomics of the date palm (*Phoenix dactylifera* L.). Nat Biotechnol 29(6):521–527

Al-Jibouri AAM, Adham KM (1990) Biochemical classification of date palm male cultivars. J Hortic Sci 65:725–729

Al-Khalifa NS, Askari E (2003) Molecular phylogeny of date palm (*Phoenix dactylifera* L.) cultivars from Saudi Arabia by DNA fingerprinting. Theor Appl Genet 107:1266–1270

Al-Khalifa NS, Askari E (2007) Early detection of genetic variation in date palms propagated from tissue culture and offshoots by DNA fingerprinting. Acta Hortic 736:105–112

Allen E, Xie Z, Gustafson AM, Carrington JC (2005) MicroRNA-directed phasing during trans-acting siRNA biogenesis in plants. Cell 121:207–221

Al-Ruqaishi IA, Davey M, Alderson P, Mayes S (2008) Genetic relationships and genotype tracing in date palms (*Phoenix dactylifera* L.) in Oman, based on microsatellite markers. Plant Genet Resour: Charact Util 6:70–72

Aranzana MJ, Kim S, Zhao K, Bakker E, Horton M, Jakob K, Lister C, Molitor J, Shindo C, Tang C, Toomajian C, Traw B, Zheng H, Bergelson J, Dean C, Marjoram P, Nordborg M (2005) Genome-wide association mapping in *Arabidopsis* identifies previously known flowering time and pathogen resistance genes. PLoS Genet 1:e60

Ashburner GR (1995) Genetic markers for coconut palms. In: Oropeza C, Howard FW, Ashburner GR (eds) Lethal yellowing: research and practical aspects. Kluwer Academic Publishers, Dordrecht, pp 173–186

Ashburner GR (1999) The application of molecular markers to coconut genetic improvement. In: Oropeza C, Verdeil JL, Ashburner GR, Cardena R, Santamaria JM (eds) Current advances in coconut biotechnology. Kluwer Academic Publishers, Dordrecht, pp 33–44

Ashburner GR, Thompson WK, Halloran GM (1997) RAPD analysis of South Pacific coconut palm populations. Crop Sci 37:992–997

Asmono D, Hartana A, Guhardja E, Yahya S (1993) Genetic diversity and similarity of 35 coconut populations based on isoenzyme banding pattern analysis. Bul Pusat Penelitian Kepala Sawit 1:39–54

Asmussen CB, Baker WJ, Dransfield J (2000) Phylogeny of the palm family (*Arecaceae*) based on *rps16* intron and *trnL-trnF* plastid DNA sequences. In: Wilson KL, Morrison DA (eds) Systematics and evolution of monocots. CSIRO Publishing, Collingwood, pp 525–537

Asmussen CB, Dransfield J, Deickmann V, Barfod AS, Pintaud J-C, Baker WJ (2006) A new subfamily classification of the palm family (*Arecaceae*): evidence from plastid DNA. Bot J Linn Soc 151:15–38

Atwell S, Huang YS, Vilhjalmsson BJ, Willems G, Horton M, Li Y, Meng D, Platt A, Tarone AM, Hu TT, Jiang R, Muliyati NW, Zhang X, Amer MA, Baxter I, Brachi B, Chory J, Dean C, Debieu M, De Meaux J, Ecker JR, Faure N, Kniskern JM, Jones JDG, Michael T, Nemri A, Roux F, Salt DE, Tang C, Todesco M, Traw MB, Weigel D, Marjoram P, Borevitz JO, Bergelson J, Nordborg M (2010) Genome-wide association study of 107 phenotypes in *Arabidopsis thaliana* inbred lines. Nature 465:627–631

Azeqour M, Majhourat K, Baaziz M (2002) Morphological variations in isozyme polymorphism of date palm clones from in vitro culture acclimated and established in South Morocco. Euphytica 123:57–66

Baaziz M, Saaidi M (1988) Preliminary identification of date palm cultivars by esterase isoenzymes and peroxidase activities. Can J Bot 66:89–93

Balick MJ, Beck HT (eds) (1990) Useful palms of the world, a synoptic bibliography. Columbia University Press, New York, 724 pp

Barakat A, Tan Han D, Benslimae A-A, Rode A, Bernardi G (1999) The gene distribution in the genomes of pea, tomato and date palm. FEBS Lett 463:139–142

Barrett HC (1973) Date breeding and improvement in North America. Fruit Varieties J 27:50–55

Barrow SC (1998) A monograph of Phoenix L (Palmae: Coryphoideae). Kew Bull 53:513–575

Baudouin L, Lebrun P (2002) The development of a microsatellite kit for use with coconuts. IPGRI, Rome, 66 p

Baudouin L, Lebrun P (2009) Coconut (*Cocos nucifera* L.) DNA studies support the hypothesis of an ancient Austronesian migration from Southeast Asia to America. Genet Resour Crop Evol 56:257–262

Baudouin L, Lebrun P, Rognon F, Ritter E (2005) Use of molecular markers for coconut improvement: status and prospects. In: Batugal P, Ramanatha R, Oliver J (eds) Coconut genetic resources. International Plant Genetic Resources Institute – Regional Office for Asia, the Pacific and Oceania (IPGRI-APO), Serdang, pp 268–281

Baudouin L, Lebrun P, Konan JL, Ritter E, Berger A, Billotte N (2006) QTL analysis of fruit components in the progeny of a Rennell Island Tallcoconut (*Cocos nucifera* L.) individual. Theor Appl Genet 112:258–268

Baudouin L, Lebrun P, Berger A, Myrie W, Been B, Dollet M (2008) The Panama Tall and the Maypan hybrid coconut in Jamaica: did genetic contamination cause a loss of resistance to lethal yellowing? Euphytica 161:353–360

Beal JM (1937) Cytological studies in the genus Phoenix. Bot Gaz 99:400–407

Beaulieu JM, Moles AT, Leitch IJ, Bennett MD, Dickie JB, Knight CA (2007) Correlated evolution of genome size and seed mass. New Phytol 173:422–437

Ben Abdallah A, Stiti K, Lepoivre P, du Jardin P (2000) Identification de cultivars de palmier dattier (*Phoenix dactylifera* L.) par l'amplification aléatoire d'AND (RAPD). Cah Agric 9:103–107

Bendiab K, Baaziz M, Brakez Z, Sedra MH (1993) Correlation of isozyme polymorphism and Bayoud-disease resistance in date palm cultivars and progeny. Euphytica 65:23–32

Bendiab K, Maaziz M, Majourhat M (1998) Preliminary date palm cultivar composition of Moroccan palm groves as revealed by leaf isoenzyme phenotypes. Biochem Syst Ecol 26:71–82

Bennaceur M, Lanaud C, Chevallier MH, Bounaga N (1991) Genetic diversity of the date palm (*Phoenix dactylifera* L.) from Algeria revealed by enzyme markers. Plant Breed 107:56–69

Benoit H (1979) Isozymic variation within and between coconut populations. Paper presented during the 50th session of the FAO technical working party on coconut production, protection and processing, Manila, 3–8 Dec 1979

Benoit H, Ghesquière M (1984) Electrophrèse, compte rendu cocotier. IV. Le déterminisme génétique. CIRAD-IRHO, Montpellier

Benslimane AA, Rose A, Quétier F, Hartmann C (1994) Characterization of two minicircular plasmid-like DNAs isolated from date-palm mitochondria. Curr Genet 26:535–541

Benslimane AA, Hartmann C, Ouenzar B, Rode A (1996) Intramolecular recombination of a mitochondrial minicircular plasmid-like DNA of date-palm mediated by a set of short direct-repeat sequences. Curr Genet 29:591–593

Beulé T, Camps C, Debiesse S, Tranchant C, Dussert S, Sabau X, Jaligot E, Alwee SSRS, Tregear JW (2010) Transcriptome analysis reveals differentially expressed genes associated with the *mantled* homeotic flowering abnormality in oil palm (*Elaeis guineensis*). Tree Genet Genomes. doi:10.1007/s11295-010-0323-9

Billotte N, Risterucci AM, Barcelos E, Noyer JL, Amblard P, Baurens FC (2001) Development, characterisation, and across-taxa utility of oil palm (*Elaeis guineensis* Jacq.) microsatellite markers. Genome 44:413–425

Billotte N, Marseillac N, Brottier P, Noyer J-L, Jacquemond-Collet J-P, Moreau C, Couvreur T, Chevallier M-H, Pintaus J-C, Risterucci A-M (2004) Nuclear microsatellite markers for the date palm (*Phoenix dactylifera* L.): characterization and utility across the genus Phoenix and in other palm genera. Mol Ecol Notes 4:256–258

Billotte N, Marseillac N, Risterucci AM, Adon B, Brottier P, Baurens FC, Singh R, Herrán A, Asmady H, Billot C, Amblard P, Durand-Gasselin T, Courtois B, Asmono D, Cheah SC, Rohde W, Ritter E, Charrier A (2005) Microsatellite-based high density linkage map in oil palm (*Elaeis guineensis* Jacq.). Theor Appl Genet 110:754–765

Billotte N, Jourjon MF, Marseillac N, Berger A, Flori A, Asmady H, Adon B, Singh R, Nouy B, Potier F, Cheah SC, Rohde W, Ritter E, Courtois B, Charrier A, Mangin B (2010) QTL detection by multi-parent linkage mapping in oil palm (*Elaeis guineensis* Jacq.). Theor Appl Genet 120:1673–1687

Bonaldo MF, Lennon G, Soares MB (1996) Normalization and subtraction: two approaches to facilitate gene recovery. Genome Res 6:791–806

Borrone JW, Kuhn DN, Schnell RJ (2004) Isolation, characterization, and development of WRKY genes as useful genetic markers in *Theobroma cacao*. Theor Appl Genet 109:495–507

Bouchireb N, Clark MS (1997) The application of biotechnology to date palm culture. In: Watanabe K, Pehu E (eds) Plant biotechnology and plant genetic resources for sustainability and productivity. RG Landes Co., Austin, pp 183–195

Budiman MA, Singh R, Low ETL, Nunberg A, Citek R, Rohlfing T, Bedell JA, Lakey ND, Martienssen RA, Cheah SC (2005) Sequencing of the oil palm genespace. In: Proceedings of PIPOC 2005 international palm oil congress. Agriculture, biotechnology and sustainability conference, Kuala Lumpur, 25–29 Sept 2005, pp 628–639

Canto-Canché B, Quintal-Salazar E, Villanueva MA (1992) Biochemical markers of variety in *Cocos nucifera* L. from Yucatan. Turrialba 42:375–381

Cao BR, Chao C-CT (2002) Identification of date cultivars in California using AFLP markers. HortScience 37:966–968

Cardeña R, Oropeza C, Zizumbo D (1998) Leaf proteins as markers useful in the genetic improvement of coconut palms. Euphytica 102:81–86

Cardeña R, Ashburner GR, Oropeza C (2003) Identification of RAPDs associated with resistance to lethal yellowing of the coconut (*Cocos nucifera* L.) palm. Sci Hortic 98:257–263

Carpenter JB, Elmer HS (1978) Pests and diseases of the date palm. Agriculture handbook no. 527. US Department of Agriculture, Washington, DC, 42 pp

Carpenter JB, Ream CL (1976) Date palm breeding, a review. Date Grow Inst Rep 53:25–33

Carpio CB (1982) Biochemical studies of *Cocos nucifera* L. Kalikasan Philipp J Biol 11:319–338

Castilho A, Vershinin A, Heslop-Harrison JS (2000) Repetitive DNA and the chromosomes in the genome of oil palm (*Elaeis guineensis*). Ann Botany 85:837–844

Cha TS, Shah FH (2001) Kernel-specific cDNA clones encoding three different isoforms of seed storage protein glutelin from oil palm *Elaeis guineensis*. Plant Sci 160:913–923

Chan PL, Ma LS, Low ETL, Elyana MS, Ooi LCL, Cheah SC, Singh R (2010) Normalized embryoid cDNA library of oil palm (*Elaeis guineensis*). Electron J Biotechnol 13:14

Chao CCT, Krueger RR (2007) The date palm (*Phoenix dactylifera* L.): overview of biology, uses, and cultivation. HortScience 42:1077–1082

Chao CCT, Krueger RR (2008) *Phoenix dactylifera*. In: Janick J, Paull RE (eds) Encyclopedia of fruits and nuts. CABI, Cambridge, pp 138–150

Cheah SC, Siti Nor Akmar A, Ooi LCL, Rahimah AR, Maria M (1993) Detection of DNA variability in the oil palm using RFLP probes. In: Proceedings of the 1991 PORIM international palm oil conference: agriculture (module 1), Malaysia, pp 144–150

Cheah SC, Ooi LCL, Rahimah AR (1996) Quality control process for oil palm tissue culture using DNA probes. PORIM information series no. 40. Palm Oil Research Institute of Malaysia (PORIM), Bangi

Cheah SC, Low ETL, Rajinder S, Fatimah HAZ, Mardhiah MZ (2003). Palm genes. MPOB information series no. 186. Malaysia Palm Oil Board (MPOB), Bangi

Chevalier A (1934) La patrie des divers Elaeis, les especes et les varieties. Rev Bot Appl Agric Trop 14:187–196

Ching A, Caldwell KS, Jung M, Dolan M, Smith OS, Tingey S, Morgante M, Rafalski AJ (2002) SNP frequency, haplotype structure and linkage disequilibrium in elite maize inbred lines. BMC Genet 3:19. doi:10.1186/1471-2156-3-19

Choong Y, Shah H, Rajanaidu N, Zakri H (1996) Isoenzyme variation of Zairean oil palm (*Elaeis guineensis* Jacq.) germplasm collection. Elaeis 8:45–53

Cochard B, Adon B, Rekima S, Billotte N, de Chenon RD, Koutou A, Nouy B, Omore A, Purba AR, Glazsmann JC, Noyer JL (2009) Geographic and genetic structure of African oil palm diversity suggests new approaches to breeding. Tree Genet Genomes 5:493–504

Cockram J, White J, Zuluaga DL, Smith D, Comadran J, Macaulay M, Luo Z, Kearsey MJ, Werner P, Harrap D, Tapsell C, Liu H, Hedley PE, Stein N, Schulte D, Steuernagel B, Marshall DF, Thomas WTB, Ramsay L, Mackay I, Balding DJ, The Agoueb Consortium, Waugh R, O'sullivan DM (2010) Genome-wide association mapping to candidate polymorphism resolution in the unsequenced barley genome. Proc Natl Acad Sci USA 107:21611–21616

Cohen Y, Gurevich V, Korchinsky R (2007) Molecular and phenotypic characterization of somaclonal variation in date palm off-types originated from tissue culture. Acta Hortic 738:417–423

Comstock RR, Robinson HF (1952) Estimation of average dominance of genes. In: Gowen JW (ed) Heterosis. Iowa State College Press, Iowa, pp 494–516

Corley RHV (1983) Potential productivity of tropical perennial crops. Exp Agric 19:217–237

Corley RHV (1998) What is the upper limit to oil extraction ratio? In: Proceedings of the 1996 International oil palm conference "Oil and kernel production in oil palm - a global perspective" (eds N Rajanaidu, IE Henson and BS Jalani), Palm Oil Research Institute of Malaysia, pp 256–269

Corley RHV, Lee CH (1992) The physiological basis for genetic improvement of oil palm in Malaysia. Euphytica 60:179–184

Corner EJH (1966) The natural history of palms. Weidenfeld and Nicolson, London

Corniquel B, Mercier L (1994) Date palm (*Phoenix dactylifera* L.) cultivar identification by RFLP and RAPD. Plant Sci 101:163–172

Corniquel B, Mercier L (1997) Identification of date palm (*Phoenix dactylifera* L.) cultivars by RFLP: partial characterization of a cDNA probe that contains a sequence encoding a zinc finger motif. Int J Plant Sci 158:152–156

Darlington SD, Wylie AP (1955) Chromosome atlas of flowering plants. George Allen & Unwin Ltd., London

Dasanayaka PN, Everard JMDT, Karunanayaka EH, Nandadasa HG (2009) Analysis of coconut (*Cocos nucifera* L.) diversity using microsatellite markers with emphasis on management and utilisation of genetic resources. J Natl Sci Found Sri Lanka 37:99–109

de Blank S (1952) A reconnaissance of the American oil palm. Trop Agric Trinidad 29:90–101
DeMason DA, Sekhar KNC (1988) The breeding system in the date palm (*Phoenix dactylifera* L.) and its recognition by early cultivators. Adv Econ Bot 6:20–35
Devakumar K, Jayadev K, Rajesh MK, Chandrasekar A, Manimekalai R, Kumaran PM, Parthasarathy VA (2006) Assessment of the genetic diversity of Indian coconut accessions and their relationship to other cultivars, using microsatellite markers. Plant Genet Resour Newsl FAO-Bioversity 145:38–45
Devanand PS, Chao CT (2003a) Identification of genetic strains of 'Medjool' and 'Deglet Noor' date palm cultivars in California using amplified fragment length polymorphism (AFLP) markers. Acta Hortic 623:333–340
Devanand PS, Chao CT (2003b) Genetic variations within 'Medjool' and 'Deglet Noor' date (*Phoenix dactylifera* L.) cultivars in California detected by fluorescent-AFLP markers. J Hortic Sci Biotechnol 78:405–409
Diaz S, Pire C, Ferrer J, Bonete MJ (2003) Identification of *Phoenix dactylifera* L. varieties based on amplified fragment length polymorphism (AFLP) markers. Cell Mol Biol Lett 8:891–899
Djerbi M (1983) Diseases of the date palm. FAO regional project for the control of Bayoud in North Africa, Baghdad, 112 pp
Dong J, Chen C, Chen Z (2003) Expression profiles of the *Arabidopsis* WRKY gene superfamily during plant defense response. Plant Mol Biol 51:21–37
Donnelly RJ, Yadav J, Zaid A (2007) Sequencing the date palm genome. Acta Hortic 736:113–115
Dransfield J, Uhl NW, Asmussen CB, Baker WJ, Harley MM, Lewis CE (2008) Genera palmarum: the evolution and classification of palms. Kew Publishing, Kew, 744 p
Duran Y, Rohde W, Kullaya A, Goikoetxea P, Ritter E (1997) Molecular analysis of East African Tall coconut genotypes by DNA marker technology. J Genet Breed 51:279–288
El Modafar C (2010) Mechanisms of date palm resistance to Bayoud disease: current state of knowledge and research prospects. Physiol Mol Plant Pathol 74:287–294
El-Assar AM, Krueger RR, Devanand PS, Chao C-CT (2005) Genetic analysis of Egyptian date (*Phoenix dactylifera* L.) accessions using AFLP markers. Genet Resour Crop Evol 52:601–607
Elhoumaizi MA, Devanand PS, Fang J, Chao C-CT (2006) Confirmation of 'Medjool' date as a landrace variety through genetic analysis of 'Medjool' accessions in Morocco. J Am Soc Hortic Sci 131:403–407
Elshibli S (2009) Genetic diversity and adaptation of date palm (*Phoenix dactylifera* L.). Ph.D. dissertation, University of Helsinki
Elshibli S, Korpelainen H (2008a) Microsatellite markers reveal high genetic diversity in date palm (*Phoenix dactylifera* L.) germplasm from Sudan. Genetica 134:251–260
Elshibli S, Korpelainen H (2008b) Excess heterozygosity and scarce genetic differentiation in the populations of *Phoenix dactylifera* L.: human impact or ecological determinants. Plant Genet Resour: Charact Util 7:95–104
Elshibli S, Korpelainen H (2009) Biodiversity of date palms (*Phoenix dactylifera* L.) in Sudan: chemical, morphological and DNA polymorphisms of selected cultivars. Plant Genet Resour: Charact Util 7:194–203
Eulgem T, Somssich IE (2007) Networks of WRKY transcription factors in defense signaling. Curr Opin Plant Biol 10:366–371
Eulgem T, Rushton PJ, Robatzek S, Somssich LE (2000) The WRKY superfamily of plant transcription factors. Trends Plant Sci 5:199–206
Fang J-G, Chao CT (2007) Methylation-sensitive amplification polymorphism in date palms (*Phoenix dactylifera* L.) and their off-shoots. Plant Biol 9:526–533
Fernando WMU (1995) Patterns of isozyme variation in *Cocos nucifera* L. Proc Annu Sess Sri Lanka Assoc Adv Sci 51:84–86
Fernando WMU, Gajanayake G (1997) Patterns of isozyme variations in Coconut (*Cocos nucifera* L.) populations used for breeding improved varieties. Plant Recherche et Développement 4:256–263
Finnegan EJ, Peacock WJ, Dennis ES (2000) DNA methylation, a key regulator of plant development and other processes. Curr Opin Genet Dev 10:21723

Garcia D, Gerald JNF, Berger F (2005) Maternal control of integument cell elongation and zygotic control of endosperm growth are coordinated to determine seed size in *Arabidopsis*. Plant Cell 17:52–60

Gaut BS, Muse SV, Clark WD, Clegg MT (1992) Relative rates of nucleotide substitution at the *rbcl* locus of monocotyledonous plants. J Mol Evol 35:292–303

Gaut BS, Morton BR, McCaig BC, Clegg MT (1996) Substitution rate comparisons between grasses and palms: synonymous rate differences at the nuclear gene *adh* parallel rate differences at the plastid gene *rbcL*. Proc Nat Acad Sci USA 93:10274–10279

Geethalakshiui P, Niral V, Parthasarathy VA (2004) Allozyme variation in populations of dwarf coconut cultivars. J Plant Crops 3:13–15

Geethalakshiui P, Parthasarathy YA, Niral V (2005) Genetic diversity among coconut (*Cocos nucifera* L.) genotypes using isozymes. Asian J Plant Sci 4:678–683

Ghesquiere M (1984) Enzyme polymorphism in oil palm I: genetic control of nine enzyme systems. Oleagineux 39:561–574

Ghesquiere M (1985) Enzyme polymorphism in oil palm (*Elaeis guineensis* Jacq.). Oleagineux 40:529–540

Ghesquire M, Barcelos E, Santos MDM, Amblard P (1987) Enzymatic polymorphism in *Elaeis oleifera* H.B.K. Analysis of populations in the Amazon Basin. Oleagineux 42:143–153

Gómez-Vidal S, Salinas J, Tena M, Lopez-Llorca LV (2009) Proteomic analysis of date palm (*Phoenix dactylifera* L.) responses to endophytic colonization by entomophagous fungi. Electrophoresis 30:2996–3005

González-Pérez MA, Sosa PA (2009) Hybridization and introgression between the endemic *Phoenix canariensis* and the introduced *P. dactylifera* in the Canary Islands. Open For Sci J 7:78–85

González-Pérez MA, Caujapé-Castells J, Sosa PA (2004) Molecular evidence of hybridisation between the endemic *Phoenix canariensis* and the widespread *P. dactylifera* with random amplified polymorphic DNA (RAPD) markers. Plant Syst Evol 247:165–175

Gruezo WS, Harries HC (1984) Self-sown, wild-type coconuts in the Philippines. Biotropica 16:140–147

Gunn B (2004) The phylogeny of the Cocoeae (Arecaceae) with emphasis on *Cocos nucifera*. Ann Missouri Bot Gard 91:505–522

Guo HS, Xie Q, Fei JF, Chua NH (2005) MicroRNA directs mRNA cleavage of the transcription factor NAC1 to downregulate auxin signals for *Arabidopsis* lateral root development. Plant Cell 17:1376–1386

Gurevich V, Lavi U, Cohen Y (2005) Genetic variation in date palms propagated from offshoots and tissue culture. J Am Soc Hortic Sci 130:46–53

Hahn WJ (2002a) A molecular phylogenetic study of the Palmae (Arecaceae) based on *atpB*, *rbcL*, and 18S nrDNA sequences. Syst Biol 51:92–112

Hahn WJ (2002b) A phylogenetic analysis of the Arecoid line of palms based on plastid DNA sequence data. Mol Phylogenet Evol 23:189–204

Hamwieh A, Farah J, Moussally S, Al-Sham'aa K, Almer K, Khierallah H, Udupa S, Lababidi S, Malek JA, Aouine M, Baum M (2010) Development of 1000 microsatellite markers across the date palm (*Phoenix dactylifera* L.) genome. Acta Hortic 882:269–277

Hamza H, Elbekky M, Ben Abederrahim MA, Ferchichi A (2011) Molecular and morphological analysis of the date palm (*Phoenix dactylifera* L.) subpopulations in the Southern of Tunisia. Span J Agric Res 9:484–493

Hardon JJ, Tan GY (1969) Interspecific hybrids in the genus *Elaeis* I and II. Euphytica 18:372–380

Hardon JJ, Rao V, Rajanaidu N (1985) A review of oil palm breeding. In: Rusell GE (ed) Progress in plant breeding. Butterworths, London, pp 139–163

Harries H (1978) The evolution, dissemination and classification of *Cocos nucifera* L. Bot Rev 44:265–319

Harries HC (1995) In: Smartt J, Simmonds NW (eds) Evolution of crop plants. Longman Scientific and Technical, New York, pp 389–395

Harries HC (2001) The coconut palm (*Cocos nucifera* L). In: Last FT (ed) Tree crop ecosystems. Elsevier, New York, pp 321–338
Harries HC, Paull RE (2008) *Cocos nucifera*. In: Janick J, Paull RE (eds) Encyclopedia of fruits and nuts. CABI, Cambridge, pp 107–118
Harries H, Baudouin L, Cardeña R (2004) Floating, boating and introgression: molecular techniques and the ancestry of coconut palm populations on Pacific islands. Ethnobot Res App 2:37–53. http://www.ethnobotanyjournal.org/vol2/i1547-3465-02-037.pdf
Hartana A, Novarianto H, Asmono D (1993) Analisis keragaman dan pewarisan pola pita isozim tanaman kelapa. Jurnal Matematika Sains 1:63–76
Hartley CWS (1977) The development of groves. In: The oil palm, 2nd edn. Longman, London, pp 77–96
Hartley CWS (1988) The botany of oil palm. In: The oil palm, 3rd edn. Longman, London, pp 47–94
Hayati A, Wickneswari R, Maizura I, Rajanaidu N (2004) Genetic diversity of oil palm (*Elaeis guineensis* Jacq.) germplasm collections from Africa: implications for improvement and conservation of genetic resources. Theor Appl Genet 108:274–1284
Hemerly AS, Ferreira PJE, Van Montagu M, Engler G, Inze D (1993) cdc2a expression in Arabidopsis is linked with the competence for cell division. Plant Cell 5:1711–1723
Henderson SA, Billotte N, Pintaus J-C (2006) Genetic isolation of Cape Verde Island *Phoenix atlantica* (Arecaceae) revealed by microsatellite markers. Conserv Genet 7:213–223
Herran A, Estioko L, Becker D, Rodriguez MJB, Rohde W, Ritter E (2000) Linkage mapping and QTL analysis in coconut (*Cocos nucifera* L). Theor Appl Genet 101:292–300
Hewezi T, Howe P, Maier TR, Baum TJ (2008) *Arabidopsis* small RNAs and their targets during cyst nematode parasitism. Mol Plant Microbe Interact 21:1622–1634
Ho CL, Kwan YY, Choi MC, Tee SS, Ng WH, Lim KA, Lee YP, Ooi SE, Lee WW, Tee JM, Tan SH, Kulaveerasingam H, Alwee SSRS, Abdullah MO (2007) Analysis and functional annotation of expressed sequence tags (ESTs) from multiple tissues of oil palm (*Elaeis guineensis* Jacq.). BMC Genomics 8:381
Huang X, Wei X, Sang T, Zhao Q, Feng Q, Zhao Y, Li C, Zhu C, Lu T, Zhang Z, Li M, Fan D, Guo Y, Wang A, Wang L, Deng L, Li W, Lu Y, Weng Q, Liu K, Huang T, Zhou T, Jing Y, Li W, Lin Z, Buckler ES, Qian Q, Zhang QF, Li J, Han B (2010) Genome-wide association studies of 14 agronomic traits in rice landraces. Nat Genet 42:961–969
Jack PL, Dimitrijevic TAF, Mayes S (1995) Assessment of nuclear, mitochondrial and chloroplast RFLP markers in oil palm (*Elaeis guineensis*). Theor Appl Genet 90:643–649
Jacquemard JC, Meunier J, Bonnet F (1982) Genetic study of the reproduction of an *Elaeis guineensis* oil palm cross. In: Puspharajah E, Chew PS (eds) The oil palm in agriculture in the eighties, vol 1. International Society of Planters (ISP), Kuala Lumpur, pp 19–32
Jay M, Bourdeix R, Potier F, Sanlaville C (1988) Premiers résultats de l'étude des polyphénols foliaires du cocotier. Oléagineux 44:151–161
Johnson D (1998) Tropical palms. Food and Agriculture Organization (FAO) forestry report, Non-wood forests products no. 10. FAO, Rome
Johnson C, Cullis TA, Cullis MA, Cullis CA (2009) DNA markers for variety identification in date palm (*Phoenix dactylifera* L.). J Hortic Sci Biotechnol 84:591–594
Jouannic S, Argout X, Lechauve F, Fizames C, Borgel A, Morcillo F, Aberlenc-Bertossi F, Duval Y, Tregear J (2005) Analysis of expressed sequence tags from oil palm (*Elaeis guineensis*). FEBS Lett 579:2709–2714
Jubrael JMS, Udupa SM, Baum M (2005) Assessment of AFLP-based genetic relationships among date palm (*Phoenix dactylifera* L.) varieties of Iraq. J Am Soc Hortic Sci 130:442–447
Kaeppler SM, Kaeppler HF, Rhee Y (2000) Epigenetic aspects of somaclonal variation in plants. Plant Mol Biol 43:179–188
Karp A (1999) The use of polymorphic microsatellites for assessing genetic diversity in coconut. In: Oropeza C, Verdeil JL, Ashburner GR, Cardena R, Santamaria JM (eds) Current advances in coconut biotechnology. Kluwer Academic Publishers, Dordrecht, pp 121–129

Khemvong S, Suvachittanont W (2005) Molecular cloning and expression of a cDNA encoding 1-deoxy-D-xylulose-5-phosphate synthase from oil palm in *Elaeis guineensis* Jacq. Plant Sci 169:571–578

Knutzon DS, Lardizabal KD, Nelsen JS, Bleibaum JL, Davies HM, Metz JC (1995) Cloning of a coconut endosperm cDNA encoding a 1-acyl-sn-alycerol-3-phosphate acyltransferase that accepts medium-chain-length substrates. Plant Physiol 109:999–1006

Krueger R (2011) Date palm germplasm. In: Jain SM, Al-Khayari JM, Johnson D (eds) Date palm biotechnology. Springer, Berlin, pp 313–336

Kularatne RS, Shah FH, Rajanaidu N (2000) Investigation of genetic diversity in African natural oil palm populations and Deli using AFLP markers. In: Proceedings of the international symposium on oil palm genetic resources and their utilisation. Malaysian Palm Oil Board, Bangi, pp 497–555

Kumar N, Modi AR, Singh AS, Gajera BB, Patel AR, Patem MP, Subhash N (2010) Assessment of genetic fidelity of micropropagated date palm (*Phoenix dactylifera* L.) plants by RAPD and ISSR markers assay. Physiol Mol Biol Plants 16:207–213

Kunert KJ, Vorster J, Bey E, Cullis CA (2000) Representational difference analysis as a DNA-based quality assurance program for date palm micropropagation. In: Proceedings of the date palm international symposium, Windhoel, Namibia, pp 72–80

Kunert KJ, Baaziz M, Cullis CA (2003) Techniques for determination of true-to-type date palm (*Phoenix dactylifera* L.) plants: a literature review. Emirates J Agric Res 15:1–16

Lacze P, Giraud-Henry I, Brackpool A (2000) Molecular fingerprinting of date palm cultivars using AFLP. In: Proceedings of the date palm international symposium, Windhoek, Namibia, Feb 2000, pp 89–95

Latiff A (2000) The biology of the genus *Elaeis*. In: Basiron Y, Jalani BS, Chan KW (eds) Advances in oil palm research, vol 1. Malaysian Palm Oil Board (MPOB), Bangi, pp 19–38

Lebrun P, Grivet L, Baudouin L (1998a) The spread and domestication of the coconut palm in the light of RFLP markers. Dissémination et domestication du cocotier à la lumière des marqueurs RFLP. Plant Recherche et développement 5:233–245

Lebrun P, N'Cho YP, Seguin M, Grivet L, Baudouin L (1998b) Genetic diversity in coconut (*Cocos nucifera* L.) revealed by restriction fragment length polymorphism (RFLP) markers. Euphytica 101:103–108

Lebrun P, Grivet L, Baudouin L (1999) Use of RFLP markers to study the diversity of the coconut palm. In: Oropeza C, Verdeil JL, Ashburner GR, Cardena R, Santamaria JM (eds) Current advances in coconut biotechnology. Kluwer Academic Publishers, Dordrecht, pp 73–89

Lebrun P, Baudouin L, Bourdeix R, Konan JL, Barker JH, Aldam C, Herran A, Ritter E (2001) Construction of a linkage map of the Rennell Island Tall coconut type (*Cocos nucifera* L.) and QTL analysis for yield characters. Genome 44:962–970

Lebrun P, Berger A, Hodgkin T, Baudouin L (2005) Biochemical and molecular methods for characterizing coconut diversity. In: Batugal P, Ramanatha R, Oliver J (eds) Coconut genetic resources. International Plant Genetic Resources Institute – Regional Office for Asia, the Pacific and Oceania (IPGRI-APO), Serdang, pp 225–247

Lebrun P, Baudouin L, Myrie W, Berger A, Dollet M (2008) Recent lethal yellowing outbreak: why is the Malayan Yellow Dwarf coconut no longer resistant in Jamaica? Tree Genet Genomes 4:125–131

Li D, Fan YM (2008) Construction and characterization of a cDNA library from the pulp of coconut (*Cocos nucifera* L.). Agric Sci China 7:1071–1076

Li D, Zheng Y, Wan L, Zhu X, Wang Z (2009) Differentially expressed microRNAs during solid endosperm development in coconut (*Cocos nucifera* L.). Sci Hortic 122:666–669

Lin HC, Morcillo F, Dussert S, Tranchant-Dubreuil C, Tregear JW, Tranbarger TJ (2009) Transcriptome analysis during somatic embryogenesis of the tropical monocot *Elaeis guineensis*: evidence for conserved gene functions in early development. Plant Mol Biol 70:173–192

Low ETL, Alias H, Boon SH, Shariff EM, Tan CYA, Ooi LCL, Cheah SC, Raha AR, Wan KL, Singh R (2008) Oil palm (*Elaeisguineensis* Jacq.) tissue culture ESTs: identifying genes associated with callogenesis and embryogenesis. BMC Plant Biol 8:62

Luo M, Dennis ES, Berger F, Peacock WJ, Chaudhury A (2005) MINISEED3 (MINI3), a WRKY family gene, and HAIKU2 (IKU2), a leucine-rich repeat (LRR) KINASE gene, are regulators of seed size in *Arabidopsis*. Proc Natl Acad Sci USA 102:17531–17536

Madon M, Clyde MM, Cheah SC (1998) Cytological analysis of *Elaeis guineensis* and *Elaeis oleifera* chromosomes. JOPR 10:68–91

Madon M, Phoon LQ, Clyde MM, Mohd DA (2008) Application of flow cytometry for estimation of nuclear DNA content in *Elaeis*. JOPR 20:447–452

Maizura I (1999) Genetic variability of oil palm (*Elaeis guineensis* Jacq) germplasm collections using restriction fragment length polymorphism (RFLP) markers. Ph.D. dissertation, National University Malaysia (Universiti Kebangsaan Malaysia)

Maizura I, Rajanaidu N, Zakri AH, Cheah SC (2006) Assessment of genetic diversity in oil palm (*Elaeis guineensis* Jacq.) using restriction fragment length polymorphism (RFLP). Genet Resour Crop Evol 53:187–195

Malek JA (2010) Next generation DNA sequencing applied to the date palm tree (*Phoenix dactylifera* L.). Acta Hortic 882:249–252

Manimekalai R, Nagarajan P (2006a) Interrelationships among coconut (*Cocos nucifera* L.) accessions using RAPD technique. Genet Resour Crop Evol 53:1137–1144

Manimekalai R, Nagarajan P (2006b) Assessing genetic relationships among coconut (*Cocos nucifera* L.) accessions using inter simple sequence repeat markers. Sci Hortic 108:49–54

Manimekalai R, Nagarajan P (2010) Bulk line analysis in coconut (*Cocos nucifera* L.) for inferring relationship between Talls, Dwarfs and Niu Leka dwarf forms. Indian J Plant Genet Resour 23:77–81

Marqués J, Fadda ZGN, Duran-Vila N, Flores R, Bové JM, Darós JA (2008) A set of novel RNAs transcribed from the chloroplast genome accumulates in date palm leaflets affected by brittle leaf disease. Phytopthology 98:337–344

Martinez MC, Jorgensen JE, Lawton MA, Lamb CJ, Doerner PW (1992) Spatial pattern of cdc2 expression in relation to meristem activity and cell proliferation during plant development. Proc Natl Acad Sci USA 89:7360–7364

Martinez RT, Baudouin L, Berger A, Dollet M (2010) Characterization of the genetic diversity of the Tall coconut(*Cocos nucifera* L.) in the Dominican republic using microsatellite (SSR) markers. Tree Genet Genomes 6:73–81

Mauro-Herrera M, Meerow AW, Borrone JW, Kuhn DN, Schnell RJ (2006) Ten informative markers developed from WRKY sequences in coconut (*Cocos nucifera*). Mol Ecol Notes 6:904–906

Mauro-Herrera M, Meerow AW, Borrone JW, Kuhn DN, Schnell RJ (2007) Usefulness of WRKY gene-derived markers for assessing genetic population structure: an example with Florida coconut cultivars. Sci Hortic 115:19–26

Mauro-Herrera M, Meerow AW, Perera L, Russell J, Schnell RJ (2010) Ambiguous genetic relationships among coconut (*Cocos nucifera* L.) cultivars: the effects of outcrossing, sample source and size, and method of analysis. Genet Resour Crop Evol 57:203–217

Mayes S, James C, Horner SF, Jack PL, Corley RHV (1996) The application of restriction fragment length polymorphism for the genetic fingerprinting of oil palm (*Elaeis guineensis*). Mol Breed 2:175–180

Mayes S, Jack PL, Marshall DF, Corley RHV (1997) Construction of a RFLP genetic linkage map for oil palm (*Elaeis guineensis* Jacq.). Genome 40:116–122

Mayes S, Jack PL, Corley RHV (2000) The use of molecular markers to investigate the genetic structure of an oil palm breeding program. Heredity 85:288–293

Meerow AW, Wisser RJ, Brown SJ, Kuhn DN, Schnell RJ, Broschat TK (2003) Analysis of genetic diversity and population structure within Florida coconut (*Cocos nucifera* L.) germplasm using microsatellite DNA, with special emphasis on the Fiji Dwarf cultivar. Theor Appl Genet 106:715–726

Meerow AW, Noblick L, Borrone JW, Couvreur TLP, Mauro-Herrera M, Hahn WJ, Kuhn DN, Nakamura K, Oleas NH, Schnell RJ (2009) Phylogenetic analysis of seven WRKY genes across the palm subtribe Attaleinae (Arecaceae) identifies *Syagrus* as sister group of the coconut. PLoS One 4:e7353. doi:10.1371/journal.pone.0007353

Men AE, Wilson P, Siemering K, Forrest S (2008) Sanger DNA sequencing. In: Janitz M (ed) Next-generation genome sequencing: towards personalized medicine. Wiley-VCH, Weinheim, pp 3–11

Michielse CB, Rep M (2009) Pathogen profile update: *Fusarium oxysporum*. Mol Plant Pathol 10:311–324

Mohd Noor Mahat (1992) Analisa polimorfisme isoenzim esterase, glutamat-oksaloasetat, transaminase dan malat dehidrogenase dalam koleksi germplasma kelapa sawit Zaire. Dissertation, Universiti Kebangsaan Malaysia

Mohsenzadeh S, Pasalari H (2010) Identification of a male-specific RAPD marker in *Phoenix dactylifera*. J Hortic Sci Biotechnol 85:144–146

Montero-Cortés M, Rodríguez-Paredes F, Burgeff C, Pérez-Nuñez T, Córdova I, Oropeza C, Verdeil J-L, Sáenz L (2010) Characterisation of a cyclin-dependent kinase (CDKA) gene expressed during somatic embryogenesis of coconut palm. Plant Cell Tissue Organ Cult 102:251–258

Moore HE Jr (1963) An annotated checklist of cultivated palms. Principes 7:119–182

Moore HE Jr (1971) Additions and corrections to "An annotated checklist of cultivated palms". Principes 15:102–106

Morcillo F, Hartmann C, Duval Y, Tregear JW (2001) Regulation of 7S globulin gene expression in zygotic and somatic embryos of oil palm. Physiol Plant 112:233–243

Morcillo F, Gagneur C, Adam H, Richaud F, Singh R, Cheah SC, Rival A, Duval Y, Tregear JW (2006) Somaclonal variation in micropropagated oil palm. Characterization of two novel genes with enhanced expression in epigenetically abnormal cell lines and in response to auxin. Tree Physiol 26:585–594

Morcillo F, Gallard A, Pillot M, Jouannic S, Aberlenc-Bertossi F, Collin M, Verdeil JL, Tregear JW (2007) EgAP2-1, an AINTEGUMENTA-like (AIL) gene expressed in meristematic and proliferating tissues of embryos in oil palm. Planta 226:1353–1362

Moretzsohn MC, Ferreira MA, Amaral ZPS, Coelho PJA, Grattapaglia D, Ferreira ME (2002) Genetic diversity of Brazilian oil palm (*Elaeis oleifera* HBK) germplasm collected in the Amazon Forest. Euphytica 124:35–45

Namsi A, Marqués J, Fadda Z, Takrouni ML, Ban Mahamoud O, Zouba A, Darós FR, Bové JM, Duran-Vila N (2006) Diagnosis of "maladie des feuilles cassantes" or brittle leaf disease of date palms by detection of associated chloroplast encoded double stranded RNAs. Mol Cell Probes 20:366–370

Namsi A, Montarone M, Serra P, Ben Mahamoud O, Takrouni ML, Zouba A, Khoualdia O, Bové JM, Duran-Vila N (2007) Manganese and brittle leaf disease of date palms. J Plant Pathol 89:125–136

NCBI (2010) dbEST release 120110: summary by organism. http://www.ncbi.nlm.nih.gov/dbEST/dbEST_summary.html. Accessed 1 Dec 2010

Nei M (1978) Estimation of average heterozygosity and genetic distance from a small number of individuals. Genetics 89:583–590

Niral V, Geethalakshmi P, Parthasarathy VA (2007) Intrapopulation allelomorphism in tall and dwarf populations of the coconut. Acta Bot Croat 66:35–42

Nixon RW (1959) Pollination, breeding, and selection of date palms. First FAO international technical meeting on date palm production and processing, Tripoli. FAO, Rome, 5–11 Dec 1959

Nixon RW, Furr JR (1965) Problems and progress in date breeding. Date Grow Inst Rept 42:2–5

Noorhariza MZ, Ismanizan I, Rozana R, Ting NC, Singh R (2010) Development and characterization of *Elaeis oleifera* microsatellite markers. Sains Malaysiana 39:909–912

Norziha A, Rafii MY, Maizura I, Ghizan S (2008) Genetic variation among oil palm parent genotypes and their progenies based on microsatellite markers. JOPR 20:533–541

Nurniwalis AW, Suhaimi N, Siti Nor Akmar A, Aminah S, Manaf MA (2008) Gene discovery via expressed sequence tags from the oil palm (*Elaeis guineensis Jacq.*) mesocarp. JOPR Spec Issue Malaysia-MIT Biotechnol Partnersh Program (Oil Palm Metabol Eng) 2:87–96

Oboh BO, Fakorede MAB (1989) Optimum time for yield evaluation and selection in the oil palm (*Elaeis guineensis* Jacq.). Oleagineux 44:509–513

Ooi LCL, Ong PW, Ithnin M, Singh R (2010) Discovery of oil palm SNPs and their application in the characterization of oil palm genetic resources and breeding materials. In: Proceedings of the national biotechnology seminar 2010, Kuala Lumpur, 24–26 May 2010

Ori N, Cohen AR, Etzioni A, Brand A, Yanai O, Shleizer S, Menda N, Amsellem Z, Efroni I, Pekker I, Alvarez JP, BlumE ZD, Eshed Y (2007) Regulation of LANCEOLATE by miR319 is required for compound-leaf development in tomato. Nat Genet 39:787–791

Ouenzar B, Trifi M, Boucharine B, Hartmann C, Marrakchi M, Benslimae AA, Rode A (2001) A mitochondrial molecular marker of resistance to Bayoud disease in date palm. Theor Appl Genet 103:366–370

Ould Mohamed Salem A, Trifi M, Sakka H, Rhouma A, Marrakchi M (2001a) Genetic inheritance analysis of four enzymes in date-palm (*Phoenix dactylifera* L.). Genet Resour Crop Evol 48:361–368

Ould Mohamed Salem A, Trifi M, Salhi-Hannachi A, Rhouma A, Marrakchi N (2001b) Genetic variability analysis of Tunisian date-palm (*Phoenix dactylifera* L.) cultivars. J Genet Breed 55:269–278

Ould Mohamed Salem A, Rhouma S, Zehdi S, Marrakchi M, Trifi M (2007) Molecular characterization of Mauritanian date palm cultivars using plasmid-like DNAs markers. Biol Plant 51:169–172

Parida SK, Kalia SK, Kaul S, Dalal V, Hemaprabha G, Selvi A, Pandit A, Singh A, Gaikwad K, Sharma TR, Srivastava PS, Singh NK, Mohapatra T (2009) Informative genomic microsatellite markers for efficient genotyping applications in sugarcane. Theor Appl Genet 118:327–338

Perera L (2002) Chloroplast DNA variation in coconut is opposite to its nuclear DNA variation. CORD XVIII:56–72

Perera L, Russell JR, Provan J, McNichol JW, Powell W (1998) Evaluating genetic relationships between indigenous coconut (*Cocos nucifera* L.) accessions from Sri Lanka by means of AFLP profiling. Theor Appl Genet 96:545–550

Perera L, Russell JR, Provan J, Powell W (1999) Identification and characterization of microsatellite loci in coconut (*Cocos nucifera* L.) and the analysis of coconut populations in Sri Lanka. Mol Ecol 8:344–346

Perera L, Russell JR, Provan J, Powell W (2000) Use of microsatellite DNA markers to investigate the level of genetic diversity and population genetic structure of coconut (*Cocos nucifera* L.). Genome 43:15–21

Perera L, Russell JR, Provan J, Powell W (2003) Studying genetic relationships among coconut varieties/populations using microsatellite markers. Euphytica 132:121–128

Purba AR, Noyer JL, Baudouin L, Perrier X, Hamon S, Lagoda PJL (2000) A new aspect of genetic diversity of Indonesian oil palm (*Elaeis guineensis* Jacq.) revealed by isoenzyme and AFLP markers and its consequences for breeding. Theor Appl Genet 101:956–961

Rabinowicz PD, Schutz K, Dedhia N, Yordan C, Parnell LD, Stein L, Mccombie WR, Martienssen RD (1999) Differential methylation of genes and retrotransposons facilitates shotgun sequencing of the maize genome. Nature Genet 23:305–308

Rajanaidu N (1985) *Elaeis oleifera* collection in Central and South America. In: Proceedings of international workshop on oil palm germplasm and utilisation. Palm Oil Research Institute of Malaysia (PORIM), Bangi, 26–27 Mar 1985, pp 84–94

Rajanaidu N (1990) Major developments in oil palm (*Elaeis guineensis*) breeding. In: Proceedings of the 12th plenary meeting of association for the taxonomic study of the flora of tropical Africa (AETFAT). Mitteilungen aus dem Staatsinstitut für Allgemeine Botanik, Hamburg, pp 39–52

Rajanaidu N, Sapurah R, Rao V, Zakri AH, Md. Noor Embi (1993) Isoenzyme variation in Nigerian oil palm (*Elaeis guineensis*) germplasm. In: Proceedings of the 1993 ISOPB international symposium on recent development in oil palm tissue culture and biotechnology, pp 209–216

Rajanaidu N, Kushiari A, Jalani BS (1999) Systems approach to the management of oil palm breeding and biotechnology. In: Darus A, Chan KW, Sharifah SRSA (eds) Proceedings of the 1999 PORIM international palm oil congress (PIPOC). Palm Oil Research Institute of Malaysia (PORIM), Bangi, 1–6 Feb 1999, pp 79–89

Rajanaidu N, Kushairi A, Rafii M, Mohd Din A, Maizura I, Jalani BS (2000a) Oil palm breeding and genetic resources. In: Basiron Y, Jalani BS, Chan KW (eds) Advances in oil palm research, vol 1. Malaysian Palm Oil Board (MPOB), Bangi, pp 171–237

Rajanaidu N, Maizura I, Cheah SC (2000b) Screening of oil palm natural populations using RAPD and RFLP molecular markers. In: Proceedings of the international symposium on oil palm genetic resources and their utilisation. Malaysian Palm Oil Board, Bangi, pp 556–584

Rance KA, Mayes S, Price Z, Jack PL, Corley RHV (2001) Quantitative trait loci for yield components in oil palm (*Elaeis guineensis* Jacq.). Theor Appl Genet 103:1302–1310

Read RW (1966) New chromosome counts in the Palmae. Principes 10:55–61

Ream CL, Carpenter JB (1975) Date palm breeding in California. In: Third FAO technical conference on the improvement of date production, processing, and marketing, Baghdad. FAO, Rome, 30 Nov–4 Dec 1975

Rhouma S, Zehdi-Azouzi S, Dakhlaoui-Dkhil S, Ould Mohaned Salem A, Othmani A, Cherif E, Marrakchi M, Trifi M (2010) Genetic variation in the Tunisian date palm (*Phoenix dactylifera* L.). In: Ramawat KG (ed) Desert plants. Springer, Berlin, pp 355–369

Riedel M, Riederer M, Becker D, Herran A, Kullaya A, Arana-López G, Peña-Rodríguez L, Billotte N, Sniady V, Rohde W, Ritter E (2009) Cuticular wax composition in *Cocos nucifera* L.: physicochemical analysis of wax components and mapping of their QTLs onto the coconut molecular linkage map. Tree Genet Genomes 5:53–69

Riju A, Arunachalam V (2009) Interspecific differences in single nucleotide polymorphisms (SNPs) and indels in expressed sequence tag libraries of oil palm *Elaeis guineensis* and *E. oleifera*. Nature proceedings. http://hdl.handle.net/10101/npre.2009.3593.2

Riju A, Chandrasekar A, Arunachalam V (2007) Mining for single nucleotide polymorphisms and insertions/deletions in expressed sequence tag libraries of oil palm. Bioinformation 2:128–131

Ritter E, Rodriguez MJB, Herran A, Estoiko L, Becker D, Rohde W (2000) Analysis of quantitative trait loci (QTL) based on linkage maps in coconut (*Cocos nucifera* L.). In: Arenciba A (ed) Plant genetic engineering towards the third millennium. Elsevier Science BV, Amsterdam, pp 42–48

Rival A, Beule T, Barre P, Hamon S, Duval Y, Noirot M (1997) Comparative flow cytometric estimation of nuclear DNA content in oil palm (*Elaeis guineensis* Jacq.) tissue cultures and seed-derived plants. Plant Cell Rep 16:884–887

Rival A, Jaligot E, Beule T, Finnegan EJ (2008) Isolation and expression analysis of genes encoding MET, CMT, and DRM methyltransferases in oil palm (*Elaeis guineensis* Jacq.) in relation to the 'mantled' somaclonal variation. J Exp Bot 59:3271–3281

Rivera R, Edwards KJ, Barker JHA, Arnold GM, Ayad G, Hodgkin T, Karp A (1999) Isolation and characterization of polymorphic microsatellites in *Cocos nucifera* L. Genome 42:668–675

Rohde W, Kullaya A, Rodriguez J, Ritter E (1995) Genome analysis of *Cocos nucifera* L. by PCR amplification of spacer sequences separating a subset of *Copia*-like *Eco*R1 repetitive elements. J Genet Breed 49:179–186

Rohde W, Becker D, Kullaya A, Rodriguez J, Herran A, Ritter E (1999) Analysis of coconut germplasm biodiversity by DNA marker technologies and construction of a genetic linkage map. In: Oropeza C, Verdeil JL, Ashburner GR, Cardena R, Santamaria JM (eds) Current advances in coconut biotechnology. Kluwer Academic Publishers, Dordrecht, pp 99–120

Roowi SH, Ho CL, Alwee SSRS, Abdullah MO, Napis S (2010) Isolation and characterization of differentially expressed transcripts from the suspension cells of oil palm (*Elaeis guineensis* Jacq.) in response to different concentration of auxins. Mol Biotechnol 46:1–19

Rozana R, Low ETL, Ooi LCL, Singh R (2005) PalmDNABase, an automated workflow for sequence analysis. In: Proceedings of the biotechnology of plantation commodities: biotechnology – emerging frontier for plantation commodities, Seri Kembangan, Selangor, 9–11 Aug 2005, pp 456–564

Saïdi MN, Gargouri-Bouzid R, Rayanni M, Drira N (2009) Optimization of RNA isolation from brittle leaf disease affected date palm leaves and construction of a subtractive DNA library. Mol Biotechnol 41:63–68

Saidi MN, Ladouce N, Hadri R, Grima-Pettenati J, Drira N, Gargouri-Bouzid R (2010) Identification and characterization of differentially expressed ESTs in date palm leaves affected by brittle leaf disease. Plant Sci 179:325–332

Saitou N, Nei M (1987) The neighbor-joining method: a new method for reconstructing phylogenetic trees. Mol Biol Evol 4:406–425

Saker MM, Moursy HA (1999) Molecular characterization of Egyptian date palm cultivars: RAPD fingerprints. Arab J Biotechnol 2:71–78

Saker MM, Bekheet SA, Taha HS, Fahmy AS, Moursy HA (2000) Detection of somaclonal variations in tissue culture-derived date palm plants using isoenzyme analysis and RAPD fingerprints. Biol Plant 43:347–351

Saker MM, Adawy SS, Mohamed AA, El-Itriby HA (2006) Monitoring of cultivar identity in tissue culture-derived date palms using RAPD and AFLP analysis. Biol Plant 50:198–204

Sakka H, Trifi M, Ould Mohamed Salem A, Rhouma A, Marrakchi M (2000) Rapid construction of a random genomic library from date palm (*Phoenix dactylifera* L.). Plant Mol Biol Rep 17:1–7

Sakka H, Zehdi S, Ould Mohamed Salem A, Rhouma A, Marrakchi M, Trifi M (2004) Genetic polymorphism of plastid DNA in Tunisian date-palm germplasm (*Phoenix dactylifera* L.) detected with PCR-RFLP. Genet Resour Crop Evol 51:479–487

Salman M, Al-Jibouri M, Al-Quadhy K, Omar S (1988) Isozyme and chromosomal analysis of tissue culture derived date palms. Date Palm J 6:401–411

Sapurah R (1990) Variasi isoenzim germplasma kelapa sawit (*Elaeis guineensis* Jacq.) dari Nigeria. Dissertation, Universiti Kebangsaan Malaysia

Sarkar SK (1957) Sex chromosomes in palms. Genet Iber 9:133–142

Schultes RE (1990) Taxonomic nomenclature and ethnobotanic notes on *Elaeis*. Elaeis 2:172–187

Schultes RE, Raffauf RF (1990) The healing forest. Dioscorides Press, Portland, pp 500

Sedra MH, Lashermes P, Trouslot P, Combes M-C, Hamon S (1998) Identification and genetic diversity analysis of date palm (*Phoenix dactylifera* L.) varieties from Morocco using RAPD markers. Euphytica 103:75–82

Sghaier-Hammami B, Drira N, Jorrín-Novo JV (2009a) Comparative 2-DE proteomic analysis of date palm (*Phoenix dactylifera* L.) somatic and zygotic embryos. J Proteomics 73:161–177

Sghaier-Hammami B, Valledor L, Drira N, Jorrín-Novo JV (2009b) Proteomic analysis of the development and germination of date palm (*Phoenix dactylifera* L.) zygotic embryos. Proteomics 9:2543–2554

Shah FH, Rasid O, Simons AJ, Dunsdon A (1994) The utility of RAPD markers for the determination of genetic variation in oil palm (*Elaeis guineensis*). Theor Appl Genet 89:713–718

Shalini KV, Manjunatha S, Lebrun P, Berger A, Baudouin L, Pirany N, Ranganath RM, Theertha PD (2007) Identification of molecular markers associated with mite resistance in coconut (*Cocos nucifera* L.). Genome 50:35–42

Shendure J, Ji H (2008) Next-generation DNA sequencing. Nat Biotechnol 26:1135–1145

Siljac-Yakovlev S, Benmalek S, Cerbah M, Coba de la Peña T, Bounaga N, Brown SC, Sarr A (1996) Chromosomal sex determination and heterochromatin structure in date palm. Sex Plant Reprod 9:127–132

Singh R, Noorhariza MZ, Ting NC, Rozana R, Tan SG, Low ETL, Maizura I, Cheah SC (2008a) Exploiting an oil palm EST database for the development of gene-derived SSR markers and their exploitation for assessment of genetic diversity. Biologia 63:227–235

Singh R, Tan SG, Panandam J, Abdul Rahman R, Cheah SC (2008b) Identification of cDNA-RFLP markers and their use for molecular mapping in oil palm (*Elaeis guineensis*). Asia Pac J Mol Biol Biotechnol 16:53–63

Singh R, Tan SG, Panandam JM, Abdul Rahman R, Ooi LCL, Low ETL, Sharma M, Jansen J, Cheah SC (2009) Mapping quantitative trait loci (QTLs) for fatty acid composition in an interspecific cross of oil palm. BMC Plant Biol 9:114

Soh AC, Kee KK, Goh KJ, Ang BN, Ooi LH (2008) *Elaeis guineensis*. In: Janick J, Paull RE (eds) Encyclopedia of fruits and nuts. CABI, Cambridge, pp 118–127

Sunkar R, Kapoor A, Zhu JK (2006) Posttranscriptional induction of two Cu/Zn superoxide dismutase genes in *Arabidopsis* is mediated by downregulation ofmiR398 and important for oxidative stress tolerance. Plant Cell 18:2051–2065

Tanksley SD, Orton TJ (eds) (1983) Isozymes in plant genetics and breeding (Developments in plant genetics and breeding, parts 1A & 1B). Elsevier, New York

Tay HH (1989) Analisis polimofisme isoenzim esterase dalam koleksi germplasma kelapa sawit Zaire. Dissertation, Universiti Kebangsaan Malaysia

Teulat B, Aldam C, Trehin R, Lebrun P, Barker JHA, Arnold GM, Karp A, Baudouin L, Rognon F (2000) An analysis of genetic diversity in coconut (*Cocos nucifera*) populations from across the geographic range using sequence-tagged microsatellites (SSRs) and AFLPs. Theor Appl Genet 100:764–771

Thuc LV, Ky H, Ooi SE, Napis S, Ishak Z, Namasivayam P (2008) Molecular characterization of an unknown protein from the ESTs of oil palm (*Elaeis guineensis* Jacq.) cell suspension culture. Pertanika J Sci Technol 16:189–199

Ting NC, Cheah SC, Zamzuri I, Tan SG, Faridah QZ, Maizura I, Singh R (2006) Statistical mapping of quantitative trait loci controlling the time to first callusing in oil palm (*Elaeis guineensis* Jacq.) tissue culture. Pertanika J Trop Agric Sci 29:35–45

Ting NC, Noorhariza MZ, Rozana R, Low ETL, Maizura I, Cheah SC, Singh R (2010) SSR mining in oil palm EST database: application in oil palm germplasm diversity studies. J Genet 89:135–145

Tisserat B, Torres AM (1979) Isozymes as genetic indicators in date palms. Date Grow Inst Rep 54:24–28

Torres AM, Tisserat B (1980) Leaf isozymes as genetic markers in date palms. Am J Bot 67:162–167

Tregear JW, Morcillo F, Richaud F, Berger A, Singh R, Cheah SC, Hartmann C, Rival A, Duval Y (2002) Characterization of a defensin gene expressed in oil palm inflorescences: induction during tissue culture and possible association with epigenetic somaclonal variation events. J Exp Bot 53:1387–1396

Trifi M, Rhouma A, Marrakchi M (2000) Phylogenetic relationships in Tunisian date-palm (*Phoenix dactylifera* L.) germplasm collection using DNA amplification fingerprinting. Agronomie 20:665–671

Triki MA, Zouba A, Khoualdia O, Ben Mahamoud O, Takrouni ML, Garnier M, Bové JM, Montarose M, Pouper A, Flores R, Darós JA, Fadda ZGN, Moreno P, Duran-Vila N (2003) "Maladie de feuilles cassantes" or brittle leaf disease of date palms in Tunisia: biotic or abiotic disease? J Plant Pathol 85:71–79

Ülker B, Somssich LE (2004) WRKY transcription factors: from DNA binding towards biological function. Curr Opin Plant Biol 7:491–498

Upadhyay A, Jayadev K, Manimekalai R, Parthasarathy VA (2004) Genetic relationship and diversity in Indian coconut accessions based on RAPD markers. Sci Hortic 99:353–362

Vorster BJ, Kunert KJ, Cullis CA (2002) Use of representational difference analysis for the characterization of sequence differences between date palm varieties. Plant Cell Rep 21:271–275

Wadt LHO, Sakiyama NS, Pereira MG, Tupinamba EA, Ribeiro FE, Aragao WM (1999) RAPD markers in the genetic diversity study of the coconut palm. In: Oropeza C, Verdeil JL, Ashburner GR, Cardena R, Santamaria JM (eds) Current advances in coconutbiotechnology. Kluwer Academic Publishers, Dordrecht, pp 89–97

Waterson JM (1953) Observations on the influence of some ecological factors on the incidence of oil palm diseases in Nigeria. J West Afr Inst Oil Palm Res 1:24–59

Wrigley G (1995) Date palm (*Phoenix dactylifera* L.). In: Smartt J, Simmonds NW (eds) The evolution of crop plants. Longman, London, pp 399–403

Wu K-L, Guo Z-J, Wang H-H, Li J (2005) The WRKY family of transcription factors in rice and *Arabidopsis* and their origins. DNA Res 12:9–26

Xie Z, Zhang Z-L, Zou X, Huang J, Ruas P, Thompson D, Shen QJ (2005) Annotations and functional analyses of the rice WRKY gene superfamily reveal positive and negative regulators of abscisic acid signaling in aleurone cells. Plant Physiol 137:176–189

Xu L, Ye R, Zheng Y, Wang Z, Zhou P, Lin Y, Li D (2010) Isolation of the endosperm-specific LPAAT gene promoter from coconut (*Cocos nucifera* L.) and its functional analysis in transgenic rice plants. Plant Cell Rep 29:1061–1068

Yang M, Zhang X, Liu G, Chen K, Yun Q, Zhao D, Al-Mssallem IS, Yu J (2010) The complete chloroplast genome sequence of date palm (*Phoenix dactylifera* L.). PLoS One 5(9):e12762. doi:10.1371/journal.pone.0012762

Younis RAA, Ismail OM, Soliman SS (2008) Identification of sex-specific DNA markers for date palm (*Phoenix dactylifera* L.) using RAPD and ISSR techniques. Res J Agric Biol Sci 4:278–284

Zehdi S, Trifi M, Ould Mohamed Salem A, Rhouma A, Marrakchi M (2002) Survey of inter simple sequence repeat polymorphisms in Tunisian date palms (*Phoenix dactylifera* L.). J Genet Y Breed 56:77–83

Zehdi S, Trifi M, Billotte N, Marrakchi M, Pintaud JC (2004) Genetic diversity of Tunisian date palms (*Phoenix dactylifera* L.) revealed by nuclear microsatellite polymorphism. Hereditas 141:278–287

Zehdi S, Pintaud JD, Billotte N, Ould Mohamed Salem A, Sakka H, Rhouma A, Marrakchi M, Trifi M (2006) Etablissement d'une clé d'intentification variétale chez le palmier dattier (*Phoenix dactylifera* L.) par les marquers microsatellites. Plant Genet Resour Newsl 145:11–18

Zeven AC (1965) The origin of the oil palm. J Nig Inst Oil Palm Res 4:218–225

Zeven AC (1967) The semi-wild oil palm and its industry in Africa. Agricultural research reports no. 698. Agricultural University, Wageningen

Zhang Y, Wang L (2005) The WRKY transcription factor superfamily: its origin in eukaryotes and expansion in plants. BMC Evol Biol 5:1

Zhu YL, Song QJ, Hyten DL, Van Tassell CP, Matukumalli LK, Grimm DR, Hyatt SM, Fickus EW, Young ND, Cregan PB (2003) Single-nucleotide polymorphisms in soybean. Genetics 163:1123–1134

Zizumbo-Villarreal D, Ruiz-Rodriguez M, Harries H, Colunga-GarcíaMarín P (2006) Population genetics, lethal yellowing disease, and relationships among Mexican and imported coconut ecotypes. Crop Sci 46:2509–2516

Index

R.J. Schnell and P.M. Priyadarshan (eds.), *Genomics of Tree Crops*,
DOI 10.1007/978-1-4614-0920-5, © Springer Science+Business Media, LLC 2012

H

MIX
Papier aus verantwortungsvollen Quellen
Paper from responsible sources
FSC® C105338

If you have any concerns about our products,
you can contact us on
ProductSafety@springernature.com

In case Publisher is established outside the EU,
the EU authorized representative is:
Springer Nature Customer Service Center GmbH
Europaplatz 3, 69115 Heidelberg, Germany

Printed by Libri Plureos GmbH
in Hamburg, Germany